Cell-based Immunotherapies for Cancer

Alok K. Mishra • Sunil K. Malonia
Editors

Cell-based Immunotherapies for Cancer

Springer

Editors
Alok K. Mishra
Department of Hematopoietic Biology & Malignancy, Division of Cancer Medicine
The University of Texas MD Anderson Cancer Center
Houston, TX, USA

Sunil K. Malonia
Department of Molecular, Cell and Cancer Biology
UMass Chan Medical School
Worcester, MA, USA

ISBN 978-3-031-88694-2 ISBN 978-3-031-88695-9 (eBook)
https://doi.org/10.1007/978-3-031-88695-9

This Springer imprint is published by the registered company Springer Nature Switzerland AG
The registered company address is: Gewerbestrasse 11, 6330 Cham, Switzerland

This book on Cell-Based Immunotherapies for Cancer is dedicated with deep gratitude and respect:

To the memory of our cherished mentor and colleague, the late Dr. Michael R. Green of UMass Chan Medical School — a distinguished cellular and molecular biologist whose groundbreaking discoveries in gene regulation and cancer biology profoundly shaped the field of cancer research. His unwavering pursuit of scientific excellence, intellectual brilliance, and generous spirit has left an enduring legacy that continues to guide and inspire.

To our parents, whose constant encouragement, boundless love, and steadfast belief have been the foundation of our growth and passion for science.

To all researchers, clinicians, and caregivers whose relentless work inspires, supports, and contributes to the advancement of cancer research and patient care: your unwavering dedication drives scientific progress and fosters hope for a more promising future in the fight against cancer.

Alok K. Mishra, PhD
Houston, TX, USA
Sunil K. Malonia, PhD
Worcester, MA, USA

Preface

Cell-based immunotherapies represent one of the most promising frontiers in modern cancer treatment, offering new hope for patients through innovative approaches that harness the body's immune system. This comprehensive volume explores the latest developments in this rapidly evolving field, bringing together insights from leading experts across immunology, oncology, and cellular therapy.

As we witness the remarkable transformation of cancer treatment paradigms through immunotherapy, this book serves as a vital resource for researchers, clinicians, and practitioners navigating this complex landscape. The carefully curated chapters within this volume address crucial aspects of cell-based immunotherapies, from fundamental biological mechanisms to clinical applications, providing readers with both theoretical foundations and practical insights.

Our contributors bring diverse expertise from different academic institutions and industry, offering unique perspectives on various aspects of cellular immunotherapy. The content spans from basic research to translational medicine, including detailed discussions of CAR-T cell therapy, tumor-infiltrating lymphocytes, natural killer cells, and emerging cellular platforms. Each chapter has been thoughtfully structured to balance scientific depth with clinical relevance, making this work valuable for both seasoned experts and those new to the field.

This book aims to serve not only as a comprehensive reference but also as a catalyst for future innovations in cancer immunotherapy. By highlighting both established approaches and emerging technologies, we hope to inspire the next generation of breakthrough treatments that will continue to revolutionize cancer care. Our objective is to foster a deeper appreciation for the potential of cell-based therapies, not only as a formidable force against cancer but also as a paradigm shift in the way we conceptualize and approach the treatment of this multifaceted disease.

As editors of this book, we extend our sincere gratitude to all contributors for their hard work in bringing this volume to fruition. Their dedication to advancing our understanding of cell-based immunotherapies will undoubtedly impact the future of cell-based therapies for cancer.

Houston, TX, USA Alok K. Mishra
Worcester, MA, USA Sunil K. Malonia

Acknowledgments

We would like to express our heartfelt gratitude to all those who contributed to the successful completion of this book, *Cell Based Immunotherapies for Cancer*.

First and foremost, we extend our sincere thanks to all the contributing authors for their dedication, and hard work in sharing their expertise and research findings. Your contributions have significantly enriched the content of this book and will undoubtedly advance the field of cancer immunotherapy. We are deeply grateful to our editorial and production team at Springer Nature for their unwavering support, guidance, and professionalism throughout the publication process. Your commitment to excellence has been instrumental in shaping this book into its final form.

Special thanks to our colleagues, mentors, and collaborators whose continuous encouragement and constructive feedback have been a source of inspiration and motivation. Your insights and shared experiences have greatly contributed to the depth and quality of this work.

We would also like to acknowledge the Department of Molecular Cell and Cancer biology, UMass Chan Medical School and the Department of Hematopoietic Biology and Malignancy at UT, MD Anderson Cancer Center, Houston, TX, for support in our research endeavors, enabling us to explore and contribute to the rapidly evolving field of cancer immunotherapy.

Finally, we express our heartfelt appreciation to our families and friends for their patience, understanding, and unwavering support throughout this journey. Your encouragement has been our driving force in bringing this book to fruition.

Houston, TX, USA — Alok K. Mishra
Worcester, MA, USA — Sunil K. Malonia

Contents

Editors and Contributors

About the Editors

Alok K. Mishra is a biotechnology expert specializing in immune-oncology and cancer drug discovery. He earned his PhD in Biological Sciences from Central Drug Research Institute, India, studying mycobacterial genetics and immunobiology. He completed his postdoctoral training in the laboratory of renowned cancer biologist Dr. Michael R. Green at the University of Massachusetts Chan Medical School, USA, where he investigated regulatory networks of innate immune checkpoints and oncogenic drivers in cancer. Currently, Dr. Mishra is a Research Scientist in the Department of Hematopoietic Biology and Malignancy at MD Anderson Cancer Center, USA, focusing on tumor–immune cell interactions and the development of targeted and cellular immunotherapies for solid malignancies. His research has led to the filing of two international patents on formulating novel ovarian cancer treatments and selective targeting of cancer stem cells in glioblastoma. Dr. Mishra has also published in high-impact journals and serves as a peer reviewer and associate editor for leading biomedical journals. As a corresponding editor of this book, he brings scientific expertise and a commitment to innovation, curating cutting-edge content that addresses key challenges in cancer biology and immunotherapy. His work holds significant translational potential, aiming to improve cancer patient outcomes and advance cancer immunotherapies.

Sunil K. Malonia is an accomplished cancer biologist and faculty member in the Department of Molecular, Cell, and Cancer Biology at UMass Chan Medical School, USA. He earned his PhD in Biotechnology from the National Center for Cell Science, India, where his doctoral research focused on the precise regulation of gene expression by cis-regulatory elements in cancer. Dr. Malonia completed his postdoctoral training at the Howard Hughes Medical Institute, University of Massachusetts Medical School, where he conducted research in functional genomics. His work explored the epigenetic regulation of tumor suppressor genes in cancer and the role of the F-box family ubiquitin ligases in tumorigenesis. Dr. Malonia's current research program is dedicated to the discovery of therapeutically targetable pathways in cancer. His work focuses on elucidating the molecular mechanisms driving cancer progression and immune evasion, identifying novel therapeutic targets, and translating these findings into innovative treatments. He has authored numerous research articles and reviews in high-impact journals and is the recipient of prestigious awards. As the senior editor of this book, Dr. Malonia blends scientific expertise with translational insight, enriching the fields of immuno-oncology and immunotherapy. His leadership and editorial contributions reflect a steadfast commitment to advancing immuno-oncology research and fostering innovative therapeutic approaches.

Contributors

Sovon Acharya Vapiano Laboratories Pvt. Ltd., Bhubaneswar, Odisha, India
Amity Institute of Biotechnology, Amity University, Gurgaon, Haryana, India

Shahid Banday Department of Molecular, Cell and Cancer Biology, UMass Chan Medical School, Worcester, MA, USA

Samyabrata Bhaduri Stylus Medicine, Cambridge, MA, USA

Neetu Bhardwaj Department of Medical Oncology, All India Institute of Medical Sciences, New Delhi, India

Subhasis Chattopadhyay School of Biological Sciences, National Institute of Science Education and Research, Bhubaneswar, Odisha, India

Homi Bhabha National Institute, Training School Complex, Mumbai, Maharashtra, India

Manu Dalela DBT Center of Excellence for Stem Cell Research, All India Institute of Medical Sciences, New Delhi, India

Rikesh Kumar Dubey Division of Medical Research, SRM Institute of Science and Technology, Kattankulathur, Tamil Nadu, India

Sridevi Duggirala Department of Molecular Oncology, Cancer Institute (W.I.A.), Adyar, Chennai, India

Shubham Dutta Department of Medicine, Division of Innate Immunity, UMass Chan Medical School, Worcester, MA, USA

Dibya Ghimire Massachusetts General Hospital, Boston, MA, USA

Deepesh Kumar Gupta Division of Oncology, Section of Cellular Therapy, Washington University, School of Medicine, St. Louis, MO, USA

Center of Gene and Cellular Immunotherapy, Washington University School of Medicine, St. Louis, MO, USA

Mahesh Hegde Department of Molecular, Cell and Cancer Biology, UMass Chan Medical School, Worcester, MA, USA

Krishan Gopal Jain Division of Nephrology, Indiana University School of Medicine, Indianapolis, IN, USA

Rohan H. Kamat Kodo Life Sciences, Vapi, Gujrat, India

Surojit Karmakar BRIC-National Centre for Cell Science, Pune, Maharashtra, India

Somlata Khamaru School of Biological Sciences, National Institute of Science Education and Research, Bhubaneswar, Odisha, India

Homi Bhabha National Institute, Training School Complex, Mumbai, Maharashtra, India

Pratirodh Koirala Takeda Pharmaceuticals, Cambridge, MA, USA

Alok Kumar Department of Immunology, University of Pittsburgh, Pittsburgh, PA, USA

Anil Kumar School of Engineering and Sciences, GD Goenka University, Gurugram, Haryana, India

Girdhari Lal BRIC-National Centre for Cell Science, Pune, Maharashtra, India

Madhu Malinee Department of Immunology, University of Pittsburgh, Pittsburgh, PA, USA

Sunil K. Malonia Department of Molecular, Cell and Cancer Biology, UMass Chan Medical School, Worcester, MA, USA

Alok K. Mishra Department of Hematopoietic Biology & Malignancy, Division of Cancer Medicine, The University of Texas MD Anderson Cancer Center, Houston, TX, USA

Apoorva Narain Versiti Blood Research Institute, Milwaukee, WI, USA

Niraj K. Nirala Program in Molecular Medicine, UMass Chan Medical School, Worcester, MA, USA

Romana Rashid Division of Innate Immunity, UMass Chan Medical School, Worcester, MA, USA

Munindra Ruwali Department of Education in Science and Mathematics (DESM), National Council of Educational Research and Training (NCERT), New Delhi, India

Bhubanananda Sahu Department of Ophthalmology and Visual Sciences, UMass Chan Medical School, Worcester, MA, USA

Bela G. Santani Foundation Medicine, Inc., Boston, MA, USA

Anshul Sharma Medical School of Wisconsin, University of Wisconsin, Milwaukee, WI, USA

Surajit Sinha Center for Cancer Research, National Cancer Institute, Bethesda, MD, USA

Reshmi Suresh BRIC-National Centre for Cell Science, Pune, Maharashtra, India

Ritesh Thakare Department of Molecular, Cell and Cancer Biology, UMass Chan Medical School, Worcester, MA, USA

Kshyama Subhadarsini Tung School of Biological Sciences, National Institute of Science Education and Research, Bhubaneswar, Odisha, India

Homi Bhabha National Institute, Training School Complex, Mumbai, Maharashtra, India

Bhawna Yadav Division of Infectious Disease and Immunology, UMass Chan Medical School, Worcester, MA, USA

Sourabh Yadav BRIC-National Centre for Cell Science, Pune, Maharashtra, India

Cellular Immune Responses in Tumor Microenvironment

Somlata Khamaru, Kshyama Subhadarsini Tung, and Subhasis Chattopadhyay

Abstract Cancer is a systemic disease at a constant tug of war with the host immune system. The tumor microenvironment (TME) continually alters immune cells and modulates the immune response favorable for tumor cell survival, growth and metastasis. Tumor cells use various immune evasion mechanisms during cancer progression and are broadly classified into three phases: elimination, equilibrium, and escape. Throughout these phases, the degree of tumor immunogenicity is reformed, and systemic immunosuppression is developed to counteract host antitumor effector responses. A meticulous balance between effector and tolerogenic immune responses determines tumor fate. The surviving tumor cells advance toward cancer progression while developing peripheral immune tolerance and recruitment of immunosuppressive cells. TME induces pro-tumorigenic M2 polarization of tumor-associated macrophages (TAMs), which promote tumor progression. Immunosuppressive signals from the TME impair dendritic cell (DC)-mediated antigen presentation to naive T cells, thereby inducing T-cell tolerance and progressing tumor growth. Cancer cells also limit the effector functions of tumor-infiltrating lymphocytes (TILs) through the imposition of multifactorial constraints such as elevated acidity, immunosuppressive metabolites and enzymes, nutrient deficiency, hypoxia, and ionic imbalance. Altogether, these factors result in T-cell anergy and exhaustion. Additionally, the tumor milieu skews natural T regulatory cells (nTregs) and recruits inducible Tregs (iTregs) that further inhibit the effector function of antitumor agents, cytotoxic T cells. Furthermore, cancer cells in the TME impede antitumor function via the activation of immunological checkpoints. These suppressor cells further restrict antitumor immunity via an intricate mesh of immunosuppressive cytokines and chemokines. This chapter explores the various aspects of cell-mediated immunity and their differential roles in the TME and cancer progression, which might help develop anticancer immunotherapies.

S. Khamaru · K. S. Tung · S. Chattopadhyay (✉)
School of Biological Sciences, National Institute of Science Education and Research, Bhubaneswar, Odisha, India

Homi Bhabha National Institute, Training School Complex, Mumbai, Maharashtra, India
e-mail: subho@niser.ac.in

A. K. Mishra, S. K. Malonia (eds.), *Cell-based Immunotherapies for Cancer*,
https://doi.org/10.1007/978-3-031-88695-9_1

Keywords Tumor microenvironment (TME) · Tumor · Cancer · Immunotherapy · Regulatory T cells (Tregs) · Immune surveillance · Immunosuppression

1 Components of the Immune System

1.1 Organs of the Immune System

The immune system encompasses an intricate network of organs, cells, and proteins. Its primary role is to safeguard the body from infections while simultaneously ensuring the protection of the body's own cells. Irregularities within the immune system can result in allergic diseases, immunodeficiencies, and autoimmune disorders [1, 2]. The immune system comprises primary and secondary lymphoid organs and diverse cells that contribute to the innate and adaptive immune systems. Primary lymphoid organs such as the thymus and bone marrow are among the most crucial components of the immune system. The secondary lymphoid organs include the spleen, tonsils, lymph vessels and nodes, adenoids, skin, and liver, which aid in the proper functionality of the immune system [1, 2].

The *bone marrow* consists of hematopoietic and adipose tissues, along with stem cells—hematopoietic stem cells (HSCs) and mesenchymal stem cells (MSCs). HSCs play a significant role in producing diverse blood cell types through hematopoiesis. While neutrophils, monocytes, dendritic cells (DCs), eosinophils, basophils, mast cells (MCs), megakaryocytes, and erythrocytes are generated from myeloid progenitor cells, standard lymphoid progenitor cells give rise to pre-T cells, pre-B cells, B cells, natural killer (NK) cells, and innate lymphoid cells. The *thymus* serves as the site for the education of T cells, providing an ideal environment for developing T cells from hematopoietic progenitor cells. The bone marrow and thymus collectively function as the central lymphoid tissues responsible for generating and initially screening lymphocytes. The *lymphatic system* is a vital component of the circulatory system. It comprises an intricate network of vessels known as the lymphatic vessels. It helps in the transportation of white blood cells to and from lymph nodes located within the body's various regions. It facilitates the transportation of antigen-presenting cells to the lymph nodes, where they initiate immune responses. Lymph nodes are widely distributed throughout the body, such as in the armpits and abdominal area, and are interconnected via lymphatic vessels. These nodes serve as strongholds for various immune cells, including B and T cells, functioning as filters or traps for foreign particles and playing crucial roles in the efficient functioning of the immune system. The primary role of the spleen is that of a blood filter. In its white pulp, it generates antibodies and removes bacteria and blood cells coated with antibodies through circulation via both blood and lymph nodes. The *palatine* and *nasopharyngeal tonsils* are lymphoepithelial tissues located near the oropharynx and nasopharynx. These immune-competent tissues act as the immune system's initial defense against foreign pathogens that are either ingested or inhaled. The *skin* serves as a vital interface with the external environment, constituting the first line of defense against external factors. It acts as an anatomical barrier that safeguards against pathogens and external damage, creating a protective boundary between the body's internal and external environments. Langerhans cells within the skin are part of the adaptive immune system. The *liver* is a multifunctional organ, including its role in the reticuloendothelial system, where it houses numerous immunologically active cells, effectively acting as a "sieve" that filters antigen carried to it through the portal system [1, 2].

1.2 Cells of the Immune System and Their Characteristic Functions

Leukocytes, commonly known as white blood cells, can be differentiated into five distinct types, all of which originate from a multipotent cell known as a hematopoietic stem cell situated in the bone marrow. Within innate leukocytes, phagocytes, mast cells, eosinophils, basophils, and natural killer cells specialize in recognizing and combating pathogens. Additionally, the adaptive immune system consists of a specific subset of leukocytes called lymphocytes. B cells and T cells, which are derived from hematopoietic stem cells in the bone marrow, represent the primary types of lymphocytes. These cells play a central role in the identification and elimination of pathogens and serve as crucial mediators in initiating the adaptive immune response.

1.2.1 Neutrophils

Neutrophils and macrophages function as phagocytes that traverse the body for invading pathogens. Neutrophils are primarily found circulating in the bloodstream. They are the most abundant type of phagocytes because they can quickly migrate in an amoeboid manner. In addition to $CD49d^{-}CD11b^{-}$ resting neutrophils, there are two distinct subsets of neutrophils in mice. Type 1 neutrophils (N1) and type 2 neutrophils (N2) differ in terms of their cytokine and chemokine production, leading to variations in the stimulation of macrophage activity as well as the expression of toll-like receptors (TLRs) and surface antigens [3]. Distinct neutrophil phenotypes have also been verified in mouse models of tumors. Inhibiting transforming growth factor beta (TGF-β) has been demonstrated to shift the differentiation of mouse neutrophils toward an anti-tumorigenic phenotype (N1). This results in increased production of proinflammatory cytokines and chemokines, reduced levels of arginase, and a heightened ability to kill tumor cells in vitro. Alternatively, in the tumor environment, TGF-β tends to support the accumulation of pro-tumorigenic N2 neutrophils in mice, thereby promoting tumor growth [4]. Neutrophils play a central role in the initial innate immune response against a wide range of bacterial and fungal pathogens, acting swiftly before the more intricate processes involving humoral and lymphocyte components of acquired immunity can occur during infection [1, 5, 6]. They carry out a multitude of other distinctive roles, including transmigration phagocytosis, degranulation, priming, adhesion, chemotaxis, oxygen radical production, and extracellular trap formation. They also perform an array of biochemical processes, such as phospholipid metabolism, GTPase activation, ion channel regulation, and adhesion molecule regulation, in addition to specialized functions and vital molecules that contribute to host defense against infections, facilitate the mediation and resolution of inflammation, and modulate cytokine/chemokine responses in the immune system [7].

1.2.2 Macrophages

Macrophages reside within different tissues and have a diverse range of functions depending on their site of residence. In general, they produce various chemical messengers, act as scavengers responsible for eliminating worn-out cells and debris, and serve as antigen-presenting cells (APCs), triggering the activation of the adaptive immune system [1, 5, 6]. Macrophages can, at times, contribute to anti-inflammatory responses, tissue healing, and maintaining balance within the body. Still, in certain situations, they can also support inflammation and the growth of tumors [8]. Broadly, macrophages can be categorized into two distinct subsets: classically activated subsets are referred to as M1 macrophages, and alternatively activated subsets are known as M2 macrophages. M1 macrophages can be polarized in vitro via T helper (Th) type 1 (Th1) cytokines such as colony-stimulating factor (GM-CSF), tumor necrosis factor alpha (TNF-α), and interferon gamma (IFN-γ). M1 macrophages express proinflammatory cytokines, including interleukin (IL)-1β, IL-6, IL-12, IL-23, and TNF-α. In contrast, M2 macrophages are polarized by Th2 cytokines such as IL-4 and IL-13, leading them to produce anti-inflammatory cytokines such as IL-10 and transforming growth factor beta (TGF-β). The polarization of macrophages can be altered in response to various stimuli [9]. M2 macrophages, in conjunction with a minority of M1 macrophages, are collectively referred to as "tumor-associated macrophages" (TAMs). TAMs represent the most heterogeneous immune cell population within the tumor microenvironment (TME) and are pivotal in supporting tumor growth. They contribute to various aspects of tumorigenesis, including immunosuppression in the TME; the promotion of tumor cell proliferation, invasion, and migration; the stimulation of angiogenesis; the facilitation of metastasis; and the development of resistance to therapeutic interventions [10]. The presence of tumor-associated macrophages (TAMs) is of significant prognostic importance and is linked to adverse clinical outcomes in multiple cancer types. Elevated TAM density is positively associated with advanced tumor stage, higher (tumor, node, metastasis) TNM stage, more severe histological grade, the presence of lymphovascular invasion, and metastasis in various cancers. These findings suggest that TAMs can serve not only as diagnostic and prognostic indicators but also as predictive tools for evaluating the clinical outcomes of cancer patients. As promising novel targets, researchers are increasingly exploring strategies to modulate TAMs for therapeutic purposes, including inhibiting macrophage recruitment, blocking macrophage survival, and inducing macrophage repolarization [11].

1.2.3 Dendritic Cells

Dendritic cells are found in tissues that are in direct contact with the external environment, such as the skin, nose, lungs, stomach, and intestines, and function as phagocytes. These cells play crucial roles in bridging the gap between bodily tissues and the innate and adaptive immune systems [1, 5, 6]. Dendritic cells (DCs) play a crucial role in conveying information in the presence of pathogens to the adaptive immune system, thereby triggering durable and antigen-specific responses. To accomplish this, DCs meticulously coordinate the activity of the proteolytic machinery within both the endosomal–lysosomal system (comprising cathepsins and other lysosomal hydrolases) and the cytosol (proteasome) and endoplasmic reticulum (ER). This orchestration facilitates the partial degradation of pathogen-derived proteins, yielding antigenic peptides. These peptides are subsequently loaded onto major histocompatibility complex (MHC) class I or class II molecules. The resulting peptide–MHC complexes are then transported to the cell membrane, where they are presented to specific T cells. Activating these T cells leads to the proliferation and development of potent effector cells, such as cytotoxic T cells, or cells that contribute to the overall progression of immune responses, such as helper T cells. The effectiveness and precision of T-cell responses are further augmented and shaped by the expression of ligands (e.g., CD80 and CD86) on DCs, which bind to co-stimulatory molecules on T cells, working in tandem with the peptide–MHC–specific T-cell receptor. Notably, DCs also produce various stimulatory cytokines (e.g., interleukin-12, or IL-12) essential for optimal T-cell stimulation [12]. While other cell types, such as macrophages and B cells, can present antigens, DCs stand out due to their remarkable efficiency, as they are uniquely responsible for initiating all antigen-specific immune responses. Within the TME, the resemblance of numerous tumor antigens, including proteins with point mutations, to naturally occurring self-proteins, coupled with their persistent exposure to DCs under noninflammatory conditions, suggests that antigen presentation by DCs exposed to tumors could lead to the induction of tolerance, likely through the generation of regulatory T cells (Tregs). Moreover, as cancer advances, robust immunosuppressive mechanisms may emerge, strongly inhibiting the activity of effector T cells. Some of these mechanisms can be triggered in DCs by factors within the TME, possibly originating from other infiltrating myeloid cells. These mechanisms may involve the expression of PD-L1 and PD-L2 (ligands for the T-cell checkpoint receptor programmed death-1, or PD-1, which induces T-cell exhaustion), TGF-β (favoring Treg production), and cytosolic enzymes such as indoleamine-2,3-dioxygenase (IDO) and arginase (which generate immunosuppressive metabolites). Although not yet fully understood in terms of mechanisms, it is increasingly evident that tumors employ strategies to resist immune recognition by restricting both T-cell and DC functionality [13].

1.2.4 Mast Cells

Mast cells, located in connective tissues and mucous membranes, regulate inflammation and are associated with allergies and anaphylaxis. Human mast cells have two phenotypes: mucosal and connective tissue mast cells [1, 5, 6]. They play roles in vasodilation, immune responses, angiogenesis, and venom detoxification. In diseases such as allergies, asthma, anaphylaxis, gastrointestinal disorders, malignancies, and cardiovascular diseases, mast cells are reportedly involved [14]. They act primarily through immunoglobulin (Ig) E (IgE)–mediated allergic reactions via the FcεRI receptor. Mature B cells, which respond to $CD4^+$ Th2 cells, initiate IgE antibody production. Initially, B cells generate IgM and IgD, which switch to IgE upon activation by antigens and interact with cytokines such as IL-4. IgE predominantly binds to FcεRI on mast cells. When mast cells contact an antigen, cross-linking FcεRI triggers granule release [15]. Mast cells also have Fc receptors for IgA and IgG, as well as receptors for adenosine, C3a, chemokines, cytokines, and pathogen-associated molecular patterns (PAMPs). Toll-like receptors (TLRs) on mast cells contribute to activation and immune responses [16]. In promoting angiogenesis, MCs release pro-angiogenic factors (vascular endothelial growth factor-VEGF, basis fibroblast growth factor-bFGF, TGF-beta, TNF-alpha, and IL-8) and proteases. Mast cell–released histamine increases microvasculature permeability, facilitating angiogenesis. The evidence suggests that mast cells enhance angiogenesis in the context of tumor growth [17].

1.2.5 Basophils and Eosinophils

Basophils, eosinophils, and neutrophils release chemical mediators to defend against parasites and contribute to allergic reactions such as asthma [1, 5, 6]. Basophils, which have a functional TLR2 receptor, are involved in innate immunity and can be activated independently of IgE secretion [18, 19]. They are primary sources of IL-4 in peripheral blood mononuclear cells (PBMCs) activated by allergens and helminth parasites [20, 21]. Increased basophil numbers are identified in cutaneous and pulmonary late-phase allergic responses and in the lungs of asthma patients [22]. Recent murine models suggest a direct role for basophils in the antigen presentation for Th2 responses [23]. Eosinophils release proinflammatory mediators, including granule-stored cationic proteins, newly synthesized eicosanoids, and cytokines [19]. Eosinophils release granule-stored cationic proteins, eicosanoids, and cytokines, including TGF-β, IL-3, IL-4, IL-5, IL-8, IL-10, IL-12, IL-13, IL-16, IL-18, TNF-α, CCL5, and CCL11. These preformed cytokines are rapidly released upon degranulation [24]. Eosinophils exhibit immunomodulatory activity through cytokine secretion, antigen presentation, and indolamine-2,3-dioxygenase expression, leading to kynurenine production with anti-Th1 activity [25].

1.2.6 Natural Killer (NK) Cells

Natural killer (NK) cells eliminate tumor cells or virus-infected cells through direct interaction, triggering granule release for target cell apoptosis and cytokine secretion [1, 26]. NK cell activity is regulated by co-activating (NKp46, NKG2D, CD16, and LFA-1) and co-inhibitory (NKG2A, KLRG1, and TIGIT) receptors that recognize specific ligands. Activation via these receptors allows NK cells to eliminate targets directly or indirectly through proinflammatory cytokine secretion [27, 28]. NK cells crucially control viral infections such as Cytomegalovirus (CMV), influenza, hepatitis C, and HIV-1, which collaborate with CD^{8+} T cells. Tumors often downregulate MHC I, evading cytotoxic T cells and hindering adaptive immune responses [29]. Human NK cells specifically identify the ligand platelet-derived growth factor DD (PDGF-DD) through the activating receptor NKp44. This interaction enables NK cells to restrict tumor growth by stimulating the secretion of IFN-γ, TNF-α, and chemokines. Cytokine production induces tumor cell cycle arrest [30, 31]. Cancer cells use various mechanisms to evade NK cells, including increasing MHC I to inhibit NK cell function, decreasing NKG2D ligand expression for impaired recognition, and upregulating inhibitory cytokines such as IL-10 or TGF-β within tumors [32, 33].

1.2.7 T Lymphocytes

To identify "nonself" targets such as pathogens, T cells require the processing and presentation of antigens in conjunction with a "self" receptor known as a major histocompatibility complex (MHC) molecule. There are two primary subtypes of T cells: cytotoxic T cells (T_Cs or CTLs), which are responsible for eliminating infected or dysfunctional cells, including those infected by viruses or other pathogens, and helper T cells (Th), including Th1, Th2, Th9, Th17, and T regulatory cells (Tregs), which oversee both innate and adaptive immune responses. Th cells play a pivotal role in determining the specific immune responses that the body mounts against a particular pathogen. Unlike T_Cs, they lack cytotoxic activity and do not directly eliminate infected cells or clear pathogens. A minor subtype, γδ T cells, can recognize intact antigens that are not bound to MHC receptors [1, 5, 6]. T cells encompass an extensive array of T-cell clones, each featuring a distinct receptor for antigens. Within this diverse repertoire, the presence of T cells with receptors capable of recognizing the body's own antigens, known as self-reactive T cells, is inevitable. These self-reactive T cells can potentially trigger harmful autoimmunity. Therefore, it is crucial to have mechanisms in place to restrain such T cells. This control involves two mechanisms that not only prevent autoimmunity but also maintain protective immunity: the shaping of the T-cell repertoire in the thymus and the regulation of T cells in peripheral tissues. To achieve potent immunity against foreign antigens while preventing harmful autoimmunity in peripheral tissues, the immune system must not only regulate the intensity and type of immune response but also distinguish between self and nonself. Intrinsic homeostatic mechanisms govern the control of intensity and type, whereas the discrimination between self and nonself is mediated primarily by Tregs [34]. Rapid and precise identification of tumor-specific antigens by immune T cells forms the basis of cancer immunology and immunotherapy. When dendritic cells (DCs) come into contact with tumor-specific antigens released into their environment, they phagocytose these antigens and present them on the surface through MHC class II molecules. Upon encountering MHC class II molecules, $CD4^+$ T cells release IFN-γ, which, in turn, triggers the elimination of tumor cells by host cells. IFN-γ promotes the secretion of two CXC chemokines, Mig and interferon gamm-inducible protein 10a (IP-10). These chemokines exert antiangiogenic effects by disrupting the tumor vasculature, inhibiting growth and necrosis [35–37]. Furthermore, IFN-γ activates macrophages to release TNF-α and nitric oxide [38]. CTLs are crucial in the immune response against cancer and form the foundation of successful cancer immunotherapy. They closely bind to antigenic peptides presented by antigen-presenting cells (APCs) or other targeted cells, connecting with MHC I molecules to eliminate target antigens or cells [39]. Additionally, CTLs engage with tumor cells, inducing apoptosis by creating pores facilitated by mechanical force. This involves death-inducing granules containing granzymes, cathepsin C, and perforin [40]. In addition, T-cell-mediated cancer immunotherapy encompasses adoptive T-cell therapy (ACT), chimeric antigen receptor (CAR) T-cell therapy, immune checkpoint blockade, and therapeutic vaccination targeting $CD4^+$ T cells [37].

1.2.8 B Lymphocytes

B cells have antigen-specific receptors, such as surface antibodies, allowing them to recognize pathogens directly. Each B-cell lineage expresses a unique antibody, forming the complete array of B-cell antigen receptors representing all possible antibodies [1, 5, 6]. Humoral immunity involves memory B and plasma cells, which are crucial for high-affinity, isotype-switched antibody production [41]. B-cell activation and differentiation are influenced by antigen nature, toll-like receptor signals, and cytokine/co-stimulatory signals [42]. In addition to humoral immunity, B cells contribute to cellular immunity by acting as antigen-presenting cells (APCs), producing inflammatory cytokines, and constituting a specific subgroup called Bregs, which secrete IL-10 to modulate immune responses [43, 44]. B cells maintain lymphoid organ architecture, notably the spleen, and foster tertiary lymphoid tissues in chronic inflammation. B cells' "cellular" functions play substantial roles in autoimmunity and allograft rejection [45, 46]. B cells mediate antigen presentation for tumor-specific $CD4^+$ and $CD8^+$ T cells, contributing to T-cell activation, especially in virus-induced tumors [47]. For example, in a syngeneic B16 melanoma cell line transfer system, the depletion of B cells through anti-CD20 antibody treatment resulted in a twofold increase in tumor volume. This depletion also led to impaired production of IFN-γ and TNF-α by both $CD4^+$ and $CD8^+$ T cells [48]. A recent study further elucidated the role of B-cell antigen presentation in tumor immunity. T follicular helper (TFH) cells, known for their involvement in B-cell maturation and activation, play a pivotal role. The interaction between TFH cells and germinal center B cells (GC B cells) activates the latter, promoting their differentiation into short-term living plasma cells, long-term living plasmablasts, and memory B cells that release antibodies with antitumoral properties, triggering antibody-dependent cellular phagocytosis (ADCP), complement-dependent cytotoxicity (CDC), and antibody-dependent cellular cytotoxicity (ADCC). These mechanisms play crucial roles in the effectiveness of antibody-based cancer therapies [49–51]. Bregs modulate immune responses in the TME by secreting anti-inflammatory cytokines such as IL-10, IL-35, and TGF-β, suppressing T cells and promoting regulatory T cell (Treg) development [52].

2 Innate and Adaptive Immune Systems Maintain Immune Homeostasis

The immune system can be categorized into two primary defensive mechanisms: innate and adaptive immunity. Innate immunity serves as the initial line of defense against invading pathogens. It is an antigen-independent, nonspecific (or with limited specificity) defense mechanism that the host deploys promptly or within hours of encountering an antigen. Unlike adaptive immunity, the innate immune response lacks immunologic memory, meaning that it cannot recognize the same pathogen upon future exposure. In contrast, adaptive immunity is antigen-dependent and antigen-specific. Consequently, there is a time delay between the exposure to the antigen and the peak of the immune response in adaptive immunity. Notably, adaptive immunity is characterized by immunologic memory, enabling the host to mount a rapid and efficient immune response upon subsequent exposure to the antigen. Synergistically, all immune cells from both the innate and adaptive immune systems work together to safeguard the host from any invading endogenous or exogenous pathogens [2, 53–55].

2.1 Defensive Barriers to Innate Immunity

Innate immunity comprises four types of defensive barriers: anatomic (skin and mucous membrane), physiologic (temperature, low pH, and chemical mediators), endocytic and phagocytic mechanisms, and inflammatory responses [2].

2.1.1 Phagocytosis and Endocytosis

Phagocytosis and endocytosis play crucial roles, with phagocytes falling into two main cell types: neutrophils and macrophages. These cell types share a fundamental function—engulfing (phagocytosing) microbes and eliminating them through various bactericidal pathways [2, 56]. Moreover, neutrophils are equipped with granules and enzyme pathways that contribute to the elimination of pathogenic microbes. In contrast to neutrophils, which have a shorter lifespan, macrophages are long-lived cells that not only engage in phagocytosis but also play a crucial role in presenting antigens to T cells. Another component of innate immunity, the complement system, constitutes a biochemical cascade designed to recognize and opsonize (coat) bacteria and other pathogens. This process makes pathogens susceptible to phagocytosis, wherein immune cells engulf microbes, clear cell debris, and directly eliminate specific pathogens and infected cells [1, 53]. The phagocytic capability of the innate immune response facilitates the removal of deceased cells or antigen–antibody complexes and eradicates foreign substances present in the organs, tissues, blood, and lymph. Additionally, it has the potential to initiate the adaptive immune response by mobilizing and activating antigen-presenting cells (APCs) [1, 2].

2.1.2 Pathogen-Associated Molecular Patterns (PAMPs)

Innate immune cells identify conserved molecular patterns commonly present in most pathogens, referred to as pathogen-associated molecular patterns (PAMPs). These patterns encompass diverse elements, such as cell wall components and genetic materials from bacteria, fungi, viruses, and other pathogens. The innate immune system recognizes PAMPs through a set of germline-coded and evolutionarily conserved proteins known as pathogen-recognition receptors (PRRs). PRRs, which are cell-surface receptors for pathogens, engage with PAMPs, facilitating processes such as phagocytosis, activating inflammatory signaling pathways, initiating cell death, and triggering complement or coagulation cascades [1, 2, 5, 6, 53, 56].

Toll-like receptors (TLRs) represent a significant category of pathogen-recognition receptors (PRRs). These molecules are present in both innate and adaptive immune cells, as well as various tissue cells, such as endothelial cells, epithelial cells, and fibroblasts. Upon binding to their respective microbial ligands, TLRs initiate several responses, including the activation of phagocytes for direct elimination of pathogens and the release of proinflammatory cytokines and antimicrobial peptides. Moreover, these receptors trigger the activation of dendritic cells, playing a crucial role in initiating adaptive immune responses. The binding of ligands to TLRs initiates the nuclear factor kappa B (NF-κB) signaling pathway. This pathway acts as a central regulator, inducing inflammation and promoting the expression of various molecules, including chemokines, cytokines, adhesion molecules, matrix metalloproteases, nitric oxide synthases, and enzymes [53–57].

2.1.3 Danger-Associated Molecular Patterns (DAMPs)

In addition to toll-like receptors (TLRs), several other tissue-related factors associated with inflammation can serve as danger signals. These signals activate phagocytes and dendritic cells, leading to the elimination of internalized pathogens and the initiation of subsequent adaptive immune responses. These factors include heat-shock proteins, lectins, cytokines, chemokines, components of the extracellular matrix, and various cell-surface molecules, including lipids derived from necrotic cells [5, 6, 57].

2.1.4 Antimicrobial Proteins and Peptides

The innate immune system includes a variety of antimicrobial proteins and peptides. Some of these molecules include lysozyme and cathepsin G, as well as smaller antimicrobial peptides such as defensins and cathelicidins and skin-specific antimicrobial agents such as dermcidin and psoriasin. These proteins and peptides contribute to the elimination of pathogens through both direct cell lysis and the induction of chemotaxis in monocytes, dendritic cells, and T cells [1, 2, 55].

2.1.5 Cytokines and Chemokines

An essential function of innate immunity is the swift mobilization of immune cells to infection and inflammation sites through the production of cytokines and chemokines—small proteins involved in cell-to-cell communication and recruitment. Cytokine production within the context of innate immunity involves various defense mechanisms throughout the body, activating localized cellular responses to infections or injuries. Notably, critical inflammatory cytokines released during the initial response to bacterial infections include tumor necrosis factor (TNF), interleukin 1 (IL-1), and interleukin 6 (IL-6). These cytokines play pivotal roles in initiating cell recruitment and local inflammation, which are crucial for eliminating numerous pathogens and contributing to the development of fever. The dysregulated production of these inflammatory cytokines is frequently linked to inflammatory or autoimmune diseases, emphasizing their importance as therapeutic targets [1, 5, 6, 53].

2.2 Lymphocytes

T and B lymphocytes constitute the cellular components of the adaptive immune system. The defining characteristics of the adaptive immune response include adaptability and memory. This adaptability is rooted in the unique way T and B cells recognize antigens. In contrast to innate immune cells, which rely on a predetermined set of inherited receptors, T and B cells undergo a process of gene recombination to generate novel and distinct antigen receptors referred to as T- and B-cell receptors (TCRs and BCRs). This innovative mechanism enables T and B cells to produce antigen receptors capable of recognizing a broad spectrum of antigens, providing them with unparalleled versatility in antigen recognition [57, 58].

2.2.1 B Lymphocytes

B cells possess a unique ability to directly recognize antigens, independent of antigen-presenting cells (APCs), owing to distinctive antibodies displayed on their cell surfaces. Primarily, B cells generate antibodies against foreign antigens. In specific situations, B cells can also function as APCs. Upon activation by foreign antigens through the binding of B-cell receptors (BCRs), B cells proliferate and differentiate into either antibody-secreting plasma cells or long-lasting memory B cells. Given their crucial role in antibody production, B cells play a central role in the humoral or antibody-mediated immune response, in contrast to the cell-mediated immune responses orchestrated by T cells and APCs [1, 5, 6, 57, 58].

2.2.2 T Lymphocytes

T cells depend on the participation of antigen-presenting cells (APCs), typically dendritic cells, but also on macrophages, B cells, fibroblasts, and epithelial cells, to recognize a specific antigen. APCs exhibit a set of proteins known as the major histocompatibility complex (MHC) on their surface. Class I MHC molecules present peptides from endogenous sources, whereas class II MHC molecules on APCs present peptides from exogenous sources to T cells [1, 2, 5, 6, 55, 58].

The MHC–antigen complex initiates the activation of the T-cell receptor (TCR), inducing the secretion of cytokines by T cells, thereby exerting additional control over the immune response. This process of antigen presentation prompts T cells to primarily differentiate into either CTLs or Th cells. CTLs play a central role in eliminating cells infected by foreign agents such as viruses and in targeting tumor cells displaying appropriate antigens. Their activation is triggered by the interaction between their TCRs, and peptides bound to MHC class I molecules, leading to the clonal expansion of cytotoxic T cells, resulting in the generation of effector and memory cells [1, 55, 57, 58].

2.2.2.1 T Helper Type 1 Cells

CD4+ Th cells play crucial roles in initiating and optimizing the immune response. The activation of Th cells occurs through the recognition of antigens bound to class II MHC molecules by their TCRs. Once activated, Th cells release cytokines that influence various cell types, including the APCs that trigger their activation and memory Th cells. APCs can induce various kinds of Th cell responses, with Th1, Th2, and Th17 responses being the most common. The Th1 response is characterized by the secretion of IFN-γ, which enhances the bactericidal capabilities of macrophages, enhances antiviral immunity, and strengthens immunity against other intracellular pathogens. Additionally, Th1-derived cytokines contribute to the differentiation of B cells, resulting in the production of opsonizing antibodies that increase phagocyte efficiency [1, 2, 5, 6, 57, 58].

2.2.2.2 T Helper Type 2 Cells

Conversely, the Th2 response is characterized by the secretion of cytokines (IL-4, 5, and 13) that participate in the generation of immunoglobulin E (IgE)-producing B cells. It also contributes to the development and recruitment of mast cells and eosinophils, which are crucial for mounting effective responses against various parasites. Furthermore, Th2 cytokines increase the production of specific forms of IgG antibodies that assist in combating bacterial infections [5, 57, 58].

2.2.2.3 T Helper Type 17 Cells

Th17 cells are identified by their generation of IL-17 family cytokines and are linked to sustained inflammatory responses, especially in chronic infections and diseases [5].

2.2.2.4 T Regulatory Cells

Among the $CD4^+$ T-cell population, a distinct subgroup known as regulatory T cells (Tregs) plays a crucial role in modulating the immune response. The existence of Tregs was a controversial topic in the early 1980s, due to limited or no knowledge about these suppressor cells [59]. This notion was changed after the identification of human $CD4^+$ T-cell clones suppressing autologous cytotoxic antitumor immune responses [60, 61]. Tregs are now considered essential regulators for immune responses, having a considerable impact during both cancer and autoimmunity. Treg cells have the ability to inhibit and mitigate immune responses, potentially acting as mechanisms to control abnormal reactions to self-antigens and the development of autoimmune disorders. Additionally, Treg cells can aid in resolving typical immune responses as pathogens or antigens are cleared from the system. These cells also play pivotal roles in establishing "immune tolerance" toward specific foreign antigens [1, 5, 6, 58].

In its entirety, our immune system has emerged as one of the most dynamic systems within our bodies. It is tasked with constantly maintaining the delicate balance of immune homeostasis amid the continuous cycles of cell proliferation and death across numerous cell types and locations. Moreover, various challenges, including responses to diverse pathogens, medications, and cancers, must be faced, and equilibrium can be subsequently restored after these challenges are resolved, or adaptation to persistent conditions can be achieved by recalibrating homeostatic settings to minimize tissue damage.

3 Cancer Cells Are Immune to Cycles of Tumor Elimination, Equilibrium, and Escape

Cancer immune surveillance is regarded as a crucial host protection process aimed at inhibiting carcinogenesis and maintaining cellular homeostasis. The concept of cancer immunoediting is commonly visualized through three phases, referred to as the "three Es of cancer immunoediting": elimination, equilibrium, and escape [62]. In the elimination phase, immune effector cells such as NK cells identify and eliminate cancer cells, restoring affected tissues to their normal functional state. If the immune system does not entirely eradicate all cancer cells, a phase of equilibrium ensues. During this phase, the immune system is believed to control the growth of tumor cells without complete elimination. Within the equilibrium phase, two potential outcomes may unfold. First, the immune system may eventually eliminate all tumor cells, resulting in an outcome that is functionally indistinguishable from the elimination phase. Second, prolonged interaction between the immune system and tumors may gradually "edit" or modify the phenotype of developing tumors. This leads to the selection of tumors that have evolved into a less immunogenic state. Tumors that are no longer susceptible to immune attack then progress into the third phase of the immunoediting process, termed "escape." The onset of clinical cancer symptoms generally correlates with the occurrence of this stage [62–64]. The cascade of events is depicted in an illustrative diagram below (Fig. 1).

3.1 Elimination

Cancer immune surveillance involves collaborative efforts between the immune system and intrinsic tumor suppressor mechanisms to eliminate developing tumor cells. Both innate and adaptive immune responses target tumor cells. Innate responses activate cells such as NK, NKT, and γδ T cells through inflammatory cytokines released by tumor cells and surrounding macrophages and stromal cells. These cytokines recruit additional immune cells that produce proinflammatory cytokines such as IL-12 and IFN-γ [62, 63]. NK cells use perforin (pfp), Fas Ligand (FasL), and Tumor necrosis factor (TNF)-related apoptosis-inducing ligand (TRAIL) to eliminate tumor cells, releasing tumor antigens (TAs) that initiate adaptive immune responses [65, 66]. In reciprocal interactions with dendritic cells (DCs), NK cells aid in DC maturation and migration to tumor-draining lymph nodes (TDLNs), enhancing antigen presentation to naive T cells for the clonal expansion of cytotoxic T lymphocytes (CTLs). These TA-specific T lymphocytes mobilize to the primary tumor site, directly targeting and eliminating tumor cells while producing cytotoxic IFN-γ [67, 68].

The elimination process can be divided into four distinct phases as follows:

3.1.1 Identification of Tumor Cells by Cells of the Innate Immune System and Restricted Cytotoxic Activity

Advancing solid tumors induce proinflammatory signals in the stromal environment, recruiting innate immune cells to the tumor site and leading to regulated tumor cell elimination via pathways such as IFN-γ signaling [64, 66, 69].

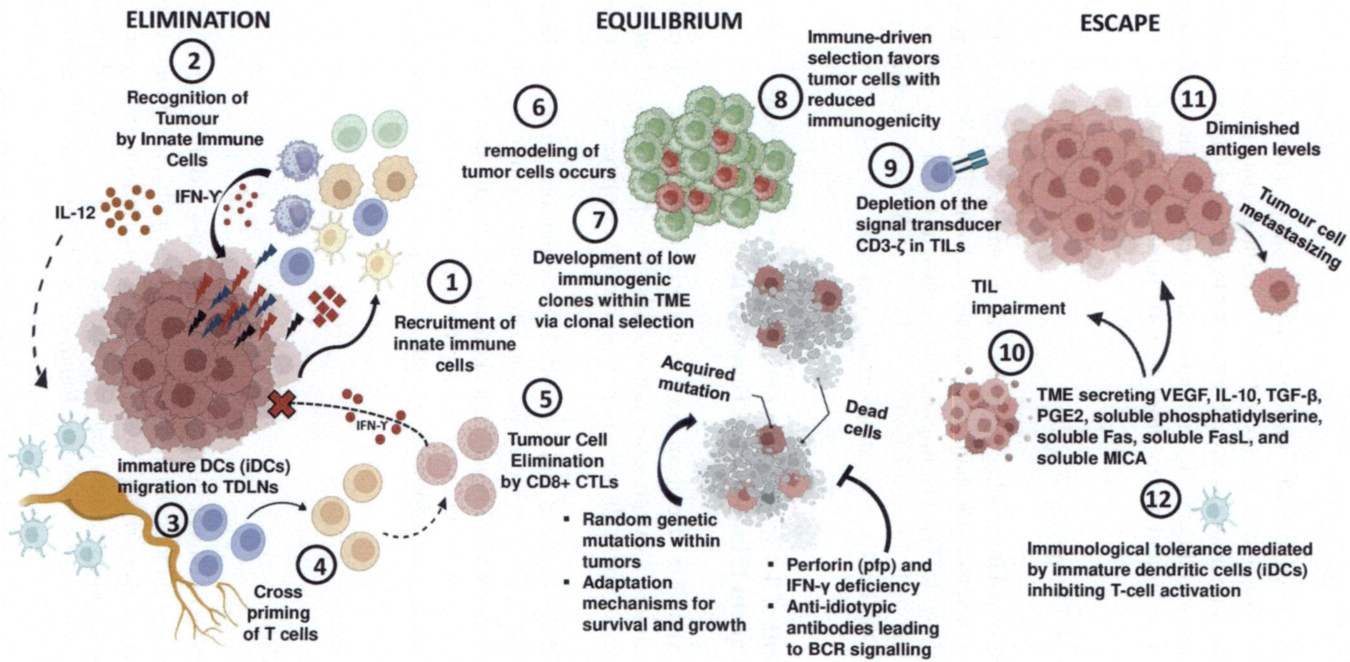

Fig. 1 Illustration of the immune-editing steps in the tumor microenvironment (TME) leading to tumor initiation and cancer progression. Cancer immune surveillance operates in three phases: elimination, equilibrium, and escape. In elimination, immune cells such as NK cells eliminate cancer cells, restoring tissues to normal. Equilibrium follows if not all cancer cells are eradicated, with two possible outcomes: eventual elimination or immune-edited, less immunogenic tumors. Escape, the third phase, indicates that tumors are no longer susceptible to immune attack, which is correlated with the onset of clinical symptoms

3.1.2 Maturation and Migration of DCs and Their Role in Cross-Priming for T Cells

IFN-γ displays limited cytotoxicity, primarily through antiproliferative and antiangiogenic actions that promote apoptosis [63, 64]. Specific chemokines from tumors and neighboring non-tumorous tissues inhibit new blood vessel formation and continuously induce tumor cell death. Inflammatory-matured immature dendritic cells (iDCs) migrate to TDLNs and phagocytose necrotic tumor cells [63–65].

3.1.3 Induction of TA-specific T Cells

In the TME, NK cells and macrophages produce IL-12 and IFN-γ, enhancing tumor cell death through cytotoxic mechanisms such as perforin, TRAIL, and reactive oxygen species (ROS). In TDLNs, migrated DCs present tumor antigens to naive $CD4^+$ T cells, driving their differentiation into $CD4^+$ T cells, which further develop into TA-specific $CD8^+$ T cells, promoting clonal expansion [63, 64].

3.1.4 Recruitment of T Cells Specific to the TA to the Tumor Site and Subsequent Elimination of Tumor Cells

TA-specific $CD4^+$ and $CD8^+$ T cells migrate to the primary tumor site, where CTLs eliminate the remaining TA-expressing tumor cells. This process is enhanced by secreted IFN-γ but also favors the selection of tumor cells with reduced immunogenicity [63, 64].

Significantly, while localized and controlled inflammation can trigger immune responses, excessive inflammatory responses can paradoxically facilitate tumor advancement under stable conditions. This phenomenon might, in part, result from anti-inflammatory responses within antigen-presenting cells, which release cytokines such as IL-10 and transforming growth factor-β (TGF-β), suppressing the activation of effector cells.

3.2 *Equilibrium*

The subsequent phase of cancer immunoediting is the equilibrium stage, involving ongoing remodeling of tumor cells that become less susceptible to immune effector cells. This dynamic process results in the immune-driven selection of tumor cells with reduced immunogenicity, contributing to the paradox of tumor development in individuals with intact immune systems. Random genetic mutations within tumors may occur, generating less immunogenic and genetically unstable variants. Immune selection pressure tends to favor the proliferation of tumor cell clones with a non-immunogenic phenotype [62–65, 69].

Experiments with mice lacking various effector molecules have shown varying levels of immune selection pressure. Lymphomas in mice lacking perforin (pfp) are more immunogenic than those in mice deficient in IFN-γ, suggesting that pfp has a more significant influence on the immune selection of lymphoma cells than IFN-γ [70, 71]. In contrast, sarcomas induced by methylcholanthrene (MCA) in mice with IFN-γ receptor deficiencies were identified as highly immunogenic [72]. Chemically induced sarcomas in nude and severe combined immunodeficiency (SCID) mice show greater immunogenicity than similar tumors in immunocompetent mice [73].

These findings suggest that initial tumor cells induced in normal mice, which are selected through T cell–mediated processes, have adapted for survival and growth in hosts with functional T-cell systems. This adaptive process leads to the elimination of highly immunogenic tumor cells, enabling non-immunogenic tumor cells to thrive [62, 63]. The decrease in immunogenicity does not correlate with reduced MHC class I expression. Notably, perforin-mediated cytotoxicity in T cells more effectively eliminates lymphoma cells than epithelial tumor cells do, whereas IFN-γ-mediated cytotoxicity predominantly targets mesenchymal tumor cells, such as sarcomas [74, 75]. Second, higher immunogenicity in tumors from immunodeficient mice than in those from immunocompetent mice suggests reduced immune selection pressure. T cell–mediated elimination seems adapted to highly immunogenic tumors induced chemically and virally [76]. Conversely, immune selection pressure favors the survival and growth of tumor variants with lower immunogenicity in the TME. In cases of spontaneously arising tumors over time, this immunogenic sculpting results in the production of fewer immunogenic tumors than chemically and virally induced tumors do [77]. Further investigations suggest that in an immune-competent host, residual tumor cells may persist, and their clinical dormancy could be linked to immune system actions [62]. A regulatory system of anti-idiotypic antibodies controls this mechanism by activating B-cell receptor (BCR) signaling. This BCR signaling inhibits cell proliferation, ultimately leading to apoptosis or a state of dormancy [78].

Clinical observations show that tumors can stay dormant in patients for extended periods, with relapse occurring even after prolonged remission, spanning a decade or more. The tumor remission phase offers insight into the immunological mechanisms regulating tumors. Despite successful therapy and extended remission, cancer recurrence remains common [79]. Recent research has shown that individuals without clinically detectable disease for more than two decades still have disseminated circulating tumor cells. Most cancer types are associated with disseminated disease that, following treatment, may persist as minimal residual disease [80, 81].

3.3 Escape

3.3.1 Alterations in Signal Transduction Molecules on Effector Cells

Owing to a lack of TA recognition, there is a crucial mechanism involving alterations in effector molecules to increase immune system recognition and activation. Depletion of the signal transducer CD3-ζ in TILs is linked to immune evasion, accompanied by the impact of immunosuppressive cytokines and local impairment of TILs [82].

3.3.2 Tumor-Derived Soluble Factors

Tumors employ immune editing to evade immune control, exerting selective pressure in the TME and potentially promoting malignancy. Various tumor-released soluble factors, including VEGF, IL-10, TGF-β, prostaglandin E2 (PGE2), soluble phosphatidylserine, soluble Fas, soluble FasL, and soluble MICA, contribute to the development of local and regional immunosuppressive networks. When these factors are initially localized at the primary tumor site, their immunosuppressive effects can extend to nearby lymph nodes and the spleen, facilitating invasion and metastasis [65, 66, 83]. The detailed role of these factors is discussed later in the chapter.

3.3.3 Immunological Ignorance and Tolerance in Tumors

The effectiveness of a tumor-specific immune response is influenced by factors such as TA levels and the maturation stages of antigen-presenting cells, especially DCs. While solid tumors such as sarcomas and carcinomas have tumor-specific antigens, making them potential targets for immune effector T cells, the overall efficacy of immune surveillance against these tumors is relatively limited [84].

Tumor cells introduced as individual cell suspensions can induce a protective cytotoxic T-cell response. However, when transplanted as small fragments, tumors often exhibit increased proliferation. This may be due to the modulation of tumor antigen levels within the TME, where tumor cells are mixed with nontumor cells, including bone marrow-derived cells such as iDCs, as well as non-bone marrow-derived cells such as fibroblasts, endothelial cells, and the extracellular matrix (ECM) [85]. Adequate TA levels can trigger an immune response facilitated by mature DCs through cross-priming. Despite sufficient antigen levels, iDCs hinder DC maturation and T-cell activation, promoting immunological tolerance. Tumor immune evasion may not only result from diminished antigen levels but also involve immunological tolerance because iDCs inhibit T-cell activation [86].

Despite considerable research, the exact mechanisms through which cancer cells evade the immune system remain unclear. Nevertheless, it is theorized that the interaction between the TME and the molecules and substances produced by cancer cells influences the development, maturation, and subsequent specialization of immune cells, shaping the immune landscape within the TME.

4 Contexture of the Immune System in the Tumor Microenvironment (TME): Impact of Immune Training in the TME on the Immune Evasion of Tumor Cells

The ability of cancer to evade the immune system presents a significant obstacle in the development of effective anticancer therapeutic approaches. Several factors contribute to the persistence of tumors, even in individuals with intact immune systems. As previously discussed, immune editing plays a crucial role in tumor evasion, allowing tumors to remain dormant within patients for extended periods, cycling through phases of "equilibrium" and "senescence" before resurfacing. Additionally, tumors exploit various immunological processes, such as targeting regulatory T-cell functions or their secretions, manipulating antigen presentation, altering the production of immune-suppressive substances, promoting tolerance, and inducing immune diversion [87]. Moreover, tumor progression is significantly influenced by tumor heterogeneity and metastasis. A pictorial representation of the immune terrain and its influence on cancer progression is given below (Fig. 2).

4.1 Regulatory Cells

The substantial immune suppression observed in the TME, which is mainly orchestrated by $CD4^{+}CD25^{+}$ $FoxP3^{+}$ Tregs or other suppressive cell types, represents a major mechanism of tumor immune evasion and presents a formidable hurdle for tumor immunotherapy [88]. Recent research has revealed that $CD4^{+}$ $NRP1^{+}$ Tregs play a role in promoting angiogenesis and contributing to tumor progression [89]. Various studies have demonstrated that Tregs derived from tumors possess greater suppressive capabilities than naturally occurring Tregs do [90]. Chemokines produced by tumor cells play a role in recruiting Tregs into the TME [91]. Evidence indicates that the release of TGF-β by diverse cell types, including tumor cells, induces the conversion of $CD4^{+}$ T cells into suppressive Tregs within the TME [92]. Therefore, eliminating Tregs through the use of anti-CD25 monoclonal antibodies (mAbs) or other approaches might increase tumor rejection [93]. Myeloid cells, particularly myeloid-derived suppressor cells (MDSCs), modulated DCs, and alternatively activated M1 and M2 macrophages, play a role in establishing an inflammatory microenvironment and can act as facilitators of tumor initiation, angiogenesis, and metastasis [94, 95]. Moreover, a detrimental cycle might commence, where increased levels of inflammatory mediators confer resistance to apoptosis in MDSCs. These apoptotic signals are otherwise regulated by T cells within intricate interaction networks [95]. Consequently, $CD11b^{+}Gr1^{+}$ MDSCs inhibit the antitumor immune response facilitated by $CD8^{+}$ T cells, and one potential mechanism for this inhibition involves the downregulation of the TCR ζ-chain [96]. MDSCs exhibiting this phenotype tend to accumulate in locations such as melanoma lesions, which are closely associated with the inflammatory environment, implying that the inflammatory

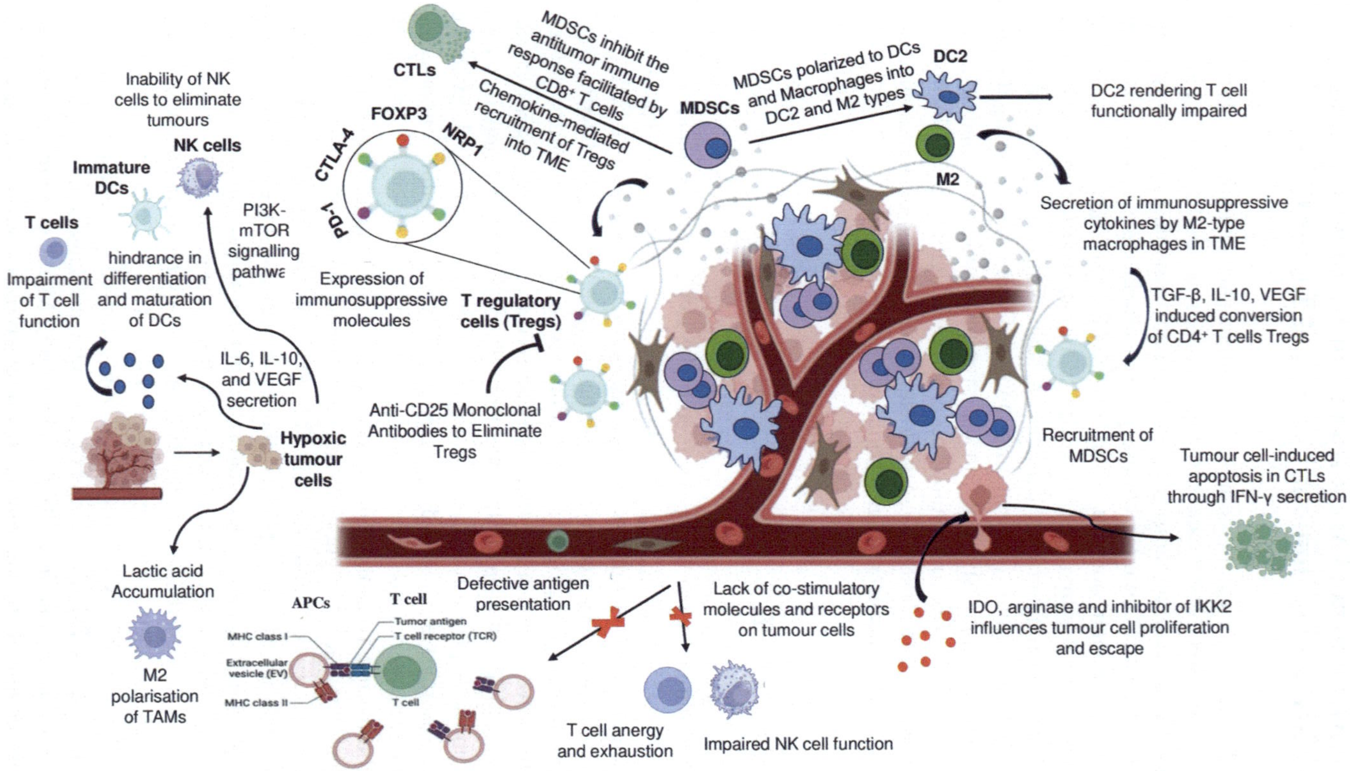

Fig. 2 Schematic representation of the composition of the immune system within the TME. The evasion of the immune system by cancer hinders effective anticancer therapies. Immune editing allows tumors to persist, cycling through "equilibrium" and "senescence" phases. Tumors manipulate regulatory T-cell function, antigen presentation, immune-suppressive substances, tolerance, and immune diversion. Tumor heterogeneity and metastasis significantly impact the progression

microenvironment of tumors promotes MDSC recruitment and immunosuppressive activity. The mitigation of chronic inflammatory mediators through pharmacological interventions can decrease MDSC levels and alleviate immunosuppression [97]. Macrophages expressing an M2 phenotype marked by $CD11b^{+}F4/80^{+}$ phenotypes produce increased levels of TGF-β, IL-10, and VEGF, fostering tumor growth [98–100]. Furthermore, various tumor-derived factors and gangliosides have been noted to impact the phenotype of DCs. These immature DCs, which are functionally impaired, exhibit diminished expression of CD80, CD86, and CD40 and increased IDO levels, thereby contributing to the inhibition of T cell–mediated immunity [101].

4.2 Defective Antigen Presentation

A well-established and crucial mechanism by which tumors evade immune surveillance is the decrease in components of the antigen-processing machinery that impact the MHC class I pathway. This involves the suppression of proteasome subunits such as latent membrane protein (LMP)2 and LMP7, transporter associated with antigen processing (TAP) protein, and tapasin [102–105]. As a result, the expression of TAs is diminished, potentially leading to an increased incidence of tumors and metastasis. This is attributed to the inability of CTLs to recognize the target antigens displayed on the surface of tumor cells [106].

4.3 Immune-Suppressive Mediators

As previously noted, tumors employ tactics to elude immune surveillance, with one strategy involving the impairment of CTLs through the production of various immune-suppressive cytokines. These cytokines can originate from cancer cells or noncancerous cells within the TME, including immune and epithelial cells. A key mediator of immune evasion is TGF-β. Additionally, immune-suppressive cytokines such as TNF-α, IL-1, IL-6, colony-stimulating factor (CSF)-1, IL-8, IL-10, and IFNs play significant roles in fostering cancer growth [107, 108]. In addition to immune-suppressive cytokines, factors such as VEGF, which are produced by tumors, impede the differentiation of progenitor cells into DCs, thereby affecting the efficient uptake and presentation of antigens [107, 108]. VEGF, IL-10, and TGF-β also impede the maturation of DCs. Immature DCs are tolerogenic and lack the presentation of antigens in the proper context with the required co-stimulation for T-cell activation [109]. It was reported that noncytolytic $CD4^{+}$ T cells produce type 2 cytokines (IL-4 and IL-10) upon stimulation by autologous APCs or tumor cells in an MHC class II–dependent way, which inhibited T-cell activation and suppressed APC co-stimulatory molecules CD80 and CD86. Neutralizing IL-10 in the tumor cell supernatant removed this inhibition, giving future directions for designing APC-based tumor vaccines like DC-based cancer vaccines [110, 111]. Moreover, factors such as tumor gangliosides and receptor-binding cancer-associated surface antigens (RCAS1) play a role in tumor progression [112]. Additional research has revealed a correlation between RCAS1 expression and the apoptosis of TILs. Similarly, ganglioside antigens, whether present on the cell surface or released from it, are recognized for their ability to inhibit CTL and DC function [113–115]. Immunosuppressive enzymes such as IDO, arginase, and inhibitors of nuclear factor kappa B kinase (IKK)2 may play a significant role in tumor progression by directly influencing tumor cell proliferation or by inducing T-cell tolerance and suppression [116, 117].

4.4 *Tolerance and Immune Deviation*

The concept of immunological ignorance and tolerance in tumors has been previously discussed. In essence, most tumor cells lack the expression of co-stimulatory molecules, potentially resulting in the induction of anergy or tolerance in T cells when they interact with the T-cell receptor without adequate co-stimulation [118]. Tumors escape immune responses by altering the Th1-to-Th2 balance, a phenomenon referred to as immune deviation. This transition is influenced by TGF-β and IL-10 [119]. It has been shown that Th1 conditioning with interleukin (IL)-12 and anti-IL-4 antibody induced better and longer-living CTL response and prevented the expansion of $CD4^+$ T cells that have downregulatory activity [120]. This paved the pathway for developing chimeric antigen receptor (CAR) T-cell therapy against cancer that is genetically altered in a lab to enable it to locate and destroy cancer cells more effectively [121]. Moreover, the manifestation of inhibitory molecules such as programmed cell death (PD)-L1/B7H1 by tumors has been demonstrated to lead to the elimination or inactivation of tumor-reactive cells [122]. Additionally, evidence indicates that the suppression of death receptors hinders the killing of tumor cells by death ligands mediated by both CTLs and NK cells [123, 124]. Studies involving CTLs and antibody blocking have demonstrated enhanced antitumor effector activity, which is primarily mediated through CD54 [125]. Moreover, reports suggest the participation of the p53 tumor suppressor gene in regulating tumor cell death via CTLs [126]. As a result, factors that contribute to tolerance and immune deviation play crucial roles in the evasion of the immune system by cancer.

4.5 *Apoptosis*

Several studies have demonstrated that cancer cells induce apoptosis in tumor-specific CTLs. IFN-γ can initiate apoptosis in these specific T cells, compromising the effectiveness of antitumor immunity. In situations with a low tumor burden, the administration of both anti-CTLA-4 and anti-PD-1 therapy antibodies activates T cells, causing increased intra-tumoral IFN-γ secretion. This heightened IFN-γ, in turn, results in T-cell death, contributing to the escape of the tumor from immune surveillance [127]. It has also been reported that tumor antigens, which are weakly immunogenic self-molecules, typically result in tumor-specific T cells characterized by low precursor frequencies and low TCR affinity. This is because those tumor-specific T cells with high avidity undergo deletion during the thymic selection process. Moreover, the TME hampers the antigen presentation process, leading to inadequate priming and boosting of T cells. Despite the infiltration of effector T cells into the TME, a complex immunosuppressive network involving cancer cells, inflammatory cells, stromal cells, and cytokines regulates their activity. Among these TME components, cancer cells, inflammatory cells, and suppressive cytokines play pivotal roles in shaping the phenotype and function of T cells. These components ultimately drive T cells to terminally differentiate into "exhausted" T cells [128]. These exhausted T cells display increased expression of inhibitory receptors such as PD-1, LAG-3, TIM-3, CTLA-4, BTLA, and TIGIT. Consequently, there is a decrease in the production of effector cytokines and a diminished ability to eradicate cancer. T-cell exhaustion may also be linked to impaired generation of memory T cells, ultimately leading to the ultimate phase of physical deletion, wherein severely exhausted T cells undergo apoptosis in the TME [129, 130].

4.6 Hypoxia

In addition to the unique functions of immunosuppressive regulators such as IL-10, TGF-β, and VEGF, hypoxia within and around tumor vessels also plays a role in the metastatic spread of cancer cells, which is dependent on hypoxia-inducible factor (HIF), VEGF, and inducible nitric oxide synthase (iNOS) [131]. A recent investigation suggested that hypoxia profoundly influences the immune microenvironment, providing tumor cells with the means to elude immune surveillance and elimination [132]. The A2A receptor (A2AR), a G protein-coupled receptor with heightened sensitivity to adenosine, is present on diverse immune cells, such as T cells, NK cells, and macrophages. Tumor cells exploit the adenosine–A2AR pathway to hinder immune cell responses, aiding in their evasion of immune surveillance. This pathway triggers apoptosis in T lymphocytes, hampers the differentiation of effector cells, and reduces the production of effector T cells and IFN-γ [133]. In the hypoxic TME, the regulation of B-cell transcription and metabolism is governed primarily by HIF-1α and the myelocytomatosis virus oncogene cellular homolog (MYC) [134]. Several studies have shown that hypoxia induces the upregulation of cytokines, including IL-6, IL-10, and VEGF, which hinders the differentiation and maturation of DCs and impairs T-cell function [135]. Hypoxia-induced activation of the phosphatidylinositol 3 kinase (PI3K)–mTOR signaling pathway inhibits the ability of NK cells to eliminate tumor cells [136]. Moreover, the accumulation of lactic acid mediated by hypoxia, under the regulation of HIF-1α, increases VEGF expression and promotes M2-like polarization of TAMs, thereby maintaining the immunosuppressive state of the TME [137].

5 Development of an Immunosuppressive Tumor Microenvironment

As tumor cells and their stroma advance, the multifaceted immunosuppressive mechanisms intensify. Molecules released by tumors mold the TME and instigate immunosuppression, weakening robust antitumor immune responses. An illustrative representation of the interplay of tumor cells, stromal components, and released molecules, collectively fostering conditions that hinder the effectiveness of the immune system in combating tumors, is given below (Fig. 3).

5.1 Immature DCs

Inadequate activation of DCs in the TME contributes to immunosuppression. When TME DCs fail to mature, they present tumor antigens, leading to T-cell anergy or tolerance [109, 118]. Tolerogenic dendritic cells (tDCs) in the TME, characterized by reduced CD80 and CD86 expression and elevated PD-L1 and CTLA-4 levels, resemble immature DCs [109]. tDCs express various immunomodulatory factors and immunosuppressive molecules induced by factors such as pathogen-derived products, VEGF, cytokines from tumors (e.g., TGF-β and IL-10), and IL-10-secreting Tregs. Immature DCs that phagocytose apoptotic/necrotic DCs can transform into tDCs with increased TGF-β secretion [138, 139]. IL-10-releasing tDCs hinder cytotoxic and T helper cell activation, promote type 1 regulatory T cell (Treg) differentiation via the ILT4/HLA-G pathway, and induce $Foxp3^+$ Tregs from naive $CD4^+$ T cells [65].

5.2 Inhibition of TILs

TILs denote lymphocytes that travel to the tumor location in reaction to molecular cues. These categories include diverse T-cell types, such as Tregs, along with innate lymphoid cells (ILCs), such as NK cells and NKT cells [140]. The secretion of chemokines by both tumor cells and the surrounding stroma plays a crucial role in shaping the infiltration of immune cells, regulating tumor cell proliferation, and influencing the likelihood of metastasis [141]. However, the penetration of TILs into the central region of the tumor is notably diminished because of compromised blood vessels. These vessels exhibit reduced expression of receptors essential for the attachment, rolling, and migration of TILs through the vascular walls [142]. Moreover, the presence of M2 macrophages and other immune cells with immunosuppressive properties can modify the pattern of chemokine secretion. This collective effect not only draws more immunosuppressive cells but also impedes the

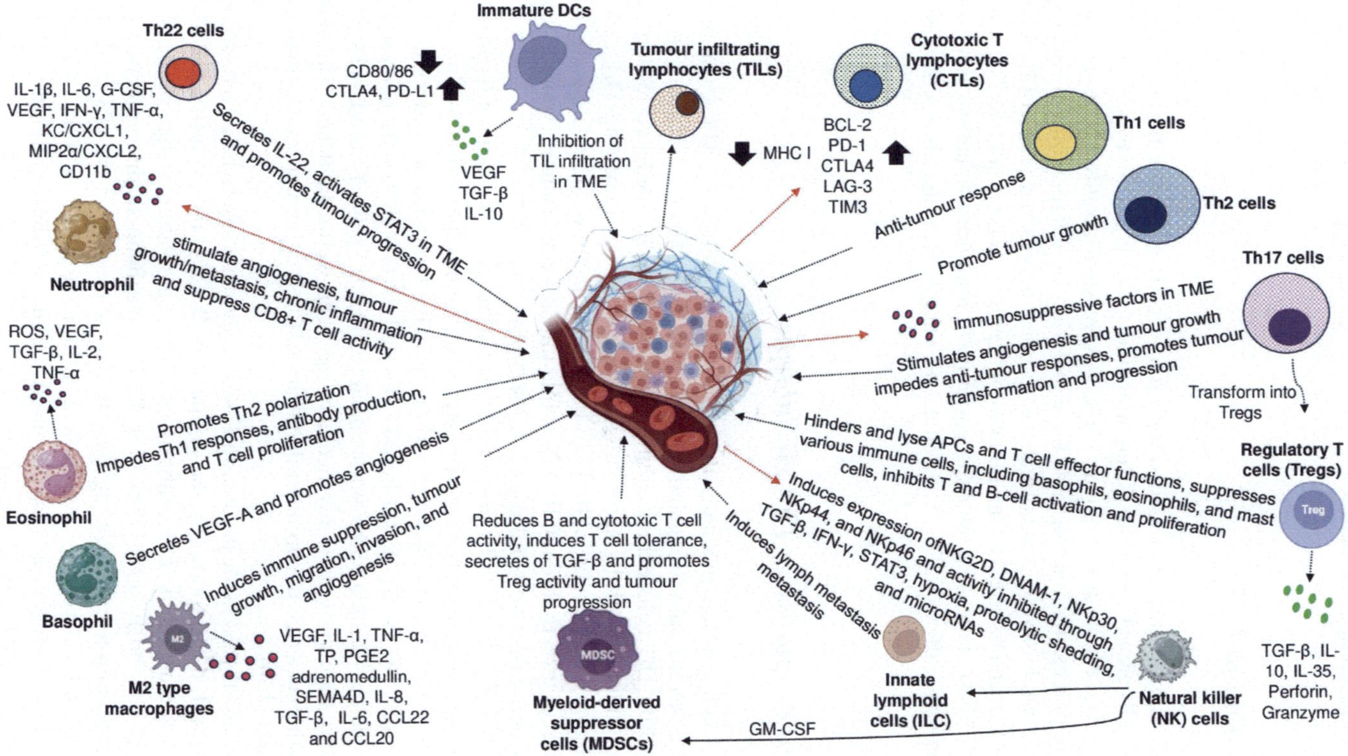

Fig. 3 Illustration showing the formation of an immunosuppressive TME. As tumor cells and the surrounding stroma continue to advance, the sophistication of immunosuppressive mechanisms further intensifies. The molecules released by tumors play an active role in molding the TME, which is involved in motion processes that induce immunosuppression. This orchestrated immunosuppressive environment, in turn, weakens the vigor of effective antitumor immune responses. Consequently, the establishment of such a microenvironment provides favorable conditions for the progression and growth of tumors

infiltration of TILs into the tumor site [143]. The inhibition of TIL penetration into the TME is discussed in detail below.

5.3 *Inhibition of CTLs*

Upon encountering mature DCs loaded with tumor antigens in the lymph nodes, naive $CD8^+$ T cells are activated and differentiate into CTLs [118]. To evade destruction by CTLs, tumor cells downregulate MHC I expression, acquire mutations hindering antigen processing, and increase the expression of antiapoptotic molecules such as BCL-2 while showing abnormal Fas receptor expression [105, 144]. As a self-regulatory mechanism, CTLs upregulate checkpoint receptors (PD-1, CTLA-4, LAG-3, and TIM-3), increasing their susceptibility to inhibitory signaling. Binding to these receptors induces CTL exhaustion and anergy, limiting their effectiveness [145].

5.4 *Immunosuppressive Th Cell Subtypes*

Naive $CD4^+$ T cells, which encounter tumor antigens presented by APCs, can differentiate into various subsets, including Th1, Th2, Th9, Th17, Th22, and Tregs [146, 147]. While Th1 cells exhibit antitumor responses, Th2-mediated responses may promote tumor growth by supporting angiogenesis and impairing Th1 cell–mediated immunity. Th17 cells in the TME have dual functions: stimulating angiogenesis and tumor growth through IL-17A while impeding antitumor responses via immunosuppressive adenosine release [146, 147].

Moreover, Th17 cells have the potential to transition into Treg cells in response to microenvironmental factors, contributing to immunosuppression within the tumor. Notably, an elevated presence of Th17 cells in the tumor is correlated with an unfavorable prognosis [148]. Nevertheless, elevated Th17 cell counts in ovarian tumors have been associated with improved survival rates [149]. The link between chronic inflammation and tumor advancement has been well-established. Th2 and Th17 cells both play a role in chronic inflammation, fostering tumor transformation and progression [150].

Th22 cells are tasked with producing IL-22, a cytokine that shares structural similarities with IL-10. Importantly, the synthesis of IL-22 is not exclusive to Th22 cells; it is also generated by Th17 subsets and NK cells. Additionally, γδ T cells express both IL-22 and IL-17A [151–153]. IL-22 contributes to the promotion of tumor cell proliferation by activating STAT3. The activation of STAT3 tends to induce immunosuppressive activities in diverse immune cells, such as DCs and macrophages [154]. Consequently, an increased presence of tumor-infiltrating Th22 cells has been associated with tumor progression.

Tregs, which are extensively found in the TME, are closely linked to a poorer survival rate. The differentiation of Tregs typically occurs through various pathways. In the thymus, naive $CD4^+$ T cells can differentiate into $CD25^+$ $Foxp3^+$ Tregs, commonly known as natural Tregs. Additionally, in peripheral tissues, these cells can differentiate into $CD25^-$ Tregs, which include various subtypes, such as IL-10-producing Tr1 cells, TGF-β-secreting Th3 cells, and $Foxp3^+$ cells. This collection is collectively referred to as inducible Tregs, or iTregs [146, 147]. Later, the distinctive roles of various other Treg subtypes in the TME and cancer progression are discussed. Generally, Tregs infiltrate the TME through signaling molecules, including chemokines and their respective receptors (such as CCL1–CCR8, CCL5–CCR5, CCL22–CCR4, CCL28–CCR10, and CXCL12–CXCR4) [155]. Once they are present, Tregs contribute to promoting tumor growth and metastasis by releasing immunosuppressive cytokines, particularly TGF-β, IL-10, and IL-35, thus contributing to various suppressive functions of Tregs [156].

Recent findings have shown that Tregs hinder the ability of DCs to function as APCs by removing peptide–MHC II complexes from DCs. This indirectly suppresses T-cell activation by reducing the number of mature DCs and transforming them into tolerogenic counterparts [157]. Tregs directly suppress T-cell activation and proliferation by generating soluble CD25, sequestering IL-2, and releasing IL-10 and TGF-β [158, 159]. Tregs interfere with T-cell metabolism via adenosine and cAMP through CD39- and CD73-expressing Tregs. They release IL-35, which induces the expression of inhibitory receptors (PD-1, LAG-3, and TIM3) and fosters exhaustion in intra-tumoral TILs [160–162]. Tregs also directly and indirectly suppress NK cell function, inhibiting NKG2D-mediated NK cell cytotoxicity through a TGF-β-dependent mechanism [163]. Tregs suppress various immune cells, including basophils, eosinophils, and mast cells, and inhibit B-cell activation and proliferation [164]. Tregs release perforin and granzymes upon T-cell receptor engagement, which target and lyse cytolytic T cells and APCs [165].

5.5 NK Cells

NK cells induce tumor cell apoptosis by releasing cytolytic granules containing perforin and granzymes. Their killing capacity is regulated by the induction of the inhibitory receptors, including NKG2D, DNAM-1, NKp30, NKp44, and NKp46. Tumor cells downregulate surface ligands, hindering recognition by the immune system, especially NK cells, through factors such as TGF-β, IFN-γ, STAT3, hypoxia, proteolytic shedding, and microRNAs (miRNA-20a, miRNA-106b, miRNA-93, miRNA-373, and miRNA-520d) [27, 166].

Cancer cells release immunosuppressive microvesicles, including exosomes with surface NKG2DLs, disrupting NKG2D receptors and hindering immune system recognition [167, 168]. $CD11b^{high}$ $CD27^{high}$ NK cells can transition into MDSCs in the TME via factors such as GM-CSF [169]. While NK cells play a key role in tumor eradication, other ILC subpopulations (ILC1s, ILC2s, and ILC3s) have dual functions, with the potential for transdifferentiation between subsets [170, 171]. Elevated RORγt$^+$ ILC3s are associated with increased lymph node metastasis [171]. Cancer cells that release TGF-β can also transform NK cells into ILC1s within the TME for immune evasion [92].

5.6 *Tumor-Associated Macrophages*

In the tumor-associated immune stroma, TAMs play crucial roles in influencing tumor progression and are categorized into proinflammatory M1 and anti-inflammatory M2 types. M2 TAMs contribute to immune suppression, tumor migration, invasion, and angiogenesis by releasing factors such as VEGF, IL-1, TNF-α, thymidine phosphorylase (TP), PGE2, adrenomedullin, and semaphorin 4D. They also participate in ECM degradation through enzymes such as uPA, MMP-7, and MMP-9 [65, 172]. ARG-1 induced by TAMs, triggered by lactic acid, impacts tumor growth by reducing L-arginine levels and downregulating the TCR ζ chain in infiltrating T cells [173]. Furthermore, TAMs release VEGF, IL-8, and IL-6, contributing to tumor progression [174]. TAMs influence tumor cell invasiveness through CCL8 induction and secretion, leading to increased CSF-1 expression and attracting circulating monocytes expressing CCR2 upon TNF-α stimulation [175]. Additionally, TAMs release the cytokines CCL22 and CCL20, recruiting CCR4$^+$ and CCR6$^+$ nTreg cells, respectively [176]. TGF-β-secreting cells such as TAMs and tumor cells inhibit T cells and NK cells, impede DC migration, and promote the differentiation of CD4$^+$ T cells into Th2 cells and Tregs [92]. Compared with those without PD-1 expression, PD-1-expressing TAMs exhibit diminished phagocytic activity [177].

5.7 *Granulocytes*

5.7.1 Basophils

Recent research has suggested that regulatory T cells can influence the activation of basophils through IL-3 and STAT5 signaling [178]. In addition to their direct interactions with immune cells, basophils can exert indirect effects on tumor growth by generating VEGF-A, a pro-angiogenic factor. Moreover, in individuals with ovarian cancer, basophils may contribute to symptoms associated with ascites, as VEGF-A can increase capillary permeability [179]. Importantly, the occurrence of basophils in the lymph nodes draining tumors in individuals with pancreatic ductal adenocarcinoma and their diminished levels in the bloodstream of patients with colorectal cancer have been linked to an unfavorable prognosis [180].

5.7.2 Eosinophils

Eosinophils release various substances, including reactive oxygen species, growth factors (VEGF and TGF-β), cytokines (IL-2 and TNF-α), and lipid mediators (eicosanoids). They present antigens to helper T cells and promote Th2 polarization [181, 182]. Regulatory eosinophils suppress immune responses, upregulate PD-L1 expression in an IFN-γ-dependent manner, inhibiting Th1 responses, hindering B-cell antibody production, restricting T-cell proliferation, and inducing mast cell degranulation through cationic protein release [183].

5.7.3 Neutrophils

In the TME, neutrophils present two phenotypes: anti-tumorigenic N1 and pro-tumorigenic N2. N2 neutrophils play a key role in immunosuppressing advancing tumors [184]. Advanced cancer often correlates with elevated blood neutrophil counts influenced by cytokines (IL-1β, IL-6, G-CSF, and VEGF). Neutrophils are recruited to tumor tissue by cytokines (IFN-γ and TNF-α), chemokines (KC/CXCL1 and MIP2α/CXCL2 in mice), and cell adhesion molecules, notably CD11b [185].

When N2 neutrophils or tumor-associated neutrophils (TANs) infiltrate a tumor, TGF-β activates them to release molecules that impact the tumor microenvironment. They stimulate angiogenesis (VEGF), support tumor growth/metastasis (cathepsin G and ROS), maintain chronic inflammation (TNF-α, IL-1β, IL-6, and IL-12), and suppress $CD8^+$ T-cell activity [186–188]. Mice with reduced $CD11b^+$/$Ly6G^+$ TANs show decreased tumor growth and increased $CD8^+$ T-cell activation [189]. Neutrophil-derived leukotrienes selectively promote tumor cell proliferation [190]. N2 TANs indirectly support tumor cell survival by attracting macrophages and regulatory T cells (Tregs) into the microenvironment [188].

5.8 Myeloid-Derived Suppressor Cells

MDSCs, including myeloid progenitor cells and immature myeloid cells, induce $CD4^+$ T-cell tolerance via MHC class II [191]. They diminish immune cell responsiveness to interferons (IFNs) by reducing STAT1 phosphorylation in tumor-bearing mice [192]. During B-cell responses, MDSCs regulate this process, as demonstrated by reduced bone marrow B cells in lung cancer model mice, which are linked to MDSC infiltration into the TME, which is mediated by IL-7 and increased STAT5 activation [193]. MDSCs promote Tregs through tumor-associated antigens, IL-10, IFN-γ, and CTLA-4 [194].

Tumor-derived exosomes containing TGF-β and PGE2 induce MDSCs, fostering tumor progression by recruiting $CD11b^+Gr\text{-}1^+$ cells [195, 196]. Under inflammatory conditions, Tregs enhance MDSC differentiation and function by releasing TGF-β, amplifying the inhibitory role of MDSCs [197]. Mast cells release histamine in the

TME, binding to histamine receptor 1 on MDSCs and inhibiting T-cell proliferation through increased Arg1 and iNOS production [198]. MDSCs possess potent immunosuppressive properties, hindering antitumor immunity.

6 Phenotypic and Functional Switching of Immune Cells in the TME

The TME shapes the immune response through the phenotypic, functional, and metabolic characteristics of TILs and TIMs. This customization aims to facilitate tumor immune evasion, metastasis, and recurrence, imposing selective pressure during early cancer development. Ongoing immunoediting occurs within immune-infiltrated regions, and TILs may not universally recognize TAs.

6.1 Tissue-Resident Memory (TRM) Cells

Among TILs, TRM cells can be categorized into two types, each of which potentially predicts clinical outcomes. The first type, marked by Ki67, CD103, PD-1, TIM-3, and ICOS, is associated with poor prognosis [199]. Conversely, the second type, with a more favorable cytolytic profile, is linked to a favorable prognosis. Two $CD8^+$ TIL subpopulations with gene expression resembling that of memory cells have been identified: one with circulating precursors and the other with tissue-resident precursors in juxta-tumor tissue. These precursors differentiate into terminally differentiated cells, which often exhibit dysfunction or exhaustion [200].

6.2 Mucosal-Associated Invariant T (MAIT) Cells

MAIT cells, a subset of proinflammatory, cytotoxic CD8+ effector memory T cells, have been linked to the promotion of tumor initiation, growth, and metastasis in a lung cancer mouse model [200]. Tumor cells expressing MR1 activate MAIT cells, leading to a reduction in NK-cell effector function, which is partially mediated by IL-17A [201].

6.3 $CD8^+$ TILs

$CD8^+$ TILs express exhaustion-associated molecules, including PD-1, LAG-3, TIM3, TIGIT, CD244, and CD160 [202]. CD39, initially identified as an activation marker, breaks down extracellular ATP and ADP into AMP, which is then converted to adenosine by CD73. Adenosine binding to A2A receptors inhibits T-cell activation and NK cell cytotoxicity [203]. The expression of migratory receptors and their ligands is disrupted in various TIL subsets. In cancer, CXCR6 is linked to dysfunctional T cells, and its increase following checkpoint blockade treatment is countered by TCF1 [204, 205]. $CD8^+$ T cells with high PD-1 expression show a distinct profile with reduced effector cytokine production and CXCL13. CXCL13 recruits immune cells to tertiary lymphoid structures (TLSs), suggesting that B cells may contribute to TLS formation [206].

6.4 Tumor-Infiltrated B Cells

There are two subtypes of tumor-infiltrating B cells: naive-like B cells and plasma cell–like B cells. Naive-like B cells decrease in advanced non-small cell lung cancer (NSCLC), with their reduced presence linked to an unfavorable prognosis owing to their growth-suppressive impact. Conversely, plasma cell–like B cells initially inhibit cancer cell growth in early-stage NSCLC but can promote growth in advanced stages. A high density of B cells within TLSs can counterbalance the negative effect of a high density of Tregs on patient survival [207].

6.5 T Follicular Helper cells (TFH)

Tfh cells in TLSs collaborate with B cells and are memory $CD4^+$ T cells that express PD-1, ICOS, and CXCR5. The development of TLSs is influenced by CXCL13 signaling, which recruits B cells through Tfh cells [208]. The abundance of Tfh cells in tumors tends to increase, possibly driven by elevated levels of TGF-β, promoting their differentiation [209].

6.6 Tregs (Tumor-Infiltrating Tregs, or ti-Tregs)

Tregs suppress antitumor immune responses, creating an immunosuppressive TME and hindering effector T-cell activity, promoting tumor growth. These cells exhibit potent suppressive features, marked by increased expression of immune checkpoint markers such as IL1R2, PDL1, PDL2, and CCR8 on their surface [210]. Elevated expression of genes such as *LAYN, MAGEH1,* or *CCR8* in these cells is associated with unfavorable prognoses. IRF4 Tregs, which are known for expressing suppressive molecules in NSCLC, are linked to multiple exhausted T-cell subpopulations and poor prognostic outcomes [211].

6.7 NK Cells

NK cells possess activating and inhibitory receptors, chemokine receptors, and regulatory/cytotoxic molecules such as CD16 (FcγRIII receptor) and CD56 (neural cell adhesion molecule) [212]. In the TME, NK cells exhibit decreased cytotoxicity, as evidenced by reduced granzyme B and IFN-γ production and diminished CD57 expression. Local TGF-β production can induce an ILC1 phenotype in NK cells, which is characterized by reduced cytotoxic activity [213, 214].

6.8 DCs

Tumors have the capacity to impede the function of DCs or manipulate the TME to attract immune-suppressive DCs. In the case of NSCLC, DCs exhibit increased levels of B7-H3 (CD276), a co-inhibitory receptor, leading to ineffective activation of T lymphocytes [215]. Compared with the healthy lungs, early-stage lung adenocarcinoma has a notable decrease in $CD141^+$ DCs, which are recognized for their preferential interaction with CD8+ T cells [216]. Conversely, there is an increase in $CD1c^+$ DCs within tumors, and these DCs express cytokines such as IL-6, IL-8, and IL-1β, which are typically associated with $CD14^+$ monocytes rather than DCs in the normal lung [217].

6.9 TAMs

Significant heterogeneity exists in TAMs, particularly between the invasive margin and the tumor center. The invasive margin tends to exhibit a greater density of M2 TAMs. Poor overall survival in cancer patients has been associated with lower numbers of M1 TAMs or the closer proximity of M2 TAMs to tumor cells [218]. TRMs contribute to tumor progression through both direct and indirect mechanisms. They can induce epithelial-to-mesenchymal transition (EMT), promoting the invasiveness of cancer cells [219]. Additionally, they play a role in expanding Tregs, providing protection to tumor cells from $CD8^+$ T lymphocyte–mediated killing. Certain macrophages expressing MARCO were found to inhibit the activation of CTLs and NK cells, with this inhibition linked to increased Treg proliferation and the production of IL-10 [220]. Cancer cells release succinate, a TCA cycle byproduct, which activates the succinate receptor (SUCNR1) signaling pathway, polarizing macrophages into TAMs via the PI3K/HIF-1α axis [221].

6.10 *TANs*

The presence of TANs in the TME is a significant predictor of an unfavorable outcome. An elevated intra-tumoral ratio of neutrophils to $CD8^+$ T cells indicates a poor prognosis, impacting both recurrence-free survival and overall survival [222]. The N2 phenotype of neutrophils is characterized by increased expression of factors that promote tumor growth, such as elevated production of IL-8/CXCL8 and reduced production of TNF-α or CXCL10 [223, 224]. The polarization of TANs is significantly influenced by TGF-β and IFN-β [225]. In a model involving acute radiation-induced lung injury, locally activated neutrophils play a crucial role in creating a tumor-supportive microenvironment in the lung. This support was mediated through enhanced regenerative Notch signaling, subsequently amplifying a stemness phenotype [226]. Additionally, neutrophils can facilitate the lung colonization of cancer cells with metastatic potential by producing leukotrienes derived from arachidonate 5-lipoxygenase. These leukotrienes aid in the selective expansion of metastasis-initiating cancer cells, allowing them to colonize distant tissues [190]. Furthermore, neutrophil extracellular traps, which form due to chronic lung inflammation resulting from exposure to tobacco smoke, can convert disseminated dormant cancer cells into rapidly growing metastases [227].

7 Tug of War Between Antitumor and Immunosuppressive Immune Responses

The ultimate fate of tumors is molded by the collective interplay between immune cells combating the tumor, those promoting its growth, and the associated cytokines, all of which function within the TME. Therefore, gaining insight into the intricate interactions and reciprocal control of these complex immune networks within the tumor environment, instead of solely focusing on individual components, is the approach most likely to lead to the development of effective and targeted cancer treatments.

Under normal circumstances, inflammation serves the purpose of safeguarding specific areas of tissue that are either infected or injured. This is achieved by summoning the necessary cells to address the issue while also isolating the area to prevent the infection from spreading. Once the problem is resolved, the tissue usually returns to its normal functioning state. However, there are instances in which inflammation persists and fails to resolve. This lingering inflammation can, in turn, stimulate the growth, survival, and development of blood vessels to support tumor cells. Intriguingly, accumulating evidence suggests that unresolved chronic inflammation plays a pivotal role in initiating, promoting, transforming, and facilitating the spread of several human cancers [107, 228, 229]. An illustrative representation of the different tumor and antitumor responses in the TME and their collective roles in cancer progression is depicted below (Fig. 4).

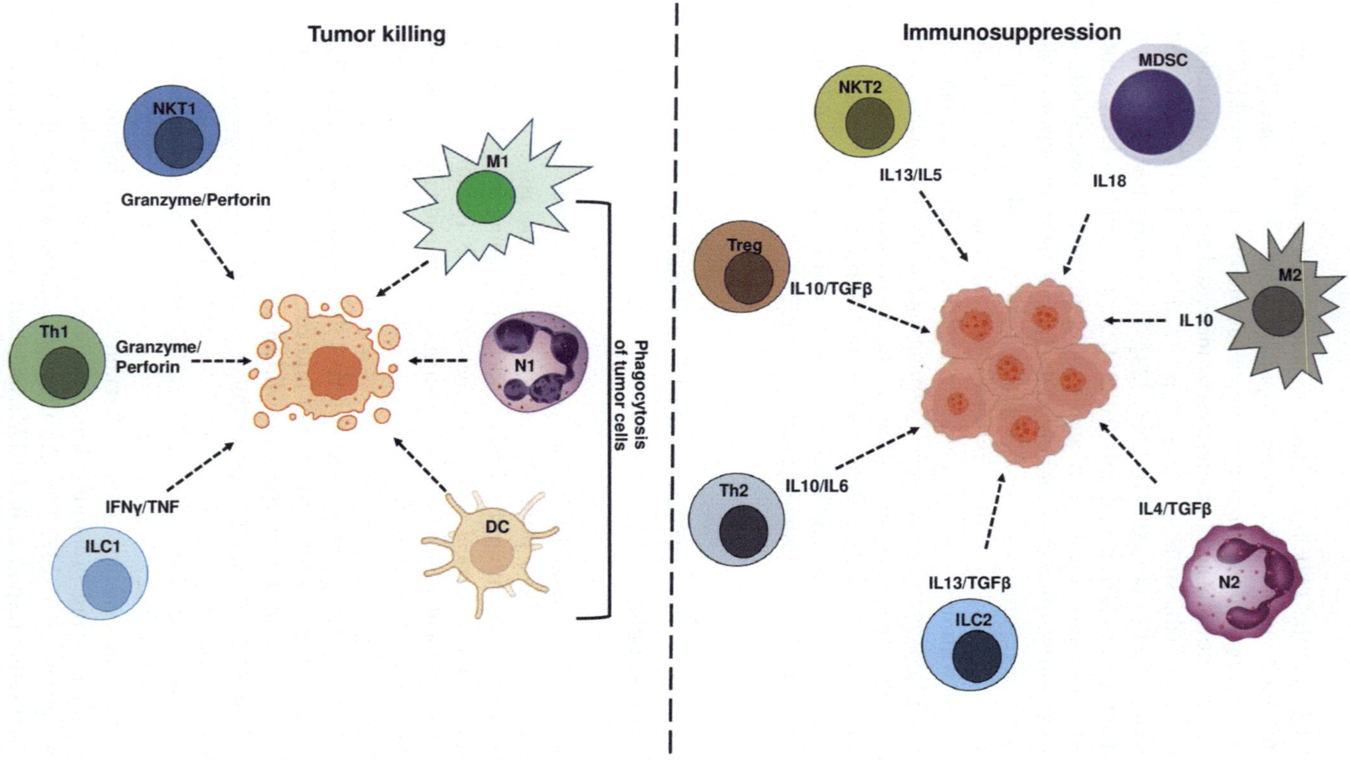

Fig. 4 Dynamic interplay between antitumor and immunosuppressive immune responses. The "tug of war" between antitumor and immunosuppressive immune responses revolves around the dynamic interactions of immune cells and cytokines within the TME. On the one hand, immune cells, such as cytotoxic T cells and natural killer cells, actively work to recognize and eliminate tumor cells, aiming to achieve an effective antitumor response. On the other hand, immunosuppressive mechanisms, which are mediated by regulatory T cells, myeloid-derived suppressor cells, and cytokines such as transforming growth factor beta (TGF-β) and interleukin-10 (IL-10), counteract these antitumor effects. This intricate interplay determines the fate of the immune response against the tumor

7.1 *Controversial Roles of Inflammatory and Suppressive Cytokines*

7.1.1 Role of Inflammatory Cytokines

IL-6 not only stimulates the proliferation of cancer cells but also inhibits their apoptosis by activating STAT3. Additionally, IL-6 signaling can impact the differentiation of T-cell subsets, especially when combined with other cytokines, such as TGF-β [230]. In carcinogen-induced liver cancer, IL-6 plays a critical role in tumor development. Furthermore, IL-6 acts as an angiogenic factor, further contributing to the progression of cancer [231]. Although TNF is a key inflammatory mediator in various models and diseases, it primarily acts through NF-κB induction [232–234]; however, in diethyl nitrosamine–induced hepatocellular carcinoma (HCC), TNF plays a pro-tumorigenic role. Ablation of TNF receptor I in mice significantly reduces obesity-enhanced HCC development [235]. Similarly, IL-1β, which activates NF-κB, similar to TNF, is genetically linked to gastric cancers [236]. In the interplay between cancer and inflammatory cells, STAT3 and NF-κB establish a reciprocal positive feedback loop, promoting cancer progression [237].

7.1.2 Role of Immunosuppressive Cytokines

TGF-β in the TME is primarily immunosuppressive, dampening anticancer immunity and influencing $CD4^+$ $CD25^+$ Tregs. Its effects on immune cells are context-dependent, influenced by the microenvironment and other factors, and play a vital yet intricate role in carcinogenesis and tumor progression [238]. TGF-β inhibits IFN-γ production by Th1 and $CD8^+$ T cells, promoting $Foxp3^+$ Treg generation and Th17 cell differentiation. IL-10, produced by $CD25^+$ Tregs, regulates immune responses. Taken together, these effects likely favor cancer growth [239, 240]. During chronic infection, such as microbially induced colitis, IL-10 downregulates inflammation and prevents carcinogenesis in mouse models. IL-10 inhibits NF-κB activation, hindering angiogenesis. However, it can also activate STAT3, contributing to tumor development by promoting cell proliferation and survival [241, 242].

7.2 *Controversial Roles of Pro-tumorigenic and Anti-tumorigenic Cells*

7.2.1 Role of M1- and M2-type TAMs

In the initial stages of tumor development, M1-type macrophages can infiltrate the tumor site. When activated by inflammatory signals, these M1 macrophages release proinflammatory cytokines and chemokines such as CXCL19 and CXCL10. This process is designed to attract and support the development and differentiation of Th1, Th17, and NK cells [174, 176]. In contrast, as the tumor progresses or is within the hypoxic regions of the TME, TAMs tend to transition toward an M2-type phenotype [172, 176]. These M2-like TAMs secrete specific factors that promote Th2 cell differentiation and recruitment. Moreover, M2 macrophages express a unique array of cytokines and chemokines, such as CCL17, CCL22, and CCL24, which facilitate the recruitment and differentiation of Tregs [243]. Moreover, M2 macrophages contribute to tissue repair, remodeling, and angiogenesis through the generation of inhibitory molecules such as VEGF and EGF [244].

7.2.2 Role of MDSCs

MDSCs constitute a diverse group of immature myeloid cells that have not yet committed to differentiating into macrophages, dendritic cells, or granulocytes [245]. These MDSCs play a critical role in exerting suppressive functions by regulating T-cell responses, which are achieved through the secretion of molecules such as NO, ROS, and TGF-β [245]. Furthermore, MDSCs promote Treg induction and facilitate anti-inflammatory responses [246]. MDSCs also secrete elevated levels of IL-6 and respond to inflammatory cytokines such as IL-6, IL-1β, and IFN-γ [191]. Additionally, their expansion can also be induced by immunosuppressive cytokines, including VEGF, produced by macrophages. Upon migrating to suitable locations, myeloid cells typically undergo differentiation. However, the specific cytokines and cellular factors present within the TME hinder this differentiation process. Instead, they promote the expansion and activation of immature myeloid cells [191].

7.2.3 Role of Helper T Cells

$CD4^+$ Th1 cells and $CD8^+$ T cells that produce IFN-γ are known to play crucial roles in restraining tumor growth by inhibiting and eliminating tumor cells. These antitumor effector immune cells are typically generated in the draining lymph nodes but can also proliferate within the TME, even in the presence of potential immunosuppressive factors. However, not all T cells exhibit antitumor activity [247]. Evidence from various cancer types, including pancreatic, breast, cervical, endometrial, non-small cell lung, ovarian, and colorectal cancers, suggests an increased presence of $CD25^+$ Treg cells [248]. These Tregs are often associated with increased secretion of the immunosuppressive cytokine TGF-β, which contributes to immune suppression in the context of cancer. For example, studies have shown an increase in the frequency of $CD4^+$ $Foxp3^+$ Tregs in both experimental murine models and human patients with HCC [249].

7.2.4 Role of T Helper Type 17 Cells

Th17 cells are a subset of T helper cells that have been linked to various autoimmune disorders and chronic inflammatory conditions. When naive $CD4^+$ T cells encounter specific signals, such as those mediated by TGF-β and IL-6, they tend to differentiate into the Th17 T-cell subset. This differentiation process involves the activation of STAT3 and results in the formation of suppressor Th17 cells [250]. Furthermore, while known for its role in maintaining and expanding Th17 inflammatory cells, IL-23 has recently been identified as a pro-carcinogenic cytokine [251]. It has been reported to promote inflammation and angiogenesis within the TME while simultaneously impeding the infiltration of $CD8^+$ T cells. The presence of Th17 cells in the TME may antagonize and counteract the tumor-induced immunosuppressive effects of $CD4^+$ Th1 cells, which produce IFN-γ [252]. Moreover, a study in a murine B16 melanoma tumor model demonstrated that IL-17 signaling plays a crucial role in tumor development [253]. IL-17 signaling directly affects both tumor cells and stromal cells, causing the induction of IL-6, which subsequently activates STAT3. This cross talk between Th17 cells and the TME impedes the differentiation and function of Th1 cells that produce IFN-γ, thereby contributing to tumor growth [254].

8 Inhibition of Naive T-cell Activation: Induction of Peripheral T-cell Tolerance Leading to T-cell Anergy, Death, and Exhaustion

T-cell activation and differentiation are intricately regulated. Under specific circumstances in which naive T cells encounter antigens, these processes may lead to functional memory development or various types of T-cell dysfunction, including exhaustion, tolerance, anergy, or senescence. In the TME, T cells play a crucial role as integral parts of immune cell infiltration. Upon encountering antigens and receiving stimulation via TCRs, T cells differentiate into effector T cells or CTLs, which are capable of eradicating the antigen. Nevertheless, if T cells previously exposed to an antigen are subjected to it consistently over an extended period, substantial alterations can manifest in their activation and differentiation processes. This may culminate in a condition of T-cell dysfunction or exhaustion [255].

Research in a murine model has demonstrated that dysfunctional exhaustion of tumor-specific T cells can occur early in the course of tumor initiation and antigen exposure [256]. Dysfunctional $CD8^+$ T cells alleviate crucial effector functions, including their ability to carry out cytotoxic functions and proliferate [255]. For example, an analysis using single-cell RNA sequencing of T cells within tumor tissues from melanoma and lung cancer patients revealed the expression of genes such as PDCD1 and LAG-3, which are linked to T-cell dysfunction [257]. Notably, blocking the PD-1 or PD-L1 pathway can effectively restore the function of T cells, highlighting its pivotal role in T-cell dysfunction [258]. Conversely, a subset of PD-1$^+$CD38$^+$CD8$^+$ T cells constitutes a population of dysfunctional cells that do not respond effectively to anti-PD-1 therapy [259].

Moreover, MDSCs employ various mechanisms to dampen the immune system, releasing elevated levels of ARG-1, iNOS, ROS, and COX2 [260]. For example, MDSCs utilize ARG-1 and iNOS, either individually or in combination, to impede the response of CD8$^+$ T cells to antigens. Notably, ARG-1 expression has been shown to decrease the production of the CD3ζ-chain, thereby compromising T-cell function [261]. Furthermore, NO alone can hinder CD8$^+$ T cells by inhibiting the phosphorylation and activation of the JAK3 and STAT5 transcription factors and can induce T-cell apoptosis [262]. Increased levels of ROS in the TME have been linked to limiting T-cell growth and differentiation and ultimately fostering T-cell exhaustion. Both MDSCs and other cells produce ROS, which may interact with T cells, inducing oxidative stress and resulting in decreased responsiveness of CD8$^+$ T cells in cancer [263]. Notably, MDSCs produce high levels of IDO, an enzyme that catabolizes tryptophan and generates kynurenine. Tryptophan depletion and kynurenine production hinder the clonal expansion of activated T cells [264]. Experimental studies have shown that IDO-mediated tryptophan depletion is a critical pathway through which MDSCs mediate the suppression of CD8$^+$ T-cell proliferation and survival within tumor-infiltrated areas, leading to their homing to the lymph nodes [265]. Moreover, reports indicate that STAT3-dependent IDO expression mediates the immunosuppressive effects of MDSCs in breast cancer, significantly inhibiting CD8$^+$ T-cell proliferation and their secretion of IFN-γ [266]. Previous studies have also demonstrated that CD11b$^+$CD33$^+$ MDSCs in tumor tissues from NSCLC patients express the surface ectonucleotidases CD39 and CD73 [267]. Moreover, TGF-β stimulates the expression of CD39 and CD73, thereby inhibiting the proliferation and function of autologous CD8$^+$ T cells [268]. Consequently, MDSCs play a crucial role in suppressing CD8$^+$ T-cell proliferation, leading to CD8$^+$ T-cell exhaustion within the TME.

TAMs contribute to immunosuppression through diverse mechanisms, two of which involve T-cell dysfunction: (1) TAMs may induce the expression of PD-L1 in monocytes, which subsequently binds to PD-1 receptors on the surface of CD8$^+$ T cells, leading to the initiation of T-cell exhaustion [269]. (2) TAMs secrete various immunosuppressive cytokines and factors, including IL-10, TGF-β, and ROS, contributing to exhaustion and dysfunction in CD8$^+$ TILs [270].

Cancer-associated fibroblasts (CAFs), which constitute the most abundant stromal cell population, release immunomodulatory substances within the TME and are known as suppressors of T cell–mediated immune responses. Recent research has revealed a novel role for CAFs, demonstrating their direct inhibition of antitumor T-cell reactions through a process involving immune checkpoint activation. One key mechanism involves the upregulation of Fas/FasL on CD8$^+$ T cells and PD-1/PD-L2 on CAFs, which contributes to the impairment of tumor-specific CD8$^+$ T-cell function [271].

The immunosuppressive cytokines present in the TME may originate from various sources, including tumor cells, MDSCs, or CAFs. These cytokines play pivotal roles in T-cell exhaustion and dysfunction [272]. TGF-β, produced by TAMs, exerts its influence by inducing TIM-3, PD-1, and CTLA-4 expression in T cells while also inhibiting the secretion of IFN-γ and granzyme B in a dose-dependent manner [273]. The administration of an anti-TGF-β antibody has been demonstrated to restore impaired T-cell cytotoxic function in malignant pleural effusion (MPE) [274]. Furthermore, the blockade of TGF-β signaling in $CD8^+$ T cells, achieved through the use of dominant negative receptors, can enhance the function of exhausted T cells [275]. In a study involving ovarian carcinoma patients, tumor-infiltrating follicular regulatory T cells (Tfr) exhibited significantly increased expression of IL-10. This upregulation of IL-10 was inversely correlated with the secretion of IFN-γ in $CD8^+$ tumor-infiltrating TILs [276].

9 Inhibition of TIL Infiltration and Subsequent Loss of Function

In various cancers, the TME affects TILs through disruptions in ion levels, hypoxia, and acidity. T cells predominantly use glycolysis for energy production during immune surveillance. Tumor cells demand high metabolic resources for growth, transitioning to aerobic glycolysis (the "Warburg effect") for a competitive energy advantage [277].

TME-induced changes in amino acid metabolism and cholesterol levels impair effector functions in TILs. In the tumor core, irregular vascularization and hypoxia attract Tregs, hindering effector T-cell activity and thus promoting tumor growth [278].

Under hypoxic conditions, LDH expression increases in cells, leading to excessive lactic acid production. Tumor cells enhance lactate transporters, facilitating lactate efflux. These effects support tumor and Treg survival but inhibit the infiltration of CTLs through multiple mechanisms, including reduced proliferation, cytokine production, cytotoxicity, motility, and energy production in the acidic TME [279].

Sustained metabolic stress and TME-induced hypoxia suppress mitochondrial biogenesis in TILs, reducing the mitochondrial mass. This leads to decreased cytokine production (e.g., TNF-α and IFN-γ) and increased expression of co-inhibitory molecules (TIM-3, LAG-3, and PD-1). Elevated PD-1 in exhausted TILs disrupts mitochondrial function, hindering metabolic reprogramming, promoting cellular senescence, and culminating in terminal exhaustion [280]. Decreased peroxisome proliferator-activated receptor gamma coactivator 1-alpha (PGC1-α) in antigen-stimulated T cells leads to persistent Akt signaling, disrupting T-cell metabolic adaptation and function [281]. Additionally, reduced PGC1-α in TILs increases mitochondrial ROS, initiating the nuclear factor of activated T cells (NFAT) signaling pathways and contributing to T-cell exhaustion [282].

Within the TME, dying or tumor cells release intracellular potassium (K), which accumulates in the nearby tumor surroundings. This results in significantly elevated extracellular potassium (K^+) levels within the interstitial fluid of the tumor, hindering the effectiveness of T cells in combating tumors [283].

10 Classification of T Regulatory Cells and Their Importance in the TME and Cancer Progression

Tregs play different roles in cancer pathogenesis than in autoimmune diseases, where they serve a protective function. It has been shown that natural T regulatory cells (nTregs) primarily exert their suppressive effects through mechanisms involving direct cell-to-cell contact mediated by membrane-bound molecules. In contrast, iTregs predominantly utilize contact-independent mechanisms, relying mainly on cytokines such as IL-10 and TGF-β for their functionality.

10.1 Natural Tregs

The immune system comprises diverse Treg subsets: nTregs, $CD8^+$ Tregs, Tr1 cells, Th3 cells, and NKT cells. nTregs, which originate in the thymus and consistently express high levels of CD25 (IL-2 receptor), form a specific subtype of $CD4^+$ T cells [284]. The identification of FoxP3 marked a key milestone in distinguishing nTregs from other $CD4^+$ T cells. FoxP3 not only serves as a Treg marker but also regulates Treg function. Continuous FoxP3 expression is vital for maintaining the suppressive capacity of Tregs [285]. Additional markers for nTregs include CTLA-4, CD62L, GITR, OX40, FR4, neuropilin-1, and CD103 [286]. Neuropilin-1 is a crucial marker of Treg functionality and stability and is implicated in processes such as angiogenesis and the promotion of tumor growth in various cancers [287].

10.2 Induced Tregs

iTregs, which originate from peripheral naive T cells, are induced by signals in the TME, including tumor-specific antigens, TGF-β, and various soluble molecules. The TCR repertoire of intra-tumoral iTregs is variable but shares similarities with that of circulating Tregs. TCRs from Tregs respond to autologous tumors and mutated neoantigens, indicating their activation and expansion in response to tumor antigens in the TME [288, 289].

In addition to nTregs and iTregs from naive T cells, IL-$17A^+Foxp3^+$ Tregs and ex-Th17 $Foxp3^+$ Tregs also contribute to the tumor-associated iTreg pool. These iTregs inhibit the antitumor immune responses of effector T (Teff) cells, NK cells, and DCs through diverse mechanisms, fostering tumor progression [290].

Tregs can also be classified on the basis of their characteristics in the TME.

Tregs produce inhibitory cytokines (IL-10, TGF-β, and IL-35) that suppress immune responses through various pathways, hindering the function of $CD8^+$ T cells and DCs via membrane-bound TGF-β and thereby modulating antitumor immune activity [289].

Tregs can eliminate effector cells via granzyme A and perforin. While nTregs use granzyme A and perforin and lack granzyme B production, iTregs express elevated levels of granzyme B [291]. Tregs maintain memory T-cell quiescence by suppressing effector and proliferation programs via CTLA-4 [292]. Canonical NK cells are susceptible to Treg suppression, unlike adaptive NK cells, which express CD57 and FcεRγ-NKG2C [293]. Reg-mediated immunosuppression involves canonical NK cells downregulating TIM-3 and upregulating inhibitory receptors such as PD-1 and IL-1R8. Tregs produce IL-37, which binds to IL-1R8 and induces phenotypic changes in canonical NK cells, diminishing their function [294]. Tregs inhibit IFN-γ secretion by $CD8^{+}$ T cells, preventing SREBP1-mediated fatty acid synthesis in M2-like TAMs [295].

Tregs impact effector cell function through metabolic modulation. Tregs compete for IL-2 with effector T cells, leading to reduced IL-2 levels in the TME and hindering effector T-cell growth [147]. When expressed with the enzymes CD39 and CD73, Tregs generate adenosine from extracellular ATP or ADP, exerting inhibitory effects through the adenosine receptor A2A on effector cells [203]. IL-27 signaling in Tregs, which promotes tumorigenic properties, upregulates CD39 [296]. Tregs disrupt effector T-cell metabolism by transferring cAMP through gap junctions [297].

Tregs impact DC regulation in the TME. CTLA-4-expressing Tregs interact with DCs, downregulating cooperative signaling and inducing DC tolerance. Tregs, via LAG-3, inhibit MHC II expression on DCs, inducing tolerance and suppressing T-cell activity through IDO action [86]. Tregs induce anergy in CD4-negative NKT cells and hinder NKT cell cytotoxicity through direct contact, which is mediated by impaired DC maturation [67].

MDSCs and Tregs create positive feedback loops, reinforcing the suppressive environment. During tumor progression, MDSCs selectively increase Treg proliferation via an in vivo mechanism that relies on TGF-β [272]. Additionally, MDSCs enhance Treg generation by expressing CD73, which produces adenosine that binds to A2aR and intensifies its immunosuppressive effects [298]. MDSCs induce Treg development in vitro and in tumor-bearing mice via the secretion of IL-10 and IFN-γ by MDSCs [299]. Increased IDO expression in the TME lowers tryptophan levels and elevates kynurenine levels [300]. Reduced T-cell activation and increased Treg production occur in chronic lymphocytic leukemia (CLL) due to elevated IDO levels in monocytic MDSCs [301]. Activated $Gr\text{-}1^{+}$ $CD115^{+}$ M-MDSCs in a mouse colon carcinoma model upregulated MHC II and released IL-10 and TGF-β, thereby promoting Treg development [302]. MDSCs influence Tregs through key cell-surface interactions, including CD40/CD40L, PD-1/PD-L1, and CD80/CTLA-4 interactions, with CD40 expression playing a crucial role in T-cell tolerance and Treg accumulation [303]. In an ovarian cancer mouse model, MDSCs increase CD80 expression through direct interactions with tumor cells. CD80 then binds to CTLA-4 on Tregs, enhancing the immunosuppressive functions of Tregs. This reciprocal modulation amplifies MDSC expansion and suppressive functions through a TGF-β-dependent mechanism [304]. IL-35, which is secreted by nTregs, enhances MDSC immunosuppressive capability through the PD-L1 pathway, with the combined signals of PD-L1 and CD169 being crucial for the induction of $IL\text{-}35^{+}$ Tregs [305].

11 Conclusion

In conclusion, the intricate interplay between cell-mediated immunity and the TME plays a pivotal role in shaping the course of cancer progression. Understanding the dynamic interactions between immune cells and malignant cells within the TME has not only shed light on the mechanisms underlying tumor evasion but also opened new avenues for therapeutic intervention. Harnessing the power of cell-mediated immunity through immunotherapies such as immune checkpoint inhibitors and adoptive cell therapies holds immense promise in the fight against cancer. As our comprehension of the TME continues to deepen, tailoring treatment strategies to bolster cell-mediated immunity represents a crucial step toward more effective and personalized cancer therapies, offering hope to countless individuals battling this formidable disease.

Competing Financial Interest The authors declare that they have no competing financial interests.

References

1. Owen J, Punt J, Stranford S, Jones P. Kuby immunology [Internet]. 2013 [cited 2023 Sep 12]. Available from: http://www.roswellpark.org/sites/default/files/jiang_11232015mir511.pdf.
2. Parkin J, Cohen B. An overview of the immune system. Lancet [Internet]. 2001 [cited 2023 Sep 12];357(9270):1777–89. Available from: https://pubmed.ncbi.nlm.nih.gov/11403834/.
3. Tsuda Y, Takahashi H, Kobayashi M, Hanafusa T, Herndon DN, Suzuki F. Three different neutrophil subsets exhibited in mice with different susceptibilities to infection by methicillin-resistant Staphylococcus aureus. Immunity. 2004;21(2):215–26.
4. Fridlender ZG, Sun J, Kim S, Kapoor V, Cheng G, Ling L, et al. Polarization of tumor-associated neutrophil phenotype by TGF-β: "N1" versus "N2" TAN. Cancer Cell. 2009;16(3):183–94.
5. Immunology e-book. Google Books [Internet]. [cited 2023 Sep 12]. Available from: https://books.google.co.in/books?hl=en&lr=&id=GIDRDwAAQBAJ&oi=fnd&pg=PP1&ots=Y0srj1lDg8&sig=ICDyZ_7FUN2S3MyTrfzMxSUWmvE&redir_esc=y#v=onepage&q&f=false.
6. Abbas AK, Lichtman AH, Pillai S. Cellular and molecular immunology e-book. Google Books [Internet]. [cited 2023 Sep 12]. Available from: https://books.google.co.in/books?hl=en&lr=&id=RWYWBAAAQBAJ&oi=fnd&pg=PP1&ots=fv3ah-xFQY&sig=i8RKJMSq7KlTpPpYBPdN94k3ei8&redir_esc=y#v=onepage&q&f=false.
7. Malech HL, Deleo FR, Quinn MT. The role of neutrophils in the immune system: an overview. Methods Mol Biol. 2014;1124:3–10. https://doi.org/10.1007/978-1-62703-845-4_1. PMID: 24504942; PMCID: PMC6777345.
8. Zhang C, Yang M, Ericsson AC. Function of macrophages in disease: current understanding on molecular mechanisms. Front Immunol. 2021;12:620510.
9. Yao Y, Xu XH, Jin L. Macrophage polarization in physiological and pathological pregnancy. Front Immunol. 2019;10:434399.
10. Kumari N, Choi SH. Tumor-associated macrophages in cancer: recent advancements in cancer nanoimmunotherapies. J Exp Clin Cancer Res [Internet]. 2022 [cited 2023 Nov 7];41(1):1–39. Available from: https://jeccr.biomedcentral.com/articles/10.1186/s13046-022-02272-x.

11. Zhou J, Tang Z, Gao S, Li C, Feng Y, Zhou X. Tumor-associated macrophages: recent insights and therapies. Front Oncol. 2020;10:507662.
12. Howard CJ, Charleston B, Stephens SA, Sopp P, Hope JC. The role of dendritic cells in shaping the immune response. Anim Health Res Rev [Internet]. 2004 [cited 2023 Nov 10];5(2):191–5. Available from: https://pubmed.ncbi.nlm.nih.gov/15984324/.
13. Mellman I. Dendritic cells: master regulators of the immune response. Cancer Immunol Res [Internet]. 2013 [cited 2023 Nov 10];1(3):145–9. Available from: https://doi.org/10.1158/2326-6066.CIR-13-0102.
14. Galli SJ, Nakae S, Tsai M. Mast cells in the development of adaptive immune responses. Nat Immunol. 2005;6(2):135–42.
15. Siraganian RP. Mast cell signal transduction from the high-affinity IgE receptor. Curr Opin Immunol. 2003;15(6):639–46.
16. da Silva EZM, Jamur MC, Oliver C. Mast cell function: a new vision of an old cell. J Histochem Cytochem [Internet]. 2014 [cited 2023 Nov 10];62(10):698–738. Available from: https://journals.sagepub.com/doi/full/10.1369/0022155414545334.
17. Norrby K. Mast cells and angiogenesis. APMIS [Internet]. 2002 [cited 2023 Nov 10];110(5):355–71. Available from: https://onlinelibrary.wiley.com/doi/full/10.1034/j.1600-0463.2002.100501.x.
18. Chirumbolo, S., Bjørklund, G., Sboarina, A., & Vella, A. (2018). The role of basophils as innate immune regulatory cells in allergy and immunotherapy. Human vaccines & immunotherapeutics, 14(4), 815–831. https://doi.org/10.1080/21645515.2017.1417711.
19. Stone, K. D., Prussin, C., & Metcalfe, D. D. (2010). IgE, mast cells, basophils, and eosinophils. The Journal of allergy and clinical immunology, 125(2 Suppl 2), S73–S80. https://doi.org/10.1016/j.jaci.2009.11.017.
20. Raza F, Babasyan S, Larson EM, Freer HS, Schnabel CL, Wagner B. Peripheral blood basophils are the main source for early interleukin-4 secretion upon in vitro stimulation with Culicoides allergen in allergic horses. PLoS One [Internet]. 2021 [cited 2023 Nov 11];16(5):e0252243. Available from: https://journals.plos.org/plosone/article?id=10.1371/journal.pone.0252243.
21. van Panhuys, N., Prout, M., Forbes, E., Min, B., Paul, W. E., & Le Gros, G. (2011). Basophils are the major producers of IL-4 during primary helminth infection. Journal of immunology (Baltimore, Md. : 1950), 186(5), 2719–2728. https://doi.org/10.4049/jimmunol.1000940.
22. Iype, J., & Fux, M. (2021). Basophils Orchestrating Eosinophils' Chemotaxis and Function in Allergic Inflammation. Cells, 10(4), 895. https://doi.org/10.3390/cells10040895.
23. Ma, T. X., & Gao, Y. (2019). New insights into roles of basophils in initiating T helper type 2 immunity. Chinese herbal medicines, 12(1), 14–18. https://doi.org/10.1016/j.chmed.2019.09.006.
24. Bandeira-Melo, C., & Weller, P. F. (2005). Mechanisms of eosinophil cytokine release. Memorias do Instituto Oswaldo Cruz, 100 Suppl 1(Suppl 1), 73–81. https://doi.org/10.1590/s0074-02762005000900013.
25. Kita H. (2011). Eosinophils: multifaceted biological properties and roles in health and disease. Immunological reviews, 242(1), 161–177. https://doi.org/10.1111/j.1600-065X.2011.01026.x.
26. Sharrock J. Natural killer cells and their role in immunity. EMJ Allergy Immunol [Internet]. 2019 [cited 2023 Nov 11];4(1):108–16. Available from: https://www.emjreviews.com/allergy-immunology/article/natural-killer-cells-and-their-role-in-immunity/.
27. Nielsen, N., Pascal, V., Fasth, A. E., Sundström, Y., Galsgaard, E. D., Ahern, D., Andersen, M., Baslund, B., Bartels, E. M., Bliddal, H., Feldmann, M., Malmström, V., Berg, L., Spee, P., & Söderström, K. (2014). Balance between activating NKG2D, DNAM-1, NKp44 and NKp46 and inhibitory CD94/NKG2A receptors determine natural killer degranulation towards rheumatoid arthritis synovial fibroblasts. Immunology, 142(4), 581–593. https://doi.org/10.1111/imm.12271.
28. Cook KD, Waggoner SN, Whitmire JK. NK cells and their ability to modulate T cells during virus infections. Crit Rev Immunol. 2014;34(5):359–88. https://doi.org/10.1615/critrevimmunol.2014010604. PMID: 25404045; PMCID: PMC4266186.

29. Rosenberg, J., & Huang, J. (2018). CD8+ T Cells and NK Cells: Parallel and Complementary Soldiers of Immunotherapy. Current opinion in chemical engineering, 19, 9–20. https://doi.org/10.1016/j.coche.2017.11.006.
30. Parodi, M., Favoreel, H., Candiano, G., Gaggero, S., Sivori, S., Mingari, M. C., Moretta, L., Vitale, M., & Cantoni, C. (2019). NKp44-NKp44 Ligand Interactions in the Regulation of Natural Killer Cells and Other Innate Lymphoid Cells in Humans. Frontiers in immunology, 10, 719. https://doi.org/10.3389/fimmu.2019.00719.
31. Barrow AD, Edeling MA, Trifonov V, Luo J, Goyal P, Bohl B, et al. Natural killer cells control tumor growth by sensing a growth factor. Cell. 2018;172(3):534–548.e19.
32. Liu, H., Wang, S., Xin, J., Wang, J., Yao, C., & Zhang, Z. (2019). Role of NKG2D and its ligands in cancer immunotherapy. American journal of cancer research, 9(10), 2064–2078.
33. Paul S, Lal G. The molecular mechanism of natural killer cells function and its importance in cancer immunotherapy. Front Immunol. 2017;8:290728.
34. Jiang H, Chess L. Regulation of immune responses by T cells [Internet]. 2006 [cited 2023 Nov 13];354(11):1166–76. https://doi.org/10.1056/NEJMra055446. Available from: https://www.nejm.org/doi/full/10.1056/NEJMra055446.
35. Bogen B, Fauskanger M, Haabeth OA, Tveita A. CD4+ T cells indirectly kill tumor cells via induction of cytotoxic macrophages in mouse models. Cancer Immunol Immunother. 2019;68(11):1865–73.
36. Schietinger A, Philip M, Liu RB, Schreiber K, Schreiber H, KSJ of E, 2010 undefined. Bystander killing of cancer requires the cooperation of CD4+ and CD8+ T cells during the effector phase. J Exp Med. 2010. rupress.org [Internet]. [cited 2023 Nov 13]. Available from: https://rupress.org/jem/article-abstract/207/11/2469/40645.
37. Ahmed H, Mahmud AR, Siddiquee MFR, Shahriar A, Biswas P, Shimul MEK, et al. Role of T cells in cancer immunotherapy: opportunities and challenges. Cancer Pathog Ther. 2023;1(2):116–26.
38. Mumberg D, Monach PA, Wanderling S, Philip M, Toledano AY, Schreiber RD, et al. CD4+ T cells eliminate MHC class II-negative cancer cells in vivo by indirect effects of IFN-γ. Proc Natl Acad Sci USA. 1999;96(15):8633–8.
39. Farhood B, Najafi M, Mortezaee K. CD8+ cytotoxic T lymphocytes in cancer immunotherapy: a review. J Cell Physiol. 2019;234(6):8509–21.
40. Dolina JS, Van Braeckel-Budimir N, Thomas GD, Salek-Ardakani S. CD8+ T cell exhaustion in cancer. Front Immunol. 2021;12:715234.
41. Hoffman, W., Lakkis, F. G., & Chalasani, G. (2016). B Cells, Antibodies, and More. Clinical journal of the American Society of Nephrology : CJASN, 11(1), 137–154. https://doi.org/10.2215/CJN.09430915.
42. Nutt SL, Hodgkin PD, Tarlinton DM, Corcoran LM, DTNR, 2015 undefined. The generation of antibody-secreting plasma cells. Nat Rev Immunol. 2015. nature.com [Internet]. [cited 2023 Nov 13]. Available from: https://www.nature.com/articles/nri3795.
43. Chalasani G, Rothstein D. Non-antibody mediated roles of B cells in allograft survival. Curr Transplant Rep. 2014;1(3):155–65.
44. Lund FE, Randall TD. Immunology TRNR, 2010 undefined. Effector and regulatory B cells: modulators of CD4+ T cell immunity. Nat Rev Immunol. 2010. nature.com [Internet]. [cited 2023 Nov 13]. Available from: https://www.nature.com/articles/nri2729.
45. Ngo VN, Cornall RJ, Cyster JG, medicine JCTJ of experimental, 2001 undefined. Splenic T zone development is B cell dependent. J Exp Med. 2001. rupress.org [Internet]. [cited 2023 Nov 13]. Available from: https://rupress.org/jem/article-abstract/194/11/1649/39085.
46. Fu YX, Huang G, Wang Y, Chaplin DD, experimental DCTJ of, 1998 undefined. B lymphocytes induce the formation of follicular dendritic cell clusters in a lymphotoxin α–dependent fashion. J Exp Med. 1998. rupress.org [Internet]. [cited 2023 Nov 13]. Available from: https://rupress.org/jem/article-abstract/187/7/1009/7559.
47. Schultz KR, Klarnet JP, Gieni RS, Hayglass KT, Greenberg PD. The role of B cells for in vivo T cell responses to a friend virus-induced leukemia. Science (1979). 1990;249(4971):921–3.

48. DiLillo DJ, Yanaba K, Tedder TF. Immunology TTTJ of, 2010 undefined. B cells are required for optimal CD4+ and CD8+ T cell tumor immunity: therapeutic B cell depletion enhances B16 melanoma growth in mice. J Immunol. 2010. journals.aai.org [Internet]. [cited 2023 Nov 13]. Available from: https://journals.aai.org/jimmunol/article/184/7/4006/83912.
49. Cui C, Wang J, Fagerberg E, Chen PM, Connolly KA, Damo M, Cheung JF, Mao T, Askari AS, et al., 2021 undefined. Neoantigen-driven B cell and CD4 T follicular helper cell collaboration promotes anti-tumor CD8 T cell responses. Cell. 2021. cell.com [Internet]. [cited 2023 Nov 13]. Available from: https://www.cell.com/cell/article/S0092-8674(21)01322-2/fulltext.
50. Hu Q, Hong Y, Qi P, Lu G, Mai X, Xu S, He X, Guo Y, Gao L, Jing Z, Wang J, Cai T, Zhang Y. Atlas of breast cancer infiltrated B-lymphocytes revealed by paired single-cell RNA-sequencing and antigen receptor profiling. Nat Commun. 2021. nature.com [Internet]. [cited 2023 Nov 13]. Available from: https://www.nature.com/articles/s41467-021-22300-2.
51. Bruhns P, of TJ of the AS, 2012 undefined. Properties of mouse and human IgG receptors and their contribution to disease models. Blood. 2012. ashpublications.org [Internet]. [cited 2023 Nov 13]. Available from: https://ashpublications.org/blood/article-abstract/119/24/5640/30328.
52. Tan R, Nie M, Long W. The role of B cells in cancer development. Front Oncol. 2022;12:958756.
53. Marshall JS, Warrington R, Watson W, Kim HL. An introduction to immunology and immunopathology. Allergy Asthma Clin Immunol [Internet]. 2018 [cited 2023 Sep 13];14(2):1–10. Available from: https://aacijournal.biomedcentral.com/articles/10.1186/s13223-018-0278-1.
54. Huntington ND, Gray DHD. Immune homeostasis in health and disease. Immunol Cell Biol. 2018;96(5):451–2.
55. Clark R, Kupper T. Old meets new: the interaction between innate and adaptive immunity. J Invest Dermatol. 2005;125(4):629–37.
56. Turvey SE, Broide DH. Innate immunity. J Allergy Clin Immunol. 2010;125(Suppl 2):S24–32.
57. Vivier E, Malissen B. Innate and adaptive immunity: specificities and signaling hierarchies revisited. Nat Immunol [Internet]. 2005 [cited 2023 Sep 13];6(1):17–21. Available from: https://www.nature.com/articles/ni1153.
58. Bonilla FA, Oettgen HC. Adaptive immunity. J Allergy Clin Immunol. 2010;125(Suppl 2):S33–40.
59. Möller G. Do suppressor T cells exist? Scand J Immunol [Internet]. 1988 [cited 2024 Oct 25];27(3):247–50. Available from: https://onlinelibrary.wiley.com/doi/full/10.1111/j.1365-3083.1988.tb02344.x.
60. Mukherji B, Guha A, Chakraborty NG, Sivanandham M, Nashed AL, Sporn JR, et al. Clonal analysis of cytotoxic and regulatory T cell responses against human melanoma. J Exp Med [Internet]. 1989 [cited 2024 Oct 25];169(6):1961–76. Available from: https://pubmed.ncbi.nlm.nih.gov/2471770/.
61. Beyer M, Schultze JL. Regulatory T cells in cancer. Blood [Internet]. 2006 [cited 2024 Oct 25];108(3):804–11. Available from: https://doi.org/10.1182/blood-2006-02-002774.
62. Teng MWL, Swann JB, Koebel CM, Schreiber RD, Smyth MJ. Immune-mediated dormancy: an equilibrium with cancer. J Leukoc Biol [Internet]. 2008 [cited 2023 Sep 13];84(4):988–93. Available from: https://onlinelibrary.wiley.com/doi/full/10.1189/jlb.1107774.
63. Kim R, Emi M, Tanabe K. Cancer immunoediting from immune surveillance to immune escape. Immunology [Internet]. 2007 [cited 2023 Sep 13];121(1):1–14. Available from: https://onlinelibrary.wiley.com/doi/full/10.1111/j.1365-2567.2007.02587.x.
64. Sandoval-Valencia RR, Sandoval-Valencia RR. Cancer immunotherapy: role of the immune system in malignant transformation. Rev Méd Hosp Gen (Méx) [Internet]. 2019 [cited 2023 Sep 13];82(4):198–207. Available from: http://www.scielo.org.mx/scielo.php?script=sci_arttext&pid=S2524-177X2019000400198&lng=es&nrm=iso&tlng=en.
65. Labani-Motlagh A, Ashja-Mahdavi M, Loskog A. The tumor microenvironment: a milieu hindering and obstructing antitumor immune responses. Front Immunol. 2020;11:525594.
66. Swann JB, Smyth MJ. Immune surveillance of tumors. J Clin Invest. 2007;117(5):1137–46.

67. Ferlazzo G, Morandi B. Cross-talks between natural killer cells and distinct subsets of dendritic cells. Front Immunol. 2014;5:159.
68. Chandrasekaran S, King MR. Microenvironment of tumor-draining lymph nodes: opportunities for liposome-based targeted therapy. Int J Mol Sci [Internet]. 2014 [cited 2023 Sep 14];15(11):20209–39. Available from: https://www.mdpi.com/1422-0067/15/11/20209/htm.
69. Kim, S. K., & Cho, S. W. (2022). The Evasion Mechanisms of Cancer Immunity and Drug Intervention in the Tumor Microenvironment. Frontiers in pharmacology, 13, 868695. https://doi.org/10.3389/fphar.2022.868695.
70. Street SEA, Cretney E, Smyth MJ. Perforin and interferon-γ activities independently control tumor initiation, growth, and metastasis. Blood [Internet]. 2001;97(1):192–7. Available from: https://www.sciencedirect.com/science/article/pii/S0006497120481261.
71. Bolitho P, Street SEA, Westwood JA, Edelmann W, MacGregor D, Waring P, et al. Perforin-mediated suppression of B-cell lymphoma. Proc Natl Acad Sci [Internet]. 2009;106(8):2723–8. Available from: https://www.pnas.org/doi/abs/10.1073/pnas.0809008106.
72. Wakita D, Chamoto K, Ohkuri T, Narita Y, Ashino S, Sumida K, et al. IFN-γ-dependent type 1 immunity is crucial for immunosurveillance against squamous cell carcinoma in a novel mouse carcinogenesis model. Carcinogenesis [Internet]. 2009;30(8):1408–15. Available from: https://doi.org/10.1093/carcin/bgp144.
73. Schreiber TH, Podack ER. A critical analysis of the tumour immunosurveillance controversy for 3-MCA-induced sarcomas. Br J Cancer [Internet]. 2009;101(3):381–6. Available from: https://doi.org/10.1038/sj.bjc.6605198.
74. Smyth MJ, Thia KYT, Street SEA, MacGregor D, Godfrey DI, Trapani JA. Perforin-mediated cytotoxicity is critical for surveillance of spontaneous lymphoma. J Exp Med [Internet]. 2000;192(5):755–60. Available from: https://doi.org/10.1084/jem.192.5.755.
75. Jorgovanovic D, Song M, Wang L, Zhang Y. Roles of IFN-γ in tumor progression and regression: a review. Biomark Res [Internet]. 2020;8(1):49. Available from: https://doi.org/10.1186/s40364-020-00228-x.
76. Teng MWL, Galon J, Fridman WH, Smyth MJ. From mice to humans: developments in cancer immunoediting. J Clin Invest [Internet]. 2015;125(9):3338–46. Available from: https://doi.org/10.1172/JCI80004.
77. Beatty GL, Gladney WL. Immune escape mechanisms as a guide for cancer immunotherapy. Clin Cancer Res [Internet]. 2015;21(4):687–92. Available from: https://doi.org/10.1158/1078-0432.CCR-14-1860.
78. Bosch R, Mora A, Cuellar C, Ferrer G, Gorlatov S, Nomdedéu J, et al. FcγRIIb-BCR coligation inhibits BCR signaling in chronic lymphocytic leukemia. Haematologica [Internet]. 2020;106(1):306–9. Available from: https://haematologica.org/article/view/9696.
79. Santos-de-Frutos K, Djouder N. When dormancy fuels tumour relapse. Commun Biol [Internet]. 2021;4(1):747. Available from: https://doi.org/10.1038/s42003-021-02257-0.
80. Badia-Ramentol J, Linares J, Gómez-Llonin A, Calon A. Minimal residual disease, metastasis and immunity. Biomolecules [Internet]. 2021;11(2). Available from: https://www.mdpi.com/2218-273X/11/2/130.
81. Tachtsidis A, McInnes LM, Jacobsen N, Thompson EW, Saunders CM. Minimal residual disease in breast cancer: an overview of circulating and disseminated tumour cells. Clin Exp Metastasis [Internet]. 2016;33(6):521–50. Available from: https://doi.org/10.1007/s10585-016-9796-8.
82. Zhao Q, Jiang Y, Xiang S, Kaboli PJ, Shen J, Zhao Y, et al. Engineered TCR-T cell immunotherapy in anticancer precision medicine: pros and cons. Front Immunol [Internet]. 2021;12. Available from: https://www.frontiersin.org/articles/10.3389/fimmu.2021.658753.
83. Prieto J, Melero I, Sangro B. Immunological landscape and immunotherapy of hepatocellular carcinoma. Nat Rev Gastroenterol Hepatol [Internet]. 2015;12(12):681–700. Available from: https://doi.org/10.1038/nrgastro.2015.173.
84. Mukherjee AG, Wanjari UR, Namachivayam A, Murali R, Prabakaran DS, Ganesan R, et al. Role of immune cells and receptors in cancer treatment: an immunotherapeutic

approach. Vaccines (Basel) [Internet]. 2022;10(9). Available from: https://www.mdpi.com/2076-393X/10/9/1493.
85. Fricke I, Gabrilovich DI. Dendritic cells and tumor microenvironment: a dangerous liaison. Immunol Invest [Internet]. 2006;35(3–4):459–83. Available from: https://doi.org/10.1080/08820130600803429.
86. Fucikova J, Palova-Jelinkova L, Bartunkova J, Spisek R. Induction of tolerance and immunity by dendritic cells: mechanisms and clinical applications. Front Immunol [Internet]. 2019;10. Available from: https://www.frontiersin.org/articles/10.3389/fimmu.2019.02393.
87. Escors D. Tumour immunogenicity, antigen presentation, and immunological barriers in cancer immunotherapy. Maiuri MC, editor. New J Sci [Internet]. 2014;2014:734515. Available from: https://doi.org/10.1155/2014/734515.
88. Kim JH, Kim BS, Lee SK. Regulatory T cells in tumor microenvironment and approach for anticancer immunotherapy. Immune Netw [Internet]. 2020;20(1). Available from: https://doi.org/10.4110/in.2020.20.e4.
89. Campos-Mora M, Contreras-Kallens P, Gálvez-Jirón F, Rojas M, Rojas C, Refisch A, et al. CD4+Foxp3+T regulatory cells promote transplantation tolerance by modulating effector CD4+ T cells in a neuropilin-1-dependent manner. Front Immunol [Internet]. 2019;10. Available from: https://www.frontiersin.org/articles/10.3389/fimmu.2019.00882.
90. Tanchot C, Terme M, Pere H, Tran T, Benhamouda N, Strioga M, et al. Tumor-infiltrating regulatory T cells: phenotype, role, mechanism of expansion in situ and clinical significance. Cancer Microenviron [Internet]. 2013;6(2):147–57. Available from: https://doi.org/10.1007/s12307-012-0122-y.
91. Li C, Jiang P, Wei S, Xu X, Wang J. Regulatory T cells in tumor microenvironment: new mechanisms, potential therapeutic strategies and future prospects. Mol Cancer [Internet]. 2020;19(1):116. Available from: https://doi.org/10.1186/s12943-020-01234-1.
92. Batlle E, Massagué J. Transforming growth factor-β signaling in immunity and cancer. Immunity [Internet]. 2019;50(4):924–40. Available from: https://www.sciencedirect.com/science/article/pii/S1074761319301414.
93. Ohue Y, Nishikawa H. Regulatory T (Treg) cells in cancer: can Treg cells be a new therapeutic target? Cancer Sci [Internet]. 2019;110(7):2080–9. Available from: https://onlinelibrary.wiley.com/doi/abs/10.1111/cas.14069.
94. Neophytou CM, Pierides C, Christodoulou MI, Costeas P, Kyriakou TC, Papageorgis P. The role of tumor-associated myeloid cells in modulating cancer therapy. Front Oncol [Internet]. 2020;10. Available from: https://www.frontiersin.org/articles/10.3389/fonc.2020.00899.
95. Gabrilovich DI, Ostrand-Rosenberg S, Bronte V. Coordinated regulation of myeloid cells by tumours. Nat Rev Immunol [Internet]. 2012;12(4):253–68. Available from: https://doi.org/10.1038/nri3175.
96. Wu Y, Yi M, Niu M, Mei Q, Wu K. Myeloid-derived suppressor cells: an emerging target for anticancer immunotherapy. Mol Cancer [Internet]. 2022;21(1):184. Available from: https://doi.org/10.1186/s12943-022-01657-y.
97. van Geffen C, Heiss C, Deißler A, Kolahian S. Pharmacological modulation of myeloid-derived suppressor cells to dampen inflammation. Front Immunol [Internet]. 2022;13. Available from: https://www.frontiersin.org/articles/10.3389/fimmu.2022.933847.
98. Yao Y, Xu XH, Jin L. Macrophage polarization in physiological and pathological pregnancy. Front Immunol [Internet]. 2019;10. Available from: https://www.frontiersin.org/articles/10.3389/fimmu.2019.00792.
99. Kumari N, Choi SH. Tumor-associated macrophages in cancer: recent advancements in cancer nanoimmunotherapies. J Exp Clin Cancer Res [Internet]. 2022;41(1):68. Available from: https://doi.org/10.1186/s13046-022-02272-x.
100. Ye X-z, Xu S-l, Xin Y-h, Yu S-c, Ping Y-f, Chen L, et al. Tumor-associated microglia/macrophages enhance the invasion of glioma stem-like cells via TGF-β1 signaling pathway. J Immunol [Internet]. 2012;189(1):444–53. Available from: https://doi.org/10.4049/jimmunol.1103248.

101. Vinay DS, Ryan EP, Pawelec G, Talib WH, Stagg J, Elkord E, et al. Immune evasion in cancer: mechanistic basis and therapeutic strategies. Semin Cancer Biol [Internet]. 2015;35:S185–98. Available from: https://www.sciencedirect.com/science/article/pii/S1044579X1500019X.
102. Johnsen AK, Templeton DJ, Sy MS, Harding CV. Deficiency of transporter for antigen presentation (TAP) in tumor cells allows evasion of immune surveillance and increases tumorigenesis1. J Immunol [Internet]. 1999;163(8):4224–31. Available from: https://doi.org/10.4049/jimmunol.163.8.4224.
103. Massa C, Wang Y, Marr N, Seliger B. Interferons and resistance mechanisms in tumors and pathogen-driven diseases-focus on the major histocompatibility complex (MHC) antigen processing pathway. Int J Mol Sci [Internet]. 2023;24(7). Available from: https://www.mdpi.com/1422-0067/24/7/6736.
104. Seliger B, Ritz U, Soldano F. Molecular mechanisms of HLA class I antigen abnormalities following viral infection and transformation. Int J Cancer [Internet]. 2006;118(1):129–38. Available from: https://onlinelibrary.wiley.com/doi/abs/10.1002/ijc.21312.
105. Dhatchinamoorthy K, Colbert JD, Rock KL. Cancer immune evasion through loss of MHC class I antigen presentation. Front Immunol [Internet]. 2021;12. Available from: https://www.frontiersin.org/articles/10.3389/fimmu.2021.636568.
106. Ke CH, Chiu YH, Huang KC, Lin CS. Exposure of immunogenic tumor antigens in surrendered immunity and the significance of autologous tumor cell-based vaccination in precision medicine. Int J Mol Sci [Internet]. 2023;24(1). Available from: https://www.mdpi.com/1422-0067/24/1/147.
107. Zhao H, Wu L, Yan G, Chen Y, Zhou M, Wu Y, et al. Inflammation and tumor progression: signaling pathways and targeted intervention. Signal Transduct Target Ther [Internet]. 2021;6(1):263. Available from: https://doi.org/10.1038/s41392-021-00658-5.
108. Basheer AS, Abas F, Othman I, Naidu R. Role of inflammatory mediators, macrophages, and neutrophils in glioma maintenance and progression: mechanistic understanding and potential therapeutic applications. Cancers (Basel) [Internet]. 2021;13(16). Available from: https://www.mdpi.com/2072-6694/13/16/4226.
109. Švajger U, Rožman P. Induction of tolerogenic dendritic cells by endogenous biomolecules: an update. Front Immunol [Internet]. 2018;9. Available from: https://www.frontiersin.org/articles/10.3389/fimmu.2018.02482.
110. Chakraborty NG, Li L, Sporn JR, Kurtzman SH, Ergin MT, Mukherji B. Emergence of regulatory CD4+ T cell response to repetitive stimulation with antigen-presenting cells in vitro: implications in designing antigen-presenting cell-based tumor vaccines. J Immunol [Internet]. 1999 [cited 2024 Oct 25];162(9):5576–83. Available from: https://doi.org/10.4049/jimmunol.162.9.5576.
111. Najafi S, Mortezaee K. Advances in dendritic cell vaccination therapy of cancer. Biomed Pharmacother. 2023;164:114954.
112. Wu AA, Drake V, Huang HS, Chiu SC, Zheng L. Reprogramming the tumor microenvironment: tumor-induced immunosuppressive factors paralyze T cells. Oncoimmunology [Internet]. 2015 [cited 2023 Nov 13];4(7):1–14. Available from: https://www.tandfonline.com/doi/abs/10.1080/2162402X.2015.1016700.
113. Toyoshima T, Nakamura S, Kumamaru W, Kawamura E, Ishibashi H, Hayashida JN, et al. Expression of tumor-associated antigen RCAS1 and its possible involvement in immune evasion in oral squamous cell carcinoma. J Oral Pathol Med [Internet]. 2006;35(6):361–8. Available from: https://onlinelibrary.wiley.com/doi/abs/10.1111/j.1600-0714.2006.00442.x.
114. Birklé S, Zeng G, Gao L, Yu RK, Aubry J. Role of tumor-associated gangliosides in cancer progression. Biochimie [Internet]. 2003;85(3):455–63. Available from: https://www.sciencedirect.com/science/article/pii/S0300908403000063.
115. Nakamura Y, Yamazaki K, Oizumi S, Nakashima M, Watanabe T, Dosaka-Akita H, et al. Expression of RCAS1 in human gastric carcinoma: a potential mechanism of immune escape. Cancer Sci [Internet]. 2004;95(3):260–5. Available from: https://onlinelibrary.wiley.com/doi/abs/10.1111/j.1349-7006.2004.tb02213.x.

116. Tran AP, Ali Al-Radhawi M, Kareva I, Wu J, Waxman DJ, Sontag ED. Delicate balances in cancer chemotherapy: modeling immune recruitment and emergence of systemic drug resistance. Front Immunol [Internet]. 2020;11. Available from: https://www.frontiersin.org/articles/10.3389/fimmu.2020.01376.
117. Yi M, Xu L, Jiao Y, Luo S, Li A, Wu K. The role of cancer-derived microRNAs in cancer immune escape. J Hematol Oncol [Internet]. 2020;13(1):25. Available from: https://doi.org/10.1186/s13045-020-00848-8.
118. Nurieva R, Wang J, Sahoo A. T-cell tolerance in cancer. Immunotherapy [Internet]. 2013;5(5):513–31. Available from: https://doi.org/10.2217/imt.13.33.
119. Taylor A, Verhagen J, Blaser K, Akdis M, Akdis CA. Mechanisms of immune suppression by interleukin-10 and transforming growth factor-β: the role of T regulatory cells. Immunology [Internet]. 2006;117(4):433–42. Available from: https://onlinelibrary.wiley.com/doi/abs/10.1111/j.1365-2567.2006.02321.x.
120. Chattopadhyay S, Chakraborty NG. Continuous presence of Th1 conditions is necessary for longer lasting tumor-specific CTL activity in stimulation cultures with PBL. Hum Immunol. 2005;66(8):884–91.
121. Sterner RC, Sterner RM. CAR-T cell therapy: current limitations and potential strategies. Blood Cancer J. 2021[Internet] [cited 2024 Oct 25];11(4):1–11. Available from: https://www.nature.com/articles/s41408-021-00459-7.
122. Sanmamed MF, Chen L. Inducible expression of B7-H1 (PD-L1) and its selective role in tumor site immune modulation. Cancer J [Internet]. 2014;20(4). Available from: https://journals.lww.com/journalppo/fulltext/2014/07000/inducible_expression_of_b7_h1__pd_l1__and_its.4.aspx.
123. Tuomela K, Ambrose AR, Davis DM. Escaping death: how cancer cells and infected cells resist cell-mediated cytotoxicity. Front Immunol [Internet]. 2022;13. Available from: https://www.frontiersin.org/articles/10.3389/fimmu.2022.867098.
124. Prager I, Watzl C. Mechanisms of natural killer cell-mediated cellular cytotoxicity. J Leukoc Biol [Internet]. 2019;105(6):1319–29. Available from: https://jlb.onlinelibrary.wiley.com/doi/abs/10.1002/JLB.MR0718-269R.
125. Slavin-Chiorini DC, Catalfamo M, Kudo-Saito C, Hodge JW, Schlom J, Sabzevari H. Amplification of the lytic potential of effector/memory CD8+ cells by vector-based enhancement of ICAM-1 (CD54) in target cells: implications for intratumoral vaccine therapy. Cancer Gene Ther [Internet]. 2004;11(10):665–80. Available from: https://doi.org/10.1038/sj.cgt.7700741.
126. Meslin F, Thiery J, Richon C, Jalil A, Chouaib S. Granzyme B-induced cell death involves induction of p53 tumor suppressor gene and its activation in tumor target cells*. J Biol Chem [Internet]. 2007;282(45):32991–9. Available from: https://www.sciencedirect.com/science/article/pii/S0021925820613096.
127. Pai CCS, Huang JT, Lu X, Simons DM, Park C, Chang A, Tamaki W, Liu E, Roybal KT, Seagal J, Chen M, Hagihara K, Wei XX, DuPage M, Kwek SS, Oh DY, Daud A, Tsai KT, Wu C, Zhang L, Fasso M, Sachidanandam R, Jayaprakash A, Lin I, Casbon AJ, Kinsbury GA, Fong L. Clonal deletion of tumor-specific T cells by interferon-γ confers therapeutic resistance to combination immune checkpoint blockade. 2019. cell.com [Internet]. 2019 [cited 2023 Nov 10];50:477–92.e8. Available from: https://www.cell.com/immunity/pdf/S1074-7613(19)30029-9.pdf.
128. Baitsch L, Fuertes-Marraco SA, Legat A, Meyer C, Speiser DE, CMT in, 2012 undefined. The three main stumbling blocks for anticancer T cells. Trends Immunol. 2012. cell.com. 2012 [cited 2023 Nov 11]. Available from: https://www.cell.com/trends/immunology/fulltext/S1471-4906(12)00037-3?large_figure=true.
129. Iwai Y, Ishida M, Tanaka Y, Okazaki T, Honjo T, Minato N. Involvement of PD-L1 on tumor cells in the escape from host immune system and tumor immunotherapy by PD-L1 blockade. Proc Natl Acad Sci USA. 2002;99(19):12293–7.
130. Jiang Y, Li Y, Zhu B. T-cell exhaustion in the tumor microenvironment. Cell Death Dis [Internet]. 2015 [cited 2023 Nov 11];6. Available from: www.nature.com/cddis.

131. Morfoisse F, Kuchnio A, Frainay C, Gomez-Brouchet A, Delisle MB, Marzi S, et al. Hypoxia induces VEGF-C expression in metastatic tumor cells via a HIF-1α-independent translation-mediated mechanism. Cell Rep [Internet]. 2014;6(1):155–67. Available from: https://www.sciencedirect.com/science/article/pii/S2211124713007560.
132. Chen G, Wu K, Li H, Xia D, He T. Role of hypoxia in the tumor microenvironment and targeted therapy. Front Oncol. 2022 Sep 23;12:961637. https://doi.org/10.3389/fonc.2022.961637. PMID: 36212414; PMCID: PMC9545774.
133. Leone RD, Emens LA. Targeting adenosine for cancer immunotherapy. J Immunother Cancer. 2018;6(1):57.
134. Burrows N, Maxwell PH. Hypoxia and B cells. Exp Cell Res. 2017;356(2):197–203.
135. Chang WH, Lai AG. The hypoxic tumour microenvironment: a safe haven for immunosuppressive cells and a therapeutic barrier to overcome. Cancer Lett. 2020;487:34–44.
136. Torres N, Regge MV, Secchiari F, Friedrich AD, Spallanzani RG, Raffo Iraolagoitia XL, et al. Restoration of antitumor immunity through anti-MICA antibodies elicited with a chimeric protein. J Immunother Cancer. 2020;8(1):e000233.
137. Chiu DKC, Tse APW, Xu IMJ, Di Cui J, Lai RKH, Li LL, et al. Hypoxia inducible factor HIF-1 promotes myeloid-derived suppressor cells accumulation through ENTPD2/CD39L1 in hepatocellular carcinoma. Nat Commun. 2017;8(1):517.
138. Li H, Shi B. Tolerogenic dendritic cells and their applications in transplantation. Cell Mol Immunol [Internet]. 2015;12(1):24–30. Available from: https://doi.org/10.1038/cmi.2014.52.
139. Mohapatra A, Sathiyamoorthy P, Park IK. Metallic nanoparticle-mediated immune cell regulation and advanced cancer immunotherapy. Pharmaceutics [Internet]. 2021;13(11). Available from: https://www.mdpi.com/1999-4923/13/11/1867.
140. Riera-Domingo C, Audigé A, Granja S, Cheng WC, Ho PC, Baltazar F, et al. Immunity, hypoxia, and metabolism–the ménage à trois of cancer: implications for immunotherapy. Physiol Rev [Internet]. 2020;100(1):1–102. Available from: https://doi.org/10.1152/physrev.00018.2019.
141. Bremnes RM, Al-Shibli K, Donnem T, Sirera R, Al-Saad S, Andersen S, et al. The role of tumor-infiltrating immune cells and chronic inflammation at the tumor site on cancer development, progression, and prognosis: emphasis on non-small cell lung cancer. J Thorac Oncol. 2011;6(4):824–33.
142. Harjunpää H, Asens ML, Guenther C, Fagerholm SC. Cell adhesion molecules and their roles and regulation in the immune and tumor microenvironment. Front Immunol. 2019;10:448153.
143. Sarkar T, Dhar S, Sa G. Tumor-infiltrating T-regulatory cells adapt to altered metabolism to promote tumor-immune escape. Curr Res Immunol. 2021;2:132–41.
144. Zeytun A, Hassuneh M, Nagarkatti M, Nagarkatti PS. Fas-Fas ligand–based interactions between tumor cells and tumor-specific cytotoxic T lymphocytes: a lethal two-way street. Blood [Internet]. 1997;90(5):1952–9. Available from: https://doi.org/10.1182/blood.V90.5.1952.
145. Anderson, A. C., Joller, N., & Kuchroo, V. K. (2016). Lag-3, Tim-3, and TIGIT: Co-inhibitory Receptors with Specialized Functions in Immune Regulation. Immunity, 44(5), 989–1004. https://doi.org/10.1016/j.immuni.2016.05.001.
146. Adeegbe DO, Nishikawa H. Natural and induced T regulatory cells in cancer. Front Immunol. 2013;4:190.
147. Paluskievicz CM, Cao X, Abdi R, Zheng P, Liu Y, Bromberg JS. T regulatory cells and priming the suppressive tumor microenvironment. Front Immunol. 2019;10:2453.
148. Gamal, W., Sahakian, E., & Pinilla-Ibarz, J. (2023). The role of Th17 cells in chronic lymphocytic leukemia: friend or foe?. Blood advances, 7(11), 2401–2417. https://doi.org/10.1182/bloodadvances.2022008985.
149. Marchenko, S., Piwonski, I., Hoffmann, I., Sinn, B. V., Kunze, C. A., Monjé, N., Pohl, J., Kulbe, H., Schmitt, W. D., Darb-Esfahani, S., Braicu, E. I., von Brünneck, A. C., Sehouli, J., Denkert, C., Horst, D., Jöhrens, K., & Taube, E. T. (2023). Prognostic value of regulatory T cells and T helper 17 cells in high grade serous ovarian carcinoma. Journal of cancer research and clinical oncology, 149(6), 2523–2536. https://doi.org/10.1007/s00432-022-04101-2.

150. Bailey SR, Nelson MH, Himes RA, Li Z, Mehrotra S, Paulos CM. Th17 cells in cancer: the ultimate identity crisis. Front Immunol. 2014;5:102119.
151. Yang X, Zheng SG. Interleukin-22: a likely target for treatment of autoimmune diseases. Autoimmun Rev. 2014 Jun;13(6):615–20. https://doi.org/10.1016/j.autrev.2013.11.008. Epub 2014 Jan 10. PMID: 24418299; PMCID: PMC3966954.
152. Dudakov, J. A., Hanash, A. M., & van den Brink, M. R. (2015). Interleukin-22: immunobiology and pathology. Annual review of immunology, 33, 747–785. https://doi.org/10.1146/annurev-immunol-032414-112123.
153. Lindahl H, Olsson T. Interleukin-22 influences the Th1/Th17 axis. Front Immunol. 2021;12:618110.
154. Pickert G, Neufert C, Leppkes M, Zheng Y, Wittkopf N, Warntjen M, et al. STAT3 links IL-22 signaling in intestinal epithelial cells to mucosal wound healing. J Exp Med [Internet]. 2009 [cited 2023 Sep 15];206(7):1465–72. Available from: https://pubmed.ncbi.nlm.nih.gov/19564350/.
155. Sarkar, T., Dhar, S., & Sa, G. (2021). Tumor-infiltrating T-regulatory cells adapt to altered metabolism to promote tumor-immune escape. Current research in immunology, 2, 132–141. https://doi.org/10.1016/j.crimmu.2021.08.002.
156. Mirlekar B. Tumor promoting roles of IL-10, TGF-β, IL-4, and IL-35: its implications in cancer immunotherapy. SAGE Open Med [Internet]. 2022;10:20503121211069012. Available from: https://doi.org/10.1177/20503121211069012.
157. Li, R., Li, H., Yang, X., Hu, H., Liu, P., & Liu, H. (2022). Crosstalk between dendritic cells and regulatory T cells: Protective effect and therapeutic potential in multiple sclerosis. Frontiers in immunology, 13, 970508. https://doi.org/10.3389/fimmu.2022.970508.
158. Levings MK, Bacchetta R, Schulz U, Roncarolo MG. The role of IL-10 and TGF-beta in the differentiation and effector function of T regulatory cells. Int Arch Allergy Immunol [Internet]. 2002 [cited 2023 Sep 15];129(4):263–76. Available from: https://pubmed.ncbi.nlm.nih.gov/12483031/.
159. Schmidt A, Oberle N, Krammer PH. Molecular mechanisms of treg-mediated T cell suppression. Front Immunol. 2012 Mar 21;3:51. https://doi.org/10.3389/fimmu.2012.00051. PMID: 22566933; PMCID: PMC3341960.
160. Vigano S, Alatzoglou D, Irving M, Ménétrier-Caux C, Caux C, Romero P, et al. Targeting adenosine in cancer immunotherapy to enhance T-cell function. Front Immunol. 2019;10:447851.
161. Jiang X, Wu X, Xiao Y, Wang P, Zheng J, Wu X, et al. The ectonucleotidases CD39 and CD73 on T cells: the new pillar of hematological malignancy. Front Immunol [Internet]. 2023 [cited 2023 Sep 15];14. Available from: https://pmc/articles/PMC9911447/.
162. Allard, B., Longhi, M. S., Robson, S. C., & Stagg, J. (2017). The ectonucleotidases CD39 and CD73: Novel checkpoint inhibitor targets. Immunological reviews, 276(1), 121–144. https://doi.org/10.1111/imr.12528.
163. Geng X, Li M, Cui B, Lu C, Liu X, Zhang P, Liu B, Ma C, Shen Y, Lu Z. CD4+CD25+ Foxp3+ regulatory T cells suppress NKG2D-mediated NK cell cytotoxicity in peripheral blood. Medicine (Baltimore). 2019 May;98(22):e15722. https://doi.org/10.1097/MD.0000000000015722. PMID: 31145286; PMCID: PMC6708973.
164. Zhang, J., Zou, Y., Chen, L., Xu, Q., Wang, Y., Xie, M., Liu, X., Zhao, J., & Wang, C. Y. (2022). Regulatory T Cells, a Viable Target Against Airway Allergic Inflammatory Responses in Asthma. Frontiers in immunology, 13, 902318. https://doi.org/10.3389/fimmu.2022.902318.
165. Grossman WJ, Verbsky JW, Barchet W, Colonna M, Atkinson JP, Ley TJ. Human T regulatory cells can use the perforin pathway to cause autologous target cell death. Immunity. 2004;21(4):589–601.
166. Regis S, Caliendo F, Dondero A, Bellora F, Casu B, Bottino C, et al. Main NK cell receptors and their ligands: regulation by microRNAs. AIMS Allergy Immunol [Internet]. 2018 [cited 2023 Sep 15];2(2):98–112. Available from: http://www.aimspress.com/article/doi/10.3934/Allergy.2018.2.98.

167. Taghikhani A, Farzaneh F, Sharifzad F, Mardpour S, Ebrahimi M, Hassan ZM. Engineered tumor-derived extracellular vesicles: potentials in cancer immunotherapy. Front Immunol. 2020;11:472095.
168. Olejarz W, Dominiak A, Żołnierzak A, Kubiak-Tomaszewska G, Lorenc T. Tumor-Derived Exosomes in Immunosuppression and Immunotherapy. J Immunol Res. 2020 May 22;2020:6272498. https://doi.org/10.1155/2020/6272498. PMID: 32537468; PMCID: PMC7261328.
169. Park YJ, Song B, Kim YS, Kim EK, Lee JM, Lee GE, et al. Tumor microenvironmental conversion of natural killer cells into myeloid-derived suppressor cells. Cancer Res [Internet]. 2013 [cited 2023 Sep 15];73(18):5669–81. Available from: https://doi.org/10.1158/0008-5472.CAN-13-0545.
170. De Salvo C, Buela KA, Pizarro TT. Cytokine-mediated regulation of innate lymphoid cell plasticity in gut mucosal immunity. Front Immunol [Internet]. 2020 [cited 2023 Sep 15];11. Available from: https://pmc/articles/PMC7794016/.
171. Irshad S, Flores-Borja F, Lawler K, Monypenny J, Evans R, Male V, et al. RORγt+ innate lymphoid cells promote lymph node metastasis of breast cancers. Cancer Res [Internet]. 2017 [cited 2023 Sep 15];77(5):1083–96. Available from: https://pubmed.ncbi.nlm.nih.gov/28082403/.
172. Lin Y, Xu J, Lan H. Tumor-associated macrophages in tumor metastasis: biological roles and clinical therapeutic applications. J Hematol Oncol. 2019 [Internet]. 2019 [cited 2022 Dec 25];12(1):1–16. Available from: https://jhoonline.biomedcentral.com/articles/10.1186/s13045-019-0760-3.
173. Grzywa, T. M., Sosnowska, A., Matryba, P., Rydzynska, Z., Jasinski, M., Nowis, D., & Golab, J. (2020). Myeloid Cell-Derived Arginase in Cancer Immune Response. Frontiers in immunology, 11, 938. https://doi.org/10.3389/fimmu.2020.00938.
174. Ge Z, Ding S. The Crosstalk Between Tumor-Associated Macrophages (TAMs) and Tumor Cells and the Corresponding Targeted Therapy. Front Oncol. 2020 Nov 3;10:590941. https://doi.org/10.3389/fonc.2020.590941. PMID: 33224886; PMCID: PMC7670061.
175. Hao Q, Vadgama JV, Wang P. CCL2/CCR2 signaling in cancer pathogenesis. Cell Commun Signal. 2020 [Internet]. 2020 [cited 2023 Sep 15];18(1):1–13. Available from: https://biosignaling.biomedcentral.com/articles/10.1186/s12964-020-00589-8.
176. Chen, Y., Song, Y., Du, W., Gong, L., Chang, H., & Zou, Z. (2019). Tumor-associated macrophages: an accomplice in solid tumor progression. Journal of biomedical science, 26(1), 78. https://doi.org/10.1186/s12929-019-0568-z.
177. Zhang, H., Liu, L., Liu, J., Dang, P., Hu, S., Yuan, W., Sun, Z., Liu, Y., & Wang, C. (2023). Roles of tumor-associated macrophages in anti-PD-1/PD-L1 immunotherapy for solid cancers. Molecular cancer, 22(1), 58. https://doi.org/10.1186/s12943-023-01725-x.
178. Das M, Stephen-Victor E, Bayry J. Regulatory T cells do not suppress rather activate human basophils by IL-3 and STAT5-dependent mechanisms. Oncoimmunology. 2020 Jun 5;9(1):1773193. https://doi.org/10.1080/2162402X.2020.1773193. PMID: 32923130; PMCID: PMC7458635.
179. Herr D, Sallmann A, Bekes I, Konrad R, Holzheu I, Kreienberg R, et al. VEGF induces ascites in ovarian cancer patients via increasing peritoneal permeability by downregulation of Claudin 5. Gynecol Oncol [Internet] 2012 [cited 2023 Sep 15];127(1):210–6. Available from: https://pubmed.ncbi.nlm.nih.gov/22579791/.
180. Chauhan, J., Stavraka, C., Grandits, M., Palhares, L. C. G. F., Josephs, D. H., Lacy, K. E., Spicer, J., Bax, H. J., & Karagiannis, S. N. (2022). Clinical and Translational Significance of Basophils in Patients with Cancer. Cells, 11(3), 438. https://doi.org/10.3390/cells11030438.
181. Long H, Liao W, Wang L, Lu Q. A player and coordinator: the versatile roles of eosinophils in the immune system. Transfus Med Hemother [Internet]. 2016 [cited 2023 Sep 15];43(2):96–108. Available from: https://doi.org/10.1159/000445215.
182. Elieh Ali Komi D, Bjermer L. Mast cell-mediated orchestration of the immune responses in human allergic asthma: current insights. Clin Rev Allergy Immunol. 2018 [Internet]. 2018

[cited 2023 Sep 15];56(2):234–47. Available from: https://link.springer.com/article/10.1007/s12016-018-8720-1.
183. Ghaffari, S., & Rezaei, N. (2023). Eosinophils in the tumor microenvironment: implications for cancer immunotherapy. Journal of translational medicine, 21(1), 551. https://doi.org/10.1186/s12967-023-04418-7.
184. Masucci, M. T., Minopoli, M., & Carriero, M. V. (2019). Tumor Associated Neutrophils. Their Role in Tumorigenesis, Metastasis, Prognosis and Therapy. Frontiers in oncology, 9, 1146. https://doi.org/10.3389/fonc.2019.01146.
185. Uribe-Querol, E., & Rosales, C. (2015). Neutrophils in Cancer: Two Sides of the Same Coin. Journal of immunology research, 2015, 983698. https://doi.org/10.1155/2015/983698.
186. Editors A, Kumar Tripathi J, Pal Singh Bhalla M, Dutta A, Bhagat S, Paul S, et al. Neutrophils in cancer and potential therapeutic strategies using neutrophil-derived exosomes. Vaccines [Internet]. 2023 [cited 2023 Sep 15];11(6):1028. Available from: https://www.mdpi.com/2076-393X/11/6/1028/htm.
187. Donne R, Lujambio A. The liver cancer immune microenvironment: therapeutic implications for hepatocellular carcinoma. Hepatology [Internet]. 2023 [cited 2023 Sep 15];77(5):1773–96. Available from: https://onlinelibrary.wiley.com/doi/full/10.1002/hep.32740.
188. Jaillon S, Ponzetta A, Di Mitri D, Santoni A, Bonecchi R, Mantovani A. Neutrophil diversity and plasticity in tumour progression and therapy. Nat Rev Cancer [Internet]. 2020 [cited 2023 Sep 15];20(9):485–503. Available from: https://www.nature.com/articles/s41568-020-0281-y.
189. Liu, Y., O'Leary, C. E., Wang, L. S., Bhatti, T. R., Dai, N., Kapoor, V., Liu, P., Mei, J., Guo, L., Oliver, P. M., Albelda, S. M., & Worthen, G. S. (2015). CD11b+Ly6G+ cells inhibit tumor growth by suppressing IL-17 production at early stages of tumorigenesis. Oncoimmunology, 5(1), e1061175. https://doi.org/10.1080/2162402X.2015.1061175.
190. Wculek, S. K., & Malanchi, I. (2015). Neutrophils support lung colonization of metastasis-initiating breast cancer cells. Nature, 528(7582), 413–417. https://doi.org/10.1038/nature16140.
191. Gabrilovich, D. I., & Nagaraj, S. (2009). Myeloid-derived suppressor cells as regulators of the immune system. Nature reviews. Immunology, 9(3), 162–174. https://doi.org/10.1038/nri2506.
192. Trikha, P., & Carson, W. E., 3rd (2014). Signaling pathways involved in MDSC regulation. Biochimica et biophysica acta, 1846(1), 55–65. https://doi.org/10.1016/j.bbcan.2014.04.003.
193. Yang, Z., Guo, J., Weng, L., Tang, W., Jin, S., & Ma, W. (2020). Myeloid-derived suppressor cells-new and exciting players in lung cancer. Journal of hematology & oncology, 13(1), 10. https://doi.org/10.1186/s13045-020-0843-1.
194. Yang, Y., Li, C., Liu, T., Dai, X., & Bazhin, A. V. (2020). Myeloid-Derived Suppressor Cells in Tumors: From Mechanisms to Antigen Specificity and Microenvironmental Regulation. Frontiers in immunology, 11, 1371. https://doi.org/10.3389/fimmu.2020.01371.
195. Xiang, X., Poliakov, A., Liu, C., Liu, Y., Deng, Z. B., Wang, J., Cheng, Z., Shah, S. V., Wang, G. J., Zhang, L., Grizzle, W. E., Mobley, J., & Zhang, H. G. (2009). Induction of myeloid-derived suppressor cells by tumor exosomes. International journal of cancer, 124(11), 2621–2633. https://doi.org/10.1002/ijc.24249.
196. Li, N., Wang, Y., Xu, H., Wang, H., Gao, Y., & Zhang, Y. (2021). Exosomes Derived from RM-1 Cells Promote the Recruitment of MDSCs into Tumor Microenvironment by Upregulating CXCR4 via TLR2/NF-κB Pathway. Journal of oncology, 2021, 5584406. https://doi.org/10.1155/2021/5584406.
197. Zhao, F., Gong, W., Song, J., Shen, Z., & Cui, D. (2022). The paradoxical role of MDSCs in inflammatory bowel diseases: From bench to bedside. Frontiers in immunology, 13, 1021634. https://doi.org/10.3389/fimmu.2022.1021634.
198. Martin, R. K., Saleem, S. J., Folgosa, L., Zellner, H. B., Damle, S. R., Nguyen, G. K., Ryan, J. J., Bear, H. D., Irani, A. M., & Conrad, D. H. (2014). Mast cell histamine promotes the immunoregulatory activity of myeloid-derived suppressor cells. Journal of leukocyte biology, 96(1), 151–159. https://doi.org/10.1189/jlb.5A1213-644R.

199. Szabo, P. A., Miron, M., & Farber, D. L. (2019). Location, location, location: Tissue resident memory T cells in mice and humans. Science immunology, 4(34), eaas9673. https://doi.org/10.1126/sciimmunol.aas9673.
200. Gueguen P, Metoikidou C, Dupic T, Lawand M, Goudot C, Baulande S, et al. Contribution of resident and circulating precursors to tumor-infiltrating CD8+ T cell populations in lung cancer. Sci Immunol [Internet]. 2021 [cited 2023 Sep 15];6(55):5778. Available from: https://www.science.org/doi/10.1126/sciimmunol.abd5778.
201. Li YR, Zhou K, Wilson M, Kramer A, Zhu Y, Dawson N, et al. Mucosal-associated invariant T cells for cancer immunotherapy. Mol Ther. 2023;31(3):631–46.
202. Dolina, J. S., Van Braeckel-Budimir, N., Thomas, G. D., & Salek-Ardakani, S. (2021). CD8+ T Cell Exhaustion in Cancer. Frontiers in immunology, 12, 715234. https://doi.org/10.3389/fimmu.2021.715234.
203. Timperi, E., & Barnaba, V. (2021). CD39 Regulation and Functions in T Cells. International journal of molecular sciences, 22(15), 8068. https://doi.org/10.3390/ijms22158068.
204. Mabrouk, N., Tran, T., Sam, I., Pourmir, I., Gruel, N., Granier, C., Pineau, J., Gey, A., Kobold, S., Fabre, E., & Tartour, E. (2022). CXCR6 expressing T cells: Functions and role in the control of tumors. Frontiers in immunology, 13, 1022136. https://doi.org/10.3389/fimmu.2022.1022136.
205. Wang, B., Wang, Y., Sun, X., Deng, G., Huang, W., Wu, X., Gu, Y., Tian, Z., Fan, Z., Xu, Q., Chen, H., & Sun, Y. (2021). CXCR6 is required for antitumor efficacy of intratumoral CD8+ T cell. Journal for immunotherapy of cancer, 9(8), e003100. https://doi.org/10.1136/jitc-2021-003100.
206. Qin, M., Jin, Y., & Pan, L. Y. (2021). Tertiary lymphoid structure and B-cell-related pathways: A potential target in tumor immunotherapy. Oncology letters, 22(6), 836. https://doi.org/10.3892/ol.2021.13097.
207. Wang, S. S., Liu, W., Ly, D., Xu, H., Qu, L., & Zhang, L. (2019). Tumor-infiltrating B cells: their role and application in anti-tumor immunity in lung cancer. Cellular & molecular immunology, 16(1), 6–18. https://doi.org/10.1038/s41423-018-0027-x.
208. Hou Y, Cao Y, Dong L, Huang Y, Zhang Z, Bi Y, et al. Regulation of follicular T helper cell differentiation in antitumor immunity. Int J Cancer [Internet]. 2023 [cited 2023 Sep 15];153(2):265–77. Available from: https://onlinelibrary.wiley.com/doi/full/10.1002/ijc.34368.
209. Yoshitomi, H., & Ueno, H. (2021). Shared and distinct roles of T peripheral helper and T follicular helper cells in human diseases. Cellular & molecular immunology, 18(3), 523–527. https://doi.org/10.1038/s41423-020-00529-z.
210. Lee G. R. (2017). Phenotypic and Functional Properties of Tumor-Infiltrating Regulatory T Cells. Mediators of inflammation, 2017, 5458178. https://doi.org/10.1155/2017/5458178.
211. Alvisi, G., Brummelman, J., Puccio, S., Mazza, E. M., Tomada, E. P., Losurdo, A., Zanon, V., Peano, C., Colombo, F. S., Scarpa, A., Alloisio, M., Vasanthakumar, A., Roychoudhuri, R., Kallikourdis, M., Pagani, M., Lopci, E., Novellis, P., Blume, J., Kallies, A., Veronesi, G., … Lugli, E. (2020). IRF4 instructs effector Treg differentiation and immune suppression in human cancer. The Journal of clinical investigation, 130(6), 3137–3150. https://doi.org/10.1172/JCI130426.
212. Bernardini G, Antonangeli F, Bonanni V, Santoni A. Dysregulation of chemokine/chemokine receptor axes and NK cell tissue localization during diseases. Front Immunol. 2016;7:222981.
213. Curio, S., & Belz, G. T. (2022). The unique role of innate lymphoid cells in cancer and the hepatic microenvironment. Cellular & molecular immunology, 19(9), 1012–1029. https://doi.org/10.1038/s41423-022-00901-1.
214. Ben-Shmuel A, Biber G, Barda-Saad M. Unleashing natural killer cells in the tumor microenvironment–the next generation of immunotherapy? Front Immunol. 2020;11:491312.
215. Zhou, W. T., & Jin, W. L. (2021). B7-H3/CD276: An Emerging Cancer Immunotherapy. Frontiers in immunology, 12, 701006. https://doi.org/10.3389/fimmu.2021.701006.
216. Gu, F. F., Zhang, K., Ma, L. L., Liu, Y. Y., Li, C., Hu, Y., Yang, Q. F., Liang, J. Y., Zeng, Y. L., Wang, Y., & Liu, L. (2020). The Superior Ability of Human BDCA3+ (CD141+) Dendritic

Cells (DCs) to Cross-Present Antigens Derived From Necrotic Lung Cancer Cells. Frontiers in immunology, 11, 1267. https://doi.org/10.3389/fimmu.2020.01267.
217. Heger, L., Hofer, T. P., Bigley, V., de Vries, I. J. M., Dalod, M., Dudziak, D., & Ziegler-Heitbrock, L. (2020). Subsets of CD1c+ DCs: Dendritic Cell Versus Monocyte Lineage. Frontiers in immunology, 11, 559166. https://doi.org/10.3389/fimmu.2020.559166.
218. Yang, M., McKay, D., Pollard, J. W., & Lewis, C. E. (2018). Diverse Functions of Macrophages in Different Tumor Microenvironments. Cancer research, 78(19), 5492–5503. https://doi.org/10.1158/0008-5472.CAN-18-1367.
219. Dudas J, Ladanyi A, Ingruber J, Steinbichler TB, Riechelmann H. Epithelial to Mesenchymal Transition: A Mechanism that Fuels Cancer Radio/Chemoresistance. Cells. 2020 Feb 12;9(2):428. https://doi.org/10.3390/cells9020428. PMID: 32059478; PMCID: PMC7072371.
220. Fleur L La, Botling J, He F, Pelicano C, Zhou C, He C, et al. Targeting MARCO and IL37R on immunosuppressive macrophages in lung cancer blocks regulatory T cells and supports cytotoxic lymphocyte function. Cancer Res [Internet]. 2021 [cited 2023 Sep 15];81(4):956–67. Available from: https://pubmed.ncbi.nlm.nih.gov/33293426/.
221. Wu JY, Huang TW, Hsieh YT, Wang YF, Yen CC, Lee GL, et al. Cancer-derived succinate promotes macrophage polarization and cancer metastasis via succinate receptor. Mol Cell [Internet]. 2020 [cited 2023 Sep 15];77(2):213–27.e5. Available from: https://pubmed.ncbi.nlm.nih.gov/31735641/.
222. Shen M, Hu P, Donskov F, Wang G, Liu Q, Du J. Tumor-associated neutrophils as a new prognostic factor in cancer: a systematic review and meta-analysis. PLoS One. 2014 Jun 6;9(6):e98259. https://doi.org/10.1371/journal.pone.0098259. PMID: 24906014; PMCID: PMC4048155.
223. Yan, M., Zheng, M., Niu, R., Yang, X., Tian, S., Fan, L., Li, Y., & Zhang, S. (2022). Roles of tumor-associated neutrophils in tumor metastasis and its clinical applications. Frontiers in cell and developmental biology, 10, 938289. https://doi.org/10.3389/fcell.2022.938289.
224. Sionov, R. V., Fridlender, Z. G., & Granot, Z. (2015). The Multifaceted Roles Neutrophils Play in the Tumor Microenvironment. Cancer microenvironment : official journal of the International Cancer Microenvironment Society, 8(3), 125–158. https://doi.org/10.1007/s12307-014-0147-5.
225. Fridlender ZG, Sun J, Kim S, Kapoor V, Cheng G, Ling L, Worthen GS, Albelda SM. Polarization of tumor-associated neutrophil phenotype by TGF-beta: "N1" versus "N2" TAN. Cancer Cell. 2009 Sep 8;16(3):183-94. https://doi.org/10.1016/j.ccr.2009.06.017. PMID: 19732719; PMCID: PMC2754404.
226. Nolan, E., Bridgeman, V. L., Ombrato, L., Karoutas, A., Rabas, N., Sewnath, C. A. N., Vasquez, M., Rodrigues, F. S., Horswell, S., Faull, P., Carter, R., & Malanchi, I. (2022). Radiation exposure elicits a neutrophil-driven response in healthy lung tissue that enhances metastatic colonization. Nature cancer, 3(2), 173–187. https://doi.org/10.1038/s43018-022-00336-7.
227. Albrengues, J., Shields, M. A., Ng, D., Park, C. G., Ambrico, A., Poindexter, M. E., Upadhyay, P., Uyeminami, D. L., Pommier, A., Küttner, V., Bružas, E., Maiorino, L., Bautista, C., Carmona, E. M., Gimotty, P. A., Fearon, D. T., Chang, K., Lyons, S. K., Pinkerton, K. E., Trotman, L. C., … Egeblad, M. (2018). Neutrophil extracellular traps produced during inflammation awaken dormant cancer cells in mice. Science (New York, N.Y.), 361(6409), eaao4227. https://doi.org/10.1126/science.aao4227.
228. Wang, D., & DuBois, R. N. (2015). Immunosuppression associated with chronic inflammation in the tumor microenvironment. Carcinogenesis, 36(10), 1085–1093. https://doi.org/10.1093/carcin/bgv123.
229. Kanterman J, Sade-Feldman M, Baniyash M. New insights into chronic inflammation-induced immunosuppression. Semin Cancer Biol. 2012;22(4):307–18.
230. Chonov DC, Ignatova MMK, Ananiev JR, Gulubova MV. IL-6 Activities in the Tumour Microenvironment. Part 1. Open Access Maced J Med Sci. 2019 Jul 20;7(14):2391–2398. https://doi.org/10.3889/oamjms.2019.589. PMID: 31592285; PMCID: PMC6765074.

231. Liu Q, Yu S, Li A, Xu H, Han X, Wu K. Targeting interlukin-6 to relieve immunosuppression in tumor microenvironment. Tumour Biol. 2017;39(6):1010428317712445.
232. Cruceriu D, Baldasici O, Balacescu O, Berindan-Neagoe I. The dual role of tumor necrosis factor-alpha (TNF-α) in breast cancer: molecular insights and therapeutic approaches. Cell Oncol. 2020;43(1):1–18.
233. Katanov C, Lerrer S, Liubomirski Y, Leider-Trejo L, Meshel T, Bar J, et al. Regulation of the inflammatory profile of stromal cells in human breast cancer: Prominent roles for TNF-α and the NF-κB pathway. Stem Cell Res Ther. 2015;6(1):87.
234. Laha D, Grant R, Mishra P, Nilubol N. The role of tumor necrosis factor in manipulating the immunological response of tumor microenvironment. Front Immunol. 2021;12:656908.
235. Nakagawa, H., Umemura, A., Taniguchi, K., Font-Burgada, J., Dhar, D., Ogata, H., Zhong, Z., Valasek, M. A., Seki, E., Hidalgo, J., Koike, K., Kaufman, R. J., & Karin, M. (2014). ER stress cooperates with hypernutrition to trigger TNF-dependent spontaneous HCC development. Cancer cell, 26(3), 331–343. https://doi.org/10.1016/j.ccr.2014.07.001.
236. Chaithongyot, S., Jantaree, P., Sokolova, O., & Naumann, M. (2021). NF-κB in Gastric Cancer Development and Therapy. Biomedicines, 9(8), 870. https://doi.org/10.3390/biomedicines9080870.
237. Yu, H., Pardoll, D., & Jove, R. (2009). STATs in cancer inflammation and immunity: a leading role for STAT3. Nature reviews. Cancer, 9(11), 798–809. https://doi.org/10.1038/nrc2734.
238. Shi X, Yang J, Deng S, Xu H, Wu D, Zeng Q, et al. TGF-β signaling in the tumor metabolic microenvironment and targeted therapies. J Hematol Oncol [Internet]. 2022 [cited 2023 Sep 15];15(1):1–39. Available from: https://jhoonline.biomedcentral.com/articles/10.1186/s13045-022-01349-6.
239. Wang J, Zhao X, Wan YY. Intricacies of TGF-β signaling in Treg and Th17 cell biology. Cell Mol Immunol [Internet]. 2023 [cited 2023 Sep 15];20(9):1002–22. Available from: https://www.nature.com/articles/s41423-023-01036-7.
240. Briukhovetska D, Dörr J, Endres S, Libby P, Dinarello CA, Kobold S. Interleukins in cancer: from biology to therapy. Nat Rev Cancer. 2021 [Internet]. 2021 [cited 2023 Sep 15];21(8):481–99. Available from: https://www.nature.com/articles/s41568-021-00363-z.
241. Kiesler P, Fuss IJ, Strober W. Experimental models of inflammatory bowel diseases. Cell Mol Gastroenterol Hepatol. 2015;1(2):154–70.
242. Carlini V, Noonan DM, Abdalalem E, Goletti D, Sansone C, Calabrone L, et al. The multifaceted nature of IL-10: regulation, role in immunological homeostasis and its relevance to cancer, COVID-19 and post-COVID conditions. Front Immunol. 2023;14:1161067.
243. Zamarron, B. F., & Chen, W. (2011). Dual roles of immune cells and their factors in cancer development and progression. International journal of biological sciences, 7(5), 651–658. https://doi.org/10.7150/ijbs.7.651.
244. Corliss, B. A., Azimi, M. S., Munson, J. M., Peirce, S. M., & Murfee, W. L. (2016). Macrophages: An Inflammatory Link Between Angiogenesis and Lymphangiogenesis. Microcirculation (New York, N.Y. : 1994), 23(2), 95–121. https://doi.org/10.1111/micc.12259.
245. Li, K., Shi, H., Zhang, B., Ou, X., Ma, Q., Chen, Y., Shu, P., Li, D., & Wang, Y. (2021). Myeloid-derived suppressor cells as immunosuppressive regulators and therapeutic targets in cancer. Signal transduction and targeted therapy, 6(1), 362. https://doi.org/10.1038/s41392-021-00670-9.
246. Lindau D, Gielen P, Kroesen M, Wesseling P, Adema GJ. The immunosuppressive tumour network: myeloid-derived suppressor cells, regulatory T cells and natural killer T cells. Immunology. 2013 Feb;138(2):105–15. https://doi.org/10.1111/imm.12036. PMID: 23216602; PMCID: PMC3575763.
247. Ahrends T, Borst J. The opposing roles of CD4+ T cells in anti-tumour immunity. Immunology. 2018 Apr 27;154(4):582–92. https://doi.org/10.1111/imm.12941. Epub ahead of print. PMID: 29700809; PMCID: PMC6050207.

248. Dwivedi, M., Tiwari, S., Kemp, E. H., & Begum, R. (2022). Implications of regulatory T cells in anti-cancer immunity: from pathogenesis to therapeutics. Heliyon, 8(8), e10450. https://doi.org/10.1016/j.heliyon.2022.e10450.
249. Kalathil, S. G., Lugade, A. A., Miller, A., Iyer, R., & Thanavala, Y. (2016). PD-1+ and Foxp3+ T cell reduction correlates with survival of HCC patients after sorafenib therapy. JCI insight, 1(11), e86182. https://doi.org/10.1172/jci.insight.86182.
250. Barbi, J., Pardoll, D., & Pan, F. (2013). Metabolic control of the Treg/Th17 axis. Immunological reviews, 252(1), 52–77. https://doi.org/10.1111/imr.12029.
251. Bunte, K., & Beikler, T. (2019). Th17 Cells and the IL-23/IL-17 Axis in the Pathogenesis of Periodontitis and Immune-Mediated Inflammatory Diseases. International journal of molecular sciences, 20(14), 3394. https://doi.org/10.3390/ijms20143394.
252. Xing J, Man C, Liu Y, Zhang Z, Peng H. Factors impacting the benefits and pathogenicity of Th17 cells in the tumor microenvironment. Front Immunol. 2023;14:1224269.
253. Tang, Q., Li, J., Zhu, H., Li, P., Zou, Z., & Xiao, Y. (2013). Hmgb1-IL-23-IL-17-IL-6-Stat3 axis promotes tumor growth in murine models of melanoma. Mediators of inflammation, 2013, 713859. https://doi.org/10.1155/2013/713859.
254. Wang, L., Yi, T., Kortylewski, M., Pardoll, D. M., Zeng, D., & Yu, H. (2009). IL-17 can promote tumor growth through an IL-6-Stat3 signaling pathway. The Journal of experimental medicine, 206(7), 1457–1464. https://doi.org/10.1084/jem.20090207.
255. Wherry EJ. T cell exhaustion. Nat Immunol [Internet]. 2011 [cited 2023 Sep 15];12(6):492–9. Available from: https://pubmed.ncbi.nlm.nih.gov/21739672/.
256. Schietinger, A., Philip, M., Krisnawan, V. E., Chiu, E. Y., Delrow, J. J., Basom, R. S., Lauer, P., Brockstedt, D. G., Knoblaugh, S. E., Hämmerling, G. J., Schell, T. D., Garbi, N., & Greenberg, P. D. (2016). Tumor-Specific T Cell Dysfunction Is a Dynamic Antigen-Driven Differentiation Program Initiated Early during Tumorigenesis. Immunity, 45(2), 389–401. https://doi.org/10.1016/j.immuni.2016.07.011.
257. Davidson, S., Efremova, M., Riedel, A., Mahata, B., Pramanik, J., Huuhtanen, J., Kar, G., Vento-Tormo, R., Hagai, T., Chen, X., Haniffa, M. A., Shields, J. D., & Teichmann, S. A. (2020). Single-Cell RNA Sequencing Reveals a Dynamic Stromal Niche That Supports Tumor Growth. Cell reports, 31(7), 107628. https://doi.org/10.1016/j.celrep.2020.107628.
258. Laba S, Mallett G, Amarnath S. The depths of PD-1 function within the tumor microenvironment beyond CD8+ T cells. Semin Cancer Biol. 2022;86:1045–55.
259. Verma, V., Shrimali, R. K., Ahmad, S., Dai, W., Wang, H., Lu, S., Nandre, R., Gaur, P., Lopez, J., Sade-Feldman, M., Yizhak, K., Bjorgaard, S. L., Flaherty, K. T., Wargo, J. A., Boland, G. M., Sullivan, R. J., Getz, G., Hammond, S. A., Tan, M., Qi, J., … Khleif, S. N. (2019). PD-1 blockade in subprimed CD8 cells induces dysfunctional PD-1+CD38hi cells and anti-PD-1 resistance. Nature immunology, 20(9), 1231–1243. https://doi.org/10.1038/s41590-019-0441-y.
260. Draghiciu, O., Lubbers, J., Nijman, H. W., & Daemen, T. (2015). Myeloid derived suppressor cells-An overview of combat strategies to increase immunotherapy efficacy. Oncoimmunology, 4(1), e954829. https://doi.org/10.4161/21624011.2014.954829.
261. Gabrilovich DI, Nagaraj S. Myeloid-derived suppressor cells as regulators of the immune system. Nat Rev Immunol. 2009 [Internet]. 2009 [cited 2023 Sep 15];9(3):162–74. Available from: https://www.nature.com/articles/nri2506.
262. Hu, X., Li, J., Fu, M., Zhao, X., & Wang, W. (2021). The JAK/STAT signaling pathway: from bench to clinic. Signal transduction and targeted therapy, 6(1), 402. https://doi.org/10.1038/s41392-021-00791-1.
263. Aboelella, N. S., Brandle, C., Kim, T., Ding, Z. C., & Zhou, G. (2021). Oxidative Stress in the Tumor Microenvironment and Its Relevance to Cancer Immunotherapy. Cancers, 13(5), 986. https://doi.org/10.3390/cancers13050986.
264. Prendergast, G. C., Malachowski, W. J., Mondal, A., Scherle, P., & Muller, A. J. (2018). Indoleamine 2,3-Dioxygenase and Its Therapeutic Inhibition in Cancer. International review of cell and molecular biology, 336, 175–203. https://doi.org/10.1016/bs.ircmb.2017.07.004.

265. Groth, C., Hu, X., Weber, R., Fleming, V., Altevogt, P., Utikal, J., & Umansky, V. (2019). Immunosuppression mediated by myeloid-derived suppressor cells (MDSCs) during tumour progression. British journal of cancer, 120(1), 16–25. https://doi.org/10.1038/s41416-018-0333-1.
266. Yu, H., Kortylewski, M., & Pardoll, D. (2007). Crosstalk between cancer and immune cells: role of STAT3 in the tumour microenvironment. Nature reviews. Immunology, 7(1), 41–51. https://doi.org/10.1038/nri1995.
267. De Cicco P, Ercolano G, Ianaro A. The new era of cancer immunotherapy: targeting myeloid-derived suppressor cells to overcome immune evasion. Front Immunol. 2020;11:550783.
268. Li, J., Wang, L., Chen, X., Li, L., Li, Y., Ping, Y., Huang, L., Yue, D., Zhang, Z., Wang, F., Li, F., Yang, L., Huang, J., Yang, S., Li, H., Zhao, X., Dong, W., Yan, Y., Zhao, S., Huang, B., … Zhang, Y. (2017). CD39/CD73 upregulation on myeloid-derived suppressor cells via TGF-β-mTOR-HIF-1 signaling in patients with non-small cell lung cancer. Oncoimmunology, 6(6), e1320011. https://doi.org/10.1080/2162402X.2017.1320011.
269. Pu, Y., & Ji, Q. (2022). Tumor-Associated Macrophages Regulate PD-1/PD-L1 Immunosuppression. Frontiers in immunology, 13, 874589. https://doi.org/10.3389/fimmu.2022.874589.
270. Wang J, Liu T, Huang T, Shang M, Wang X. The mechanisms on evasion of anti-tumor immune responses in gastric cancer. Front Oncol. 2022 Nov 10;12:943806. https://doi.org/10.3389/fonc.2022.943806. PMID: 36439472; PMCID: PMC9686275.
271. Raskov H, Orhan A, Christensen JP, Gögenur I. Cytotoxic CD8+ T cells in cancer and cancer immunotherapy. Br J Cancer. 2020 [Internet]. 2020 [cited 2023 Sep 15];124(2):359–67. Available from: https://www.nature.com/articles/s41416-020-01048-4.
272. Sui, H., Dongye, S., Liu, X., Xu, X., Wang, L., Jin, C. Q., Yao, M., Gong, Z., Jiang, D., Zhang, K., Liu, Y., Liu, H., Jiang, G., & Su, Y. (2022). Immunotherapy of targeting MDSCs in tumor microenvironment. Frontiers in immunology, 13, 990463. https://doi.org/10.3389/fimmu.2022.990463.
273. Li, Y., Xiang, S., Pan, W., Wang, J., Zhan, H., & Liu, S. (2023). Targeting tumor immunosuppressive microenvironment for pancreatic cancer immunotherapy: Current research and future perspective. Frontiers in oncology, 13, 1166860. https://doi.org/10.3389/fonc.2023.1166860.
274. Li L, Yang L, Wang L, Wang F, Zhang Z, Li J, et al. Impaired T cell function in malignant pleural effusion is caused by TGF-β derived predominantly from macrophages. Int J Cancer [Internet]. 2016 [cited 2023 Sep 15];139(10):2261–9. Available from: https://pubmed.ncbi.nlm.nih.gov/27459735/.
275. Kloss CC, Lee J, Zhang A, Chen F, Melenhorst JJ, Lacey SF, et al. Dominant-negative TGF-β receptor enhances PSMA-targeted human CAR T cell proliferation and augments prostate cancer eradication. Mol Ther [Internet] 2018 [cited 2023 Sep 15];26(7):1855–66. Available from: http://www.cell.com/article/S1525001618302065/fulltext.
276. Li L, Ma Y, Xu Y. Follicular regulatory T cells infiltrated the ovarian carcinoma and resulted in CD8 T cell dysfunction dependent on IL-10 pathway. Int Immunopharmacol [Internet]. 2019 [cited 2023 Sep 15];68:81–7. Available from: https://pubmed.ncbi.nlm.nih.gov/30616170/.
277. Burns, J. S., & Manda, G. (2017). Metabolic Pathways of the Warburg Effect in Health and Disease: Perspectives of Choice, Chain or Chance. International journal of molecular sciences, 18(12), 2755. https://doi.org/10.3390/ijms18122755.
278. Verma, N. K., Wong, B. H. S., Poh, Z. S., Udayakumar, A., Verma, R., Goh, R. K. J., Duggan, S. P., Shelat, V. G., Chandy, K. G., & Grigoropoulos, N. F. (2022). Obstacles for T-lymphocytes in the tumour microenvironment: Therapeutic challenges, advances and opportunities beyond immune checkpoint. EBioMedicine, 83, 104216. https://doi.org/10.1016/j.ebiom.2022.104216.
279. Payen, V. L., Mina, E., Van Hée, V. F., Porporato, P. E., & Sonveaux, P. (2020). Monocarboxylate transporters in cancer. Molecular metabolism, 33, 48–66. https://doi.org/10.1016/j.molmet.2019.07.006.

280. Saka D, Gökalp M, Piyade B, Cevik NC, Sever EA, Unutmaz D, et al. Mechanisms of T-cell exhaustion in pancreatic cancer. Cancers [Internet]. 2020 [cited 2023 Sep 15];12(8):2274. Available from: https://www.mdpi.com/2072-6694/12/8/2274/htm.
281. Kishton, R. J., Sukumar, M., & Restifo, N. P. (2017). Metabolic Regulation of T Cell Longevity and Function in Tumor Immunotherapy. Cell metabolism, 26(1), 94–109. https://doi.org/10.1016/j.cmet.2017.06.016.
282. Kim, B., Jung, J. W., Jung, J., Han, Y., Suh, D. H., Kim, H. S., Dhanasekaran, D. N., & Song, Y. S. (2017). PGC1α induced by reactive oxygen species contributes to chemoresistance of ovarian cancer cells. Oncotarget, 8(36), 60299–60311. https://doi.org/10.18632/oncotarget.19140.
283. Eil R, Vodnala SK, Clever D, Klebanoff CA, Sukumar M, Pan JH, et al. Ionic immune suppression within the tumour microenvironment limits T cell effector function. Nature [Internet]. 2016 [cited 2023 Sep 15];537(7621):539–43. Available from: https://www.nature.com/articles/nature19364.
284. Gol-Ara M, Jadidi-Niaragh F, Sadria R, Azizi G, Mirshafiey A. The role of different subsets of regulatory T cells in immunopathogenesis of rheumatoid arthritis. Arthritis. 2012;2012:805875. https://doi.org/10.1155/2012/805875. Epub 2012 Oct 24. PMID: 23133752; PMCID: PMC3486158.
285. Saleh R, Elkord E. FoxP3+ T regulatory cells in cancer: prognostic biomarkers and therapeutic targets. Cancer Lett. 2020;490:174–85.
286. Karkhah A, Javanian M, Ebrahimpour S. The role of regulatory T cells in immunopathogenesis and immunotherapy of viral infections. Infect Genet Evol. 2018;59:32–7.
287. Chuckran CA, Liu C, Bruno TC, Workman CJ, Vignali DAA. Neuropilin-1: a checkpoint target with unique implications for cancer immunology and immunotherapy. J Immunother Cancer [Internet]. 2020 [cited 2023 Sep 15];8(2):e000967. Available from: https://jitc.bmj.com/content/8/2/e000967.
288. Tay RE, Richardson EK, Toh HC. Revisiting the role of CD4+ T cells in cancer immunotherapy—new insights into old paradigms. Cancer Gene Ther. 2020 [Internet]. 2020 [cited 2023 Sep 15];28(1):5–17. Available from: https://www.nature.com/articles/s41417-020-0183-x.
289. Li C, Jiang P, Wei S, Xu X, Wang J. Regulatory T cells in tumor microenvironment: new mechanisms, potential therapeutic strategies and future prospects. Mol Cancer. 2020 [Internet]. 2020 [cited 2023 Sep 15];19(1):1–23. Available from: https://molecular-cancer.biomedcentral.com/articles/10.1186/s12943-020-01234-1.
290. Downs-Canner S, Berkey S, Delgoffe GM, Edwards RP, Curiel T, Odunsi K, et al. Suppressive IL-17A+Foxp3+ and ex-Th17 IL-17AnegFoxp3+ Treg cells are a source of tumour-associated Treg cells. Nat Commun [Internet]. 2017 [cited 2023 Sep 15];8. Available from: https://pubmed.ncbi.nlm.nih.gov/28290453/.
291. Cao X, Cai SF, Fehniger TA, Song J, Collins LI, Piwnica-Worms DR, et al. Granzyme B and perforin are important for regulatory T cell-mediated suppression of tumor clearance. Immunity. 2007;27(4):635–46.
292. Kalia V, Penny LA, Yuzefpolskiy Y, Baumann FM, Sarkar S. Quiescence of memory CD8(+) T cells is mediated by regulatory T cells through inhibitory receptor CTLA-4. Immunity [Internet]. 2015 [cited 2023 Sep 15];42(6):1116–29. Available from: https://pubmed.ncbi.nlm.nih.gov/26084026/.
293. Kared, H., Martelli, S., Tan, S. W., Simoni, Y., Chong, M. L., Yap, S. H., Newell, E. W., Pender, S. L. F., Kamarulzaman, A., Rajasuriar, R., & Larbi, A. (2018). Adaptive NKG2C+CD57+ Natural Killer Cell and Tim-3 Expression During Viral Infections. Frontiers in immunology, 9, 686. https://doi.org/10.3389/fimmu.2018.00686.
294. Sarhan D, Hippen KL, Lemire A, Hying S, Luo X, Lenvik T, et al. Adaptive NK cells resist regulatory T-cell suppression driven by IL37. Cancer Immunol Res [Internet]. 2018 [cited 2023 Sep 15];6(7):766–75. Available from: https://pubmed.ncbi.nlm.nih.gov/29784636/.
295. Sojka DK, Fowell DJ. Regulatory T cells inhibit acute IFN-γ synthesis without blocking T-helper cell type 1 (Th1) differentiation via a compartmentalized requirement for IL-10.

Proc Natl Acad Sci U S A [Internet]. 2011 [cited 2023 Sep 15];108(45):18336–41. Available from: https://www.pnas.org/doi/abs/10.1073/pnas.1110566108.

296. Park, Y. J., Ryu, H., Choi, G., Kim, B. S., Hwang, E. S., Kim, H. S., & Chung, Y. (2019). IL-27 confers a protumorigenic activity of regulatory T cells via CD39. Proceedings of the National Academy of Sciences of the United States of America, 116(8), 3106–3111. https://doi.org/10.1073/pnas.1810254116.
297. Klein, M., & Bopp, T. (2016). Cyclic AMP Represents a Crucial Component of Treg Cell-Mediated Immune Regulation. Frontiers in immunology, 7, 315. https://doi.org/10.3389/fimmu.2016.00315.
298. Roh, M., Wainwright, D. A., Wu, J. D., Wan, Y., & Zhang, B. (2020). Targeting CD73 to augment cancer immunotherapy. Current opinion in pharmacology, 53, 66–76. https://doi.org/10.1016/j.coph.2020.07.001.
299. Li, Y., He, H., Jihu, R., Zhou, J., Zeng, R., & Yan, H. (2021). Novel Characterization of Myeloid-Derived Suppressor Cells in Tumor Microenvironment. Frontiers in cell and developmental biology, 9, 698532. https://doi.org/10.3389/fcell.2021.698532.
300. Munn, D. H., & Mellor, A. L. (2016). IDO in the Tumor Microenvironment: Inflammation, Counter-Regulation, and Tolerance. Trends in immunology, 37(3), 193–207. https://doi.org/10.1016/j.it.2016.01.002.
301. Jitschin R, Braun M, Büttner M, Dettmer-Wilde K, Bricks J, Berger J, et al. CLL-cells induce IDOhi CD14+HLA-DRlo myeloid-derived suppressor cells that inhibit T-cell responses and promote TRegs. Blood [Internet]. 2014 [cited 2023 Sep 15];124(5):750–60. Available from: https://doi.org/10.1182/blood-2013-12-546416.
302. Huang B, Pan PY, Li Q, Sato AI, Levy DE, Bromberg J, et al. Gr-1+CD115+ immature myeloid suppressor cells mediate the development of tumor-induced T regulatory cells and T-cell anergy in tumor-bearing host. Cancer Res [Internet]. 2006 [cited 2023 Sep 15];66(2):1123–31. Available from: https://doi.org/10.1158/0008-5472.CAN-05-1299.
303. Haist M, Stege H, Grabbe S, Bros M. The functional crosstalk between myeloid-derived suppressor cells and regulatory T cells within the immunosuppressive tumor microenvironment. Cancers. 2021 [Internet]. 2021 [cited 2023 Sep 15];13(2):210. Available from: https://www.mdpi.com/2072-6694/13/2/210/htm.
304. Kumar, V., Patel, S., Tcyganov, E., & Gabrilovich, D. I. (2016). The Nature of Myeloid-Derived Suppressor Cells in the Tumor Microenvironment. Trends in immunology, 37(3), 208–220. https://doi.org/10.1016/j.it.2016.01.004.
305. Li, C., Jiang, P., Wei, S., Xu, X., & Wang, J. (2020). Regulatory T cells in tumor microenvironment: new mechanisms, potential therapeutic strategies and future prospects. Molecular cancer, 19(1), 116. https://doi.org/10.1186/s12943-020-01234-1.

The Evolution of Cancer Immunotherapy: A Journey from Concept to Clinical Reality

Alok K. Mishra and Sunil K. Malonia

Abstract Cancer immunotherapy has transformed the oncology landscape by harnessing the immune system's ability to recognize and eliminate malignant cells. This paradigm shift from conventional therapies to biologically targeted interventions underscores a new era in cancer therapy. From its early theoretical and experimental foundations to the groundbreaking clinical applications we see today, immunotherapy has revolutionized cancer treatment. Landmark achievements, particularly immune checkpoint inhibitors targeting the PD-1/PD-L1 axis or CTLA-4, have demonstrated durable responses across various cancers, validating the therapeutic potential of immune modulation. Building on these advances, multiple approaches such as CAR-T cells, have provided unprecedented efficacy in hematologic malignancies. At the same time, cancer vaccines and oncolytic viruses bring innovative approaches to tackling solid tumors. Despite challenges, including immune-related toxicities and resistance mechanisms, ongoing advancements in biomarkers, combination therapies, and novel immune modulators address these limitations, paving the way for more personalized and resilient treatment responses. This chapter highlights the evolution of cancer immunotherapy and its transition to engineered cell-based therapies for cancer, tracing its journey from conceptual origins to scientific milestones.

A. K. Mishra (✉)
Department of Molecular, Cell and Cancer Biology, UMass Chan Medical School, Worcester, MA, USA

Department of Hematopoietic Biology & Malignancy, Division of Cancer Medicine, The University of Texas MD Anderson Cancer Center, Houston, TX, USA
e-mail: AKMishra1@mdanderson.org

S. K. Malonia (✉)
Department of Molecular, Cell and Cancer Biology, UMass Chan Medical School, Worcester, MA, USA

Cancer Center, UMass Chan Medical School, Worcester, MA, USA
e-mail: Sunil.Malonia@umassmed.edu

A. K. Mishra, S. K. Malonia (eds.), *Cell-based Immunotherapies for Cancer*,
https://doi.org/10.1007/978-3-031-88695-9_2

Keywords Cancer · Immunotherapy · Cell-based therapy · Immunoediting · CAR-T · TCR therapy · Oncolytic virus · Bone marrow transplantation (BMT)

Abbreviations

CAR	Chimeric antigen receptor
GVHD	Graft-versus-host disease
MHC	Major histocompatibility complex
TCR	T-cell receptor
TIL	Tumor-infiltrating lymphocyte

1 Introduction

Cancer, often referred to as the "emperor of all maladies," has long been one of the most formidable challenges in medicine [1, 2]. For centuries, it was a disease that confounded physicians, with treatments limited to surgery, rudimentary forms of radiation, and later, chemotherapy. Despite the incremental progress made through these approaches, many cancers remained untreatable or recurred after initial success, leaving patients and doctors searching for more effective solutions [3–5]. One of the most promising breakthroughs in recent decades has been cancer immunotherapy. This treatment strategy seeks to harness the body's own immune system to recognize and destroy cancer cells. The emergence of immunotherapy as a viable cancer treatment is nothing short of revolutionary, fundamentally transforming the field of oncology and offering new hope to patients with cancers previously deemed untreatable [6–9].

The idea of using the immune system to fight cancer is not new; its roots can be traced back more than a century. However, in the past few decades, the field has gained significant traction, evolving from experimental and largely theoretical work to becoming a central pillar in the fight against cancer. The success of immunotherapy is a testament to the convergence of decades of research in immunology, molecular biology, and clinical oncology. It represents a treatment breakthrough and a paradigm shift in understanding the interaction between cancer and the immune system [6, 8].

The latter half of the twentieth century witnessed a series of groundbreaking discoveries that would eventually bring cancer immunotherapy to the forefront of oncology. These included identifying immune checkpoint molecules that regulate immune responses and can be exploited by cancer cells to avoid immune destruction. The development of monoclonal antibodies in the 1970s also provided a powerful new tool for targeting specific molecules on cancer cells, laying the groundwork for future immunotherapies (Table 1). The culmination of these advances was the

Table 1 Key milestones in the evolution of cancer immunotherapy

Year	Milestone	Description	Reference
1891	Coley's toxins	William Coley, regarded as the "*Father of Immunotherapy*," used bacterial toxins to stimulate the immune system in cancer patients, initiating early exploration into immunotherapy.	[10]
1976	Discovery of interferons	Interferons, proteins that help regulate immune responses, were shown to have potential in treating cancer, representing one of the first forms of cytokine therapy.	[11]
1986	FDA approval of interleukin-2 (IL-2)	IL-2 was approved for the treatment of renal cell carcinoma and melanoma, becoming one of the first immunotherapy agents. It works by enhancing the activity of T cells.	[12]
1990s	Tumor-infiltrating lymphocyte (TIL) therapy	Adoptive cell transfer using TILs showed promise in treating melanoma, marking an early success for cell-based immunotherapy.	[13]
1997	FDA approval of rituximab	The first monoclonal antibody approved for cancer treatment, *rituximab*, targets CD20 in B-cell non-Hodgkin's lymphoma, paving the way for targeted immunotherapies.	[14]
2010	FDA approval of sipuleucel-T (Provenge)	The first cancer vaccine approved for prostate cancer, *sipuleucel-T*, personalized treatment by using the patient's own immune cells to fight cancer.	[15]
2011	FDA approval of ipilimumab	*Ipilimumab*, an anti-CTLA-4 immune checkpoint inhibitor, was approved for metastatic melanoma, ushering in the era of immune checkpoint inhibitors.	[16]
2012	CAR-T-cell therapy development	Early clinical trials of chimeric antigen receptor (CAR)-T-cell therapy demonstrated remarkable efficacy in treating leukemia and lymphoma, marking the rise of engineered immune cells.	[17]
2013	Identification of PD-1/PD-L1 pathway in cancer	Research showing that the *PD-1/PD-L1* axis could be targeted to enhance immune responses against tumors led to the rapid development of PD-1/PD-L1 inhibitors.	[18]
2017	FDA approval of CAR-T therapies	The FDA approved the first CAR-T-cell therapy: *Tisagenlecleucel (Kymriah)* for pediatric acute lymphoblastic leukemia.	[19]
2018	Nobel Prize for checkpoint inhibitors	James Allison and Tasuku Honjo were awarded the Nobel Prize in Physiology or Medicine for their discovery of immune checkpoint inhibitors targeting CTLA-4 and PD-1.	[20]
2024	FDA approval for TIL therapy	The first FDA approval of TIL therapy for unresectable or metastatic melanoma.	[21]

development of immune checkpoint inhibitors in the early twenty-first century, which represented a significant turning point in cancer treatment [6–8, 22–25].

The approval of the first immune checkpoint inhibitor, ipilimumab, for treating metastatic melanoma in 2011 marked the beginning of a new era in cancer immunotherapy [9, 16, 26]. This was followed by the success of other checkpoint inhibitors, including pembrolizumab and nivolumab, in treating various cancers. These drugs, by blocking inhibitory signals that prevent immune cells from attacking cancer cells, have transformed the outlook for patients with previously considered incurable cancers. In parallel, other forms of immunotherapy, such as adoptive T-cell therapy and cancer vaccines, have shown promise in clinical trials, further expanding the therapeutic arsenal available to oncologists [9, 22, 27–31].

2 Early Beginnings: From Ancient Hypotheses to Coley's Toxins (1890s)

The concept of manipulating the immune system to treat disease can be traced as far back as ancient medical practices. However, the modern conceptualization of immunotherapy began in the late nineteenth century, when Dr. William Coley, an American surgeon, observed spontaneous tumor regressions in patients with concurrent bacterial infections. Coley postulated that stimulating the immune system could lead to cancer regression. In the 1890s, he developed "*Coley's toxins*," a mixture of *Streptococcus* and *Serratia marcescens* bacteria that he injected into cancer patients to provoke immune responses. While results were variable, Coley's work is often cited as the origin of immunotherapy as a scientific endeavor [10, 32–35]. Despite limited success and skepticism from the medical community, Coley's findings offered a glimpse into the potential of immunotherapy and paved the way for future researchers. His hypothesis that the immune system could be a crucial component in the fight against cancer was groundbreaking, though the precise mechanisms of his observations remained obscure at the time. The scientific world, however, soon shifted its focus toward surgery, chemotherapy, and radiation, relegating Coley's pioneering work to the periphery until the concept resurfaced decades later [10, 34].

3 Immune Surveillance Hypothesis (1950s–1970s)

The *immune surveillance hypothesis*, which gained prominence between the 1950s and 1970s, marked a pivotal moment in our understanding of the immune system's role in cancer prevention. The roots of this concept can be traced back to *Paul Ehrlich's* 1909 proposition, though it was not formally named at the time. Ehrlich suggested that the immune system might have the ability to detect and destroy cells that had the potential to become malignant, preventing the formation of tumors. Building on this foundational idea, in 1957, *Frank Macfarlane Burnet* and *Lewis Thomas* independently formalized the hypothesis, giving it a structured framework. They proposed that the immune system's surveillance mechanisms were not only designed to fight infections but also to monitor the body for emerging cancer cells.

Their theory posited that specific immune cells such as cytotoxic T cells, natural killer (NK) cells, and macrophages could recognize and eliminate tumor cells by identifying abnormal antigens displayed on their surface [36–40].

The hypothesis revolutionized the way scientists viewed the immune system's role in cancer, sparking the belief that a properly functioning immune system was integral in protecting the body from cancer, just as it defends against pathogens. Burnet and Thomas's work suggested that immunological mechanisms were constantly at play, guarding against the spontaneous emergence of cancerous cells [41]. This era also laid the groundwork for future immunological research, inspiring decades of study into how the immune system could be harnessed or enhanced to prevent or treat cancer, ultimately leading to the development of modern cancer immunotherapies.

3.1 The Concept of Immunoediting (2000s)

The *concept of immunoediting*, emerging in the 2000s, evolved from and expanded upon the original *immune surveillance hypothesis* by offering a more nuanced understanding of the relationship between the immune system and cancer. While immune surveillance focused primarily on the immune system's role in detecting and eliminating cancerous cells, *Robert D. Schreiber* and colleagues in the early 2000s proposed that the immune system plays a much more dynamic and complex role in shaping tumor progression. This process, known as *immunoediting*, introduced the idea that the immune system not only suppresses tumor growth but also exerts selective pressure that can lead to the emergence of more resistant and immune-evasive cancer cells over time.

Immunoediting consists of three distinct phases: *elimination*, *equilibrium*, and *escape*. In the *elimination* phase, the immune system successfully identifies and destroys nascent tumor cells, mirroring the concept of immune surveillance. However, in the *equilibrium* phase, some cancer cells survive the immune attack and enter a state of dormancy, where they are held in check by the immune system but continue to mutate and adapt. This phase can last for years and represents a battle between the immune system and the tumor as it evolves under immune pressure. Finally, in the *escape* phase, the most resilient tumor cells, having developed mechanisms to evade immune detection, begin to proliferate uncontrollably, leading to clinically detectable cancer [9, 37, 42, 43].

This refined understanding introduced by immunoediting highlighted the immune system's dual role in preventing and shaping cancer. It not only reinforced the immune system's protective role but also acknowledged the tumor's capacity to evolve under immune pressure, ultimately leading to immune resistance. The concept of immunoediting has been crucial in guiding the development of modern immunotherapies, which aim to break this cycle by boosting the immune system's ability to target resistant cancer cells or reverse the mechanisms tumors use to escape immune detection [42, 43].

3.2 Modern Cancer Immunotherapy (2010s–Present)

Modern immunotherapy represents the culmination of decades of research building on the concepts of *immune surveillance* and *immunoediting*, that has led to a transformative approach to cancer treatment. By the 2010s, it became clear that the immune system could be harnessed to prevent cancer and treat it effectively, even after tumors had developed mechanisms to evade immune detection (Table 1). Modern immunotherapy focuses on enhancing the immune system's natural ability to recognize and eliminate cancer cells, and it is now considered one of the most promising areas in oncology. These advancements are rooted in a deeper understanding of how the immune system interacts with tumors, and they leverage various strategies across six key pillars [8, 9, 22, 44, 45].

4 The Six Pillars of Modern Cancer Immunotherapy

Cancer immunotherapy has transformed over the years into a multifaceted field, with six core pillars emerging as the foundation of current and future therapeutic strategies (Fig. 1). These pillars represent different approaches to effectively harnessing the immune system to combat cancer.

4.1 The Age of Cytokine Therapy and Interferons

In the 1970s and 1980s, the discovery of cytokine proteins that regulate immune responses marked the beginning of clinical immunotherapy trials. Among these cytokines, interleukin-2 (IL-2) and interferon-alpha (IFN-α) were at the forefront. These agents were used to boost the immune system's activity against cancer cells [12, 46–48]. IL-2 became the first immunotherapy drug approved by the U.S. Food and Drug Administration (FDA) in 1992 for metastatic kidney cancer [12]. Its mechanism of action involves the activation and proliferation of T cells, which are essential for mounting an immune response. Similarly, IFN-α was approved for treating certain types of leukemia, melanoma, and Kaposi's sarcoma. These therapies produced dramatic results in some patients, particularly in those with renal cell carcinoma and melanoma, but they were often accompanied by severe toxicities that limited their widespread use [46, 48].

While important, early cytokine therapies also highlighted the challenges of harnessing the immune system: the balance between efficacy and toxicity. These treatments demonstrated that the immune system could indeed fight cancer, but the lack of specificity led to significant side effects such as systemic inflammation, fever, and organ damage. Despite these challenges, the approval of IL-2 and interferons marked a critical step in the evolution of immunotherapy, demonstrating for the first time that immunomodulation could produce durable responses in cancer patients [9, 12, 46, 48].

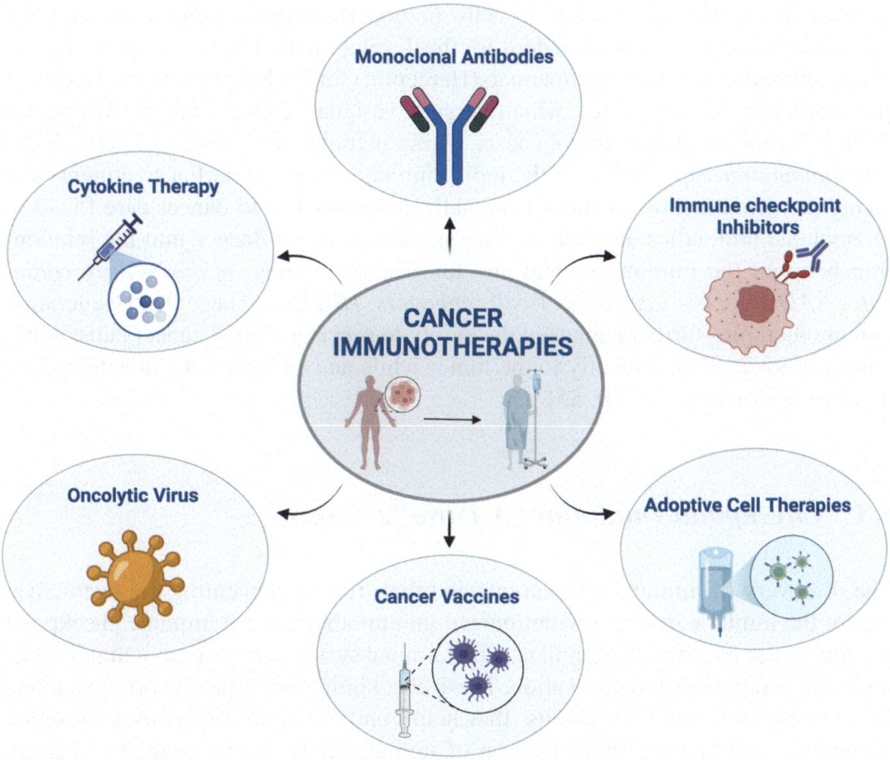

Fig. 1 Overview of six major classes of cancer immunotherapies. (1) Monoclonal Antibodies: Engineered antibodies that bind specific antigens on cancer cells, enhancing immune recognition, blocking signaling pathways, or delivering cytotoxic agents. (2) Immune Checkpoint Inhibitors: Agents that block inhibitory receptors such as PD-1, PD-L1, or CTLA-4 on T cells, restoring their ability to recognize and destroy tumor cells. (3) Adoptive Cell Therapies: Treatments including CAR-T cell therapy, where a patient's T cells are genetically modified ex vivo to express tumor-specific receptors and reinfused to target cancer cells. (4) Cancer Vaccines: Vaccines designed to elicit or amplify immune responses against tumor-associated antigens, promoting tumor-specific immunity. (5) Oncolytic Virus Therapy: Utilizes genetically engineered viruses that selectively infect and lyse cancer cells while stimulating systemic anti-tumor immune responses. (6) Cytokine Therapy: Involves administration of immune-activating cytokines, such as interleukins (e.g., IL-2) or interferons (e.g., IFN-α), to boost the immune system's capacity to combat cancer. Figure created with Biorender.com

4.2 Monoclonal Antibodies: A Targeted Revolution

The development of *monoclonal antibodies (mAbs)* in the late twentieth century represented a turning point in immunotherapy, enabling the precise targeting of cancer cells while sparing healthy tissue. Unlike cytokine therapies, monoclonal antibodies are designed to recognize and bind to specific antigens on the surface of cancer cells, marking them for destruction by the immune system [49, 50]. The first monoclonal antibody approved for cancer treatment was *rituximab* in 1997, which targeted the CD20 antigen on B cells in non-Hodgkin's lymphoma [14, 51]. Rituximab's success opened the door for the development of numerous other monoclonal antibodies, such as trastuzumab (Herceptin) for HER2-positive breast cancer and bevacizumab (Avastin), which targets vascular endothelial growth factor (VEGF) to inhibit tumor angiogenesis. These therapies introduced the era of *targeted immunotherapy*, reducing the indiscriminate effects of earlier treatments and laying the groundwork for more personalized approaches to cancer care [52–54]. Monoclonal antibodies also allowed researchers to delve deeper into the relationship between the immune system and tumors, developing *antibody–drug conjugates (ADCs)* and *bispecific T-cell engagers (BiTEs)*. These next-generation immunotherapies further enhanced the ability to precisely target cancer cells, delivering cytotoxic agents directly to the tumor while activating T cells to enhance the immune response [9, 49, 50, 55].

4.3 Checkpoint Inhibition: A Paradigm Shift

The discovery of immune checkpoints and their role in preventing the overactivation of the immune system revolutionized immunotherapy. An immune checkpoint is a molecular pathway that regulates the immune system's activity to maintain self-tolerance and prevent overactivation. These checkpoints are typically proteins found on immune cells, such as T-cells, that help control immune responses by either promoting or inhibiting the activation of immune cells. In the context of cancer, immune checkpoints can be exploited by tumor cells to evade immune detection and destruction, contributing to tumor growth and immune resistance. *Checkpoint inhibitors* were designed to block these immune brakes, thereby unleashing the full power of the immune system against cancer [7, 25, 45] (Fig. 2). This concept was particularly transformative, enabling the immune system to bypass the inhibitory signals that tumors exploit to evade immune detection and clearance. *CTLA-4*, one of the first immune checkpoints identified, became the target of *ipilimumab*, a monoclonal antibody approved in 2011 for metastatic melanoma. Ipilimumab was the first checkpoint inhibitor to demonstrate durable long-term survival in a subset of patients, igniting widespread interest in checkpoint blockade therapy [16]. Following this, inhibitors of *PD-1* and *PD-L1*, such as *pembrolizumab (Keytruda)* and *nivolumab (Opdivo)*, were developed and rapidly expanded into a range of cancers, including lung, bladder, and head and neck cancers (REF). The success of checkpoint inhibitors has been nothing short of revolutionary. Checkpoint blockade has provided long-lasting remissions for patients who previously had few or no treatment options. However, despite the remarkable efficacy in some patients, not all tumors respond to checkpoint inhibitors, and some patients develop secondary resistance. Identifying biomarkers for durable response and combining checkpoint inhibitors with other therapies remains an area of intense research to improve outcomes [7, 9, 20, 26, 45, 56–59].

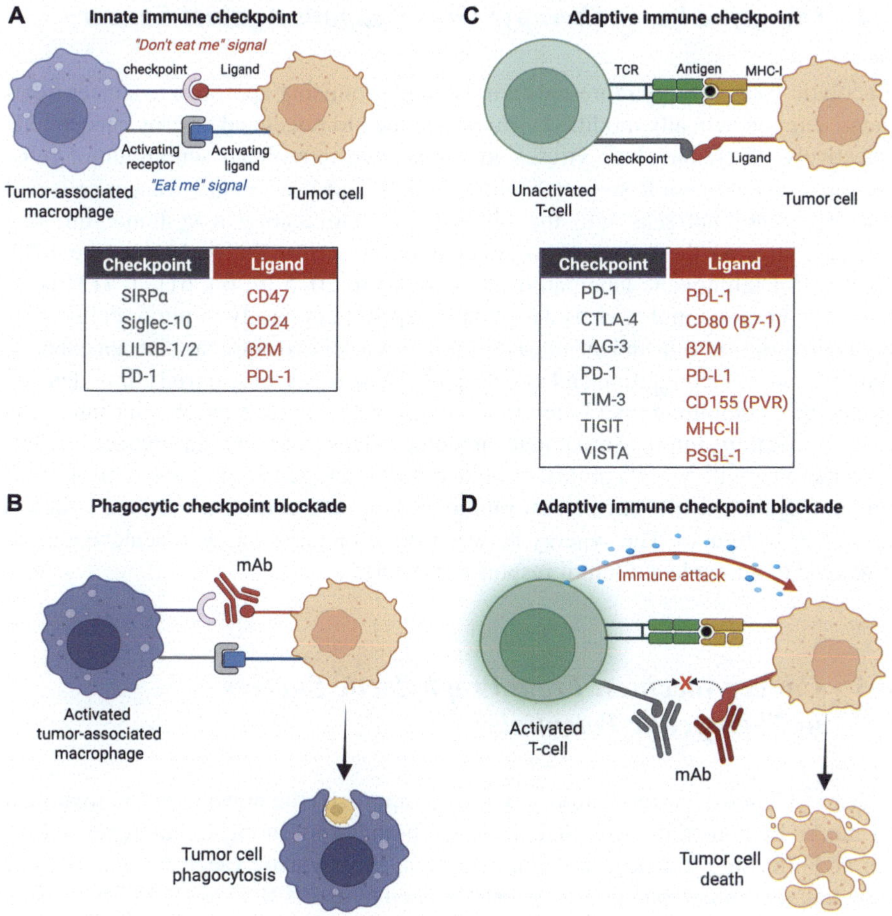

Fig. 2 The figure illustrates the roles of innate and adaptive immune checkpoints in cancer immune evasion and the therapeutic impact of checkpoint blockade. Panel A depicts innate immune checkpoints that regulate macrophagemediated phagocytosis. A tumor-associated macrophage is shown interacting with a tumor cell via inhibitory ("Don't eat me") or activating ("Eat me") signals. The associated table lists key innate immune checkpoints and their tumor cell ligands: SIRPα–CD47, Siglec-10–CD24, LILRB1/2–β2M, and PD-1–PD-L1. These inhibitory interactions suppress phagocytosis, allowing tumor cells to evade innate immune clearance. Panel B illustrates therapeutic phagocytic checkpoint blockade using monoclonal antibodies (mAbs) targeting these inhibitory pathways, thereby promoting macrophage-mediated engulfment and clearance of tumor cells. Panel C transitions to adaptive immunity, showing a T cell engaging with a tumor cell through checkpoint-ligand interactions that inhibit T-cell activation. The accompanying table lists major adaptive immune checkpoints and their ligands: PD-1–PD-L1, CTLA-4–CD80, LAG-3–β2M, TIM-3–CD155, TIGIT–MHC-II, and VISTA–PSGL-1. These pathways dampen T-cell responses, enabling tumor immune escape. Panel D demonstrates adaptive immune checkpoint blockade, in which mAbs targeting inhibitory receptors on T cells restore T-cell activation, leading to enhanced cytotoxic activity and effective tumor cell killing. Overall, the figure highlights the dual role of immune checkpoints in both innate and adaptive arms of the immune system and the therapeutic strategy of checkpoint inhibition to reinvigorate anti-tumor immunity. Figure created with Biorender.com

4.4 Oncolytic Virus Therapy: Viral Weapons Against Cancer

Oncolytic virus therapy represents a novel and promising approach to cancer treatment, using genetically modified viruses designed to infect and destroy cancer cells selectively. These oncolytic viruses are engineered to replicate within tumor cells, causing cell death while sparing healthy tissues. This process kills cancer cells and releases tumor antigens into the bloodstream, stimulating a systemic immune response against the tumor. A significant milestone in this field was the FDA approval of talimogene laherparepvec (T-VEC) in 2015 [9, 60, 61]. T-VEC is an oncolytic herpes simplex virus designed to replicate precisely in tumor cells, leading to their destruction and the release of tumor-associated antigens that can enhance immune system recognition and targeting of the cancer. Clinical trials have demonstrated that combining oncolytic virus therapy with immune checkpoint inhibitors can significantly improve treatment outcomes. This combination approach leverages the oncolytic virus's ability to induce tumor antigen release and activation of immune system, while checkpoint inhibitors help overcome the immune system's resistance to tumors. The synergy between these therapies aims to achieve a more effective and durable antitumor response [60–64].

4.5 Cancer Vaccines: From Prophylactic Success to Therapeutic Potential

Cancer vaccines have revolutionized how we approach the prevention and treatment of cancer, demonstrating significant success both in preventing certain types of cancer and offering promise as a therapeutic tool. Prophylactic vaccines such as those targeting the human papillomavirus (HPV) and hepatitis B virus (HBV) have been particularly successful in reducing the incidence of cervical and liver cancers, respectively. These vaccines work by priming the immune system to recognize and eliminate virus-infected cells before they can develop into cancer. While preventive vaccines have proven highly effective, the development of therapeutic cancer vaccines that aim to treat existing cancers by stimulating the immune system to recognize and destroy cancer cells has faced more challenges. *The first FDA-approved therapeutic vaccine, sipuleucel-T (Provenge)*, marked a significant milestone for the treatment of prostate cancer in 2010 [15]. Provenge demonstrated the ability to activate the immune system against tumor-specific antigens. However, therapeutic vaccines have not yet achieved the widespread clinical success seen in other immunotherapy approaches like checkpoint inhibitors. One major hurdle for therapeutic vaccines is the immune-suppressive tumor microenvironment, which can prevent the immune system from effectively attacking the cancer. To overcome these challenges, ongoing research is focused on optimizing vaccine formulations, enhancing antigen presentation, and combining vaccines with other immunotherapies, such as checkpoint inhibitors, to amplify the immune response. As research continues, cancer vaccines are emerging as promising therapeutic strategies, capable of complementing existing treatments and offering hope for more personalized, targeted immunotherapy options [65–69].

4.6 *Adoptive Cell Therapy: The CAR-T Revolution and Emerging Innovations*

Adoptive cell therapy (ACT) represents a cutting-edge approach in cancer immunotherapy that harnesses and enhances the body's own immune cells to target and eliminate cancer (Fig. 3). Among the various ACT strategies, chimeric antigen receptor-T-cell (CAR-T) therapy has been particularly transformative. However, other innovative methods such as T-cell receptor (TCR) therapy, engineered natural killer (NK) cells, and tumor-infiltrating lymphocyte (TIL) therapy are also advancing the field. Recently, CAR macrophages have emerged as promising new developments in ACT [9, 30, 31, 70].

5 Foundation of Cell-Based Therapies for Cancer

The foundation of cell-based immunotherapies can be traced back to the 1950s when bone marrow transplantation (BMT) emerged as a potential treatment for various blood disorders and cancers. The pioneering work in BMT provided valuable insights into cellular therapies and set the stage for future advancements [71].

Bone marrow transplantation involves the transfer of healthy hematopoietic stem cells (HSCs) from a compatible donor to a patient with a diseased or damaged bone marrow. These transplanted HSCs have the ability to repopulate the patient's bone marrow and give rise to new blood cells, thereby restoring the immune system and potentially eradicating cancer cells [71–73].

The first successful bone marrow transplant was performed in 1956 by *Dr. E. Donnall Thomas*, who later received the *Nobel Prize in Physiology or Medicine* for his groundbreaking contributions [74]. This achievement demonstrated the potential of cell-based therapies in treating hematological disorders, such as leukemia and certain immune deficiencies.

Over the following decades, researchers and clinicians refined the techniques and expanded the applications of bone marrow transplantation. They improved the understanding of matching donor and recipient characteristics, developed protocols for reducing graft-versus-host disease (GVHD), and explored alternative sources of stem cells, such as umbilical cord blood [71, 75].

The success of bone marrow transplantation laid the groundwork for other cell-based immunotherapies, including CAR-T-cell therapy. Building upon the knowledge gained from BMT, scientists sought to engineer immune cells to enhance their cancer-fighting capabilities and create a more targeted and effective treatment approach [71–73, 75]. The following section summarizes the different cell-based immunotherapies utilized in cancer treatment (Fig. 3).

5.1 CAR-T Therapy: A Game Changer

CAR-T therapy has been a game changer in cancer treatment due to its unprecedented success in treating patients with advanced, treatment-resistant blood cancers. Unlike traditional therapies such as chemotherapy and radiation, which indiscriminately target both healthy and cancerous cells, CAR-T therapy is a highly personalized approach that relies on engineering the patient's own immune cells to specifically target cancer cells. This precision enables more effective tumor destruction with fewer off-target effects. The FDA approvals of *tisagenlecleucel (Kymriah)* and *axicabtagene ciloleucel (Yescarta)* in 2017 marked a turning point, as both therapies demonstrated remarkable outcomes in patients who had exhausted all other treatments, including complete remission in some cases. CAR-T's ability to harness the body's immune system and provide lasting remission in previously incurable blood cancers has fundamentally changed the landscape of cancer treatment, offering new hope for patients with otherwise limited options. However, its transformative potential is tempered by challenges, particularly in adapting the therapy to solid tumors. Factors such as antigen heterogeneity and the immunosuppressive tumor microenvironment present new hurdles for CAR-T therapy, and its active areas of research. Despite these challenges, CAR-T therapy's precision and success in hematologic cancers have solidified its status as a groundbreaking advancement in oncology [9, 19, 76–78].

5.2 TCR Therapy

T-cell receptor (TCR) therapy involves engineering T cells to recognize and target cancer cells based on specific antigens presented by major histocompatibility complex (MHC) molecules. TCR therapy modifies T cells to express receptors that can recognize cancer-associated peptides presented by MHC molecules on cancer cell surfaces. This method can target a broader range of antigens compared to CAR-T therapy, including those derived from intracellular proteins. TCR therapy is particularly useful for targeting cancers with specific antigenic profiles not easily addressed by CARs. It has shown promise in treating various solid tumors and hematologic malignancies [79–81].

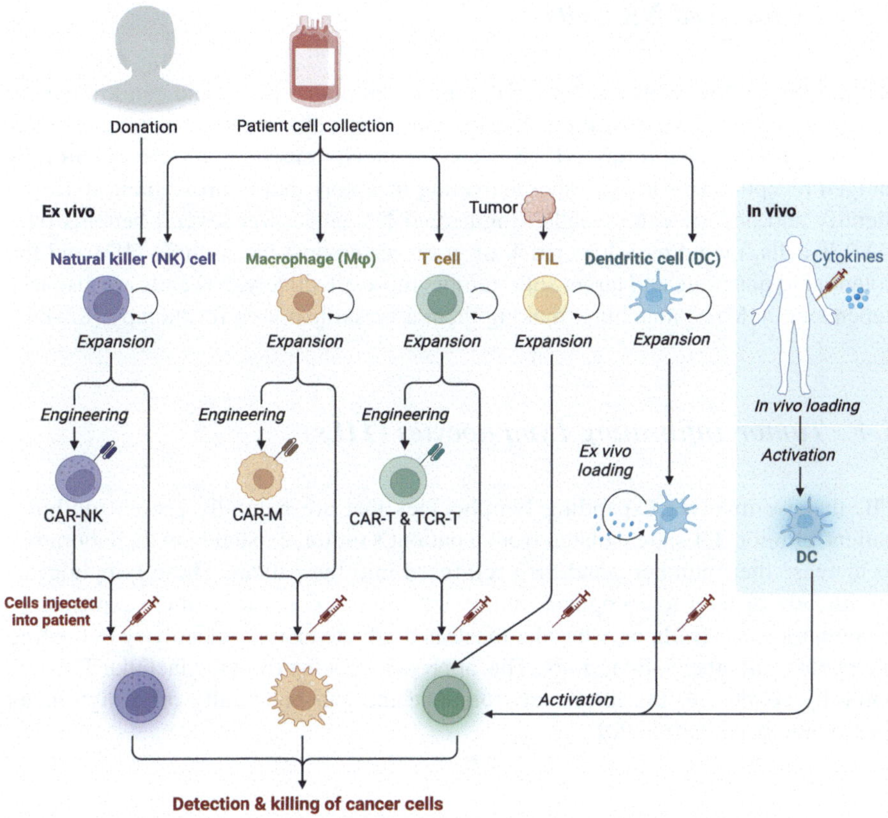

Fig. 3 Schematic representation of diverse cell therapies for cancer treatment. This figure illustrates the multi-step process of developing and administering various cell-based immunotherapies designed to target and eliminate cancer cells. The process begins with the collection of peripheral blood from both healthy donors and cancer patients, which serves as the source of key immune effector cells. Specific immune cell populations—including Natural Killer (NK) cells, macrophages (Mφ), T cells, tumor-infiltrating lymphocytes (TILs), and dendritic cells (DCs)—are isolated from the blood samples and expanded ex vivo (outside the body) under controlled laboratory conditions. Following expansion, these immune cells undergo genetic modification or functional enhancement to improve their tumor-targeting capabilities. NK cells are engineered into CAR-NK cells by introducing chimeric antigen receptors (CARs), enabling them to recognize and kill tumor cells independently of MHC presentation. Similarly, macrophages are modified into CAR-M cells to enhance their phagocytic and tumor-infiltrating functions. T cells are genetically altered into two primary types: CAR-T cells, which express synthetic receptors for surface tumor antigens, and TCR-T cells, which are equipped with engineered T cell receptors targeting intracellular antigens presented on MHC molecules. TILs, which are naturally present within tumor tissues, are isolated, expanded, and reinfused to augment the patient's existing anti-tumor immune response. Dendritic cells are prepared either by loading them ex vivo with tumor-associated antigens or by in vivo activation through cytokine stimulation, promoting their role as potent antigen-presenting cells that initiate and amplify adaptive immune responses. Once engineered or activated, these immune cells are reinfused into the patient's bloodstream, where they home to tumor sites and work to identify, attack, and eliminate malignant cells. The diagram emphasizes the critical role of genetically engineered immune cells and antigen-specific immune activation in the precise targeting and eradication of cancer cells, highlighting the potential of personalized immunotherapies in oncology. Figure created with Biorender.com

5.3 Engineered NK Cells

Natural killer (NK) cells are part of the innate immune system and can kill cancer cells without prior sensitization. Engineering NK cells involves enhancing their cancer-targeting capabilities. NK cells are genetically modified to express chimeric antigen receptors (CARs) or other activating receptors that improve their ability to identify and destroy cancer cells. Engineered NK cells offer several benefits over CAR-T cells, including a lower risk of graft-versus-host disease (GVHD) and the potential to penetrate and target solid tumors more effectively. NK cells are also less dependent on MHC matching, making them a versatile option for therapy [82–84].

5.4 Tumor-Infiltrating Lymphocytes (TILs)

TIL therapy involves expanding lymphocytes that are naturally present within a patient's tumor. TILs are isolated from a patient's tumor, expanded in the laboratory to increase their numbers, and then reinfused into the patient. These lymphocytes are already primed to recognize and attack tumor cells due to their exposure to tumor antigens. TIL therapy has demonstrated effectiveness, particularly in treating melanoma and other solid tumors. This approach benefits from the fact that TILs are naturally adapted to the tumor environment and can potentially offer significant therapeutic benefits [85–90].

5.5 CAR Macrophages: The Latest Advancement

The development of CAR macrophages represents a novel advancement in ACT, building on the inherent properties of macrophages. Macrophages, as innate immune cells, have the ability to phagocytize and destroy cancer cells. By engineering these cells to express CARs, researchers aim to enhance their ability to target and eradicate tumors. CAR macrophages can penetrate dense tumors and overcome some barriers that limit CAR-T cells' effectiveness. CAR macrophages leverage macrophages' natural tumor-infiltrating properties and phagocytic capabilities. They offer a promising approach to targeting solid tumors and dealing with the immunosuppressive tumor microenvironment. Researchers are optimizing CAR macrophage designs to improve their survival, activation, and tumor-killing efficacy. Studies are ongoing to evaluate their performance in preclinical models and early-phase clinical trials [9, 31, 91].

6 The Innovators Who Unleashed the Power of the Immune System Against Cancer

The evolution of cancer immunotherapy has been shaped by groundbreaking discoveries, many of which have been recognized with the Nobel Prize (Table 2). These visionary scientists unveiled fundamental mechanisms that enable the

Table 2 A chronological list of Nobel Prize winning discoveries that shaped the field of cancer immunotherapy

Year	Nobel laureate(s)	Discovery/contribution	Impact on cancer immunotherapy	Reference
2018	James P. Allison and Tasuku Honjo	Discovery of CTLA-4 as a negative regulator of immune responses and discovery of PD-1 and its role in immune inhibition	Development of CTLA-4 inhibitors (e.g., ipilimumab) to enhance immune response against cancer. Development of PD-1 inhibitors (e.g., pembrolizumab and nivolumab) to restore immune activity against cancer	[20]
2012	Shinya Yamanaka	Discovery of induced pluripotent stem cells (iPSCs)	Provides tools for regenerative medicine and immune system research, with potential applications in cancer immunotherapy	[92]
2011	Ralph M. Steinman	Discovery of dendritic cells and their role in immune activation	Foundation for dendritic cell vaccines and immunotherapy strategies	[93]
1987	Susumu Tonegawa	Discovery of the genetic basis for antibody diversity	Foundation for understanding immune response and development of antibody-based therapies	[94]
1984	Niels K. Jerne, Georges J.F. Köhler, and César Milstein	Theory of immune networks and development of monoclonal antibody technology	Creation of monoclonal antibodies for targeted cancer therapies and immune checkpoint inhibitors	[95]
1980	Baruj Benacerraf, Jean Dausset, and George D. Snell	Discovery of the major histocompatibility complex (MHC) and its role in immune recognition	Critical for understanding immune system interactions and developing therapies involving MHC	[96]
1960	Sir Frank Macfarlane Burnet	Discovery of acquired immunological tolerance	Burnet's discovery laid the foundation for modern immunotherapy, enabling targeted cancer treatments and immune modulation techniques	[97]
1908	Paul Ehrlich and Ilia Metchnikoff	Concept of "*magic bullets*" and side-chain theory	Influenced development of targeted therapies and antibody-based treatments	[98]

immune system to combat cancer, leading to transformative therapies that have saved countless lives. From pioneering research on immune checkpoints to revolutionary advancements in adoptive T-cell therapy, their contributions have ushered in a new era of oncology, redefining cancer treatment and patient outcomes.

Paul Ehrlich (1854–1915) Paul Ehrlich, a pioneer in immunology, first proposed that the immune system might prevent the development of cancer by eliminating cells that become malignant. Though not widely accepted in his time, Ehrlich's concept of immune surveillance laid the foundation for later theories. His idea of a "*magic bullet*" to target diseases was revolutionary, foreshadowing modern targeted therapies such as monoclonal antibodies and CAR-T cells. Ehrlich famously said, "*We must learn to shoot microbes with magic bullets*," a vision that resonates with today's cancer treatments [99, 100].

Sir Frank Macfarlane Burnet (1899–1985) Building on Ehrlich's ideas, Frank Macfarlane Burnet, along with Lewis Thomas, proposed the *immune surveillance hypothesis* in 1957. They theorized that the immune system is constantly patrolling the body, identifying and destroying emerging cancer cells before they can grow into tumors. Burnet's influential work helped frame the immune system as a key player in cancer prevention and treatment. He earned the Nobel Prize in 1960 for predicting acquired immune tolerance. He also developed the theory of clonal selection [97].

Dr. Lloyd J. Old (1933–2011) Often called the *Father of Modern Immunotherapy*, Dr. Lloyd J. Old was instrumental in linking the immune system to cancer treatment. Working alongside *Ted Boyse*, Old discovered *tumor antigens*, proteins on the surface of cancer cells that can be recognized by the immune system. Their discovery was groundbreaking, as it led to the development of cancer vaccines and other immune-based therapies. Old's visionary work helped launch the field of cancer immunology, revolutionizing cancer care [101].

Tasuku Honjo (1942–) and James P. Allison (1948–) Tasuku Honjo and James P. Allison revolutionized cancer treatment by discovering immune checkpoints PD-1 and CTLA-4 that regulate T-cell activity against tumors. Honjo's work on PD-1 led to therapies that prevent cancer cells from evading immune attack, while Allison's research on CTLA-4 demonstrated that blocking this protein unleashes T cells to destroy tumors. Together, their findings paved the way for immune checkpoint inhibitors, transforming outcomes for patients with cancers like melanoma and lung cancer. Their breakthroughs earned them the Nobel Prize in 2018, marking a new era in immunotherapy [102, 103].

Edward Donnall (Don) Thomas (1920–2012) Edward Donnall (Don) Thomas was a pioneering physician who revolutionized bone marrow transplantation. Despite early failures and skepticism from immunologists, he persisted in overcoming immunological barriers, ultimately making transplants a life-saving treatment for blood disorders. His groundbreaking work earned him the 1990 Nobel Prize in Physiology or Medicine. Thomas's pioneering efforts not only transformed the field of transplantation medicine but also provided a foundation for advancements in immunology and regenerative medicine. His legacy continues to impact countless lives worldwide.

Steven Rosenberg The advent of cancer immunotherapy in the 1970s and 1980s was driven by Steven Rosenberg at the National Cancer Institute (NCI), who developed adoptive cell transfer (ACT) therapy using tumor-infiltrating lymphocytes (TILs) to enhance the body's immune response against cancer. His research demonstrated that engineered immune cells could successfully target and destroy tumors, laying the foundation for CAR-T-cell therapy.

Carl H. June Carl June's work on *CAR-T-cell therapy* revolutionized cancer treatment, particularly for blood cancers like leukemia and lymphoma. In 2010, June's team engineered a patient's T cells to recognize and attack cancer cells, resulting in a groundbreaking remission. This success led to the FDA approval of *Kymriah* in 2017, the first CAR-T therapy for pediatric acute lymphoblastic leukemia. June's work demonstrated the immense potential of reprogramming the immune system to fight cancer, opening a new frontier in immunotherapy.

7 Challenges and Future Directions

The field of ACT is rapidly advancing, with ongoing research aimed at expanding the range of cancers that can be treated and enhancing the efficacy of these therapies. Innovations in cell engineering, improved understanding of tumor biology, and strategies to overcome limitations in existing therapies are driving the development of new approaches. As these technologies continue to evolve, they hold the potential to revolutionize cancer treatment, offering new hope for patients across various cancer types and improving overall outcomes. Despite the rapid progress, several challenges remain in cancer immunotherapy. The heterogeneity of tumors, immune escape mechanisms, and tumor microenvironment can limit the effectiveness of current therapies.

Additionally, the high cost of immunotherapies, particularly CAR-T and checkpoint inhibitors, presents significant barriers to widespread access. Researchers are exploring combination therapies integrating immunotherapy with traditional treatments such as chemotherapy, radiation, and targeted therapies. Biomarker discovery is also crucial to predicting patient durable responses and tailoring treatments more effectively. Advances in personalized medicine, artificial intelligence, and synthetic biology promise to further refine cancer immunotherapy approaches, making treatments more precise, less toxic, and potentially curative for a broader range of cancers. The future of cancer immunotherapy lies in overcoming these obstacles through continued innovation, cross-disciplinary collaboration, and a deeper understanding of the intricate relationship between cancer and the immune system [71, 79, 90, 102, 104–108].

8 Conclusion

Cancer immunotherapy has come a long way since the days of William Coley, evolving from rudimentary immune stimulation to highly sophisticated, personalized therapies that have changed the oncology landscape. As we continue to unlock the complexities of the immune system and its interactions with cancer, the future of cancer treatment looks increasingly promising. The ultimate goal is to make cancer a manageable or curable disease for all patients. This vision is steadily becoming a reality, thanks to the relentless pursuit of scientific discovery and innovation.

Acknowledgments We sincerely acknowledge the pioneering contributions of researchers whose work, due to space constraints, could not be cited but has greatly advanced cellular immunotherapy. Authors acknowledge the support from UMass Chan Medical School and the MD Anderson Cancer Center. This book chapter was language-edited with the assistance of OpenAI's ChatGPT. The final content remains the responsibility of the authors. All figures in this chapter were created with Biorender.com.

Declaration of Interests The authors declare no competing interests.

References

1. Riddell S. The emperor of all maladies. J Clin Invest. 2011;121(1):5.
2. Schwartz R. The emperor of all maladies — the beginning of the beginning. N Engl J Med. 2011;365(25):2353.
3. Chhikara BS, Parang K. Global Cancer Statistics 2022: the trends projection analysis. Chem Biol Lett. 2023;10(1):451.
4. Siegel RL, Miller KD, Wagle NS, Jemal A. Cancer statistics, 2023. CA Cancer J Clin. 2023;73(1):17.
5. Sung H, Ferlay J, Siegel RL, Laversanne M, Soerjomataram I, Jemal A, et al. Global Cancer Statistics 2020: GLOBOCAN estimates of incidence and mortality worldwide for 36 cancers in 185 countries. CA Cancer J Clin. 2021;71(3):209.
6. Chen DS, Mellman I. Oncology meets immunology: the cancer-immunity cycle. Immunity. 2013;39:1.
7. Pardoll DM. The blockade of immune checkpoints in cancer immunotherapy. Nat Rev Cancer. 2012;12:252.
8. Dobosz P, Dzieciątkowski T. The intriguing history of cancer immunotherapy. Front Immunol. 2019;10:2965.
9. Mishra AK, Ali A, Dutta S, Banday S, Malonia SK. Emerging trends in immunotherapy for cancer. Diseases [Internet]. 2022;10(3):60. Available from: https://www.mdpi.com/2079-9721/10/3/60.
10. Bickels J, Kollender Y, Merinsky O, Meller I. Coley's toxin: historical perspective. Isr Med Assoc J. 2002;4:471–2.
11. Pestka S. The interferons: 50 years after their discovery, there is much more to learn. J Biol Chem. 2007;282:20047.
12. Rosenberg SA. IL-2: the first effective immunotherapy for human cancer. J Immunol. 2014;192(12):5451.

13. Muul LM, Spiess PJ, Director EP, Rosenberg SA. Identification of specific cytolytic immune responses against autologous tumor in humans bearing malignant melanoma. J Immunol. 1987;138(3):989.
14. Leget GA, Czuczman MS. Use of rituximab, the new FDA-approved antibody. Curr Opin Oncol. 1998;10:548.
15. Cheever MA, Higano CS. PROVENGE (sipuleucel-T) in prostate cancer: the first FDA-approved therapeutic cancer vaccine. Clin Cancer Res. 2011;17:3520.
16. Lipson EJ, Drake CG. Ipilimumab: an anti-CTLA-4 antibody for metastatic melanoma. Clin Cancer Res. 2011;17:6958.
17. Mitra A, Barua A, Huang L, Ganguly S, Feng Q, He B. From bench to bedside: the history and progress of CAR T cell therapy. Front Immunol. 2023;14:1188049.
18. Ishida Y. Pd-1: its discovery, involvement in cancer immunotherapy, and beyond. Cells. 2020;9:1376.
19. Prasad V. Tisagenlecleucel – the first approved CAR-T-cell therapy: implications for payers and policy makers. Nat Rev Clin Oncol. 2018;15:11.
20. Huang PW, Chang JWC. Immune checkpoint inhibitors win the 2018 Nobel Prize. Biom J. 2019;42:299.
21. Healey N. Tumor infiltrating lymphocyte approval heralds new era for precision cancer immunotherapy FDA approval of Iovance's lifileucel for melanoma shows the promise of tumor-infiltrating lymphocytes for the treatment of solid tumors, with trials ongoing for lung cancer, cervical cancer and more. Nat Med. 2024;30:1795.
22. Rosenberg SA, Yang JC, Restifo NP. Cancer immunotherapy: moving beyond current vaccines. Nat Med. 2004;10:909.
23. Mellman I, Coukos G, Dranoff G. Cancer immunotherapy comes of age. Nature. 2011;480:480.
24. Chen DS, Mellman I. Elements of cancer immunity and the cancer-immune set point. Nature. 2017;541:321.
25. Ribas A, Wolchok JD. Cancer immunotherapy using checkpoint blockade. Science. 2018;359:1350.
26. Korman AJ, Garrett-Thomson SC, Lonberg N. The foundations of immune checkpoint blockade and the ipilimumab approval decennial. Nat Rev Drug Discov. 2022;21:163.
27. Want MY, Bashir Z, Najar RA. T cell based immunotherapy for cancer: approaches and strategies. Vaccines. 2023;11:835.
28. Kang BH, Lee HK. Dendritic cell-based immunotherapy in hot and cold tumors. Int J Mol Sci. 2022;23:7325.
29. Anderson NR, Minutolo NG, Gill S, Klichinsky M. Macrophage-based approaches for cancer immunotherapy. Cancer Res. 2021;81:1201.
30. Lee S, Kivimäe S, Dolor A, Szoka FC. Macrophage-based cell therapies: the long and winding road. J Control Release. 2016;240:527–40.
31. Mishra AK, Malonia SK. Advancing cellular immunotherapy with macrophages. Life Sci. 2023;328:121857.
32. Starnes CO. Coley's toxins [10]. Nature. 1992;360:23.
33. Tsung K, Norton JA. Lessons from Coley's toxin. Surg Oncol. 2006;15(1):25–8.
34. Carlson RD, Flickinger JC, Snook AE. Talkin' toxins: from Coley's to modern cancer immunotherapy. Toxins (Basel). 2020;12(4):241.
35. Sell S. Cancer immunotherapy: breakthrough or "deja vu, all over again"? Tumour Biol. 2017;39:101042831770776.
36. Silverstein AM, Rose NR. On the mystique of the immunological self. Immunol Rev. 1997;159:197.
37. Möller G, Möller E. The concept of immunological surveillance against neoplasia. Immunol Rev. 1976;28(1):3.
38. Burnet FM. The concept of immunological surveillance. Prog Exp Tumor Res. Fortschritte der experimentellen Tumorforschung. Progres de la recherche experimentale des tumeurs. 1970;13:1–27.

39. Chow MT, Möller A, Smyth MJ. Inflammation and immune surveillance in cancer. Semin Cancer Biol. 2012;22:23.
40. Ostrand-Rosenberg S. Immune surveillance: a balance between protumor and antitumor immunity. Curr Opin Genet Dev. 2008;18:11.
41. Corthay A. Does the immune system naturally protect against cancer? Front Immunol. 2014;5:197.
42. Wagner M, Koyasu S. Cancer immunoediting by innate lymphoid cells. Trends Immunol. 2019;40:415.
43. Dunn GP, Old LJ, Schreiber RD. The immunobiology of cancer immunosurveillance and immunoediting. Immunity. 2004;21:137.
44. Shafer P, Kelly LM, Hoyos V. Cancer therapy with TCR-engineered T cells: current strategies, challenges, and prospects. Front Immunol. 2022;13:835762.
45. Peng S, Bao Y. A narrative review of immune checkpoint mechanisms and current immune checkpoint therapy. Ann Blood. 2022;7.
46. Zhou Y, Quan G, Liu Y, Shi N, Wu Y, Zhang R, et al. The application of Interleukin-2 family cytokines in tumor immunotherapy research. Front Immunol. 2023;14:1090311.
47. Morgan DA, Ruscetti FW, Gallo R. Selective in vitro growth of T lymphocytes from normal human bone marrows. Science (1979). 1976;193(4257):1007.
48. Kirkwood J. Cancer immunotherapy: the interferon-α experience. Semin Oncol. 2002;29(3 SUPPL. 7):18.
49. Zarnani AH, Bozorgmehr M, Shabani M, Barzegar-Yarmohammadi L, Ghaemimanesh F, Jeddi-Tehrani M. Monoclonal antibodies for cancer immunotherapy. In: Cancer immunology: bench to bedside immunotherapy of cancers. Springer; 2015.
50. Weiner L, Surana R, Wang S. Antibodies and cancer therapy: versatile platforms for cancer immunotherapy. Nat Rev Immunol. 2010;10(5):317.
51. Pierpont TM, Limper CB, Richards KL. Past, present, and future of rituximab-the world's first oncology monoclonal antibody therapy. Front Oncol. 2018;8:163.
52. Garcia J, Hurwitz HI, Sandler AB, Miles D, Coleman RL, Deurloo R, et al. Bevacizumab (Avastin®) in cancer treatment: a review of 15 years of clinical experience and future outlook. Cancer Treat Rev. 2020;86:102017.
53. EUCTR2007-005182-35-HU. Treatment of HER2 positive breast cancer with chemotherapy plus trastuzumab vs chemotherapy plus trastuzumab plus bevacizumab. 2008. https://trialsearch.who.int/Trial2.aspx?TrialID=EUCTR2007-005182-35-HU.
54. Arvind K, Mayur P, Vaibhav R, Arun MK. Impact of ado-trastuzumab emtansine therapy in human epidermal growth factor receptor 2 positive metastatic breast cancer: a recent survey. Asian J Pharm. 2016;10(4):S444.
55. Zarnani AH, Jafari D, Bozorgmehr M, Shabani M, Barzegar-Yarmohammadi L, Ghaemimanesh F, et al. Monoclonal antibodies for cancer immunotherapy. In: Cancer immunology: bench to bedside immunotherapy of cancers. 2nd ed. Springer; 2020.
56. Shiravand Y, Khodadadi F, Kashani SMA, Hosseini-Fard SR, Hosseini S, Sadeghirad H, et al. Immune checkpoint inhibitors in cancer therapy. Curr Oncol. 2022;29:3044.
57. Xu F, Jin T, Zhu Y, Dai C. Immune checkpoint therapy in liver cancer. J Exp Clin Cancer Res. 2018;37:110.
58. Borgeaud M, Sandoval J, Obeid M, Banna G, Michielin O, Addeo A, et al. Novel targets for immune-checkpoint inhibition in cancer. Cancer Treat Rev. 2023;120:102614.
59. Guo Z, Zhang R, Yang AG, Zheng G. Diversity of immune checkpoints in cancer immunotherapy. Front Immunol. 2023;14:1121285.
60. Raja J, Ludwig JM, Gettinger SN, Schalper KA, Kim HS. Oncolytic virus immunotherapy: future prospects for oncology. J Immunother Cancer. 2018;6:1–13.
61. Mondal M, Guo J, He P, Zhou D. Recent advances of oncolytic virus in cancer therapy. Hum Vaccin Immunother. 2020;16:2389.
62. Fukuhara H, Ino Y, Todo T. Oncolytic virus therapy: a new era of cancer treatment at dawn. Cancer Sci. 2016;107:1373.

63. Muthukutty P, Yoo SY. Oncolytic virus engineering and utilizations: cancer immunotherapy perspective. Viruses. 2023;15:1645.
64. Cao GD, He XB, Sun Q, Chen S, Wan K, Xu X, et al. The oncolytic virus in cancer diagnosis and treatment. Front Oncol. 2020;10:1786.
65. Buonaguro L, Tagliamonte M. Selecting target antigens for cancer vaccine development. Vaccines. 2020;8:615.
66. Saxena M, van der Burg SH, Melief CJM, Bhardwaj N. Therapeutic cancer vaccines. Nat Rev Cancer. 2021;21:360.
67. Makker S, Galley C, Bennett CL. Cancer vaccines: from an immunology perspective. Immunother Adv. 2024;4:ltad030.
68. Janes ME, Gottlieb AP, Park KS, Zhao Z, Mitragotri S. Cancer vaccines in the clinic. Bioeng Transl Med. 2024;9:e10588.
69. Liu J, Fu M, Wang M, Wan D, Wei Y, Wei X. Cancer vaccines as promising immunotherapeutics: platforms and current progress. J Hematol Oncol. 2022;15:28.
70. Cabanos HF, Hata AN. Emerging insights into targeted therapy-tolerant persister cells in cancer. Cancers (Basel). 2021;13(11):2666.
71. Dazzi F, Goldman JM. Adoptive immunotherapy following allogeneic bone marrow transplantation. Annu Rev Med. 1998;49:329.
72. Cirillo M, Tan P, Sturm M, Cole C. Cellular immunotherapy for hematologic malignancies: beyond bone marrow transplantation. Biol Blood Marrow Transplant. 2018;24:433.
73. Decker WK, da Silva RF, Sanabria MH, Angelo LS, Guimarães F, Burt BM, et al. Cancer immunotherapy: historical perspective of a clinical revolution and emerging preclinical animal models. Front Immunol. 2017;8:829.
74. Storb R. Edward Donnall Thomas (1920–2012). Nature. 2012;491(7424):20777–8.
75. Zhang L, Yu J, Wei W. Advance in targeted immunotherapy for graft-versus-host disease. Front Immunol. 2018;9:1087.
76. Valiullina AK, Zmievskaya EA, Ganeeva IA, Zhuravleva MN, Garanina EE, Rizvanov AA, et al. Evaluation of CAR-T cells' cytotoxicity against modified solid tumor cell lines. Biomedicines. 2023;11(2):626.
77. D'Aloia MM, Zizzari IG, Sacchetti B, Pierelli L, Alimandi M. CAR-T cells: the long and winding road to solid tumors review-article. Cell Death Dis. 2018;9:282.
78. Safarzadeh Kozani P, Safarzadeh Kozani P, Ahmadi Najafabadi M, Yousefi F, Mirarefin SMJ, Rahbarizadeh F. Recent advances in solid tumor CAR-T cell therapy: driving tumor cells from hero to zero? Front Immunol. 2022;13:795164.
79. Rath JA, Arber C. Engineering strategies to enhance TCR-based adoptive T cell therapy. Cells. 2020;9:1485.
80. Sun Y, Li F, Sonnemann H, Jackson KR, Talukder AH, Katailiha AS, et al. Evolution of cd8+ t cell receptor (Tcr) engineered therapies for the treatment of cancer. Cells. 2021;10:2379.
81. Schendel DJ. Evolution by innovation as a driving force to improve TCR-T therapies. Front Oncol. 2023;13:1216829.
82. Rezvani K, Rouce R, Liu E, Shpall E. Engineering natural killer cells for cancer immunotherapy. Mol Ther. 2017;25:1769.
83. Karvouni M, Vidal-Manrique M, Lundqvist A, Alici E. Engineered NK cells against cancer and their potential applications beyond. Front Immunol. 2022;13:825979.
84. Marofi F, Saleh MM, Rahman HS, Suksatan W, Al-Gazally ME, Abdelbasset WK, et al. CAR-engineered NK cells; a promising therapeutic option for treatment of hematological malignancies. Stem Cell Res Ther. 2021;12:374.
85. Lin B, Du L, Li H, Zhu X, Cui L, Li X. Tumor-infiltrating lymphocytes: warriors fight against tumors powerfully. Biomed Pharmacother. 2020;132:110873.
86. Kazemi MH, Sadri M, Najafi A, Rahimi A, Baghernejadan Z, Khorramdelazad H, et al. Tumor-infiltrating lymphocytes for treatment of solid tumors: it takes two to tango? Front Immunol. 2022;13:1018962.

87. Lee S, Margolin K. Tumor-infiltrating lymphocytes in melanoma. Curr Oncol Rep. 2012;14(5):468.
88. Paijens ST, Vledder A, de Bruyn M, Nijman HW. Tumor-infiltrating lymphocytes in the immunotherapy era. Cell Mol Immunol. 2021;18:842.
89. Dushyanthen S, Beavis PA, Savas P, Teo ZL, Zhou C, Mansour M, et al. Relevance of tumor-infiltrating lymphocytes in breast cancer. BMC Med. 2015;13:202.
90. Rosenberg SA, Spiess P, Lafreniere R. A new approach to the adoptive immunotherapy of cancer with tumor-infiltrating lymphocytes. Science (1979). 1986;233(4770):1318.
91. Mantovani A, Allavena P, Marchesi F, Garlanda C. Macrophages as tools and targets in cancer therapy. Nat Rev Drug Disc. Nature Research. 2022;21:799.
92. Ochi M. Shinya Yamanaka's 2012 Nobel Prize and the radical change in orthopedic strategy thanks to his discovery of iPS cells. Acta Orthop. 2013;84:1.
93. Volchenkov R, Sprater F, Vogelsang P, Appel S. The 2011 Nobel Prize in physiology or medicine. Scand J Immunol. 2012;75(1):1–4.
94. Shampo MA, Kyle RA, Steensma DP. Susumu Tonegawa – Japan's first Nobel laureate in physiology or medicine. Mayo Clin Proc. 2011;86:e14.
95. Raju TN. The Nobel chronicles. 1984: Niels Kai Jerne, (1911–94); César Milstein (b 1926); and Georges Jean Franz Köhler (1946–95). Lancet. 2000;355(9197):75.
96. Marx JL. 1980 Nobel Prize in Physiology or Medicine. Science (1979). 1980, 210(4470):621–3.
97. Silverstein AM. The curious case of the 1960 Nobel Prize to Burnet and Medawar. Immunology. 2016;147(3):269.
98. Winau F, Westphal O, Winau R. Paul Ehrlich – in search of the magic bullet. Microbes Infect. 2004;6:786.
99. Kemeny DM, MacAry PA. Immunology: making magic bullets. BMJ. 2007;334(Suppl 1):s13.
100. Blair J. Making magic bullets. Nat Microbiol. 2017;2:1.
101. Smyth MJ. Lloyd John Old 1933–2011. Nat Immunol. 2012;13(2):103.
102. Wolchok J. Putting the immunologic brakes on cancer. Cell. 2018;175:1452.
103. Allison JP. Discovery of cancer therapy by inhibition of negative immune regulation-the 2018 Nobel Prize in Physiology or Medicine. The Nobel Assembly at Karolinska Institutet; 2018.
104. Rosenberg SA, Restifo NP, Yang JC, Morgan RA, Dudley ME. Adoptive cell transfer: a clinical path to effective cancer immunotherapy. Nat Rev Cancer. 2008;8:299.
105. Morotti M, Albukhari A, Alsaadi A, Artibani M, Brenton JD, Curbishley SM, et al. Promises and challenges of adoptive T-cell therapies for solid tumours. Br J Cancer. 2021;124:1759.
106. Turk H, Hamilton S, Siu L, June C, Ott P. Clinical trials in the evolving landscape of precision medicine. Cancer Cell. 2020;38:3–8.
107. Ruella M, Kalos M. Adoptive immunotherapy for cancer. Immunol Rev. 2014;257:1.
108. Capponi S, Daniels KG. Harnessing the power of artificial intelligence to advance cell therapy. Immunol Rev. 2023;320:147.

Tumor-Infiltrating Lymphocytes: ACTing against Solid Tumors

Anshul Sharma, Bhubanananda Sahu, and Alok K. Mishra

Abstract The emergence of cellular immunotherapy has revolutionized cancer treatment. Amidst the arsenal of immune cells, tumor-infiltrating lymphocytes (TILs) have emerged as potent warriors in the battle against cancer. From the initial recognition of tumor antigens to the complex orchestration of immune responses, the story of TILs unfolds as a promising narrative in the ever-evolving landscape of cancer therapy. This chapter explores the evolving landscape of TIL-based cancer therapy, encompassing their role in the immune microenvironment and nuanced interplay between TILs and cancer. This chapter also explores the methods of isolation of TILs, ex vivo expansion, and the transformative outcomes observed in clinical applications. Delving into the molecular intricacies of TIL–cancer interactions, we have discussed the potential and challenges inherent in harnessing TILs for the personalized treatment of diverse malignancies.

Keywords Cancer · Immunotherapy · Solid tumors · Cytokines · Cell-based therapy · Immunoediting · Immune surveillance · CAR-T

A. Sharma
Department of Molecular, Cell and Cancer Biology, UMass Chan Medical School, Worcester, MA, USA

Department of Medicine, Division of Hematology and Oncology, Medical College of Wisconsin, Milwaukee, WI, USA

B. Sahu
Department of Ophthalmology and Visual Sciences, UMass Chan Medical School, Worcester, MA, USA

A. K. Mishra (✉)
Department of Molecular, Cell and Cancer Biology, UMass Chan Medical School, Worcester, MA, USA

Department of Hematopoietic Biology & Malignancy, Division of Cancer Medicine, The University of Texas MD Anderson Cancer Center, Houston, TX, USA
e-mail: AKMishra1@mdanderson.org

A. K. Mishra, S. K. Malonia (eds.), *Cell-based Immunotherapies for Cancer*,
https://doi.org/10.1007/978-3-031-88695-9_3

Abbreviations

ACT	Adoptive cell therapy
APCs	Antigen-presenting cells
CAR-T cells	Chimeric antigen receptor T cells
CRC	Colorectal cancer
CTLs	Cytotoxic T lymphocytes
FasL	Fas ligand
FDA	Food and Drug Administration
G-CSF	Granulocyte colony-stimulating factor
HNSCC	Head and neck squamous cell carcinoma
IL-2	Interleukin-2
MHC	Major histocompatibility complex.
NSCLC	Non-small cell lung cancer
PD-1	Programmed cell death protein 1
TCR	T-cell receptor
TIL	Tumor-infiltrating lymphocytes
TIME	Tumor immune microenvironment
TME	Tumor microenvironment
TNF	Tumor necrosis factor
TRAIL	TNF-related apoptosis-inducing ligand

1 Introduction

Tumor-infiltrating lymphocytes (TILs) constitute a specialized subset of immune cells that actively penetrate the complex microenvironment of tumors [1–4]. TILs, primarily T lymphocytes, are integral components of the body's immune defense system against cancer. TILs exhibit a unique ability to recognize and engage with tumor-specific antigens expressed on the surface of cancer cells. As they infiltrate the tumor site, TILs unleash a targeted immune response, deploying cytotoxic mechanisms to eliminate malignant cells [2, 3]. The composition, density, and functional status of TILs within tumors have emerged as critical factors influencing the prognosis and treatment outcomes in various cancers [5–8]. The presence of antigen-specific TILs in tumors indicates that these cells often lack the sufficient numbers or functionality to eliminate the tumor on their own. However, their numbers can be increased to achieve the levels required for full tumor eradication in the body. In line with this, adoptive cell therapy (ACT), which involves the ex vivo expansion of TILs called TIL therapy, has been shown to contribute significantly to tumor regression [2, 8–10].

The pioneering efforts in the field of tumor-infiltrating lymphocyte (TIL) therapy were done by Dr. Steven Rosenberg's research group [4]. In 1986, Rosenberg and his team isolated TILs from murine tumors and cultivated them in vitro, amplifying their numbers in the presence of interleukin-2 (IL-2). Subsequently, these activated TILs were reintroduced into hosts bearing tumors, resulting in significant antitumor reactivity [1, 4]. Notably, their investigations revealed that combining IL-2 with the infusion of TILs could enhance the therapeutic efficacy [5].

In the field of cancer immunotherapy, there is a growing interest in exploiting the potential of TILs for therapeutic purposes. Adoptive cell therapy (ACT), a promising immunotherapeutic approach, involves the extraction, expansion, and reinfusion of autologous TILs into the patient to enhance the antitumor immune response. This personalized strategy has demonstrated remarkable success, particularly in treating advanced melanoma, underscoring the therapeutic potential inherent in manipulating tumor-infiltrating T cells [6, 7, 11–13].

TILs represent a diverse group of immune cells, including clusters of T cells, B cells, natural killer (NK) cells, and T regulatory cells (Tregs). Among these, T cells are of particular interest in cancer research due to their prominence as the second most common immune cell type found in human tumors across various cancer types. During the early stages of tumor development, if enough immunogenic antigens are present, naïve T cells are primed in the draining lymph nodes or bloodstream. This priming triggers their activation and subsequent migration to the tumor microenvironment (TME). Among the different T-cell subtypes, $CD8^+$ T cells (cytotoxic T cells or killer T cells) and $CD4^+$ T cells (helper T cells) are the most studied and well-known [3, 14].

Cytotoxic $CD8^+$ T cells (cytotoxic T lymphocytes (CTLs)) are the primary effectors in the adaptive immune system's response against cancer. CTL activation is an antigen-specific process that requires the T-cell receptor (TCR) to interact with a peptide derived from a tumor antigen bound to a major histocompatibility complex (MHC) class I molecule presented by antigen-presenting cells (APCs) or tumor cells [15, 16]. CTLs combat cancer through two main mechanisms: granule exocytosis and the death ligand/death receptor pathway. In granule exocytosis, secretory granules containing perforin and granzymes fuse with the plasma membrane and release their contents, leading to tumor cell death. In the death ligand/death receptor–mediated pathway, activated CTLs express FasL and TNF-related apoptosis-inducing ligands (TRAIL) on their surface, which can induce apoptosis in susceptible cancer cells by interacting with their corresponding death receptors [3, 14, 17, 18].

Within the complex environment of the tumor microenvironment (TME), tumor-infiltrating lymphocytes (TILs) are increasingly recognized as crucial players due to their diverse roles and clinical significance. The formation of TILs begins with circulating lymphocytes migrating across the tumor endothelium to the tumor site. However, the tumor endothelium is often compromised, which can directly inhibit T-cell function and hinder the infiltration of TILs. Notably, blood vessels carrying lymphocytes are typically absent in the tumor core, instead being concentrated in the surrounding stroma or invasive margins, indicating a directional aspect to TIL infiltration. These observations underscore the tumor endothelium's key role in regulating T-cell migration and infiltration, as well as the importance of the stroma in tumor progression. Histopathological studies of tumors have shown distinct patterns of TIL distribution, with immune cells clustering in specific regions around and within tumors, suggesting an organized and nonrandom distribution. The presence of TILs within a tumor is often associated with a favorable prognosis, as their activity reflects an ongoing immune response against the malignancy [2, 3, 8, 10].

TILs play a crucial role in immune surveillance, continuously monitoring and eliminating emerging tumor cells. Their ability to specifically target cancer cells while sparing normal tissues makes them an attractive option for adoptive cell therapy. TILs can be made available for therapeutic purposes through two ways: (1) expanding the TILs to increase their numbers and (2) engineering TILs to enhance their cancer-fighting capabilities [6, 7].

2 TILs Are Different from CAR-T Cells

Among T-cell-based immunotherapies, chimeric antigen receptor T-cell (CAR-T) therapy has garnered widespread attention, particularly for its transformative success in treating hematologic malignancies [19]. However, tumor-infiltrating lymphocyte (TIL) therapy offers unique advantages, particularly for solid tumors, underscoring the importance of understanding the differences between these two adoptive cell therapies (ACTs) [2, 3]. CAR-T cells are genetically engineered to target specific antigens, such as CD19, BCMA, CD20, CD22, CD33, and FLT3, and have achieved remarkable results in B-cell leukemias, multiple myeloma, B-cell malignancies, acute lymphoblastic leukemia (ALL), and acute myeloid leukemia (AML), where tumor cells often express well-defined and homogenous antigens. This makes CAR-T therapy an ideal choice for hematologic cancers, though it is often associated with severe side effects like cytokine release syndrome (CRS), B-cell aplasia, and neurotoxicity [19–22].

In contrast, TIL therapy capitalizes on the natural antitumor capabilities of a patient's own T cells, which are extracted from the tumor microenvironment, expanded ex vivo, and reintroduced into the patient. TILs are inherently polyclonal, recognizing a broad spectrum of tumor-associated antigens and neoantigens without the need for genetic engineering. This makes TIL therapy particularly suited for solid tumors, which are characterized by high antigenic heterogeneity and immune evasion mechanisms that challenge the monoclonal specificity of CAR-T cells. Additionally, TILs can overcome the heterogeneity of solid tumors by targeting multiple antigens simultaneously, including neoantigens unique to individual tumors [1–3, 8, 9]. Despite hurdles such as complex manufacturing processes and variability in expansion success, TIL therapy holds promise as a highly personalized approach for solid tumors (see also Table 1).

3 The TIL Therapy Process

This therapeutic strategy involves the extraction of TILs from a patient's tumor tissue, followed by their ex vivo expansion to create a robust and highly specific population of immune cells. Once expanded, these potent TILs are reintroduced into the patient's body through adoptive cell therapy (ACT), where they mount a targeted

Table 1 A comparison of TIL therapy with CAR-T-cell therapy

Feature	Tumor-infiltrating lymphocytes (TILs)	CAR-T cells
Source	Naturally occurring T cells from the tumor	Genetically engineered T cells from blood
Antigen	Broad range of naturally occurring tumor antigens, including neoantigens	Specific, predefined antigen (e.g., CD19, HER2, etc.)
Personalization	Highly personalized, patient-specific	Specific to antigens but can be used in multiple patients if antigens are common
Primary cancer types treated	Solid tumors (e.g., melanoma)	Hematologic malignancies (e.g., B-cell leukemias and lymphomas)
Expansion method	In vitro expansion using IL-2	In vitro genetic modification and expansion
Immune surveillance	Continuous monitoring and elimination of emerging tumor cells	Targeted killing of cells expressing specific antigens
Clinical success	Effective in melanoma and other solid tumors	Remarkable success in certain blood cancers
Adverse effects	Reversible side effects (e.g., cytokine release syndrome)	Potential for severe side effects (e.g., cytokine release syndrome and neurotoxicity)
Challenges	Complex and resource-intensive manufacturing process Tumor heterogeneity can affect efficacy Limited by the need for tumor resection and variability in TIL quality	Limited success in solid tumors High cost and complex genetic engineering process Complex logistics for production and administration

assault against cancer cells [18, 23]. The therapeutic approach involves the following steps (Fig. 1).

4 Isolation and Initial Expansion of Tumor-Infiltrating Lymphocytes (TILs)

The process of producing the TIL infusion product begins in the laboratory with the surgical resection of one or more tumor deposits from the patient. These tumor specimens are then carefully dissected into small fragments, each measuring approximately 2–3 millimeters in diameter. These fragments are placed into tissue culture flasks containing a specialized growth medium that is supplemented with high-dose interleukin-2 (IL-2), a cytokine essential for promoting the growth and survival of TILs. Within the culture environment, lymphocytes that are embedded in the tumor stroma begin to migrate out (extravasate) of the tissue fragments and into the surrounding medium. Over the course of 10–14 days, these lymphocytes proliferate, resulting in an initial expansion of TILs [24].

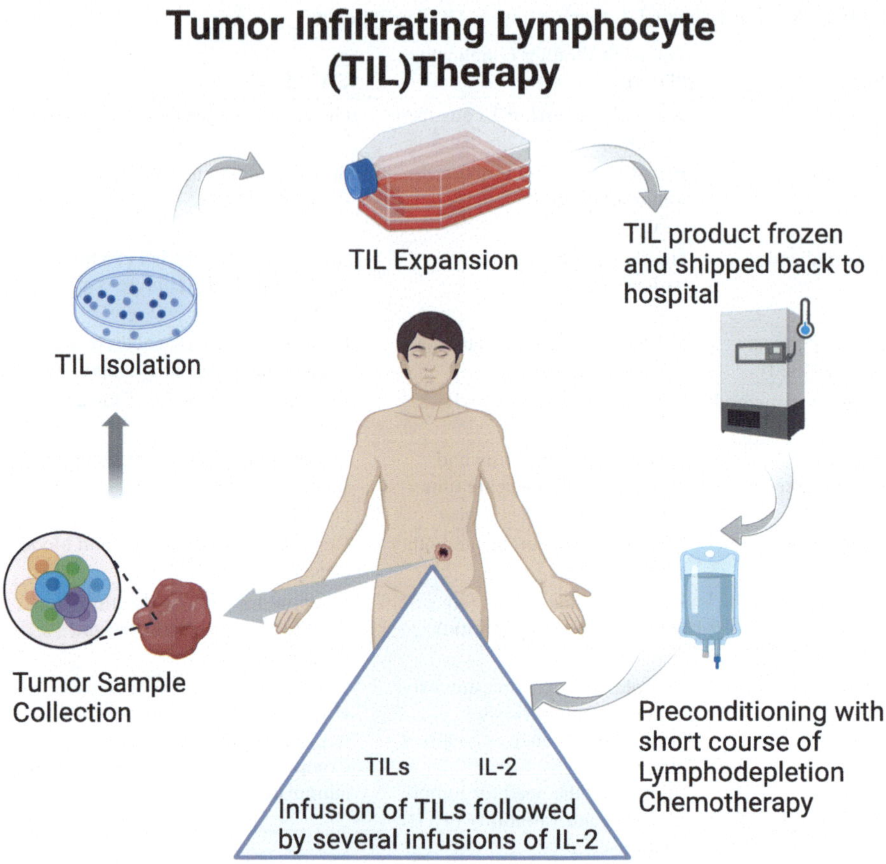

Fig. 1 Overview of the tumor-infiltrating lymphocyte (TIL) therapy process. Step-by-step process of tumor-infiltrating lymphocyte (TIL) therapy involves several key steps to enhance the immune system's ability to fight cancer. First, a tumor sample is collected from the patient to obtain immune cells, followed by the isolation of TILs from the tumor tissue in the laboratory. The isolated TILs are then expanded ex vivo to increase their numbers, after which the TIL product is cryopreserved and shipped back to the hospital for administration. Before infusion, the patient undergoes a short course of lymphodepleting chemotherapy to create an optimal environment for TIL engraftment. Finally, the expanded TILs are infused into the patient, accompanied by several doses of interleukin-2 (IL-2) to support TIL survival and function

5 Preparation of Tumor-Reactive Lymphocytes

After isolating and expanding TILs, the next step is to prepare them for therapeutic use by enhancing their tumor-reactivity. This is achieved by continuing the in vitro culture of TILs in the presence of interleukin-2 (IL-2) using multiple fragments of the resected tumor [25–27]. To identify which TILs are most effective at targeting cancer cells, whole-exome and RNA sequencing are performed on both the tumor and normal tissues [26]. This sequencing helps identify all the cancer-specific mutations that are presented on the patient's dendritic cells. These tumor antigens are then expressed by engineered antigen-presenting cells (APCs), which are used to assess the reactivity of the expanded TILs. The TILs that demonstrate high activity are selected for reinfusion back into the patient, where they will seek out and attack the tumor [26, 28, 29].

6 Large-Scale Expansion of Selected TILs

Following initial expansion, the cells undergo a rapid expansion protocol (REP) to increase their numbers exponentially. This phase typically involves culturing the TILs in large-scale bioreactors or flasks, which are designed to accommodate the significant increase in cell volume. The culture medium is supplemented not only with IL-2 but also with other stimulatory agents such as anti-CD3 antibodies and feeder cells (irradiated peripheral blood mononuclear cells). These stimulatory factors help drive robust proliferation while maintaining the TILs' cytotoxicity and specificity. Throughout the expansion process, the TILs are routinely monitored for various quality control parameters, including cell viability, growth rate, phenotypic markers, and functional assays to ensure that they retain their ability to recognize and kill cancer cells. Sterility tests are also conducted to prevent contamination, ensuring that the final product is safe for infusion into the patient. At the end of the REP, the goal is to generate billions of TILs (often around 10–100 billion cells), which are then harvested, concentrated, and prepared for infusion back into the patient [26, 29, 30].

7 Infusion of Expanded TILs

The process of delivering activated, tumor-specific tumor-infiltrating lymphocytes (TILs) to patients involves several critical steps aimed at maximizing the therapeutic potential of these immune cells. Initially, TILs are expanded in bioreactors to achieve the desired quantity and quality [31]. Before TIL infusion, patients undergo a non-myeloablative lymphodepleting regimen (usually fludarabine and cyclophosphamide) for a duration of 5–7 days. This treatment can be administered either in an inpatient or outpatient setting, depending on the treatment team's discretion. The infusion is administered through intravenous injection, usually over several hours, in a controlled clinical environment such as an infusion center or hospital, where specialized medical teams monitor the patient for immediate adverse reactions. Following the infusion, patients often receive supportive treatments to enhance the effectiveness of the TILs. A common adjunct therapy is high-dose interleukin-2 (IL-2), which promotes the proliferation and activation of the infused TILs. However, IL-2 therapy is associated with significant side effects, including capillary leak syndrome, hypotension, and flu-like symptoms, necessitating close monitoring and supportive care, such as fluids and vasopressors. The infused TILs then circulate through the bloodstream, targeting tumor sites to engage in the destruction of cancer cells [32, 33].

8 The First TIL Therapy Human Trial

Building on his findings from preclinical studies, Rosenberg extended TIL therapy to patients diagnosed with melanoma, marking a significant milestone with the publication of the first report on the successful treatment of melanoma using TIL-ACT in 1988 [24, 26].

Dr. Steven Rosenberg's seminal clinical trial marked a pivotal moment in cancer therapy. In this pioneering study, lymphocytes extracted from freshly resected melanomas were cultured and expanded in vitro. These enhanced immune cells demonstrated a remarkable ability to specifically target and destroy autologous tumor cells without harming allogeneic tumor cells or normal autologous cells. The treatment protocol involved adoptive transfer of these tumor-infiltrating lymphocytes along with interleukin-2 (IL-2), following lymphodepletion by single-dose intravenous administration of cyclophosphamide [6, 24, 26].

The trial enrolled 20 patients with metastatic melanoma, of whom 15 had not previously received IL-2. Encouragingly, objective regression of cancer was observed in 60% of these treatment-naïve patients and in 40% of those for whom previous IL-2 therapy had failed. Regression occurred in diverse metastatic sites including the lungs, liver, bone, skin, and subcutaneous tissues, with durations ranging from 2 to more than 13 months. Although IL-2 treatment induced reversible toxic effects during the brief five-day course, the overall results underscored the potential of adoptive TIL therapy as a groundbreaking approach in cancer treatment. Dr. Rosenberg's work laid crucial groundwork for advancing immunotherapy, transforming the landscape of oncology research and patient care [24, 34, 35].

9 Clinical Applications and Successes in Various Solid Tumors

After the first clinical trial of TILs in 1988 by Rosenberg, TIL therapy has demonstrated significant antitumor potential and high efficacy rates, particularly in the treatment of melanoma. Clinical trials have shown promising results with TIL therapies for advanced melanoma and certain other solid tumors as listed below (Table 2).

9.1 Melanoma

Melanoma serves as the model cancer for TIL therapy. Early studies demonstrated remarkable response rates in metastatic melanoma patients, with objective response rates (ORRs) as high as 72% when combined with non-myeloablative chemotherapy and high-dose IL-2. Lifileucel, an autologous TIL product, has shown great promise in refractory melanoma, achieving durable responses in clinical trials [13, 36, 37].

Table 2 Selected TIL candidates in and approaching the clinic

Candidate	Sponsor	Properties	Cancer type	Trial status	Identifier
BNT221	BioNTech	Neoantigen-reactive T cells from peripheral blood	Melanoma	I	NCT04625205
LN-145	Iovance Biotherapeutics	TIL	NSCLC, HNSCC	II	NCT03108495
CISH-inactivated TILs	Intima Bioscience	CISH-inactivated neoantigen-reactive engineered TILs	Solid cancers	I/II	NCT05566223
NEXTGEN-TIL	Vall d'Hebron Institute of Oncology (VHIO)	Neoantigen-reactive TILs	Solid tumors	I	NCT05141474
Lifileucel/ LN-144	Iovance Biotherapeutics	TIL	Melanoma	Approved	NCT02360579
BNT221	BioNTech	Neoantigen-reactive T cells from peripheral blood	Melanoma	I	NCT04625205
LN-145-S1	Iovance Biotherapeutics	PD-1-selected TIL	Melanoma, HNSCC	II	NCT03083873
IOV-4001	Iovance Biotherapeutics	PD-1-inactivated engineered TIL	Melanoma, NSCLC	I/II	NCT05361174
AGX-148 plus PH-762	AgonOx, Inc., Phio Pharmaceuticals	CD8-positive TILs plus PD-1 silencing siRNA	Solid tumors	I	NCT05902520
ATL001	Achilles	Neoantigen-reactive TILs	Melanoma, NSCLC	I/II	NCT03997474
ITIL-306	InstilBio	Antifolate receptor alpha CoStAR-boosted engineered TIL	Ovarian, NSCLC, RCC	I	NCT05397093
OBX-115	Obsidian Therapeutics	IL15-boosted engineered TIL (IL2-free)	Melanoma	I	NCT06060613
iTCR	NCI	TIL-derived TCRs engineered into peripheral T cells	Metastatic cancer	II	NCT03412877
LYL-845	Lyell Immunopharma	Epigenetically reprogrammed TILs	Melanoma, NSCLC, CRC	I	NCT05573035

9.2 *Cervical Cancer*

TIL therapy has been effective in metastatic cervical cancer, particularly when used in combination with checkpoint inhibitors like anti-PD-1 or anti-CTLA-4 antibodies. Clinical trials have reported ORRs of 44%, providing hope for patients who have failed conventional therapies [37, 38].

9.3 *Triple-Negative Breast Cancer (TNBC)*

TNBC, a subtype of breast cancer with limited treatment options, benefits significantly from TIL therapy due to its high immune cell infiltration. Studies show a 20% reduction in recurrence and mortality rates with increased TIL density [39].

9.4 *Non-small Cell Lung Cancer (NSCLC)*

NSCLC, characterized by high mutation loads and neoantigens, is a promising candidate for TIL therapy. Early trials combining TILs with anti-PD-1 therapy have demonstrated tumor regression in a majority of patients [39].

9.5 *Ovarian Cancer*

Despite feasibility, clinical results in ovarian cancer have been modest, primarily due to challenges like TIL exhaustion within the immunosuppressive tumor microenvironment (TME). However, combining TILs with immune checkpoint inhibitors has shown potential for improving outcomes [39].

10 FDA Approval of TIL Therapy: Lifileucel

In a landmark event over three decades in the making, the Food and Drug Administration (FDA) has approved lifileucel (Amtagvi), the first cancer treatment utilizing tumor-infiltrating lymphocytes (TILs). This approval not only marks the advent of TIL therapy for melanoma but also signifies the first cellular therapy approved for a solid tumor. The FDA's accelerated approval of lifileucel is specifically for patients with advanced melanoma that has progressed following treatment with certain immunotherapy drugs or targeted therapies. This new therapy offers a promising option for patients who have exhausted other treatment avenues.

Lifileucel's approval was based on a clinical trial sponsored by Iovance Biotherapeutics. Among the more than 70 participants treated with the approved dose of lifileucel, nearly one-third experienced a reduction in tumor size, with some achieving complete tumor responses. Approximately 40% of these patients had no progression of their cancer 1 year after receiving the treatment. Notably, more than half of the 48 patients whose cancer responded to lifileucel treatment lived at least a year without disease progression. Some responses have lasted several years, with the longest exceeding 5 years. The FDA approval of lifileucel marks a significant milestone in the treatment of melanoma, offering new hope for patients with advanced disease. As research and development continue, TIL therapy holds promise not only for melanoma but potentially for other solid tumors, underscoring its transformative potential in cancer treatment. This approval heralds the integration of cellular therapy into the mainstream of oncology, highlighting the continuous evolution of cancer treatment paradigms [13, 36].

11 Genetically Engineered or Synthetic TILs

Traditional tumor-infiltrating lymphocyte (TIL) therapy has demonstrated promise by harnessing the body's own immune cells to fight cancer, with notable success in patients with metastatic melanoma and non-small cell lung cancer. However, the efficacy of conventional TIL therapy has been limited due to several challenges. One of the main issues is the limited persistence of TILs in the patient after reinfusion, as these cells often lose their ability to function effectively over time, becoming exhausted in the harsh tumor microenvironment. Furthermore, traditional TIL therapy requires high doses of IL-2 to support TIL survival, which is associated with significant toxicity and can lead to severe side effects.

These limitations have driven the need for synthetic TILs, genetically engineered to enhance their trafficking, durability, function, and resistance to exhaustion. By modifying TILs to express specific receptors or immune-enhancing genes, these engineered cells can better target tumors, resist the immunosuppressive signals of the tumor microenvironment, and persist longer in the body. This advancement addresses the efficacy challenges seen in conventional TIL therapy and reduces the dependence on toxic IL-2 regimens, paving the way for more effective and safer treatments for a broader range of patients [40].

12 Challenges with TIL Therapy

TIL therapy faces several challenges and limitations that impact its clinical feasibility and efficacy, particularly in treating solid tumors [41]. Logistical hurdles represent a major barrier, as the process begins with the surgical resection of tumor tissue to isolate TILs [35, 42]. This step may not always be feasible for tumors located in inaccessible regions or for patients who are not candidates for surgery due to poor health or comorbidities [43, 44]. Once the tumor is resected, the preparation process involves a rapid expansion protocol (REP) that typically takes 6–8 weeks to generate sufficient TILs for reinfusion [25, 45]. This delay poses a significant challenge for patients with rapidly progressing or aggressive tumors, who may miss the optimal treatment window [46].

The immunosuppressive tumor microenvironment (TME) presents another critical challenge. Tumors create an immunosuppressive milieu characterized by the presence of immune checkpoint molecules like PD-1, CTLA-4, and LAG3, along with regulatory T cells (Tregs) and myeloid-derived suppressor cells (MDSCs). These factors suppress TIL activity, reducing their ability to exert effective antitumor responses once reinfused into the patient. This immune suppression limits the persistence and functionality of TILs within the TME [47, 48].

High costs and technical expertise further limit the accessibility of TIL therapy. The entire process requires specialized facilities compliant with good manufacturing practices (GMPs), highly trained personnel, and significant resources for cell culture and expansion. These requirements contribute to the high cost of therapy, making it less accessible to many patients compared to other immunotherapies like checkpoint inhibitors [25, 48, 49].

Variability in TIL quality and expansion is another significant limitation. Not all tumors yield sufficient TILs, and approximately one-third of TIL cultures fail to expand during ex vivo processing. Even when successful, reinfused TILs often exhibit low persistence and reduced cytotoxicity, which can compromise therapeutic outcomes. Additionally, the variability in tumor antigen presentation and TIL responsiveness adds to the unpredictability of this treatment. Last, toxicity concerns arise primarily from the use of high-dose IL-2, a critical component of TIL expansion and post-infusion support. IL-2 is associated with significant side effects, including cytokine release syndrome (CRS) and other systemic toxicities, which require close monitoring and supportive care [41]. Together, these challenges highlight the need for advancements in manufacturing processes, strategies to overcome the immunosuppressive TME, and the development of less toxic and more efficient protocols to maximize the clinical potential of TILs [27, 41].

13 Conclusion and Future Prospects

Tumor-infiltrating lymphocyte (TIL) therapy represents a critical frontier in the fight against cancer, especially in solid tumors, leveraging the natural immune surveillance capabilities of T cells to target tumors with precision and adaptability [3, 35, 44, 49]. The deepening understanding of TIL biology, coupled with advancements in therapeutic approaches, positions TIL therapy as a cornerstone of personalized cancer immunotherapy, particularly for solid tumors. While the therapy has demonstrated significant potential, its widespread adoption requires addressing several key challenges, including logistical hurdles, cost, immunosuppressive tumor microenvironments, and variability in TIL quality and persistence [2, 7, 41, 48].

Future research is actively exploring solutions to these challenges to maximize the clinical impact of TIL therapy. Efforts to reduce production timelines, such as optimizing rapid expansion protocols and adopting young TIL methodologies, aim to make the treatment more accessible to patients with aggressive diseases. Parallelly, innovations in genetic engineering, such as CRISPR-based modification, offer opportunities to enhance TIL functionality, persistence, and antigen specificity. Early studies of genetically engineered TILs show promise, particularly in targeting tumors that exhibit immune evasion mechanisms or heterogeneous antigen expression [1, 7, 41].

Declaration of Interests The authors declare no competing interests.

References

1. Zhao Y, Deng J, Rao S, Guo S, Shen J, Du F, et al. Tumor infiltrating lymphocyte (TIL) therapy for solid tumor treatment: progressions and challenges. Cancers. 2022;14:4160.
2. Lin B, Du L, Li H, Zhu X, Cui L, Li X. Tumor-infiltrating lymphocytes: warriors fight against tumors powerfully. Biomed Pharmacother. 2020;132:110873.
3. Paijens ST, Vledder A, de Bruyn M, Nijman HW. Tumor-infiltrating lymphocytes in the immunotherapy era. Cell Mol Immunol. 2021;18:842.
4. Spiess PJ, Yang JC, Rosenberg SA. In vivo antitumor activity of tumor-infiltrating lymphocytes expanded in recombinant interleukin-2. J Natl Cancer Inst. 1987;79(5):1067.
5. Rosenberg SA, Spiess P, Lafreniere R. A new approach to the adoptive immunotherapy of cancer with tumor-infiltrating lymphocytes. Science (1979). 1986;233(4770):1318.
6. Rosenberg SA, Restifo NP, Yang JC, Morgan RA, Dudley ME. Adoptive cell transfer: a clinical path to effective cancer immunotherapy. Nat Rev Cancer. 2008;8:299.
7. Morotti M, Albukhari A, Alsaadi A, Artibani M, Brenton JD, Curbishley SM, et al. Promises and challenges of adoptive T-cell therapies for solid tumours. Br J Cancer. 2021;124:1759.
8. Geukes Foppen MH, Donia M, Svane IM, Haanen JBAG. Tumor-infiltrating lymphocytes for the treatment of metastatic cancer. Mol Oncol. 2015;9:1918.
9. Bai Z, Zhou Y, Ye Z, Xiong J, Lan H, Wang F. Tumor-infiltrating lymphocytes in colorectal cancer: the fundamental indication and application on immunotherapy. Front Immunol. 2022;12:808964.
10. Dushyanthen S, Beavis PA, Savas P, Teo ZL, Zhou C, Mansour M, et al. Relevance of tumor-infiltrating lymphocytes in breast cancer. BMC Med. 2015;13:202.
11. Fu Q, Chen N, Ge C, Li R, Li Z, Zeng B, et al. Prognostic value of tumor-infiltrating lymphocytes in melanoma: a systematic review and meta-analysis. OncoImmunology. 2019;8:e1593806.
12. Lee S, Margolin K. Tumor-infiltrating lymphocytes in melanoma. Curr Oncol Rep. 2012;14(5):468.
13. Sarnaik AA, Hamid O, Khushalani NI, Lewis KD, Medina T, Kluger HM, et al. Lifileucel, a tumor-infiltrating lymphocyte therapy, in metastatic melanoma. J Clin Oncol. 2021;39(24):2656.
14. Almangush A, De Keukeleire S, Rottey S, Ferdinande L, Vermassen T, Leivo I, et al. Tumor-infiltrating lymphocytes in head and neck cancer: ready for prime time? Cancers. 2022;14:1558.
15. Hennecke J, Wiley DC. T cell receptor-MHC interactions up close. Cell. 2001;104:1.
16. Cole DK, Pumphrey NJ, Boulter JM, Sami M, Bell JI, Gostick E, et al. Human TCR-binding affinity is governed by MHC class restriction. J Immunol. 2007;178(9):5727.

17. Cullen SP, Martin SJ. Mechanisms of granule-dependent killing. Cell Death Differ. 2008;15:251.
18. Upadhyay R, Boiarsky JA, Pantsulaia G, Svensson-Arvelund J, Lin MJ, Wroblewska A, et al. A critical role for fas-mediated off-target tumor killing in t-cell immunotherapy. Cancer Discov. 2021;11(3):599.
19. Mishra AK, Ali A, Dutta S, Banday S, Malonia SK. Emerging trends in immunotherapy for cancer. Diseases [Internet]. 2022;10(3):60. Available from: https://www.mdpi.com/2079-9721/10/3/60.
20. Fried S, Avigdor A, Bielorai B, Meir A, Besser MJ, Schachter J, et al. Early and late hematologic toxicity following CD19 CAR-T cells. Bone Marrow Transplant. 2019;54(10):1643.
21. Taraseviciute A, Tkachev V, Ponce R, Turtle CJ, Snyder JM, Liggitt HD, et al. Chimeric antigen receptor T cell–mediated neurotoxicity in nonhuman primates. Cancer Discov. 2018;8(6):750.
22. Zah E, Lin MY, Anne SB, Jensen MC, Chen YY. T cells expressing CD19/CD20 bispecific chimeric antigen receptors prevent antigen escape by malignant B cells. Cancer Immunol Res. 2016;4(6):498.
23. Guedan S, Ruella M, June CH. Emerging cellular therapies for cancer. Annu Rev Immunol. 2019;37:145.
24. Rosenberg SA, Restifo NP. Adoptive cell transfer as personalized immunotherapy for human cancer. Science. 2015;348:62.
25. Rohaan MW, Van Den Berg JH, Kvistborg P, Haanen JBAG. Adoptive transfer of tumor-infiltrating lymphocytes in melanoma: a viable treatment option. J ImmunoTher Cancer. 2018;6:102.
26. Dudley ME, Wunderlich JR, Shelton TE, Even J, Rosenberg SA. Generation of tumor-infiltrating lymphocyte cultures for use in adoptive transfer therapy for melanoma patients. J Immunother. 2003;26(4):332.
27. Wu R, Forget MA, Chacon J, Bernatchez C, Haymaker C, Chen JQ, et al. Adoptive T-cell therapy using autologous tumor-infiltrating lymphocytes for metastatic melanoma: current status and future outlook. Cancer J. 2012;18:160.
28. Leko V, Rosenberg SA. Identifying and targeting human tumor antigens for T cell-based immunotherapy of solid tumors. Cancer Cell. 2020;38:454.
29. Prieto PA, Durflinger KH, Wunderlich JR, Rosenberg SA, Dudley ME. Enrichment of CD8+ cells from melanoma tumor-infiltrating lymphocyte cultures reveals tumor reactivity for use in adoptive cell therapy. J Immunother. 2010;33(5):547.
30. Itzhaki O, Hovav E, Ziporen Y, Levy D, Kubi A, Zikich D, et al. Establishment and large-scale expansion of minimally cultured young tumor infiltrating lymphocytes for adoptive transfer therapy. J Immunother. 2011;34(2):212.
31. Somerville RPT, Devillier L, Parkhurst MR, Rosenberg SA, Dudley ME. Clinical scale rapid expansion of lymphocytes for adoptive cell transfer therapy in the WAVE ®bioreactor. J Transl Med. 2012;10(1):1–11.
32. Betof Warner A, Hamid O, Komanduri K, Amaria R, Butler MO, Haanen J, et al. Expert consensus guidelines on management and best practices for tumor-infiltrating lymphocyte cell therapy. J Immunother Cancer. 2024;12(2):e008735.
33. Schwartz RN, Stover L, Dutcher J. Managing toxicities of high-dose interleukin-2. Oncology (Williston Park, N.Y.). 2002;16:11–20.
34. Rosenberg SA, Packard BS, Aebersold PM, Solomon D, Topalian SL, Toy ST, et al. Use of tumor-infiltrating lymphocytes and interleukin-2 in the immunotherapy of patients with metastatic melanoma. A preliminary report [see comments]. N Engl J Med. 1988;319(25):1676.
35. Monberg TJ, Borch TH, Svane IM, Donia M. TIL therapy: facts and hopes. Clin Cancer Res. 2023;29:3275.
36. Healey N. Tumor infiltrating lymphocyte approval heralds new era for precision cancer immunotherapy FDA approval of Iovance's lifileucel for melanoma shows the promise of tumor-infiltrating lymphocytes for the treatment of solid tumors, with trials ongoing for lung cancer, cervical cancer and more. Nat Med. 2024;30:1795.

37. Warner AB, Corrie PG, Hamid O. Tumor-infiltrating lymphocyte therapy in melanoma: facts to the future. Clin Cancer Res. 2023;29:1835.
38. Muul LM, Spiess PJ, Director EP, Rosenberg SA. Identification of specific cytolytic immune responses against autologous tumor in humans bearing malignant melanoma. J Immunol. 1987;138(3):989.
39. Kazemi MH, Sadri M, Najafi A, Rahimi A, Baghernejadan Z, Khorramdelazad H, et al. Tumor-infiltrating lymphocytes for treatment of solid tumors: it takes two to tango? Front Immunol. 2022;13:1018962.
40. Jiménez-Reinoso A, Nehme-Álvarez D, Domínguez-Alonso C, Álvarez-Vallina L. Synthetic TILs: engineered tumor-infiltrating lymphocytes with improved therapeutic potential. Front in Oncol. 2021;10:593848.
41. Zhang P, Zhang G, Wan X. Challenges and new technologies in adoptive cell therapy. J Hematol Oncol. 2023;16:97.
42. Mullinax JE, Egger ME, McCarter M, Monk BJ, Toloza EM, Brousseau S, et al. Surgical considerations for tumor tissue procurement to obtain tumor-infiltrating lymphocytes for adoptive cell therapy. Cancer J (United States). 2022;28:285–93.
43. Du S, Yan J, Xue Y, Zhong Y, Dong Y. Adoptive cell therapy for cancer treatment. Exploration. 2023;3(4):20210058.
44. Kumar A, Watkins R, Vilgelm AE. Cell therapy with TILs: training and taming T cells to fight cancer. Front Immunol. 2021;12:690499.
45. Wickström S, Lövgren T. Expansion of tumor-infiltrating lymphocytes from melanoma tumors. Methods Mol Biol. 2019;1913:105–18.
46. Besser MJ, Itzhaki O, Ben-Betzalel G, Zippel DB, Zikich D, Kubi A, et al. Comprehensive single institute experience with melanoma TIL: long term clinical results, toxicity profile, and prognostic factors of response. Mol Carcinog. 2020;59(7):736.
47. Lim AR, Rathmell WK, Rathmell JC. The tumor microenvironment as a metabolic barrier to effector T cells and immunotherapy. eLife. 2020;9:9.
48. Verma NK, Wong BHS, Poh ZS, Udayakumar A, Verma R, Goh RKJ, et al. Obstacles for T-lymphocytes in the tumour microenvironment: therapeutic challenges, advances and opportunities beyond immune checkpoint. eBioMedicine. 2022;83:104216.
49. Ward JP. Tumor infiltrating lymphocytes (TIL): from bench to bedside. In: Cancer drug discovery and development. Hayle Medical; 2022.

CAR-T Cells: Design Principles, Core Concepts, and Clinical Applications

Rohan H. Kamat, Samyabrata Bhaduri, Bhawna Yadav, Niraj K. Nirala, Alok K. Mishra, Surajit Sinha, and Sunil K. Malonia

Abstract The development of Chimeric Antigen Receptor (CAR) T-cell therapy represents a transformative advancement in cancer immunotherapy. Initially conceptualized in the early 1980s, CAR T-cell therapy emerged from efforts to engineer T cells to recognize and target tumor-specific antigens. This therapy involves the genetic modification of autologous T cells to express synthetic receptors (CARs) that target tumor antigens independently of major histocompatibility complex (MHC) presentation, providing a highly specific and personalized treatment modality. CAR-T-cell therapy has demonstrated significant success, particularly in the treatment of relapsed or refractory hematological malignancies. Despite its revolutionary impact, CAR-T therapy faces several challenges, including antigen escape, treatment-related toxicities such as cytokine release syndrome (CRS) and neurotoxicity, and limited efficacy in solid tumors. Over the years, the field has advanced with innovations in

R. H. Kamat
Kodo Life Science, Vapi, Gujarat, India

S. Bhaduri
Stylus Medicine, Cambridge, MA, USA

B. Yadav
Division of Infectious Disease and Immunology, UMass Chan Medical School, Worcester, MA, USA

N. K. Nirala
Program in Molecular Medicine, UMass Chan Medical School, Worcester, MA, USA

A. K. Mishra
Department of Hematopoietic Biology & Malignancy, Division of Cancer Medicine, The University of Texas MD Anderson Cancer Center, Houston, TX, USA

S. Sinha
Center for Cancer Research, National Cancer Institute, Bethesda, MD, USA

S. K. Malonia (✉)
Department of Molecular, Cell and Cancer Biology, UMass Chan Medical School, Worcester, MA, USA

Cancer Center, UMass Chan Medical School, Worcester, MA, USA
e-mail: Sunil.Malonia@umassmed.edu

A. K. Mishra, S. K. Malonia (eds.), *Cell-based Immunotherapies for Cancer*,
https://doi.org/10.1007/978-3-031-88695-9_4

CAR design, manufacturing processes, and combinatorial therapeutic approaches. These advancements have broadened the clinical applications of CAR-T-cell therapy, and ongoing research holds promise for overcoming current limitations and enhancing therapeutic outcomes across a wider range of cancers. This chapter provides an overview of the fundamental principles underlying CAR design, recent innovations in CAR-T technologies, and the clinical translation of CAR-T-cell therapy.

Keywords Chimeric antigen receptor · Single-chain variable fragment · Major histocompatibility complex · CAR generation

Abbreviations

B-ALL	B-cell acute lymphoblastic leukemia
BCMA	B-cell maturation antigen
CAR	Chimeric antigen receptor
CD	Cluster of differentiation
CRISPR	Clustered regularly interspaced short palindromic repeats
CRS	Cytokine release syndrome
DLBCL	Diffuse large B-cell lymphoma
EDAPT	Enara Dark Antigen Platform Technology
FcγR	Fc gamma receptor
FDA	Food and Drug Administration
GvHD	Graft-versus-host disease
ICANS	Immune effector cell–associated neurotoxicity syndrome
ICOS	Inducible T-cell co-stimulator
IL	Interleukin
ITAM	Immunoreceptor tyrosine activation motif
LBCL	Large B-cell lymphoma
MHC	Major histocompatibility complex
NFAT	Nuclear factor of activated T cells
scFv	Single-chain variable fragment
SOC	Standard of care
T-ALL	T-cell acute lymphoblastic leukemia
TCR	T-cell receptor
TMD	Transmembrane domain
TME	Tumor microenvironment

1 Introduction

The immune system plays a pivotal role in recognizing and eliminating malignant cells, a process that forms the foundation of cancer immunosurveillance. However, tumor progression often reflects the failure of this defense, driven by a complex interplay of immune evasion mechanisms and the immunosuppressive tumor microenvironment (TME). Among the central players in antitumor immunity are cytotoxic T lymphocytes (CTLs), which must be activated, expanded, trafficked to the tumor site, and differentiated into potent effector cells capable of executing cytolytic functions. Yet, in many cancers, the full potential of CTLs remains unrealized due to factors such as antigen loss, T cell dysfunction, and downregulation of major histocompatibility complex (MHC) molecules by tumor cells. These challenges have prompted intensive efforts to develop therapeutic strategies that enhance T cell-mediated immunity [1]. A major focus has been the identification of tumor-associated antigens—either self-antigens aberrantly expressed in tumors or neoantigens arising from somatic mutations—that can serve as targets for immune recognition. While antigen recognition is a prerequisite for T cell activation, it is often insufficient for effective tumor clearance, given the numerous immunoregulatory mechanisms that limit T cell function within the TME. Adoptive cell transfer (ACT) emerged as a promising immunotherapeutic approach, offering the ability to harness and amplify tumor-specific T cell responses. ACT involves isolating T cells from patients, expanding or genetically engineering them ex vivo, and reinfusing them to mediate tumor regression. Early ACT strategies, particularly those utilizing tumor-infiltrating lymphocytes (TILs), demonstrated clinical potential but were limited by the scarcity and functional exhaustion of reactive T cells [1, 2]. The advent of gene transfer technologies has since revolutionized this field, enabling the redirection of T cell specificity through the introduction of genetically encoded tumor-targeting receptors. This technological breakthrough culminated in the development of chimeric antigen receptor (CAR) T cell therapy—a form of ACT in which patient-derived T cells are engineered to express synthetic receptors that recognize tumor antigens independently of MHC presentation. CAR T cells combine the antigen specificity of monoclonal antibodies with the effector capabilities of T cells, resulting in a powerful therapeutic modality capable of targeting otherwise evasive tumor cells [3–6]. While current CAR-T therapies are primarily autologous, efforts are underway to develop allogeneic, "off-the-shelf" CAR-T cells derived from healthy donors. These products aim to improve scalability, reduce manufacturing time, and increase accessibility. If proven safe and effective, allogenic CAR T cells could greatly expand the availability of this transformative treatment. This chapter provides an overview of the fundamental concepts of, design principles, essential component ofs CARs, and the clinical application of CAR-T-cell therapy in cancer treatment.

2 Conventional T Cells Versus CAR-T Cells

During a physiological T-cell-mediated immune response, a normal T cell is activated upon binding its T-cell receptor (TCR) to a major histocompatibility complex (MHC) molecule presenting a foreign peptide antigen, along with co-stimulatory signals from an antigen-presenting cell. This interaction forms a classical immunological synapse between the T cell and the target cell, initiating intracellular

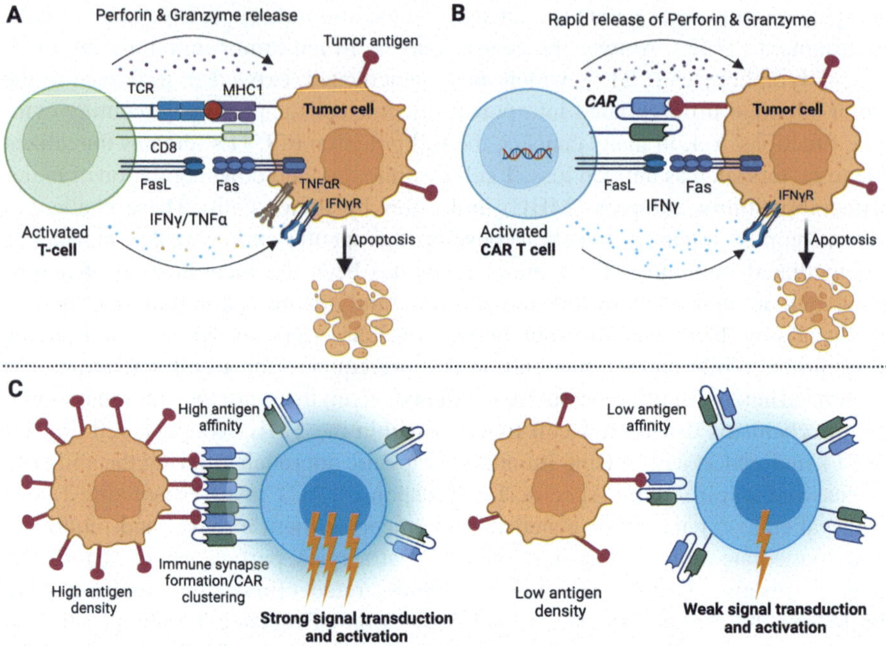

Fig. 1 (**a**) Normal T cells recognize tumors through the interaction of their T-cell receptor (TCR) with MHC I molecules presenting tumor antigen peptides. Tumor cell cytotoxicity is primarily mediated by perforin and granzyme release, along with FasL/Fas interactions and cytokine secretion (e.g., IFN-γ and TNF-α) to induce apoptosis. (**b**) CAR-T cells employ similar mechanisms but exhibit rapid release of perforin and granzyme, leading to more rapid tumor cell death. They also rely more on Fas/FasL interactions and secrete IFN-γ to enhance tumor-killing activity. (**c**) The level of CAR-T-cell activation depends on the antigen density and antigen affinity on the target tumor cell. (Figure created with BioRender.com)

signaling pathways that drive T-cell proliferation and effector functions [7, 8]. The conventional T cells employ three main cytotoxic mechanisms (Fig. 1a). The first is the perforin/granzyme pathway, which is the primary cytotoxic mechanism. Following the formation of the immunological synapse, cytotoxic effector proteins such as perforin and granzyme are released into the synaptic cleft. Perforin creates pores in the target cell membrane, allowing granzymes to enter the cell and induce apoptosis [9, 10]. The second is the death receptor pathway. Death receptors are members of the tumor necrosis factor receptor (TNFR) superfamily, characterized by a cytoplasmic "death domain" that enables cytotoxic signaling upon ligand binding. In T cells, Fas ligand (FasL) on the cell surface binds to Fas receptors on target tumor cells, triggering programmed cell death [11]. The third is the cytokine release pathway, in which activated T cells secrete cytokines such as TNF-α and IFN-γ, enhancing target cell recognition and killing efficiency [12].

CAR-T cells differ from conventional T cells in several key aspects. First, most CAR-T cells feature a chimeric receptor that directly recognizes and binds to the antigen on the surface of the target cancer cell, bypassing the need for MHC presentation (Fig. 1b). This design of CAR-T cells enables them to recognize a broader range of target antigens than conventional TCRs, with a significantly greater binding affinity (100–1000 times greater), more akin to that of antibodies than TCRs [13, 14]. Additionally, CAR clustering on the T-cell surface facilitates the formation of a strong immunological synapse, enabling more rapid CAR-T activation upon target antigen recognition [15–17]. The immunological synapse between CAR-T cells and the target tumor is often regarded as an indicator of CAR-T efficacy and potential side effects. Hyperactivation of CAR-T cells can result in toxicity due to excessive cytokine release or reduced efficacy caused by limited persistence. The extent of CAR-T-cell activation depends on target antigen density, CAR affinity, and signal transduction strength [18] (Fig. 1c). Like conventional T cells, CAR-T cells also execute effector functions against target tumor cells but more robustly. However, the interactions and the formation of synapses that drive CAR-T-cell proliferation, cytokine secretion, cytotoxicity, and persistence differ from those of conventional T cells [13, 18, 19]. CAR-T cells utilize mechanisms that induce apoptosis in target cells, similar to conventional T cells, involving all three cytotoxic pathways. However, the kinetics of the effector response are more rapid and robust (Fig. 1c). Initially, it was thought that CAR-T cells relied predominantly on the perforin/granzyme pathway rather than death receptor pathways for effective tumor cell killing. However, recent findings indicate that both pathways are involved. CAR-T cells also employ the death receptor pathway to induce cell death [18]. Compared to native T cells, CAR-T cells mediate tumor cell killing more rapidly, which is attributed to the quick degranulation and release of perforin and granzyme upon encountering the target antigen [4, 18]. It has been shown that CAR-T derived from either $CD4^+$ or $CD8^+$ T cells can efficiently kill tumor cells through this mechanism, regardless of Fas expression [20, 21].

2.1 Chimeric Antigen Receptors (CARs)

Chimeric antigen receptors (CARs) are modular proteins that are synthetically engineered to be expressed on T lymphocytes. The DNA sequences encoding CARs are designed in silico, synthesized, and cloned into a transfer vector, which is then used to introduce CAR trans gene into T cells, generating CAR-T cells. The funnel process of CAR development and generation of CAR-T cells is outlined in Fig. 2 and described below.

2.1.1 Target Antigen Identification

The initial step in this process involves identifying the tumor-specific antigen or target molecule that the CAR-T cell will recognize. It is critical to ensure that the chosen antigen is highly expressed on cancer cells and minimally expressed on healthy/normal cells to reduce off-target effects [22]. Profiling cancer cells to identify antigens that are overexpressed on cancer cells is essential for optimizing the CAR design. Advancements

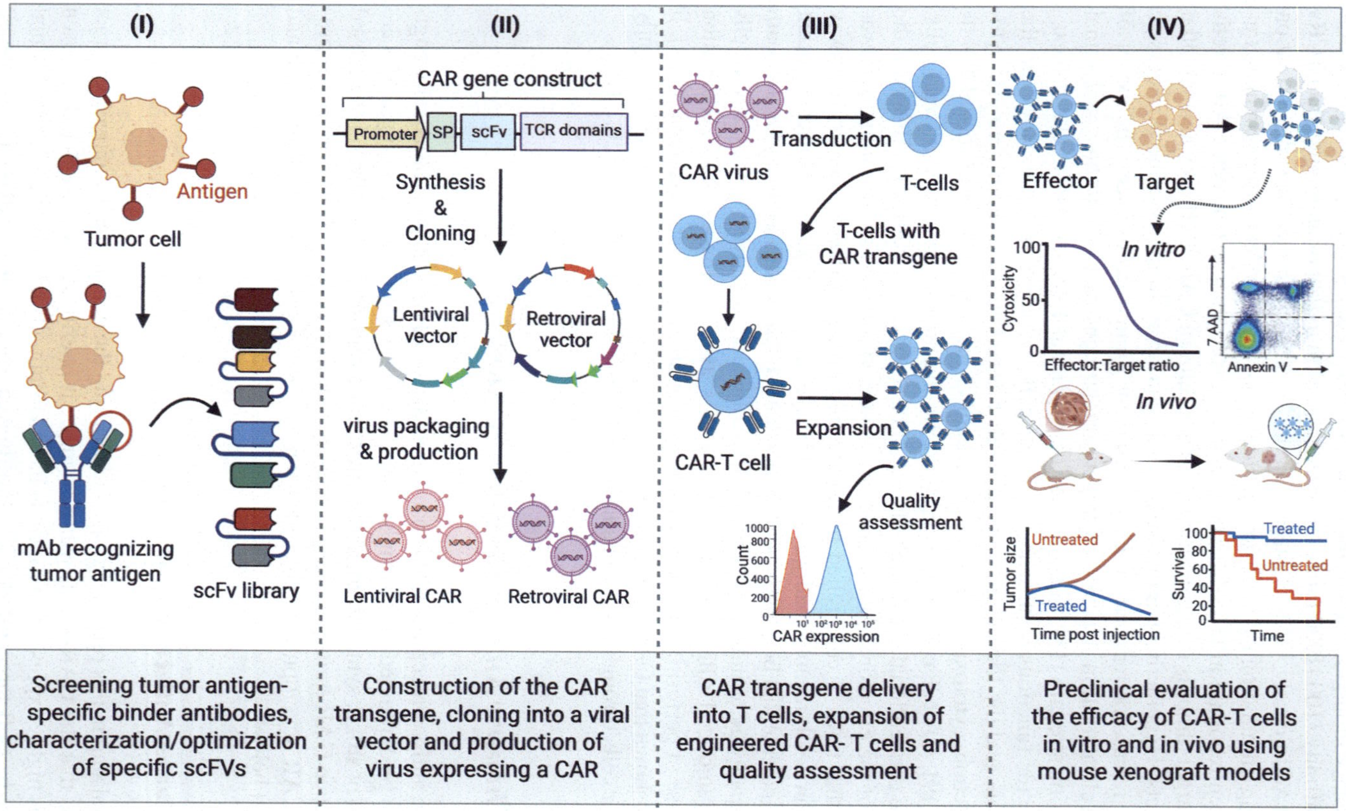

Fig. 2 The steps involved in CAR development and CAR-T-cell generation [1]. Screening for appropriate binding antibodies against the target tumor antigen. (II & III) Generating the CAR construct by assembling the CAR gene and components of the native TCR in a viral cloning vector, followed by packaging the virus expressing the CAR and transducing into T cells. (IV) In vitro and in vivo evaluation of engineered CAR-T cells. SP: signal peptide. (Figure created with BioRender.com)

in the multiomics space, including mass spectrometry-based immunopeptidomics combined with transcriptomics and ribosome profiling, have enabled identification of noncanonical tumor antigens [23]. Proteogenomic-based assays have been used to identify druggable tumor-specific antigens derived from noncoding regions of the genome [24]. Given the mode of action of CAR-T therapy, a prerequisite for the target antigen is its surface expression. A cell subtype–specific surface antigen profile (surfaceome), generated through transcriptomics and surface proteomics can be used to identify cancer-specific targets [25]. Neoantigens are novel tumor-generated antigens resulting from mutations in genomic regions; these can be identified using various NGS-based techniques. These neoantigens hold immense value due to their immunogenic properties [26]. These neoantigens are also called "dark antigens." Enara Bio is a commercial entity that has developed a dark antigen discovery platform, Enara Dark Antigen Platform Technology (EDAPT), which uses multiomics and bioinformatics to explore and validate new tumor antigens [27, 28]. When selecting a target antigen, it is crucial to consider its relevance to the disease and its role in tumor progression. The critical aspects of target selection involve three primary criteria: coverage, specificity, and stability. Coverage ensures that the target antigen is present on majority of tumor cells, while specificity aims to minimize damage to healthy tissues. Stability ensures the target's consistent expression to prevent treatment failure [22].

2.1.2 Identification of Binder Antibodies Against the Target Antigens

The process of screening binder monoclonal antibodies (mAbs) is a critical step in designing CARs, aimed at identifying candidates that bind specifically and robustly to the target antigen. Common approaches employed for this purpose include hybridoma screening, phage display, and high-throughput single B-cell sequencing technologies. Each method enables the identification of high-affinity binders tailored for the CAR's antigen-recognition domain. Hybridoma technology involves fusing antibody-producing B cells with myeloma cells to create immortalized cell lines capable of producing monoclonal antibodies. This technique remains widely used to generate specific antibodies that serve as the initial template for CAR design [29, 30]. Phage display is a versatile, high-throughput method that allows the screening of vast libraries of antibody fragments displayed on the surface of bacteriophages. This approach is particularly valuable for identifying single-chain variable fragments (scFvs) with optimal binding properties and minimal immunogenicity [31–33]. The key selection criteria include binding affinity, specificity, and functional validation to ensure compatibility with CAR design.

2.1.3 Constructing a CAR Gene Construct

The selected monoclonal antibody (mAb) sequence is engineered into a single-chain variable fragment (scFv), which serves as the antigen-recognition domain of the CAR. The scFv sequence, along with a linker sequence and various sequences derived from the transmembrane and signaling domains of native T cells, is assembled into a single gene cassette known as the CAR construct. The CAR gene is then cloned into a viral vector, which includes a signal peptide and an appropriate promoter, followed by the CAR ORF (open reading frame). Commonly used promoters for driving CAR expression through viral vectors include elongation factor 1 alpha (EF1α), cytomegalovirus (CMV), mouse stem cell virus (MSCV), and murine leukemia virus MND promoter (myeloproliferative sarcoma virus enhancer, negative control region deleted, dl587rev primer-binding site substituted. These promoters are known for their strong, constitutive expression, ensuring high levels of CAR protein production [34, 35].

2.1.4 Introducing CAR Constructs into T Cells

Efficient delivery of the CAR transgene into T cells is critical for therapeutic efficacy of CAR-T cell therapy, as it facilitates tumor antigen recognition and immune-mediated cytotoxicity. Multiple genetic engineering methods have been developed for CAR transgene delivery, each with distinct advantages and limitations in terms of efficiency, safety, duration of transgene expression, and clinical scalability [36]. Both viral and nonviral genetic engineering tools have been utilized to introduce the CAR transgene into T cells. Among these lentiviral and gamma-retroviral vectors are most widely used due to their ability to stably integrate the transgene into the host genome, resulting in long-term expression and durable therapeutic effects. Typically require a a single transduction is sufficient to achieve integration, after which the engineered cells can expand and persist, generating a potent population of cells capable of neutralizing cancer cells for extended periods [36]. Although concerns regarding insertional mutagenesis and oncogenic potential remain, longitudinal genomic studies of patients treated with viral vector-modified CAR-T cells suggest that these risks are relatively low [37]. In parallel, nonviral platforms are being actively investigated as an alternative to viral vectors. These include transposon-based systems such as Sleeping Beauty and PiggyBac, which facilitate stable integration of the CAR transgene into the host genome without relying on viral components. Additionally, transient expression systems—such as mRNA electroporation—enable non-integrative, short-term expression of the CAR, thereby reducing the risk of insertional mutagenesis [38]. These nonviral approaches may confer advantages in terms of manufacturing scalability, regulatory compliance, and improved safety profiles, particularly for applications where transient CAR expression is desirable, such as in solid tumors or to mitigate long-term toxicities.

2.1.5 In Vitro Assessment of CAR-T Cells

The engineered T cells are first analyzed for CAR transgene expression and then subsequently tested in vitro for their ability to recognize and kill antigen-expressing target cells. Various functional assays are used to assess CAR-T cell functionality, including cytotoxicity assays such as the chromium (51Cr)-release assay (51Cr assay), luciferase-mediated bioluminescence imaging (BLI) assay, impedance-based assay, and flow cytometry assays. These assays evaluate the ability of CAR-T cells to kill target cells, cytokine secretion assays to evaluate immune activation, and proliferation assays to assess the expansion of CAR-T cells upon antigen engagement. These tests are typically used to evaluate parameters such as specificity, efficacy, and potential off-target effects [39–41].

2.1.6 In Vivo Assessment of Lead CAR-T Cells

The top-performing CAR constructs from in vitro studies are selected for preclinical evaluation in animal models. This step involves tumor regression studies using xenograft, syngeneic, or humanized mouse models; monitoring toxicity and off-target effects; and assessing the persistence and expansion of CAR-T cells. The purpose of studies is to identify the lead CAR construct with the optimal balance of efficacy and safety for clinical translation [42–44]. This systematic process ensures that CAR-T cells are optimized for specificity, functionality, and safety before advancing to clinical application.

3 CARs' Basic Structure and Design Elements

CARs are modular synthetic receptors composed of four key structural components: (1) an extracellular antigen-binding domain, (2) a hinge or spacer region, (3) a transmembrane domain, and (4) one or more intracellular signaling domains (Fig. 3). CAR design has evolved over the years, with each generation resulting in improvements in efficacy, safety, and durability [45–47]. This section will discuss the fundamental principles guiding CAR design.

3.1 Antigen-Binding Domain (scFv)

The antigen-binding domain, located on the extracellular region of the CAR, is responsible for recognizing the target antigen and directing the specificity of CAR-expressing T cells. It plays a pivotal role in determining the affinity and avidity of CAR-T cells, directly influencing their overall effectiveness. In most CARs, the antigen-binding domain is derived from the variable heavy (V_H) and light (V_L) chains of monoclonal antibodies, which are connected by a flexible peptide linker to create single-chain variable fragments (scFvs). The most commonly used linker is (Gly4Ser)3, which ensures proper folding of scFv while maintaining its ability to bind antigens [45–47]. scFv sequences are commonly sourced from murine or human monoclonal antibodies. CARs can also incorporate the variable regions of

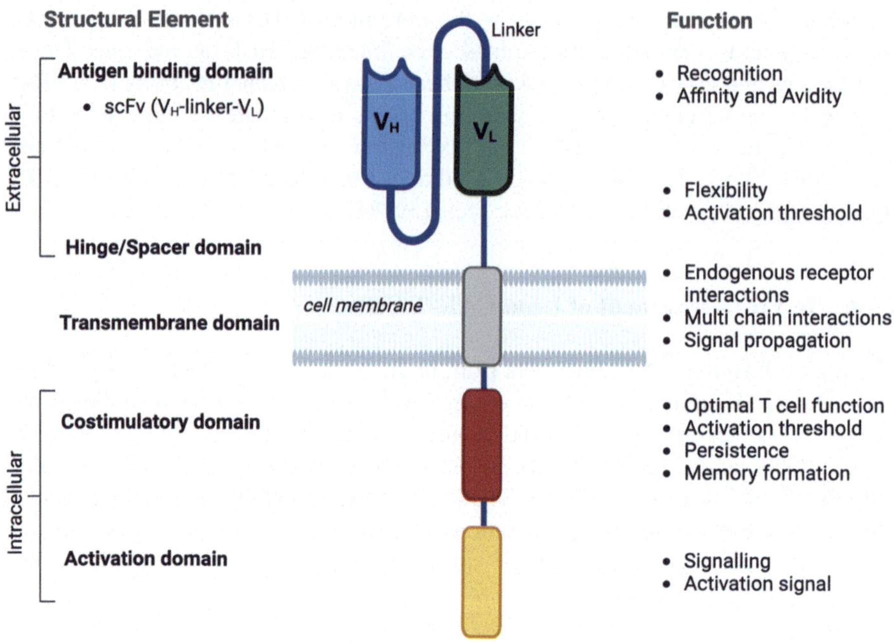

Fig. 3 Illustration showing the basic structural elements of chimeric antigen receptors (CARs) and their functions. scFv: single-chain variable fragment, V_H: variable domain of heavy chain, V_L: variable domain of light chain. (Figure created with BioRender.com)

heavy-chain-only antibodies (V_{HH} or nanobodies), which are small, stable, single-domain antibody fragments derived from camelids, offering affinities similar to those of traditional scFvs [48]. The key parameters of scFvs that influence CAR function are affinity (binding strength to the target antigen), avidity (overall binding strength due to multiple antigen–scFv interactions), the location of antigen epitope, and the accessibility of the epitope. Affinity is critical for CAR-T-cell specificity and minimizing "on-target, off-tumor" side effects, particularly when the target antigen is expressed in healthy tissues. Variations in scFv affinity, epitope specificity, and recognition of posttranslational modifications, such as glycosylation, can significantly impact therapeutic outcomes. Modulating affinity can impact CAR-T-cell signaling, proliferation, cytokine production, and persistence. Therefore, the affinities of scFvs must be carefully tuned to strike a balance between achieving a robust antitumor response and minimizing the risk of on-target, off-tumor toxicity [46, 47, 49, 50]. Furthermore, affinity thresholds are not universal and vary based on interrelated factors such as antigen density on target cells, CAR expression levels, and the location of the binding epitope. Therefore, optimizing CAR binding to its target antigen requires the consideration of additional factors, such as epitope location, target antigen density, and minimizing ligand-independent tonic signaling associated with certain scFvs [47, 49]. The avidity of scFvs is another important factor influencing CAR-T-cell response. It is affected by CAR expression, ligand density, and the affinity of individual binding domains [51, 52]. Several factors beyond merely recognizing and binding the target epitope impact CAR function. For example, the interaction between the VH and VL chains and the relative positioning of the complementarity-determining regions can influence the affinity and specificity of the CAR for target epitope [53]. A critical aspect of antigen-binding domain development is managing scFv aggregation. Aggregation or misfolding of scFvs may result from poor folding stability of the VH or VL domains or the exposure of hydrophobic residues at the VH–VL interface. Additionally, scFv linkers may impose steric constraints on VH–VL interactions, potentially leading to oligomerization [54, 55]. Aggregation can result in tonic signaling—activation in the absence of antigen—which may lead to early T-cell exhaustion [54]. To mitigate these effects, precise control of CAR expression levels in T cells is crucial. Excessive CAR expression can also promote the exchange of V_H–V_L domains between different CAR molecules, increasing the risk of aggregation on the cell surface [54]. The length and composition of the linker connecting the VH and VL domains are also critical. Shorter linkers have been associated with increased tonic signaling in certain CAR constructs, whereas longer, more flexible linkers may mitigate this risk and improve antigen-binding efficiency. Consequently, the precise design of VH–VL pairs and linker regions is fundamental for developing CAR-T-cell therapies with reduced tonic signaling and enhanced clinical efficacy [17]. In conclusion, ensuring optimal CAR-T-cell efficiency requires rigorous scFv characterization and engineering. Achieving a balance between affinity, avidity, epitope specificity, linker design, and CAR expression levels is essential for developing effective CAR-T therapies.

3.2 Hinge Domain

The hinge region, also referred to as spacer, is an extracellular component of CARs that connects the antigen-binding domain (scFv) to the transmembrane domain, providing flexibility and optimizing antigen interaction (Fig. 1). This domain influences the interaction of CAR-T cells with the target antigen and provides flexibility to the scFv, facilitating the recognition of target epitopes. The design of the spacer, including its length and composition, significantly influences CAR expression, signaling, and activation strength [56]. The optimal spacer length is determined by the location of the target epitope and the degree of steric hindrance on the target cells. For example, long spacers are often more effective at accessing membrane-proximal epitopes or glycosylated antigens [57, 58], whereas CARs bearing shorter spacers are better suited for accessing membrane-distal epitopes [59]. The spacer length has been suggested to be critical in ensuring sufficient intercellular spacing for the formation of an effective immunological synapse. This underscores the importance of tailoring spacers to suit specific epitopes. Thus, modulating spacer length can have a profound effect on the cytolytic activity and signaling of CAR-T cells [45, 59, 60]. Common hinge regions are derived from membrane-bound receptors such as CD8 or CD28 or from IgG domains (IgG1 or IgG4). However, IgG-derived spacers can interact with Fcγ receptors (FcγRs), triggering off-target innate immune activation and CAR-T-cell depletion [46, 47, 61, 62]. To mitigate this, IgG-derived spacers are often modified to minimize FcγR interactions. Alternatively, CD28 and CD8α spacers, which naturally lack FcγR interactions, are widely used [59, 60, 63].

3.3 *Transmembrane Domain*

The transmembrane domain (TMD) of CARs, a hydrophobic α-helix spanning the cell membrane, primarily anchors the receptor. TMDs in CAR structures function as critical conduits, transmitting ligand recognition signals to the cytoplasmic domain. Most CAR designs incorporate the TMD from the same protein as the adjacent hinge or signaling domains, commonly derived from endogenous T-cell proteins, such as CD4, CD8, and CD28, or the TCR-associated CD3ζ chain [64, 65]. Although the precise mechanisms of CAR signaling remain unclear, it is hypothesized that, similar to TCRs, CARs may require dimerization or interaction with accessory molecules to initiate proper signaling. Some of these TMDs can facilitate molecular interactions that promote self-association or assembly with their native T-cell counterparts. Such interactions can influence CAR expression and function, potentially limiting precise control over signaling outcomes [65–67]. For example, CD3ζ CARs can interact with TCR components, enabling both cis- and trans-signaling mechanisms [68]. The choice of TMD can thus affect CAR surface expression, stability, and interactions with endogenous signaling complexes. While this can enhance CAR function in some cases, it can also lead to unintended signaling effects, reducing precise control over CAR activation. Overall, the TMD may play a role in synapse formation and signaling, influencing CAR expression, stability, and function.

3.4 *Intracellular/Cytoplasmic Domains*

In normal T cells, the TCR–CD3 complex is responsible for transducing the critical "signal 1," which initiates the intracellular signaling cascade necessary for T-cell activation. This signal is generated upon the recognition of antigenic peptides presented by MHC present on the surface of antigen-presenting cells (APCs). However, the first signal is not sufficient for full T-cell activation. Co-stimulatory receptors, such as CD28, provide an additional "signal 2" that is crucial for sustaining the T-cell response, preventing anergy (a state of unresponsiveness), and promoting T-cell proliferation [69, 70]. In addition to CD28, other co-stimulatory receptors, such as 4-1BB, ICOS, and OX40, play important roles in regulating T-cell differentiation, metabolism, and apoptosis. These co-stimulatory signals help fine-tune the T-cell response, ensuring effective immune function [71]. In CAR-T cells, the intracellular signaling domain includes the CD3ζ cytoplasmic region, providing the primary activation signal (signal 1) upon antigen binding. To enhance CAR-T function, co-stimulatory domains are incorporated into the receptor construct in cis with CD3ζ, significantly improving proliferation, persistence, and antitumor efficacy. The type, number, and spatial arrangement of co-stimulatory domains are key factors influencing CAR-T performance [46, 47, 72].

CAR signaling relies on immunoreceptor tyrosine activation motifs (ITAMs) within the cytoplasmic CD3ζ signaling domain. ITAMs are phosphorylated upon receptor engagement, recruiting key molecules such as ZAP70 to initiate downstream signaling cascades. Although CAR-T cells form nonclassical immune synapses, they activate ZAP70 in an antigen-dependent manner [73, 74]. In conventional T cells, the TCR–CD3 complex contains multiple ITAMs essential for optimal signaling [75]. A key difference between T cells and CAR-T cells is the number of ITAMs involved in signaling and activity. The number and arrangement of ITAMs in CAR-T cells significantly impact activation, proliferation, and antitumor responses. Increasing ITAMs enhances signaling, proliferation, and cytotoxicity but may also induce tonic signaling, whereas reducing them can improve specificity for high antigen densities [76, 77]. Research on ITAM multiplicity in CAR-T cells has also revealed that the position of ITAMs within the cytoplasmic domain can influence signaling outcomes. The availability of in-trans-ITAMs (located in transmembrane regions) can enhance CAR-T-cell signaling by promoting stronger interactions with ZAP70. Alterations in the position or number of ITAMs can affect T-cell activation, differentiation, proliferation, cytokine secretion, and effector function [78, 79]. Understanding the precise role of ITAMs in CAR-T-cell signaling is essential for optimizing CAR design and improving therapeutic outcomes. In conclusion, CAR-T-cell design is influenced by a combination of factors, including the number and type of co-stimulatory domains, the order and spatial arrangement of these domains, and the number of ITAMs. These elements work together to affect signal strength, T-cell activation, persistence, and antitumor efficacy. Further studies exploring the complex interplay between these components, as well as clinical comparisons, will be crucial for refining CAR-T-cell therapies.

4 Generations of CARs

First-generation CARs were equipped with a single intracellular signaling domain, either the CD3ζ-chain or FcεRIγ (Fig. 4). These constructs closely resembled natural TCRs but had several limitations, including an inability to generate adequate levels of interleukin-2 (IL-2) [80]. Additionally, these engineered cells exhibited limited proliferation and a reduced lifespan in vivo, suggesting the need for co-stimulatory domains to improve efficacy [81, 82]. Second-generation CARs were designed to overcome the limitations of first-generation CAR-T cells, such as insufficient proliferation, low cytokine production, and short lifespan. These CARs incorporate dual signaling mechanisms, which enhance T-cell proliferation, more

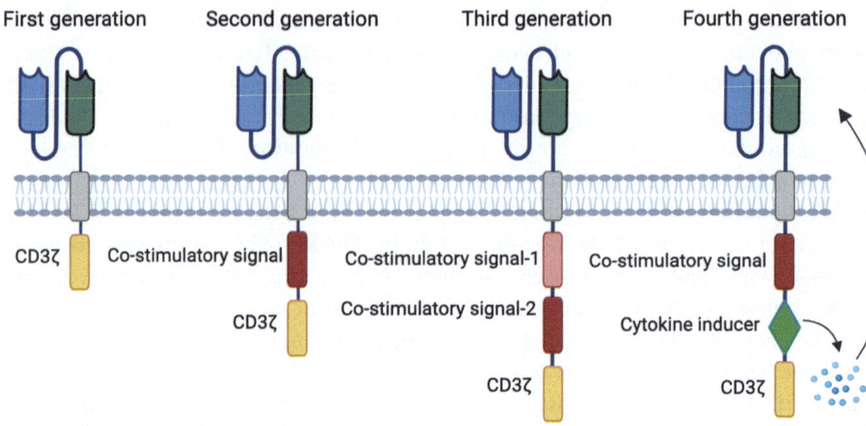

Fig. 4 Illustration showing different generations of CAR designs with different domains. (Figure created with BioRender.com)

closely mimicking natural immune responses. In addition to the intracellular CD3ζ domain, second-generation CARs include co-stimulatory molecules like CD28, CD134 (OX-40), or CD137 (4-1BB), which provide a secondary signal upon tumor antigen recognition. This results in dual signaling pathways mediated by both CD3ζ and the co-stimulatory molecules [83–85] (Fig. 4). Preclinical and clinical studies have demonstrated that these co-stimulatory signals enhance T-cell proliferation, cytotoxicity, and durability by extending their in vivo lifespan [86, 87]. Research has also highlighted the critical role of the co-stimulatory domain composition in shaping the functional effects of CAR-T cells. For example, studies suggest that 4-1BBζ-CAR-T cells may circulate longer than CD28ζ-CAR-T cells [88, 89], although the former may be prone to early exhaustion, while the latter faces the risks constitutive activation in the absence of an antigen. Consequently, CAR designs have continued to evolve to optimize co-stimulatory domains for improved performance [89]. All FDA-approved CAR-T-cell therapies are currently based on second-generation CAR-T cells. Third-generation CARs were developed by incorporating multiple co-stimulatory signaling domains, such as CD28, CD137 (4-1BB), CD134 (OX-40), NKG2D, CD27, TLR2, or inducible T-cell co-stimulators (ICOSs) within their endodomain. Examples of these constructs include combinations like CD3ζ-CD28-OX40, CD3ζ-CD28-41BB, CD3ζ-ICOS-4-1BB, and CD3ζ-TLR2-CD28 [90] (Fig. 4). Compared with second-generation CARs, third-generation CARs have shown improved T-cell proliferation, survival, and antitumor activity in preclinical studies. These constructs have demonstrated success in treating certain cancers and exhibit enhanced safety profiles, along with increased persistence and proliferative capacity of CAR T cells [91]. Fourth-generation CARs were developed based on second-generation constructs (Fig. 4). The key innovation in this generation is the addition of a constitutive or inducible expression cassette encoding a transgenic protein, such as a cytokine. These advanced CAR-T cells, known as T cells redirected for universal cytokine-mediated killing (TRUCKs), are engineered to deliver the transgenic product directly to the tumor site. This is typically achieved by incorporating a nuclear factor of activated T cells (NFAT)–responsive cassette that produces a cytokine (e.g., IL-12) when the CD3ζ-containing CAR engages its target. Engineering TRUCK CAR-T cells involves introducing two transgenic cassettes: one for the CAR structure and the other for the inducible cytokine. In preclinical models, the inclusion of a cytokine transgene significantly improved the efficacy of CAR-T-cell therapies compared with second-generation CARs [92, 93]. Fifth-generation CAR designs aim to expand the therapeutic potential of CAR-T cells by enhancing their antigen-binding capabilities, optimizing intracellular signaling, and incorporating regulatory mechanisms to improve safety and efficacy. These include dual- or multiple-antigen-targeting CARs [94, 95], switchable CARs that offer external control mechanisms to activate or deactivate CAR-T cells as needed, providing a strategy to manage toxicity [96], armored CARs designed to secrete cytokines or other immune-modulating molecules to enhance their own function and sustain activity within the immunosuppressive tumor microenvironment [97], and the universal allogenic CARs [98].

5 Autologous Versus Allogeneic CAR-T Cell Therapy

Autologous CAR-T therapy involves harvesting T cells from the patient, genetically modifying them to express a tumor antigen specific CAR and reinfusing them back into the same patient. This personalized approach reduces the risk of immune rejection by using the patient's own immune cells (Fig. 5). The majority of FDA-approved CAR-T-cell therapies and those in clinical trials currently rely on autologous T cells

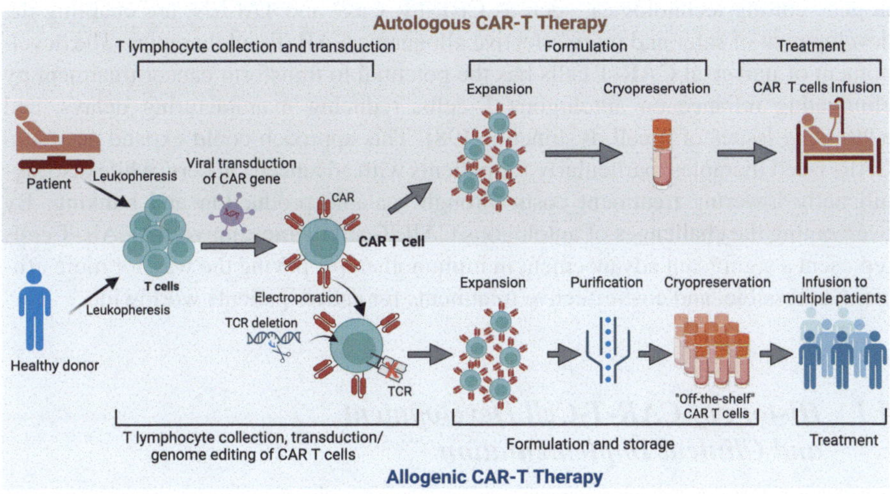

Fig. 5 Illustration showing autologous and allogeneic CAR-T-cell therapies. (Figure created with BioRender.com)

obtained from patients. Although autologous CAR-T-cell therapies have demonstrated remarkable efficacy, achieving durable remission in many patients with hematological cancers, the personalized nature of this therapy presents challenges and limitations [99]. Harvesting sufficient T cells from patients can be challenging, particularly in those with lymphopenia caused by advanced cancer or previous chemotherapy, which often reduces the number of naive T cells critical for CAR-T-cell efficacy. In pediatric patients, chemotherapy depletes naive T cells, potentially causing CAR-T production failure due to insufficient T cell numbers [100, 101]. Second, the manufacturing process for autologous CAR-T-cell products is lengthy and highly individualized, which may render it impractical for patients with advanced-stage cancer, where disease progression can occur during the production period. Additionally, CAR-T-cell production failure can arise because of the characteristics of the apheresis product [102]. In contrast, allogeneic CAR-T-cell therapy uses T cells derived from healthy donors, which are engineered to express CARs and are subsequently administered to patients. This off-the-shelf approach offers several advantages, including rapid availability, scalability, and the use of high-quality donor T cells, which may result in more robust and consistent therapeutic outcomes (Fig. 5). Allogeneic CAR-T-cell therapy also has the potential to reduce production costs significantly and streamline the treatment process [103]. However, the major challenge with this modality is the risk of immune rejection or graft-versus-host disease (GvHD), where the donor cells may attack the patient's tissues or be rejected by the patient's immune system. To mitigate these risks, donor T cells are often modified by gene editing to delete endogenous TCR or MHC molecules, thereby reducing the likelihood of adverse immune responses [104]. Both autologous and allogeneic CAR-T-cell therapies have unique advantages and limitations, and their application depends on patient-specific factors, cancer type, and logistical considerations. While autologous CAR-T-cell therapy is currently more established and widely used, ongoing advances in gene-editing technologies, such as CRISPR-Cas9 and TALEN, are enabling the development of safer and more effective allogeneic CAR-T-cell therapies. The development of universal CAR-T cells has the potential to transform cancer treatment by eliminating reliance on autologous T cells, reducing manufacturing delays, and addressing issues of T-cell dysfunction [98]. This approach could expand access to CAR-T-cell therapies, particularly for patients with advanced cancers, while also significantly lowering treatment costs through scalable production and banking. By overcoming the challenges of autologous CAR-T-cell therapy, universal CAR-T cells represent a significant advancement in immunotherapy, paving the way for more efficient, accessible, and cost-effective treatments for cancer patients worldwide.

5.1 History of CAR-T-Cell Development and Clinical Implementation

The development of CAR-T therapy spans several decades of scientific advancement. In 1987, Zelig Eshhar and colleagues at the Weizmann Institute introduced the concept of "T-bodies"—engineered T cells with antibody-like recognition capabilities. The first functional chimeric antigen receptors (CARs) were developed in 1993, combining antibody-binding domains with T-cell signaling components, enabling antigen recognition independent of MHC presentation [80, 105]. In the late 1990s and early 2000s, Michel Sadelain at Memorial Sloan Kettering Cancer Center advanced CAR technology by incorporating co-stimulatory domains, leading to the development of second-generation CARs. These modifications were crucial for improving T-cell activation, proliferation, and persistence, addressing the limitations seen in first-generation CARs [106]. These advancements were particularly pivotal for reducing the rapid T-cell exhaustion observed withearlier designs, significantly enhancing in vivo responses. Around the same time, Steven Rosenberg at the National Cancer Institute refined T-cell engineering strategies based on tumor-infiltrating lymphocytes (TILs), contributing to the broader field of adoptive cell therapy (ACT). During the early 2000s, preclinical research gained momentum, with focussed efforts on optimizing CAR constructs and assessing their efficacy in various cancer models. A significant milestone occurred in 2003 when Michel Sadelain's team at Memorial Sloan Kettering Cancer Center introduced second-generation CAR-T cells targeting CD19, a B-cell-specific antigen. This approach demonstrated substantial efficacy in a mouse models of cancer [107]. By the late 2000s and into the 2010s, clinical trials evaluating these advanced CAR-T-cell therapies began reporting promising results, marking a significant step toward their therapeutic application. The 2010s witnessed dramatic breakthroughs in CAR-T-cell therapy. In 2011, Carl June and his team at the University of Pennsylvania reported remarkable outcomes in patients with refractory chronic lymphocytic leukemia (CLL) treated with CD19-targeted CAR-T cells [108]. This success led to large-scale clinical trials, culminating in 2017, when the FDA approved Kymriah (tisagenlecleucel), the first-ever CAR-T-cell therapy, for pediatric and young adult B-cell acute lymphoblastic leukemia (B-ALL). As of December 2023, the US Food and Drug Administration (FDA) has approved six CAR-T-cell therapies for use in 12 different indications, including large B-cell lymphoma (LBCL) [109–112], mantle cell and follicular lymphoma [113, 114], and B-cell acute lymphoblastic leukemia (B-ALL) [115–117]. Clinical trials have shown that CD19 CAR-T-cell therapy significantly outperforms the standard of care (SOC) as a second-line treatment for LBCL and has demonstrated exceptional efficacy as a first-line treatment [118–120]. The versatility of CAR-T-cell technology has been further validated by expanding beyond CD19 targeting. Two BCMA-CAR-T-cell therapies, which target BCMA antigen, have been approved for multiple myeloma [121, 122] and have high response rates, with CD22-CARs in LBCL [123], CD22-CARs in B-ALL [124], CD7-CARs in T-cell acute lymphoblastic leukemia (T-ALL) [125, 126], CD22-CARs in B-ALL, and CD30-CARs in Hodgkin's lymphoma [127]. The timeline of CAR-T-cell therapy from the conceptualization phase to the commercialization phase [95, 148, 149, 150, 151, 152, 153, 154, 155, 156, 157], highlights the remarkable progress made in the field, which has transformed the landscape of cancer treatment (Fig. 6).

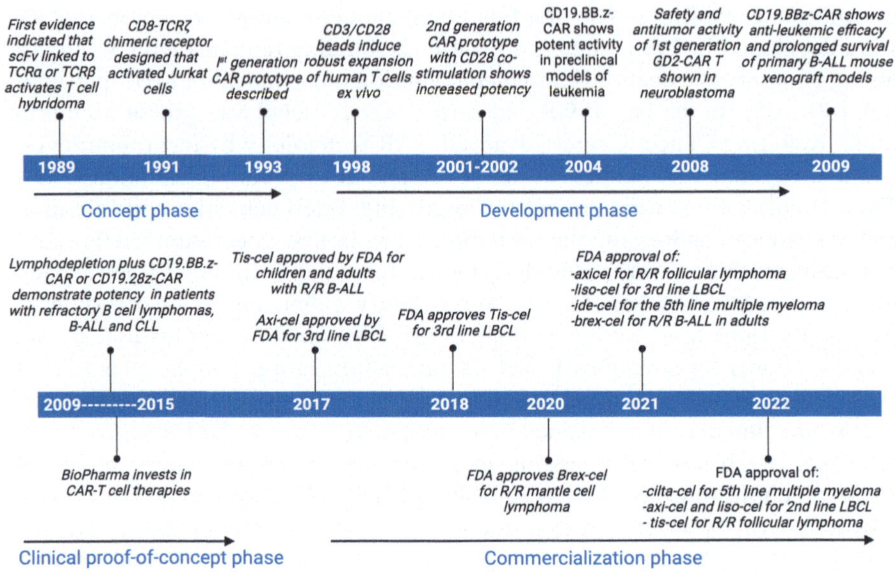

Fig. 6 Timeline of CAR-T-cell therapy development from conceptualization to commercialization [95, 148–157]

5.2 *Limitations of CAR-T-Cell Therapies and Potential Solutions*

Although CAR-T-cell therapy has revolutionized cancer treatment, several limitations hinder its broader application. The primary mechanism by which CAR-T cells eliminate tumor cells is through antigen recognition. However, antigen escape, where malignant cells lose or downregulate target antigen expression, often leads to tumor resistance and relapse [128, 129]. For instance, while CD19-targeted CAR-T cells have shown initial success in treating relapsed/refractory acute lymphoblastic leukemia (ALL), up to 70% of patients experience disease recurrence due to antigen loss [130]. Similarly, the loss of BCMA expression has been observed in multiple myeloma patients undergoing treatment with BCMA-targeted CAR-T cells [131, 132]. Comparable mechanisms of antigen escape and resistance patterns have also been observed in solid tumors. To overcome this challenge, dual or tandem CAR constructs targeting multiple antigens, such as CD19/CD22 or BCMA/CD19 in hematological malignancies and HER2/IL13Rα2 or HER2/MUC1 in solid tumors, have shown promising results [133–136]. However, solid tumor therapy presents unique challenges, such as on-target, off-tumor effects, where CAR-T cells attack normal cells expressing the same antigen. Additionally, heterogeneity in target antigen expression, even among cells within the same tumor, makes CAR-T-cell therapy more challenging. Strategies such as targeting tumor-restricted posttranslational modifications and improving antigen specificity are being actively explored to minimize toxicity [129]. Another major obstacle is CAR-T-cell trafficking and tumor infiltration, particularly in solid tumors, where physical barriers such as the tumor stroma and an immunosuppressive microenvironment limit effectiveness of CAR-T cells [137]. Approaches like localized delivery, including intraventricular or intrapleural injectionsand engineering CAR-T cells to express chemokine receptors or enzymes such as heparanase, to degrade the tumor stroma, have shown potential in enhancing tumor penetration [129]. The immunosuppressive tumor microenvironment (TME) further complicates CAR-T-cell therapy by promoting T-cell exhaustion and reducing persistence [138]. Combination therapies, such as pairing CAR-T cells with immune checkpoint inhibitors (e.g., anti-PDL1), have demonstrated improved outcomes in preclinical studies [139, 140]. Additionally, innovative strategies such as engineering "armored CARs" to secrete stimulatory cytokines could help overcome these barriers [141, 142]. Toxicities, including cytokine release syndrome (CRS), hemophagocytic lymphohistiocytosis/macrophage activation syndrome (HLH/MAS), and immune effector cell–associated neurotoxicity syndrome (ICANS), limit the widespread use of CAR-T cells. These toxicities arise from excessive cytokine release and immune activation, often necessitating interventions such as IL-6 inhibitors or corticosteroids [143–145]. To mitigate these risks, next-generation CAR-T-cell engineering focuses on optimizing antigen-binding affinity, modulating activation thresholds, and incorporating safety switches to balance efficacy and safety [38, 95, 146, 147]. These advancements are crucial for improving CAR-T-cell function and persistence, reducing off-target effects, extending the therapeutic potential of CAR-T-cell therapies to more patients and cancer types.

6 Conclusion and Future Perspectives

The integration of adoptive cell transfer with gene engineering has revolutionized cancer immunotherapy. CAR-T cell therapy represents a major milestone in personalized medicine, offering precise, potent, and persistent tumor targeting and reshaping the therapeutic landscape for cancer. CAR-T cell therapy represents a major milestone in personalized medicine, offering precise, potent, and persistent tumor targeting and reshaping the therapeutic landscape for cancer. Despite significant progress, continuous refinements in CAR designs remains necessary to further expand their therapeutic applications, offering a promising avenue for personalized cancer treatment. Over the years, CAR designs have undergone substantial advancements to enhance their efficacy and safety. While the ectodomains and transmembrane regions have remained relatively consistent since the inception of CAR-T cells, there have been major innovations in the evolution of the endodomains. Five generations of CAR-T cells have now been identified, each incorporating unique strategies to optimize T-cell activation and proliferation in response to specific antigens. Among the most promising developments is the integration of molecular switches that enable precise control over CAR activation or degradation through chemical signals such as small molecules or drugs. These innovations aim to improve both the efficacy and safety of CAR-T-cell therapy, broadening its applications in anticancer treatment. However, no universally optimal CAR design has been established, as even minor modifications can significantly impact therapeutic outcomes. While CAR-T-cell therapy has demonstrated remarkable clinical efficacy and has become a standard of care for relapsed or refractory hematologic malignancies, its full therapeutic potential remains largely untapped—particularly in the treatment of solid tumors. Key challenges include antigen loss leading to disease relapse, limited persistence of CAR-T cells, tumor immune escape mechanisms, and the immunosuppressive tumor microenvironment (TME). Moreover, severe toxicities, including on-target, off-tumor effects and cytokine release syndrome, remain significant concerns that need to be addressed. To overcome these limitations, researchers are developing novel strategies such as multi-specific CARs that target multiple antigens, allogeneic and induced pluripotent stem cell (iPSC)-derived CAR-T cells, and combination therapies that integrate checkpoint inhibitors or oncolytic viruses. A major bottleneck in the field is the lack of preclinical models that accurately recapitulate the complexities of the human immune system and TME. Current mouse models do not fully represent human cancer phenotypes, limiting the predictive power of preclinical studies. While humanized models incorporating additional immune components offer some improvements, they are still in trial stage and remain imperfect for studying critical aspects such as on-target, off-tumor toxicities. Therefore, developing more physiologically relevant preclinical in vivo models is crucial for advancing CAR-T-cell therapies and ensuring their safety and effectiveness. Looking ahead, in vivo CAR-T-cell platforms represent a groundbreaking shift by enabling direct engineering of T cells within patients through viral or nonviral delivery systems, such as lipid nanoparticles (LNPs) or mRNA-based gene transfer. This approach eliminates the need for complex ex vivo manufacturing, reduces costs, and may potentially enhance T-cell fitness. However, challenges such as precise targeting, efficient transduction, and minimization of off-target effects must be addressed through the development of more advanced gene delivery tools. As the field continues to evolve, efforts are underway to deepen our mechanistic understanding of CAR signaling and explore novel engineering strategies to enhance efficacy and safety. While most CAR T therapies in clinical use are autologous, efforts are underway to develop allogeneic, "off-the-shelf" CAR T cells derived from healthy donors. These approaches aim to improve the scalability, reduce production time, and enhance the accessibility of CAR T therapies. Allogeneic products, if proven safe and effective, represent a promising advancement in making this powerful treatment modality more broadly available. Moving forward, CAR-T-cell therapy is poised to expand beyond hematologic malignancies into solid tumors and other diseases, including autoimmune disorders. The next generation of CAR-T cells will likely incorporate improved design principles guided by computational modeling and structure-based predictions, which may reduce the need for exhaustive empirical testing. With ongoing advancements, CAR-Tl therapy is expected to become a central pillar of oncology, offering durable and potentially curative treatment options for a broader range of patients.

Acknowledgments The authors wish to acknowledge the invaluable contributions of numerous researchers in the fields of immunotherapy and CAR-T-cell therapies, whose work, whose work, due to space limitations, could not be cited in this chapter. Their pioneering efforts have significantly advanced our understanding of cellular immunotherapy. The authors also gratefully acknowledge the support provided by UMass Chan Medical School. Additionally, the authors acknowledge the assistance of ChatGPT (OpenAI) for its contribution to language editing and the enhancement of the manuscript's clarity.

Declaration of Interests The authors declare that they have no competing interests.

References

1. Ho WY, Blattman JN, Dossett ML, Yee C, Greenberg PD. Adoptive immunotherapy: engineering T cell responses as biologic weapons for tumor mass destruction. Cancer Cell. 2003;3(5):431–7.
2. Sadelain M, Rivière I, Brentjens R. Targeting tumours with genetically enhanced T lymphocytes. Nat Rev Cancer. 2003;3(1):35–45.
3. June CH, O'Connor RS, Kawalekar OU, Ghassemi S, Milone MC. CAR T cell immunotherapy for human cancer. Science. 2018;359(6382):1361–5.
4. Davenport AJ, Jenkins MR, Cross RS, Yong CS, Prince HM, Ritchie DS, et al. CAR-T cells inflict sequential killing of multiple tumor target cells. Cancer Immunol Res. 2015;3(5):483–94.
5. Stephan MT, Ponomarev V, Brentjens RJ, Chang AH, Dobrenkov KV, Heller G, et al. T cell-encoded CD80 and 4-1BBL induce auto- and transcostimulation, resulting in potent tumor rejection. Nat Med. 2007;13(12):1440–9.
6. June CH, Sadelain M. Chimeric antigen receptor therapy. N Engl J Med. 2018;379(1):64–73.
7. Huppa JB, Davis MM. T-cell-antigen recognition and the immunological synapse. Nat Rev Immunol. 2003;3(12):973–83.
8. Krogsgaard M, Huppa JB, Purbhoo MA, Davis MM. Linking molecular and cellular events in T-cell activation and synapse formation. Semin Immunol. 2003;15(6):307–15.
9. Weigelin B, Friedl P. T cell-mediated additive cytotoxicity – death by multiple bullets. Trends Cancer. 2022;8(12):980–7.
10. Stinchcombe JC, Majorovits E, Bossi G, Fuller S, Griffiths GM. Centrosome polarization delivers secretory granules to the immunological synapse. Nature. 2006;443(7110):462–5.
11. Martínez-Lostao L, Anel A, Pardo J. How do cytotoxic lymphocytes kill cancer cells? Clin Cancer Res. 2015;21(22):5047–56.
12. Seki N, Brooks AD, Carter CR, Back TC, Parsoneault EM, Smyth MJ, et al. Tumor-specific CTL kill murine renal cancer cells using both perforin and Fas ligand-mediated lysis in vitro, but cause tumor regression in vivo in the absence of perforin. J Immunol. 2002;168(7):3484–92.
13. Maus MV, June CH. Making better chimeric antigen receptors for adoptive T-cell therapy. Clin Cancer Res. 2016;22(8):1875–84.
14. Wang Z, Guo Y, Han W. Current status and perspectives of chimeric antigen receptor modified T cells for cancer treatment. Protein Cell. 2017;8(12):896–925.
15. Lim WA, June CH. The principles of engineering immune cells to treat cancer. Cell. 2017;168(4):724–40.
16. Xiong W, Chen Y, Kang X, Chen Z, Zheng P, Hsu YH, et al. Immunological synapse predicts effectiveness of chimeric antigen receptor cells. Mol Ther. 2018;26(4):963–75.
17. Benmebarek MR, Karches CH, Cadilha BL, Lesch S, Endres S, Kobold S. Killing mechanisms of chimeric antigen receptor (CAR) T cells. Int J Mol Sci. 2019;20(6):1283.

18. Korell F, Berger TR, Maus MV. Understanding CAR T cell-tumor interactions: paving the way for successful clinical outcomes. Med. 2022;3(8):538–64.
19. Neeson P, Shin A, Tainton KM, Guru P, Prince HM, Harrison SJ, et al. Ex vivo culture of chimeric antigen receptor T cells generates functional CD8+ T cells with effector and central memory-like phenotype. Gene Ther. 2010;17(9):1105–16.
20. Liadi I, Singh H, Romain G, Rey-Villamizar N, Merouane A, Adolacion JR, et al. Individual motile CD4(+) T cells can participate in efficient multikilling through conjugation to multiple tumor cells. Cancer Immunol Res. 2015;3(5):473–82.
21. Hombach A, Köhler H, Rappl G, Abken H. Human CD4+ T cells lyse target cells via granzyme/perforin upon circumvention of MHC class II restriction by an antibody-like immunoreceptor. J Immunol. 2006;177(8):5668–75.
22. Liu B, Yan L, Zhou M. Target selection of CAR T cell therapy in accordance with the TME for solid tumors. Am J Cancer Res. 2019;9(2):228–41.
23. Chong C, Coukos G, Bassani-Sternberg M. Identification of tumor antigens with immunopeptidomics. Nat Biotechnol. 2022;40(2):175–88.
24. Laumont CM, Vincent K, Hesnard L, Audemard É, Bonneil É, Laverdure JP, et al. Noncoding regions are the main source of targetable tumor-specific antigens. Sci Transl Med. 2018;10(470):eaau5516.
25. DeLucia DC, Lee JK. Identification of cell surface targets for CAR T cell immunotherapy. Methods Mol Biol. 2020;2097:45–54.
26. Xie N, Shen G, Gao W, Huang Z, Huang C, Fu L. Neoantigens: promising targets for cancer therapy. Signal Transduct Target Ther. 2023;8(1):9.
27. Abbott R, Alassadi O, Davies M, Essen MV, Freire-Benéitez V, Jefferson V, et al. 312 Identification of novel NSC lung cancer Dark Antigens™ with expression in multiple tumor types, as promising targets for immunotherapies. J Immunother Cancer. 2022;10(Suppl 2):A328.
28. Hulen TM, Crowther MD, Khan S, Schina A, Howie D, Met Ö, et al. 363 Identification of dark antigen® reactive T cell receptors from tumor-infiltrating lymphocytes. J Immunother Cancer. 2023;11(Suppl 1):A412.
29. Köhler G, Milstein C. Continuous cultures of fused cells secreting antibody of predefined specificity. Nature. 1975;256(5517):495–7.
30. Daigo K, Sugita S, Mochizuki Y, Iwanari H, Hiraishi K, Miyano K, et al. A simple hybridoma screening method for high-affinity monoclonal antibodies using the signal ratio obtained from time-resolved fluorescence assay. Anal Biochem. 2006;351(2):219–28.
31. André AS, Moutinho I, Dias JNR, Aires-da-Silva F. In vivo phage display: a promising selection strategy for the improvement of antibody targeting and drug delivery properties. Front Microbiol. 2022;13:962124.
32. Chung DH, Kong S, Young NJ, Chuo SW, Shiah JV, Connelly EJ, et al. Rare antibody phage isolation and discrimination (RAPID) biopanning enables identification of high-affinity antibodies against challenging targets. Commun Biol. 2023;6(1):1036.
33. Schofield DJ, Pope AR, Clementel V, Buckell J, Chapple S, Clarke KF, et al. Application of phage display to high throughput antibody generation and characterization. Genome Biol. 2007;8(11):R254.
34. Rad SMA, Poudel A, Tan GMY, McLellan AD. Promoter choice: who should drive the CAR in T cells? PLoS One. 2020;15(7):e0232915.
35. Ho JY, Wang L, Liu Y, Ba M, Yang J, Zhang X, et al. Promoter usage regulating the surface density of CAR molecules may modulate the kinetics of CAR-T cells in vivo. Mol Ther Methods Clin Dev. 2021;21:237–46.
36. Qin DY, Huang Y, Li D, Wang YS, Wang W, Wei YQ. Paralleled comparison of vectors for the generation of CAR-T cells. Anti-Cancer Drugs. 2016;27(8):711–22.
37. Jadlowsky JK, Hexner EO, Marshall A, Grupp SA, Frey NV, Riley JL, et al. Long-term safety of lentiviral or gammaretroviral gene-modified T cell therapies. Nat Med. 2025. https://doi.org/10.1038/s41591-024-03478-6.

38. Rossi M, Breman E. Engineering strategies to safely drive CAR T-cells into the future. Front Immunol. 2024;15:1411393.
39. Si X, Xiao L, Brown CE, Wang D. Preclinical evaluation of CAR T cell function: in vitro and in vivo models. Int J Mol Sci. 2022;23(6):3154.
40. Nouri Y, Weinkove R, Perret R. An in vitro model to assess CRS potential of CAR T cells using a tumor cell line and autologous monocytes. Curr Protoc. 2023;3(8):e864.
41. Piccinini C, Carloni S, Arienti C, Pancisi E, Fanini F, Pignatta S, et al. In vitro CAR-T cell killing: validation of the potency assay. Cancer Immunol Immunother. 2024;73(9):168.
42. Ahmed EN, Cutmore LC, Marshall JF. Syngeneic mouse models for pre-clinical evaluation of CAR T cells. Cancers (Basel). 2024;16(18):3186.
43. Duncan BB, Dunbar CE, Ishii K. Applying a clinical lens to animal models of CAR-T cell therapies. Mol Ther Methods Clin Dev. 2022;27:17–31.
44. Wu Y, Yu XZ. Modelling CAR-T therapy in humanized mice. EBioMedicine. 2019;40:25–6.
45. Srivastava S, Riddell SR. Engineering CAR-T cells: design concepts. Trends Immunol. 2015;36(8):494–502.
46. Guedan S, Calderon H, Posey AD Jr, Maus MV. Engineering and design of chimeric antigen receptors. Mol Ther Methods Clin Dev. 2019;12:145–56.
47. Jayaraman J, Mellody MP, Hou AJ, Desai RP, Fung AW, Pham AHT, et al. CAR-T design: elements and their synergistic function. EBioMedicine. 2020;58:102931.
48. Xie YJ, Dougan M, Jailkhani N, Ingram J, Fang T, Kummer L, et al. Nanobody-based CAR T cells that target the tumor microenvironment inhibit the growth of solid tumors in immunocompetent mice. Proc Natl Acad Sci USA. 2019;116(16):7624–31.
49. Duan Y, Chen R, Huang Y, Meng X, Chen J, Liao C, et al. Tuning the ignition of CAR: optimizing the affinity of scFv to improve CAR-T therapy. Cell Mol Life Sci. 2021;79(1):14.
50. Flugel CL, Majzner RG, Krenciute G, Dotti G, Riddell SR, Wagner DL, et al. Overcoming on-target, off-tumour toxicity of CAR T cell therapy for solid tumours. Nat Rev Clin Oncol. 2023;20(1):49–62.
51. Zhong S, Malecek K, Johnson LA, Yu Z, Vega-Saenz de Miera E, Darvishian F, et al. T-cell receptor affinity and avidity defines antitumor response and autoimmunity in T-cell immunotherapy. Proc Natl Acad Sci USA. 2013;110(17):6973–8.
52. Oren R, Hod-Marco M, Haus-Cohen M, Thomas S, Blat D, Duvshani N, et al. Functional comparison of engineered T cells carrying a native TCR versus TCR-like antibody-based chimeric antigen receptors indicates affinity/avidity thresholds. J Immunol. 2014;193(11):5733–43.
53. Chailyan A, Marcatili P, Tramontano A. The association of heavy and light chain variable domains in antibodies: implications for antigen specificity. FEBS J. 2011;278(16):2858–66.
54. Ajina A, Maher J. Strategies to address chimeric antigen receptor tonic signaling. Mol Cancer Ther. 2018;17(9):1795–815.
55. Gil D, Schrum AG. Strategies to stabilize compact folding and minimize aggregation of antibody-based fragments. Adv Biosci Biotechnol. 2013;4(4a):73–84.
56. Qin L, Lai Y, Zhao R, Wei X, Weng J, Lai P, et al. Incorporation of a hinge domain improves the expansion of chimeric antigen receptor T cells. J Hematol Oncol. 2017;10(1):68.
57. Guest RD, Hawkins RE, Kirillova N, Cheadle EJ, Arnold J, O'Neill A, et al. The role of extracellular spacer regions in the optimal design of chimeric immune receptors: evaluation of four different scFvs and antigens. J Immunother. 2005;28(3):203–11.
58. Wilkie S, Picco G, Foster J, Davies DM, Julien S, Cooper L, et al. Retargeting of human T cells to tumor-associated MUC1: the evolution of a chimeric antigen receptor. J Immunol. 2008;180(7):4901–9.
59. Hudecek M, Sommermeyer D, Kosasih PL, Silva-Benedict A, Liu L, Rader C, et al. The nonsignaling extracellular spacer domain of chimeric antigen receptors is decisive for in vivo antitumor activity. Cancer Immunol Res. 2015;3(2):125–35.
60. Watanabe N, Bajgain P, Sukumaran S, Ansari S, Heslop HE, Rooney CM, et al. Fine-tuning the CAR spacer improves T-cell potency. Onco Targets Ther. 2016;5(12):e1253656.

61. Hombach A, Hombach AA, Abken H. Adoptive immunotherapy with genetically engineered T cells: modification of the IgG1 Fc 'spacer' domain in the extracellular moiety of chimeric antigen receptors avoids 'off-target' activation and unintended initiation of an innate immune response. Gene Ther. 2010;17(10):1206–13.
62. Almåsbak H, Walseng E, Kristian A, Myhre MR, Suso EM, Munthe LA, et al. Inclusion of an IgG1-Fc spacer abrogates efficacy of CD19 CAR T cells in a xenograft mouse model. Gene Ther. 2015;22(5):391–403.
63. Jonnalagadda M, Mardiros A, Urak R, Wang X, Hoffman LJ, Bernanke A, et al. Chimeric antigen receptors with mutated IgG4 Fc spacer avoid fc receptor binding and improve T cell persistence and antitumor efficacy. Mol Ther. 2015;23(4):757–68.
64. Abate-Daga D, Davila ML. CAR models: next-generation CAR modifications for enhanced T-cell function. Mol Ther Oncolytics. 2016;3:16014.
65. Fujiwara K, Tsunei A, Kusabuka H, Ogaki E, Tachibana M, Okada N. Hinge and transmembrane domains of chimeric antigen receptor regulate receptor expression and signaling threshold. Cells. 2020;9(5):1182.
66. Bridgeman JS, Hawkins RE, Bagley S, Blaylock M, Holland M, Gilham DE. The optimal antigen response of chimeric antigen receptors harboring the CD3zeta transmembrane domain is dependent upon incorporation of the receptor into the endogenous TCR/CD3 complex. J Immunol. 2010;184(12):6938–49.
67. Muller YD, Nguyen DP, Ferreira LMR, Ho P, Raffin C, Valencia RVB, et al. The CD28-transmembrane domain mediates chimeric antigen receptor heterodimerization with CD28. Front Immunol. 2021;12:639818.
68. Bridgeman JS, Ladell K, Sheard VE, Miners K, Hawkins RE, Price DA, et al. CD3ζ-based chimeric antigen receptors mediate T cell activation via cis- and trans-signalling mechanisms: implications for optimization of receptor structure for adoptive cell therapy. Clin Exp Immunol. 2014;175(2):258–67.
69. Courtney AH, Lo WL, Weiss A. TCR signaling: mechanisms of initiation and propagation. Trends Biochem Sci. 2018;43(2):108–23.
70. Shah K, Al-Haidari A, Sun J, Kazi JU. T cell receptor (TCR) signaling in health and disease. Signal Transduct Target Ther. 2021;6(1):412.
71. Sharpe AH. Mechanisms of costimulation. Immunol Rev. 2009;229(1):5–11.
72. Weinkove R, George P, Dasyam N, McLellan AD. Selecting co-stimulatory domains for chimeric antigen receptors: functional and clinical considerations. Clin Transl Immunology. 2019;8(5):e1049.
73. Davenport AJ, Cross RS, Watson KA, Liao Y, Shi W, Prince HM, et al. Chimeric antigen receptor T cells form nonclassical and potent immune synapses driving rapid cytotoxicity. Proc Natl Acad Sci USA. 2018;115(9):E2068–e76.
74. Karlsson H, Svensson E, Gigg C, Jarvius M, Olsson-Strömberg U, Savoldo B, et al. Evaluation of intracellular signaling downstream chimeric antigen receptors. PLoS One. 2015;10(12):e0144787.
75. Love PE, Hayes SM. ITAM-mediated signaling by the T-cell antigen receptor. Cold Spring Harb Perspect Biol. 2010;2(6):a002485.
76. Harris DT, Kranz DM. Adoptive T cell therapies: a comparison of T cell receptors and chimeric antigen receptors. Trends Pharmacol Sci. 2016;37(3):220–30.
77. Harris DT, Hager MV, Smith SN, Cai Q, Stone JD, Kruger P, et al. Comparison of T cell activities mediated by human TCRs and CARs that use the same recognition domains. J Immunol. 2018;200(3):1088–100.
78. James JR. Tuning ITAM multiplicity on T cell receptors can control potency and selectivity to ligand density. Sci Signal. 2018;11(531):eaan1088.
79. Feucht J, Sun J, Eyquem J, Ho YJ, Zhao Z, Leibold J, et al. Calibration of CAR activation potential directs alternative T cell fates and therapeutic potency. Nat Med. 2019;25(1):82–8.
80. Eshhar Z, Waks T, Gross G, Schindler DG. Specific activation and targeting of cytotoxic lymphocytes through chimeric single chains consisting of antibody-binding domains and the

gamma or zeta subunits of the immunoglobulin and T-cell receptors. Proc Natl Acad Sci USA. 1993;90(2):720–4.
81. Brocker T, Karjalainen K. Signals through T cell receptor-zeta chain alone are insufficient to prime resting T lymphocytes. J Exp Med. 1995;181(5):1653–9.
82. Thistlethwaite FC, Gilham DE, Guest RD, Rothwell DG, Pillai M, Burt DJ, et al. The clinical efficacy of first-generation carcinoembryonic antigen (CEACAM5)-specific CAR T cells is limited by poor persistence and transient pre-conditioning-dependent respiratory toxicity. Cancer Immunol Immunother. 2017;66(11):1425–36.
83. Finney HM, Lawson AD, Bebbington CR, Weir AN. Chimeric receptors providing both primary and co-stimulatory signaling in T cells from a single gene product. J Immunol. 1998;161(6):2791–7.
84. Imai C, Mihara K, Andreansky M, Nicholson IC, Pui CH, Geiger TL, et al. Chimeric receptors with 4-1BB signaling capacity provoke potent cytotoxicity against acute lymphoblastic leukemia. Leukemia. 2004;18(4):676–84.
85. van der Stegen SJ, Hamieh M, Sadelain M. The pharmacology of second-generation chimeric antigen receptors. Nat Rev Drug Discov. 2015;14(7):499–509.
86. Krause A, Guo HF, Latouche JB, Tan C, Cheung NK, Sadelain M. Antigen-dependent CD28 signaling selectively enhances survival and proliferation in genetically modified activated human primary T lymphocytes. J Exp Med. 1998;188(4):619–26.
87. Finney HM, Akbar AN, Lawson AD. Activation of resting human primary T cells with chimeric receptors: co-stimulation from CD28, inducible co-stimulator, CD134, and CD137 in series with signals from the TCR zeta chain. J Immunol. 2004;172(1):104–13.
88. Feucht J, Sadelain M. Function and evolution of the prototypic CD28ζ and 4-1BBζ chimeric antigen receptors. Immunooncol Technol. 2020;8:2–11.
89. Asmamaw Dejenie T, Tiruneh GMM, Dessie Terefe G, Tadele Admasu F, Wale Tesega W, Chekol Abebe E. Current updates on generations, approvals, and clinical trials of CAR T-cell therapy. Hum Vaccin Immunother. 2022;18(6):2114254.
90. Tomasik J, Jasiński M, Basak GW. Next generations of CAR-T cells – new therapeutic opportunities in hematology? Front Immunol. 2022;13:1034707.
91. Ramos CA, Rouce R, Robertson CS, Reyna A, Narala N, Vyas G, et al. In vivo fate and activity of second- versus third-generation CD19-specific CAR-T cells in B cell non-Hodgkin's lymphomas. Mol Ther. 2018;26(12):2727–37.
92. Chmielewski M, Abken H. TRUCKs: the fourth generation of CARs. Expert Opin Biol Ther. 2015;15(8):1145–54.
93. Tang L, Pan S, Wei X, Xu X, Wei Q. Arming CAR-T cells with cytokines and more: innovations in the fourth-generation CAR-T development. Mol Ther. 2023;31(11):3146–62.
94. Xie B, Li Z, Zhou J, Wang W. Current status and perspectives of dual-targeting chimeric antigen receptor T-cell therapy for the treatment of hematological malignancies. Cancers (Basel). 2022;14(13):3230.
95. Labanieh L, Mackall CL. CAR immune cells: design principles, resistance and the next generation. Nature. 2023;614(7949):635–48.
96. Liu D, Zhao J, Song Y. Engineering switchable and programmable universal CARs for CAR T therapy. J Hematol Oncol. 2019;12(1):69.
97. Hawkins ER, D'Souza RR, Klampatsa A. Armored CAR T-cells: the next chapter in T-cell cancer immunotherapy. Biologics. 2021;15:95–105.
98. Zhao J, Lin Q, Song Y, Liu D. Universal CARs, universal T cells, and universal CAR T cells. J Hematol Oncol. 2018;11(1):132.
99. Cappell KM, Kochenderfer JN. Long-term outcomes following CAR T cell therapy: what we know so far. Nat Rev Clin Oncol. 2023;20(6):359–71.
100. Singh N, Perazzelli J, Grupp SA, Barrett DM. Early memory phenotypes drive T cell proliferation in patients with pediatric malignancies. Sci Transl Med. 2016;8(320):320ra3.
101. Das RK, Vernau L, Grupp SA, Barrett DM. Naïve T-cell deficits at diagnosis and after chemotherapy impair cell therapy potential in pediatric cancers. Cancer Discov. 2019;9(4):492–9.

102. Elavia N, Panch SR, McManus A, Bikkani T, Szymanski J, Highfill SL, et al. Effects of starting cellular material composition on chimeric antigen receptor T-cell expansion and characteristics. Transfusion. 2019;59(5):1755–64.
103. Jallouk AP, Sengsayadeth S, Savani BN, Dholaria B, Oluwole O. Allogeneic and other innovative chimeric antigen receptor platforms. Clin Hematol Int. 2024;6(3):61–72.
104. Depil S, Duchateau P, Grupp SA, Mufti G, Poirot L. 'Off-the-shelf' allogeneic CAR T cells: development and challenges. Nat Rev Drug Discov. 2020;19(3):185–99.
105. Gross G, Waks T, Eshhar Z. Expression of immunoglobulin-T-cell receptor chimeric molecules as functional receptors with antibody-type specificity. Proc Natl Acad Sci USA. 1989;86(24):10024–8.
106. Maher J, Brentjens RJ, Gunset G, Rivière I, Sadelain M. Human T-lymphocyte cytotoxicity and proliferation directed by a single chimeric TCRzeta/CD28 receptor. Nat Biotechnol. 2002;20(1):70–5.
107. Brentjens RJ, Latouche JB, Santos E, Marti F, Gong MC, Lyddane C, et al. Eradication of systemic B-cell tumors by genetically targeted human T lymphocytes co-stimulated by CD80 and interleukin-15. Nat Med. 2003;9(3):279–86.
108. Kalos M, Levine BL, Porter DL, Katz S, Grupp SA, Bagg A, et al. T cells with chimeric antigen receptors have potent antitumor effects and can establish memory in patients with advanced leukemia. Sci Transl Med. 2011;3(95):95ra73.
109. Neelapu SS, Locke FL, Bartlett NL, Lekakis LJ, Miklos DB, Jacobson CA, et al. Axicabtagene ciloleucel CAR T-cell therapy in refractory large B-cell lymphoma. N Engl J Med. 2017;377(26):2531–44.
110. Nastoupil LJ, Jain MD, Feng L, Spiegel JY, Ghobadi A, Lin Y, et al. Standard-of-care axicabtagene ciloleucel for relapsed or refractory large B-cell lymphoma: results from the US Lymphoma CAR T Consortium. J Clin Oncol. 2020;38(27):3119–28.
111. Abramson JS, Palomba ML, Gordon LI, Lunning MA, Wang M, Arnason J, et al. Lisocabtagene maraleucel for patients with relapsed or refractory large B-cell lymphomas (TRANSCEND NHL 001): a multicentre seamless design study. Lancet. 2020;396(10254):839–52.
112. Schuster SJ, Bishop MR, Tam CS, Waller EK, Borchmann P, McGuirk JP, et al. Tisagenlecleucel in adult relapsed or refractory diffuse large B-cell lymphoma. N Engl J Med. 2019;380(1):45–56.
113. Wang M, Munoz J, Goy A, Locke FL, Jacobson CA, Hill BT, et al. KTE-X19 CAR T-cell therapy in relapsed or refractory mantle-cell lymphoma. N Engl J Med. 2020;382(14):1331–42.
114. Fowler NH, Dickinson M, Dreyling M, Martinez-Lopez J, Kolstad A, Butler J, et al. Tisagenlecleucel in adult relapsed or refractory follicular lymphoma: the phase 2 ELARA trial. Nat Med. 2022;28(2):325–32.
115. Maude SL, Laetsch TW, Buechner J, Rives S, Boyer M, Bittencourt H, et al. Tisagenlecleucel in children and young adults with B-cell lymphoblastic leukemia. N Engl J Med. 2018;378(5):439–48.
116. Pasquini MC, Hu ZH, Curran K, Laetsch T, Locke F, Rouce R, et al. Real-world evidence of tisagenlecleucel for pediatric acute lymphoblastic leukemia and non-Hodgkin lymphoma. Blood Adv. 2020;4(21):5414–24.
117. Schultz LM, Baggott C, Prabhu S, Pacenta HL, Phillips CL, Rossoff J, et al. Disease burden affects outcomes in pediatric and young adult B-cell lymphoblastic leukemia after commercial tisagenlecleucel: a pediatric real-world chimeric antigen receptor consortium report. J Clin Oncol. 2022;40(9):945–55.
118. Kamdar M, Solomon SR, Arnason J, Johnston PB, Glass B, Bachanova V, et al. Lisocabtagene maraleucel versus standard of care with salvage chemotherapy followed by autologous stem cell transplantation as second-line treatment in patients with relapsed or refractory large B-cell lymphoma (TRANSFORM): results from an interim analysis of an open-label, randomised, phase 3 trial. Lancet. 2022;399(10343):2294–308.

119. Locke FL, Miklos DB, Jacobson CA, Perales MA, Kersten MJ, Oluwole OO, et al. Axicabtagene ciloleucel as second-line therapy for large B-cell lymphoma. N Engl J Med. 2022;386(7):640–54.
120. Neelapu SS, Dickinson M, Munoz J, Ulrickson ML, Thieblemont C, Oluwole OO, et al. Axicabtagene ciloleucel as first-line therapy in high-risk large B-cell lymphoma: the phase 2 ZUMA-12 trial. Nat Med. 2022;28(4):735–42.
121. Berdeja JG, Madduri D, Usmani SZ, Jakubowiak A, Agha M, Cohen AD, et al. Ciltacabtagene autoleucel, a B-cell maturation antigen-directed chimeric antigen receptor T-cell therapy in patients with relapsed or refractory multiple myeloma (CARTITUDE-1): a phase 1b/2 open-label study. Lancet. 2021;398(10297):314–24.
122. Munshi NC, Anderson LD Jr, Shah N, Madduri D, Berdeja J, Lonial S, et al. Idecabtagene vicleucel in relapsed and refractory multiple myeloma. N Engl J Med. 2021;384(8):705–16.
123. Zhang WY, Liu Y, Wang Y, Wang CM, Yang QM, Zhu HL, et al. Long-term safety and efficacy of CART-20 cells in patients with refractory or relapsed B-cell non-Hodgkin lymphoma: 5-years follow-up results of the phase I and IIa trials. Signal Transduct Target Ther. 2017;2:17054.
124. Shah NN, Highfill SL, Shalabi H, Yates B, Jin J, Wolters PL, et al. CD4/CD8 T-cell selection affects chimeric antigen receptor (CAR) T-cell potency and toxicity: updated results from a phase I anti-CD22 CAR T-cell trial. J Clin Oncol. 2020;38(17):1938–50.
125. Lu P, Liu Y, Yang J, Zhang X, Yang X, Wang H, et al. Naturally selected CD7 CAR-T therapy without genetic manipulations for T-ALL/LBL: first-in-human phase 1 clinical trial. Blood. 2022;140(4):321–34.
126. Zhang M, Chen D, Fu X, Meng H, Nan F, Sun Z, et al. Autologous nanobody-derived fratricide-resistant CD7-CAR T-cell therapy for patients with relapsed and refractory T-cell acute lymphoblastic leukemia/lymphoma. Clin Cancer Res. 2022;28(13):2830–43.
127. Ramos CA, Grover NS, Beaven AW, Lulla PD, Wu MF, Ivanova A, et al. Anti-CD30 CAR-T cell therapy in relapsed and refractory Hodgkin lymphoma. J Clin Oncol. 2020;38(32):3794–804.
128. Majzner RG, Mackall CL. Tumor antigen escape from CAR T-cell therapy. Cancer Discov. 2018;8(10):1219–26.
129. Sterner RC, Sterner RM. CAR-T cell therapy: current limitations and potential strategies. Blood Cancer J. 2021;11(4):69.
130. Maude SL, Teachey DT, Porter DL, Grupp SA. CD19-targeted chimeric antigen receptor T-cell therapy for acute lymphoblastic leukemia. Blood. 2015;125(26):4017–23.
131. Brudno JN, Maric I, Hartman SD, Rose JJ, Wang M, Lam N, et al. T cells genetically modified to express an anti-B-cell maturation antigen chimeric antigen receptor cause remissions of poor-prognosis relapsed multiple myeloma. J Clin Oncol. 2018;36(22):2267–80.
132. Cohen AD, Garfall AL, Stadtmauer EA, Melenhorst JJ, Lacey SF, Lancaster E, et al. B cell maturation antigen-specific CAR T cells are clinically active in multiple myeloma. J Clin Invest. 2019;129(6):2210–21.
133. Spiegel JY, Patel S, Muffly L, Hossain NM, Oak J, Baird JH, et al. CAR T cells with dual targeting of CD19 and CD22 in adult patients with recurrent or refractory B cell malignancies: a phase 1 trial. Nat Med. 2021;27(8):1419–31.
134. Shi M, Wang J, Huang H, Liu D, Cheng H, Wang X, et al. Bispecific CAR T cell therapy targeting BCMA and CD19 in relapsed/refractory multiple myeloma: a phase I/II trial. Nat Commun. 2024;15(1):3371.
135. Hegde M, Mukherjee M, Grada Z, Pignata A, Landi D, Navai SA, et al. Tandem CAR T cells targeting HER2 and IL13Rα2 mitigate tumor antigen escape. J Clin Invest. 2016;126(8):3036–52.
136. Wilkie S, van Schalkwyk MC, Hobbs S, Davies DM, van der Stegen SJ, Pereira AC, et al. Dual targeting of ErbB2 and MUC1 in breast cancer using chimeric antigen receptors engineered to provide complementary signaling. J Clin Immunol. 2012;32(5):1059–70.
137. Guzman G, Reed MR, Bielamowicz K, Koss B, Rodriguez A. CAR-T therapies in solid tumors: opportunities and challenges. Curr Oncol Rep. 2023;25(5):479–89.

138. Liu Z, Zhou Z, Dang Q, Xu H, Lv J, Li H, et al. Immunosuppression in tumor immune microenvironment and its optimization from CAR-T cell therapy. Theranostics. 2022;12(14):6273–90.
139. Luo Y, Gadd ME, Qie Y, Otamendi-Lopez A, Sanchez-Garavito JE, Brooks MM, et al. Solid cancer-directed CAR T cell therapy that attacks both tumor and immunosuppressive cells via targeting PD-L1. Mol Ther Oncol. 2024;32(4):200891.
140. Wang Y, Cho JW, Kastrunes G, Buck A, Razimbaud C, Culhane AC, et al. Immune-restoring CAR-T cells display antitumor activity and reverse immunosuppressive TME in a humanized ccRCC mouse model. iScience. 2024;27(2):108879.
141. Huang M, Deng J, Gao L, Zhou J. Innovative strategies to advance CAR T cell therapy for solid tumors. Am J Cancer Res. 2020;10(7):1979–92.
142. Liu Y, Xiao L, Yang M, Chen X, Liu H, Wang Q, et al. CAR-armored-cell therapy in solid tumor treatment. J Transl Med. 2024;22(1):1076.
143. Zhou X, Rasche L, Kortüm KM, Danhof S, Hudecek M, Einsele H. Toxicities of chimeric antigen receptor T cell therapy in multiple myeloma: an overview of experience from clinical trials, pathophysiology, and management strategies. Front Immunol. 2020;11:620312.
144. Rejeski K, Jain MD, Shah NN, Perales MA, Subklewe M. Immune effector cell-associated haematotoxicity after CAR T-cell therapy: from mechanism to management. Lancet Haematol. 2024;11(6):e459–e70.
145. Lee DW, Gardner R, Porter DL, Louis CU, Ahmed N, Jensen M, et al. Current concepts in the diagnosis and management of cytokine release syndrome. Blood. 2014;124(2):188–95.
146. Neeser A, Ramasubramanian R, Wang C, Ma L. Engineering enhanced chimeric antigen receptor-T cell therapy for solid tumors. Immunooncol Technol. 2023;19:100385.
147. Uslu U, June CH. Beyond the blood: expanding CAR T cell therapy to solid tumors. Nat Biotechnol. 2024. https://doi.org/10.1038/s41587-024-02446-2.
148. Irving BA, Weiss A. The cytoplasmic domain of the T cell receptor zeta chain is sufficient to couple to receptor-associated signal transduction pathways. Cell. 1991;64(5):891–901.
149. Levine BL, Cotte J, Small CC, Carroll RG, Riley JL, Bernstein WB, et al. Large-scale production of CD4+ T cells from HIV-1-infected donors after CD3/CD28 costimulation. J Hematother. 1998;7(5):437–48.
150. Haynes NM, Trapani JA, Teng MW, Jackson JT, Cerruti L, Jane SM, et al. Single-chain antigen recognition receptors that costimulate potent rejection of established experimental tumors. Blood. 2002;100(9):3155–63.
151. Dudley ME, Wunderlich JR, Robbins PF, Yang JC, Hwu P, Schwartzentruber DJ, et al. Cancer regression and autoimmunity in patients after clonal repopulation with antitumor lymphocytes. Science. 2002;298(5594):850–4.
152. Pule MA, Savoldo B, Myers GD, Rossig C, Russell HV, Dotti G, et al. Virus-specific T cells engineered to coexpress tumor-specific receptors: persistence and antitumor activity in individuals with neuroblastoma. Nat Med. 2008;14(11):1264–70.
153. Milone MC, Fish JD, Carpenito C, Carroll RG, Binder GK, Teachey D, et al. Chimeric receptors containing CD137 signal transduction domains mediate enhanced survival of T cells and increased antileukemic efficacy in vivo. Mol Ther. 2009;17(8):1453–64.
154. Kochenderfer JN, Wilson WH, Janik JE, Dudley ME, Stetler-Stevenson M, Feldman SA, et al. Eradication of B-lineage cells and regression of lymphoma in a patient treated with autologous T cells genetically engineered to recognize CD19. Blood. 2010;116(20):4099–102.
155. Brentjens RJ, Rivière I, Park JH, Davila ML, Wang X, Stefanski J, et al. Safety and persistence of adoptively transferred autologous CD19-targeted T cells in patients with relapsed or chemotherapy refractory B-cell leukemias. Blood. 2011;118(18):4817–28.
156. Savoldo B, Ramos CA, Liu E, Mims MP, Keating MJ, Carrum G, et al. CD28 costimulation improves expansion and persistence of chimeric antigen receptor-modified T cells in lymphoma patients. J Clin Invest. 2011;121(5):1822–6.
157. Porter DL, Levine BL, Kalos M, Bagg A, June CH. Chimeric antigen receptor-modified T cells in chronic lymphoid leukemia. N Engl J Med. 2011;365(8):725–33.

CAR-T-Cell Therapy: Success and Advancements in the Treatment of Hematological Malignancies

Mahesh Hegde

Abstract Chimeric antigen receptor (CAR) T-cell therapies have garnered attention as promising approaches for treating hematological malignancies. The remarkable success in relapsed/refractory (r/r) B-cell malignancies, specifically leukemia, lymphoma, and multiple myeloma (MM), has propelled CAR-T-cell therapy into the spotlight. Targeting molecules such as CD19 and B-cell maturation antigen (BCMA) has led to durable complete remission (CR) and a notably high objective response rate (ORR) in certain hematological malignancies. These unprecedented outcomes have spurred exploration of novel therapeutic targets for immunotherapy applications. Despite positive results in B-cell lymphoma and acute lymphoblastic leukemia (ALL), the treatment of other malignant blood disorders, such as acute myeloid leukemia (AML), remains challenging. Furthermore, the broader application of CAR-T-cell therapy is hindered by adverse effects, including cytokine release syndrome (CRS), cytopenia, infections, and nervous system complications. Addressing these challenges is crucial for expanding the use of CAR-T-cell therapy for various hematological malignancies. The present chapter delves into existing CAR-T-cell therapies designed for hematological malignancies and explores potential targets for CAR-T-cell therapy in the context of hematological malignancies.

Keywords Immunotherapy · CAR-T cells · Leukemia · Lymphoma · Myeloma · Hematology

Abbreviations

ALL	Acute lymphocytic/lymphoblastic leukemia
AML	Acute myeloid leukemia

M. Hegde (✉)
Department of Molecular, Cell and Cancer Biology, UMass Chan Medical School, Worcester, MA, USA
e-mail: mahesh.hegde@umassmed.edu

A. K. Mishra, S. K. Malonia (eds.), *Cell-based Immunotherapies for Cancer*,
https://doi.org/10.1007/978-3-031-88695-9_5

BCMA	B-cell maturation antigen
BCR	B-cell receptor
CAR	Chimeric antigen receptor
CLL	Chronic lymphocytic leukemia
CR	Complete remission
CRS	Cytokine release syndrome
DLBCL	Diffuse large B-cell lymphoma
FDA	Food and Drug Administration
FL	Follicular lymphoma
HGBL	High-grade B-cell lymphoma
HL	Hodgkin lymphoma
HSC	Hematopoietic stem cell
LSC	Leukemic stem cell
MCL	Mantle cell lymphoma
MHC	Major histocompatibility complex
MM	Multiple myeloma
NHL	non-Hodgkin lymphoma
ORR	Objective response rate
r/r	Relapsed/refractory
TCR	T-cell receptor
WHO	World Health Organization

1 Introduction

With advances in tumor immunology, a new era in cellular immunotherapy has emerged. Historically, hematological malignancies have typically been addressed through a combination of chemotherapy, radiotherapy, and hematopoietic stem cell (HSC) transplantation. Nevertheless, recent advancements have led to the development of immune-targeted therapies, such as monoclonal antibodies, bispecific antibodies, antibody–drug conjugates (ADCs), and chimeric antigen receptor (CAR) T-cell therapy. Allogeneic hematopoietic stem cell transplantation, an ancient form of cellular immunotherapy, provides the theoretical foundation for harnessing the immune system against malignancies. Its underlying mechanism, graft-versus-leukemia or graft-versus-tumor effects, where donor immune cells target abnormal cells within the patient's body, lays the groundwork for future innovations [1]. This progress was further propelled by genetic engineering, enabling the modification of T-cell receptors (TCRs) to target specific tumor antigens, forming the rational basis for CAR-T-cell therapy.

T lymphocytes (T cells) are crucial components of cell-mediated immune responses and are pivotal in monitoring and eradicating tumor or premalignant cells [2]. When T-cell activity is hindered, cancer can develop. Recognizing the ability of many cancer types to suppress anticancer immune responses, researchers have pioneered immunotherapeutic strategies aimed at bolstering a patient's immune system against cancer cells. T-cell-based adoptive immunotherapy represents a method to modify and divert T cells against malignant cells. An innovative approach within this paradigm is CAR-T-cell therapy, a recent treatment option that involves altering a patient's T cells through the introduction of a CAR construct. These engineered cells are reintroduced into the patient's bloodstream, where they initiate an assault against cancer cells [3].

The CAR itself is a hybrid protein that replaces the natural T-cell receptor. It consists of several components: the extracellular domain, which is typically derived from a monoclonal antibody and recognizes a specific antigen, is coupled via a flexible linker. The hinge region serves as the connection between the extracellular domain and the transmembrane domain, providing conformational flexibility. Within the intracellular domain, there is a co-stimulatory domain that mimics the co-stimulatory signal necessary for complete T-cell activation [4]. Scientists have developed approaches to improve the tumor recognition capabilities of adoptively stimulated cells, going beyond the conventional TCR/major histocompatibility complex (MHC)/peptide recognition [5].

Genetic engineering has given rise to these novel receptors, CARs, which recognize cancer-associated antigens, activate T cells, promote proliferation, and form a cellular memory. CAR constructs are hybrid molecules capable of substituting major functions of the TCR, including surface antigen recognition and the activation of T cells [6]. Early positive outcomes indicate that T cells engineered with CAR constructs tailored to target specific antigens effectively express CAR and display antigen-specific cytotoxicity [7]. Activated CAR-T cells perform targeted destruction of cancer cells. An inherent advantage lies in their recognition, which is not restricted to the MHC. CAR-T-cell therapy has been applied predominantly in the field of hematological malignancies, including acute lymphoblastic leukemia (ALL), chronic lymphocytic leukemia (CLL), and multiple myeloma (MM). These malignancies are more accessible for targeting because of the availability of suitable tumor antigens, marking a significant step forward in the realm of cancer immunotherapy [8–10].

CAR-T-cell therapy has emerged as a promising treatment avenue for patients diagnosed with non-Hodgkin lymphoma (NHL) that has either relapsed or proven refractory to conventional therapies [11]. In clinical trials, anti-CD19 CAR infusion exhibited high response rates in patients with B-cell malignancies, including those with conditions such as diffuse large B-cell lymphoma (DLBCL) and B-ALL [12]. Aggressive B-cell lymphoma manifests as a complex and diverse entity, exhibiting variations in both clinical presentation and biological characteristics. These cases are classified as high-grade B-cell lymphoma (HGBL) with MYC/BCL2/BCL6 rearrangements. This distinct categorization of HGBL differs from that of DLBCL and Burkitt lymphoma (BL) because of differing biological and clinical characteristics [13]. Efforts to improve outcomes for HGBL patients have included diverse strategies, such as high-dose chemotherapy (HDT) followed by autologous stem cell transplantation (ASCT), along with allogeneic transplant procedures in younger patients. Despite these endeavors, the survival rates for these individuals continue to fall below the desired levels. The approval of CAR-T-cell therapy for HGBL holds promise for patients, signifying a potential breakthrough in treatment options. Additionally, the Food and Drug Administration (FDA) has approved CAR-T-cell therapy for relapsed/refractory (r/r) mantle cell lymphoma (MCL) and r/r follicular lymphoma (FL). These approvals have demonstrated promising outcomes for patients, expanding the scope of effective treatments for these specific lymphoma subtypes.

CAR-T-cell therapy has revolutionized cancer immunotherapy, standing as a landmark alongside immune checkpoint inhibitors such as programmed death-1 (PD-1) and cytotoxic T lymphocyte-associated protein-4 (CTLA-4) blockade in combination therapies [14]. Pediatric patients often face limited options with immune checkpoint inhibitors due to the requirements of Programmed Death-Ligand 1 (PD-L1) positivity or increased tumor mutational burden. Consequently, checkpoint blockade has demonstrated restricted efficacy in treating pediatric cancers. Considering these challenges, combination therapy involving CAR-T cells has emerged as a promising and helpful approach for enhancing treatment outcomes in pediatric cancer patients [15].

The success of CAR-T-cell therapy crucially hinges on the careful selection of an optimal target that is uniformly expressed on tumor cells but is absent in healthy vital cells. CD19 is promising because of its expression in B-cell malignancies, although clinical trials have reported a 30% relapse rate, possibly related to CAR-T-cell persistence issues and antigen escape mechanisms. Antigen escape, a significant challenge, occurs not only in non-B-cell malignancies but also in solid tumors. Additionally, on-target, off-tumor toxicities pose a risk, varying across different targeted antigens. Despite these hurdles, targeting CD19 has shown promise, but the pursuit of multi-antigen targeting strategies has emerged as a potential solution. These challenges have spurred the exploration of novel targets for CAR-T-cell therapy, driving ongoing advancements in the field [16].

In this comprehensive chapter, the objective is to provide an up-to-date overview of approved CAR-T-cell treatment options specifically tailored for hematological malignancies and to summarize the potential targets for CAR-T-cell therapies currently under exploration. These advancements hold the potential to significantly reshape the landscape of autologous T-cell immunotherapy, heralding transformative changes in the field of cancer treatment.

2 CAR-T-Cell Therapy for Hematological Malignancies

The majority of CAR-T-cell therapy techniques have been applied to treat hematological malignancies, especially B-cell malignancies, where the first CAR-T-cell therapy was approved by the FDA (Fig. 1 and Table 1). The primary reason for its success in patients with hematological malignancies can be attributed to several factors.

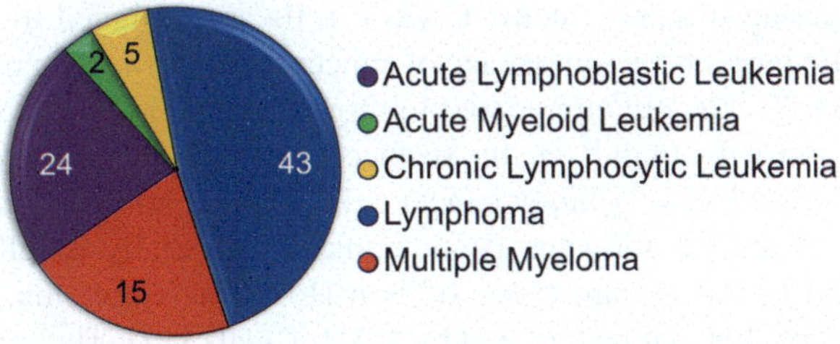

Fig. 1 A pie chart illustrating the clinical trials conducted on patients with hematological malignancies treated with CAR-T-cell therapy. The data were obtained from completed studies listed on clinicaltrials.gov and categorized according to clinical conditions (*n* = 89)

Table 1 FDA-approved CAR-T-cell therapies for hematological malignancies

Product name	Target	Clinical trial	Disease type	Month and year	References
Kymriah	CD19	ELIANA (NCT02435849)	R/R B-cell acute lymphoblastic leukemia	August 2017	[19]
Yescarta	CD19	ZUMA-1 (NCT02348216)	R/R large B-cell lymphoma	October 2017	[20]
Kymriah	CD19	JULIET (NCT02445248)	R/R large B-cell lymphoma	May 2018	[21]
Tecartus	CD19	ZUMA-2 (NCT02601313)	R/R mantle cell lymphoma	July 2020	[22]
Breyanzi	CD19	TRANSCEND NHL 001 (NCT02631044)	R/R diffuse large B-cell lymphoma	February 2021	[23]
Yescarta	CD19	ZUMA-5 (NCT03105336)	R/R follicular lymphoma	March 2021	[24]
Abecma	BCMA	KarMMa trial (NCT03361748)	R/R multiple myeloma	March 2021	[25]
Tecartus	CD19	ZUMA-3 (NCT02614066)	R/R B-cell acute lymphoblastic leukemia	October 2021	[26]
Carvykti	BCMA	CARTITUDE-1 phase 1b/2 trial (NCT03548207)	R/R multiple myeloma	February 2022	[27]
Yescarta	CD19	ZUMA-7 (NCT03391466)	R/R diffuse large B-cell lymphoma	April 2022	[28]
Kymriah	CD19	ELARA (NCT03568461)	R/R follicular lymphoma	May 2022	[29]
Breyanzi	CD19	PILOT trial (NCT03483103) and TRANSFORM trial (NCT03575351)	R/R diffuse large B-cell lymphoma	June 2022	[30]

1. Specific targeting of cancer cells: CAR-T cells are designed to recognize and target specific proteins on the surface of cancer cells. In the case of B-cell malignancies, CAR-T cells are engineered to target specific antigens, such as CD19, which is commonly found on the surface of B cells. This specificity allows CAR-T cells to selectively target cancer cells while sparing normal cells.
2. Availability of unique antigens: Blood cancers, especially B-cell malignancies such as B-ALL and certain types of non-Hodgkin lymphoma, often express unique antigens that can be targeted by CAR-T cells [17]. The presence of such distinct antigens on cancer cells enhances the specificity of CAR-T-cell therapy.
3. Rapid and potent immune response: CAR-T cells are engineered to rapidly multiply and mount a potent immune response against cancer cells. This is particularly advantageous in the treatment of hematological malignancies, where the ability to quickly target and eliminate circulating cancer cells is crucial.
4. Immunosuppressive microenvironment: Solid tumors often create an immunosuppressive microenvironment that can hinder the effectiveness of immune responses [18]. In contrast, blood cancers are typically associated with a more accessible and less immunosuppressive environment, making it easier for CAR-T cells to infiltrate and attack malignant cells.
5. Established infrastructure for blood-based therapies: The infrastructure for blood-based therapies, such as apheresis and leukapheresis (processes used to collect blood components), is well-established. This facilitates the extraction of T cells from patients, their genetic modification to express CARs, and the subsequent reinfusion of engineered CAR-T cells. Moreover, monitoring treatment efficiency is easier in patients with hematological malignancies than in those with solid tumors.
6. Clinical trial success: Initial success in clinical trials with B-cell malignancies paved the way for the approval of CAR-T-cell therapies targeting CD19, such as Kymriah and Yescarta. This success has contributed to the continued exploration and development of CAR-T-cell therapies for other types of blood cancers.

In summary, the established infrastructure for blood-based therapies streamlines various aspects of CAR-T-cell therapy development and administration, from cell collection to patient monitoring. This preexisting infrastructure contributes to the efficiency, safety, and accessibility of CAR-T-cell therapy for blood cancer patients.

2.1 *CAR-T-Cell Therapy for Acute Lymphoblastic Leukemia (ALL)*

ALL is a swiftly progressing hematological malignancy characterized by the uncontrolled proliferation of immature lymphoid cells in the bone marrow and peripheral blood. While ALL predominantly affects children, it can also impact adults. The understanding of the molecular and genetic underpinnings of ALL has significantly shaped diagnostic and therapeutic approaches. Genetic mutations in lymphoid precursor cells instigate their transformation into leukemic blasts, bypassing normal maturation and causing bone marrow failure. The identification of specific genetic abnormalities has become integral for risk stratification and treatment planning in ALL [31].

Over 75% of ALL cases originate from precursors of the B-cell lineage, whereas the remaining cases involve malignant T-cell precursors. The World Health Organization (WHO) classifies ALL on the basis of immunophenotype, cytogenetics, and molecular features, aiding in prognosis prediction and treatment customization. Subtypes encompass B-cell ALL and T-cell ALL, each distinguished by unique clinical and biological characteristics [32]. Continual research in ALL aims to enhance risk stratification, pinpoint novel therapeutic targets, and mitigate treatment-related toxicities. Ongoing clinical trials exploring new agents, combination therapies, and immunotherapeutic strategies have contributed to the constant refinement of treatment outcomes for ALL patients. Stem cell transplantation, whether autologous (utilizing the patient's own stem cells) or allogeneic (employing donor stem cells), plays a role in consolidating treatment for high-risk or relapsed cases. The decision to undergo transplantation hinges on factors such as age, comorbidities, and the availability of a suitable donor.

The landscape of ALL treatment has evolved, with combination chemotherapy remaining pivotal. This involves the administration of cytotoxic agents such as vincristine, daunorubicin, and asparaginase to eliminate leukemic cells and induce remission. Corticosteroids, such as prednisone, are standard components of induction therapy. Molecular diagnostics have identified specific genetic alterations with prognostic significance. For example, the Philadelphia chromosome resulting from the *BCR-ABL1* fusion gene indicates a poorer prognosis, but tyrosine kinase inhibitors such as imatinib have improved outcomes for patients with this mutation.

Immunotherapy has emerged as a promising avenue for ALL treatment. Monoclonal antibodies, including rituximab and inotuzumab ozogamicin, target specific antigens on leukemic cells, enhancing the ability of the immune system to recognize and eliminate them [33]. Chimeric antigen receptor T-cell therapy (CAR-T) has demonstrated remarkable success, particularly in r/r B-ALL. CAR-T cells are engineered to express receptors that target CD19, a surface antigen on B cells, facilitating the destruction of leukemic cells [19]. The clinical investigations leading to the approval of Kymriah demonstrated exceptional efficacy, marking a transformative shift in the landscape of cancer treatment. The initial approval by the FDA in August 2017 signified a milestone for CAR-T-cell therapy, specifically acknowledging the 4-1BB-based CD19 CAR-T-cell therapy Kymriah (CTL019/tisagenlecleucel from Novartis, Basel, Switzerland). This approval pertained to its application in patients up to 25 years of age with ALL who are refractory to treatment or who are experiencing a second or subsequent relapse. By targeting CD19, a B-cell surface marker, Kymriah induced well-tolerated B-cell aplasia as an off-tumor effect [34].

In a pivotal study involving patients with r/r B-ALL, a noteworthy percentage achieved complete remission (CR) following Kymriah administration. This therapeutic approach has the capacity to elicit enduring responses, instilling optimism in patients whose conventional treatment options are depleted. The clinical trial, registered under NCT02435849, included individuals aged 3–23 years ($n = 75$) without prior anti-CD19 therapy, where non-leukopenia patients (72/75) received a lymphodepletion regimen before a dose of 2.0–5.0 × 10^6 Kymriah per kg body weight. The findings revealed an overall remission rate of 81% (95% CI: 71–89) within 3 months of treatment, and all patients who responded to the treatment were determined to be negative for minimal residual disease [19].

Despite the unprecedented success of Kymriah, its implementation is not devoid of challenges. Therapy can induce severe side effects, including cytokine release syndrome (CRS) and neurological events, necessitating meticulous management. The medical community is actively refining protocols and treatment strategies to mitigate these adverse effects and enhance patient outcomes. Following its approval, multiple trials have been initiated, investigating various CAR constructs, stimulatory and co-stimulatory domains, and modified manufacturing processes. In October 2021, the FDA granted approval to Tecartus (brexucabtagene autoleucel), a CD19-based CAR-T-cell therapy, for treating r/r B-ALL. This approval was grounded in the results of the multicenter phase 2 ZUMA-3 study (NCT02614066), which assessed the efficacy and safety of Tecartus in adult patients with the specified conditions. In this study, 55 adult patients (with a median age of 40 years) exhibiting more than 5% blast cells in the bone marrow received a single Tecartus infusion at a dose of 1 × 10^6 CAR-T cells per kg body weight. Notably, the results, assessed at a median follow-up of 16.4 months, demonstrated a CR rate of 56%, with a median duration of remission of 12.8 months and a median overall survival of 18.2 months. The most commonly observed adverse events included anemia (49%), pyrexia (36%), infections (25%), cytokine release syndrome (24%), and neurological events (25%). Overall, Tecartus treatment resulted in a high rate of CR or CR with incomplete hematological recovery in adult patients with r/r B-ALL, maintaining a manageable safety profile [26]. Tecartus is the first FDA-approved CAR-T-cell therapy for MCL and the third anti-CD19-based chimeric antigen receptor T-cell therapy. While its potential therapeutic application in r/r B-ALL is evident, further clinical research is warranted to explore its suitability for patients with T-ALL.

2.2 *CAR-T-Cell Therapy for Lymphoma*

Lymphoma, a diverse group of hematological malignancies, originates in the lymphatic system and is a vital component of the immune system. This category includes Hodgkin lymphoma (HL) and non-Hodgkin lymphoma (NHL), each exhibiting distinct clinical, histological, and genetic features. HL is characterized by Reed–Sternberg cells and abnormal B lymphocytes, contributing to the distinctive histological pattern observed. In contrast, non-Hodgkin lymphomas comprise various subtypes originating from B cells, T cells, or natural killer (NK) cells, increasing the complexity of diagnostic and treatment approaches.

Understanding the molecular and genetic underpinnings of lymphomas has played a pivotal role in advancing therapeutic strategies. The WHO classification provides a comprehensive framework for categorizing lymphoid neoplasms on the basis of genetic and molecular characteristics, facilitating tailored treatment strategies and predicting disease prognosis [35]. The landscape of lymphoma treatment is continually evolving, with ongoing research aimed at identifying novel therapeutic targets and refining treatment approaches. Research efforts have focused on unraveling the intricate genetic and molecular pathways driving lymphomagenesis, enabling the development of more precise and effective therapies. Research has also explored the role of stem cell transplantation in lymphoma treatment. Autologous stem cell transplantation, which uses a patient's own stem cells, has been employed to manage certain lymphomas, especially in r/r cases. Allogeneic stem cell transplantation, involving donor stem cells, is another approach, albeit with challenges such as graft-versus-host disease.

In recent years, substantial progress has been made in developing targeted therapies for lymphomas. Monoclonal antibodies such as rituximab have revolutionized the treatment of B-cell lymphomas by specifically targeting CD20-expressing cells [36]. Immunomodulatory agents, such as lenalidomide, have demonstrated efficacy in specific lymphoma subtypes, suggesting alternative therapeutic options [37]. CAR-T-cell therapy has emerged as a groundbreaking modality, particularly for certain B-cell lymphomas. CAR-T-cell therapies targeting CD19, such as Kymriah and Yescarta, have shown remarkable success in clinical trials and have received approval for treating r/r B-cell lymphomas [38]. In May 2018, Kymriah received a second approval from the FDA for the treatment of adult patients with r/r large B-cell lymphoma. This regulatory green light was predicated on findings from the phase 2 JULIET clinical trial (NCT02445248), in which 115 assessable adult patients, with a median age of 56 years, underwent Kymriah treatment (5.0×10^8 CAR-T cells) after lymphodepletion chemotherapy or prior leukopenia. The overall response rate was approximately 61%, and the median time to the initial response was 29 days. Notably, there were no recorded deaths attributed to therapy-related causes [21].

In May 2022, Kymriah again garnered FDA approval for a novel indication—treating adult patients grappling with r/r FL after undergoing two or more lines of systemic therapy. This regulatory endorsement hinged on the outcomes of the global phase 2 ElARA trial (NCT03568461). Ninety-seven adult patients, with a median age of 57 years and facing r/r FL, received Kymriah following more than two lines of therapy or rejection of autologous stem cell transplant. The results revealed an ORR of 86% (95% CI, 78–92), with a CR rate of 66% (95% CI, 56–75), suggesting that Kymriah holds promise as a therapeutic option for those with r/r FL [29]. Yescarta, also known as axicabtagene ciloleucel, is an innovative chimeric antigen receptor T-cell (CAR-T) therapy based on the CD19 antigen that has demonstrated significant efficacy in the treatment of certain hematologic malignancies. Developed by Kite Pharma, a subsidiary of Gilead Sciences, Yescarta received approval from the US FDA in October 2017 for the treatment of adult patients with r/r large B-cell lymphoma after two or more lines of systemic therapy for primary mediastinal

B-cell lymphoma, high-grade B-cell lymphoma, transformed FL lymphoma, and DLBCL.

The pivotal clinical trial (ZUMA-1, NCT02348216) played a crucial role in establishing the efficacy and safety profile of Yescarta. In this trial, 111 lymphoma patients underwent lymphodepleting chemotherapy before receiving a single infusion of Yescarta (2×10^6 anti-CD19 CAR-T cells per kilogram of body weight). The results demonstrated a remarkable objective response rate (ORR) (82%) with a CR rate of 54%. This study reported that 42% of the patients continued to respond, with 40% continuing to achieve a CR, with a median follow-up of 15.4 months, and that a substantial proportion of patients survived (52% at 18 months of follow-up). However, similar to other CAR-T-cell therapies, Yescarta is associated with unique challenges and potential side effects. During the clinical study, neutropenia (78% of the patients), anemia (43%), and thrombocytopenia (38%) were observed. In addition, cytokine release syndrome (13%) and neurological toxicity (28%) were noted. Three of the patients died during treatment. The medical community has been actively refining protocols to mitigate these side effects and optimize the therapeutic benefits of Yescarta [20].

In March 2021, Yescarta received renewed FDA approval for the treatment of patients with r/r FL on the basis of findings from ZUMA-5 clinical trials (NCT03105336) [24]. This study included 148 adult patients with indolent non-Hodgkin lymphoma, categorized as FL (124) or marginal zone lymphoma [24], who experienced r/r disease or had undergone multiple lines of therapy at 15 medical centers in the USA and two in France. Yescarta was administered at a dose of 2×10^6 CAR-T cells/kg body weight. With a median follow-up of 17.5 months, the clinical results revealed an impressive ORR of 92% (95% CI 85–97) and a CR rate of 74%. Adverse effects, akin to CAR-T-cell therapy, were observed, including cytopenias (low blood cell counts) in 70% of patients, infections in 18%, cytokine release syndrome in 7%, and neurological events in 19% of patients.

The FDA granted Yescarta another approval in April 2022 for the treatment of r/r DLBCL, which was supported by the ZUMA-7 clinical study (NCT03391466). In this phase 3 trial, 180 adult patients with r/r large B-cell lymphoma received Yescarta, while 179 patients received standard care. The results demonstrated the superior effectiveness of Yescarta, with a 24-month survival rate of 41%, compared with 16% for standard care therapy. Moreover, the ORR was 83% with Yescarta treatment, as opposed to 50% with standard care therapy, with CR rates of 65% and 32%, respectively. Adverse effects mirrored those of CAR-T-cell therapy, with cytokine release syndrome in 6% and neurological events in 21% of patients. Overall, Yescarta showed notable enhancements in event-free survival and response compared with standard care, albeit with the anticipated occurrence of high-grade toxic effects [28].

The FDA granted approval to Tecartus, a CD19-based CAR-T-cell therapy, in July 2020 to address r/r MCL. In phase 2 of the ZUMA-2 multicenter trial (NCT02601313), 68 adult patients with mantle cell lymphoma, previously treated with a Bruton's tyrosine kinase (BTK) inhibitor either before or concurrently with immunotherapy, received a single infusion of Tecartus at a dose of 2×10^6 CAR-T cells per kg of body weight. Notably, the clinical outcomes demonstrated an impressive ORR of 93% (95% CI 84–98) and a CR rate of 67% (95% CI 53–78). After 12 months of treatment, the estimated progression-free survival and overall survival rates were 61% and 83%, respectively. However, adverse effects, including cytopenias (94% of cases), infections (32%), increased CRS (15%), and neurological events (31%), were noted. In summary, Tecartus elicited durable remission in a majority of patients with r/r MCL, and the observed toxic effects were consistent with those reported in other CAR-T-cell therapies [22]. A significant challenge in the production of Tecartus in MCL lies in the starting material, as the T cells of patients with MCL are relatively scarce, and there are elevated levels of circulating tumor cells in the peripheral blood [39]. The need to eliminate these circulating tumor cells for T-cell enrichment can pose an additional obstacle, potentially impeding the production and manufacturing processes of CAR-T cells for MCL [40].

In February 2021, the FDA granted approval to Breyanzi (lisocabtagene maraleucel), an additional CD19-based CAR-T-cell therapy, for the treatment of r/r DLBCL in patients who have undergone two or more lines of systemic therapy. Breyanzi incorporated a nonfunctional truncated epidermal growth factor receptor (tEGFR) co-expressed with a CD19-specific chimeric antigen receptor on cell surfaces, which served as a surrogate for chimeric antigen receptor expression. The multicenter clinical trial TRANSCEND NHL 001 (NCT02631044) focused on adult patients with large B-cell lymphomas, including subgroups such as high-grade B-cell lymphoma, DLBCL transformed from any indolent lymphoma, B-cell lymphoma, and FL. The study involved 269 patients divided into three groups, each administered Breyanzi at different doses. The results demonstrated the consistent safety and activity of different doses of Breyanzi. The ORR was 73% (95% CI 66.8–78.0), with a CR rate of 53% (46.8–59.4). The observed adverse effects included neutropenia (60%), anemia (37%), and thrombocytopenia (27%). CRS occurred in 42% of patients, whereas neurological events were noted in 30% of patients [23].

In June 2022, the FDA granted renewed approval for Breyanzi as a treatment option for patients with large B-cell lymphoma following their first-line therapy. Two pivotal clinical trials demonstrated the substantial efficacy of Breyanzi in this patient population. In the multicenter phase 2 PILOT trial (NCT03483103), 61 patients who had undergone their first-line therapy involving anthracycline and a CD20-targeted agent received two sequential infusions of Breyanzi, each consisting of equal target doses of $CD8^+$ and $CD4^+$ CAR-T cells, resulting in a total target dose of 100×10^6 CAR-T cells. The results revealed an overall response rate of 80% (95% CI 68–89), with a median follow-up of 12.3 months. Noteworthy adverse effects included neutropenia (48%), leukopenia (21%), and thrombocytopenia (20%). Cytokine release syndrome occurred in 38% of patients, and neurological events were observed in 31% of patients. These findings support Breyanzi's potential as a second-line treatment for patients with large B-cell lymphoma [41].

In the phase 3 TRANSFORM trial (NCT03575351), Breyanzi was compared with standard therapy as a second-line therapy for large B-cell lymphoma. Adult patients were randomly assigned at a 1:1 ratio to receive either Breyanzi (100×10^6

CAR-T cells) or standard therapy. The CR rate was 74% for Breyanzi therapy versus 43% for standard therapy. These data highlight the significant efficacy of Breyanzi compared with standard therapy, positioning Breyanzi as a preferred second-line treatment for patients with r/r large B-cell lymphoma [30].

2.3 *CAR-T-Cell Therapy for Multiple Myeloma (MM)*

MM is a hematologic malignancy characterized by the abnormal proliferation of plasma cell, a type of white blood cell responsible for antibody production. This cancer typically originates in the bone marrow and can result in various complications, such as bone lesions, anemia, renal dysfunction, and immunodeficiency [42]. MM accounts for approximately 1% of all cancers and 10% of hematologic malignancies and poses a significant medical challenge. MM is a complex hematologic malignancy with diverse clinical and genetic features [43].

Advances in molecular and genetic research have led to an enhanced understanding of the pathogenesis of MM. Myeloma cells often exhibit chromosomal abnormalities, including translocations involving the immunoglobulin (Ig) heavy chain locus. Specific genetic alterations, such as mutations in the *KRAS* and *NRAS* genes, have provided valuable insights into disease biology and potential therapeutic targets [44]. The WHO classification categorizes MM on the basis of distinct clinical and laboratory features, aiding in risk stratification and treatment decisions. Precursor conditions such as monoclonal gammopathy of undetermined significance (MGUS) and smoldering multiple myeloma (SMM) are recognized, as they may progress to symptomatic multiple myeloma [45].

Treatment strategies for MM have evolved significantly in recent years. High-dose chemotherapy followed by autologous stem cell transplantation is a standard approach for eligible patients, aiming for profound and enduring responses. Revolutionary agents, such as proteasome inhibitors (bortezomib and carfilzomib), immunomodulatory drugs (lenalidomide and pomalidomide), and monoclonal antibodies (daratumumab and elotuzumab), have transformed the therapeutic landscape. Clinical trials play a pivotal role in testing and establishing the efficacy of these novel agents, with ongoing research exploring new treatment combinations, sequencing strategies, and the integration of immunotherapeutic approaches. Specific molecular subtypes have been identified, and targeted therapies tailored to these subtypes have been actively investigated. Genomic studies have revealed the heterogeneous nature of MM, emphasizing the necessity for personalized treatment approaches facilitated by next-generation sequencing technologies [46].

Bone disease is a characteristic feature of MM, and bisphosphonates, such as zoledronic acid and pamidronate, play crucial roles in preventing skeletal-related events. These agents alleviate bone pain, fractures, and hypercalcemia associated with myeloma-induced bone lesions. The management of r/r MM poses a substantial challenge. Salvage therapies, which incorporate newer agents and immunotherapies, provide alternatives for patients who have exhausted standard treatments. Investigative efforts have focused on the potential of CAR-T-cell therapy, a promising immunotherapeutic approach, in refractory MM.

Owing to the limited expression of the CD19 antigen in MM patients [47], widely used CAR-T-cell therapies targeting CD19 are not considered suitable for their treatment. Consequently, various alternative targets have been explored within the context of MM, with a notable focus on B-cell maturation antigen (BCMA) (TNFRSF17/CD269). BCMA-targeted CAR-T-cell therapy has emerged as effective for treating MM patients. BCMA, a transmembrane glycoprotein belonging to the TNF receptor family of proteins, is selectively induced during plasma cell differentiation. Importantly, BCMA expression is low in $CD34^+$ hematopoietic stem cells, as well as in naïve and memory B cells, rendering BCMA a suitable candidate for target selection [48, 49].

The first CAR-T-cell therapy based on a BCMA target, Abecma (idecabtagene vicleucel), received FDA approval in March 2021. Phase 2 clinical studies within the KarMMa trial (NCT03361748) utilized Abecma in patients with r/r MM. Among 128 patients who had undergone treatment with four or more lines of therapy, including a proteasome inhibitor, an immunomodulatory agent, and an anti-CD38 antibody, Abecma doses ranging from 150×10^6 to 450×10^6 CAR-T cells were used. With a median follow-up of 13.3 months, the therapy achieved an ORR of 73% and a CR rate of 33%. Common toxic effects, such as neutropenia (91%), anemia (70%), thrombocytopenia (63%), cytokine release syndrome (84%), and neurological effects (18%), were observed. Overall, Abecma treatment has demonstrated patient satisfaction and instilled hope in patients with MM [25].

Carvykti (ciltacabtagene autoleucel), the second BCMA-based CAR-T-cell therapy, received approval in February 2022 for the treatment of r/r MM. A notable feature of Carvykti is its incorporation of two single-domain variable domain of heavy-chain-only antibodies (VHH) antibodies, thereby enhancing its affinity for the BCMA target. The CARTITUDE-1 multicenter phase 1b/2 clinical study (NCT03548207) aimed to evaluate the safety and clinical efficacy of Carvykti. In this study, 97 adult MM patients, with 29 in phase 1b and 68 in phase 2, who had previously received treatment with a proteasome inhibitor, an immunomodulatory agent, and an anti-CD38 antibody, were given carvacrol at a dose of 0.75×10^6 CAR-T cells/kg body weight. The results revealed, after a median follow-up of 12.4 months, an impressive ORR of 97% (95% CI 91.2–99.4; 94 out of 97 patients) and a CR rate of 67% (65 patients). Importantly, the overall survival rate was reported to be 89% (80.2–93.5%). Common adverse effects included neutropenia (95%), anemia (68%), leukopenia (61%), thrombocytopenia (60%), and lymphopenia (50%). Cytokine release syndrome occurred in 95% of the patients, whereas neurotoxicity was observed in 21%. In general, Carvykti treatment resulted in early, profound, and sustained responses in patients with MM who had undergone extensive prior treatments, and it maintained a manageable safety profile [27].

2.4 CAR-T-Cell Therapy for Acute Myeloid Leukemia (AML)

AML is a diverse and aggressive blood cancer characterized by the rapid growth of abnormal myeloid precursor cells in both the bone marrow and peripheral blood. It is the most prevalent form of acute leukemia in adults and encompasses a wide array of diseases with varying genetic and molecular characteristics. The understanding of the origins of AML has undergone significant refinement, influencing diagnostic methodologies and the formulation of targeted therapeutic approaches. AML arises from mutations that disrupt the typical differentiation process of myeloid cells, leading to the accumulation of immature blast cells. These blast cells commonly infiltrate the bone marrow, hindering the production of normal blood cells. AML presents with a spectrum of symptoms, including fatigue, shortness of breath, and increased susceptibility to infections and bleeding due to the failure of bone marrow functions [50].

Advances in molecular and genetic research have highlighted the intricate landscape of AML. Chromosomal irregularities and genetic mutations play pivotal roles in the development and prognosis of this disease. Recurrent genetic alterations, such as mutations in genes including *NRAS*, *KRAS*, FLT3, NPM1, and DNMT3A, are frequently observed and hold significance for risk stratification and treatment decisions [51]. The classification of AML by the WHO hinges on a combination of clinical, morphological, immunophenotypic, and genetic features. This classification acts as a guide for treatment strategies and prognostic evaluations. AML is broadly categorized into various subtypes, each characterized by distinct features, including de novo AML, therapy-related AML, and AML with recurrent genetic abnormalities [52].

AML research has underscored the critical role of minimal residual disease assessment, allowing for a more meticulous evaluation of treatment response and the risk of relapse. Minimal residual disease monitoring employs sensitive techniques such as flow cytometry and next-generation sequencing to identify minute quantities of residual leukemic cells. The management of AML demands a comprehensive approach that integrates chemotherapy, stem cell transplantation, and targeted therapies tailored to the patient's unique characteristics and disease profile. Supportive care measures, including blood and platelet transfusions, antimicrobial prophylaxis, and growth factor support, are crucial for handling complications associated with AML and its treatment [53].

The standard treatment protocol for AML involves intensive chemotherapy, with the aim of inducing remission by eliminating leukemic cells. Decisions regarding additional consolidation therapy, such as stem cell transplantation, are influenced by factors such as age, cytogenetic risk, and the response to initial treatment. Allogeneic stem cell transplantation, which utilizes donor cells, is a potentially curative option for eligible patients. Despite initial positive responses to treatment, the challenge of relapse looms large in AML management. Research endeavors have focused on the development of targeted therapies to enhance outcomes and minimize the adverse effects of treatment. Tyrosine kinase inhibitors such as midostaurin and gilteritinib, which target FLT3 mutations, have demonstrated promise in specific subsets of AML patients [54].

The landscape of AML treatment continues to evolve with the introduction of novel agents. Venetoclax, a BCL-2 inhibitor combined with hypomethylating agents, has exhibited efficacy, particularly in older adults with newly diagnosed AML who may not withstand intensive chemotherapy [55]. Enasidenib and ivosidenib, inhibitors of isocitrate dehydrogenase (IDH) mutations, represent approved targeted therapies for r/r AML. Immunotherapeutic approaches, including immune checkpoint inhibitors and monoclonal antibodies, are currently under investigation in clinical trials, aiming to augment the immune system's capacity to identify and eliminate leukemic cells [56].

AML is characterized by clinical heterogeneity [57]. Currently, there are no available CAR-T-cell therapy–based treatments for AML, primarily due to limitations in targeting specific malignant cells. Existing CAR-T-cell therapy targets for AML are also expressed in healthy hematopoietic stem and progenitor cells, potentially leading to bone marrow destruction [58]. Only two CAR-T-cell-based clinical trials in AML have been completed, and the outcomes are either unsatisfactory or await disclosure. Some preclinical studies in hematological and solid tumors have demonstrated the efficacy of CAR-T-cell therapy based on natural killer group 2D (NKG2D) ligands. These antigens, ligands for NK-cell-activating receptors, are found in hematologic malignancies, including lymphomas, leukemias, myelodysplastic syndromes, and myelomas [59].

Notably, human NKG2D-CAR-T cells do not react to healthy normal cells in vitro [60]. In one clinical trial (NCT02203825) investigating NKG2D ligand–based CAR-T cells (CM-CS1 T cells) in adult patients with AML, myelodysplastic syndrome with refractory anemia with excess blasts (MDS-RAEB), and MM [61, 62], up to 3×10^7 CAR-T cells were utilized. The results revealed no dose-limiting toxicity, cytokine release syndrome, or CAR-T-cell-related neurotoxicity. Furthermore, no notable autoimmune reactions were observed, and none of the grade 3 adverse events were attributed to NKG2D-CAR-T cells. However, there were limitations in NKG2D-CAR-T-cell expansion and persistence, and no robust efficacy signal was evident in this initial phase 1 study. The NKG2D-CAR approach is still in the early stages of clinical development, with unique features and strengths that require further testing.

Interleukin (IL)-3, a versatile cytokine, has its receptor, IL-3R (CD123), which serves as the α chain of the interleukin 3 receptor. CD123 is highly expressed in various hematologic malignancies, including subsets of B-lymphoid leukemia, AML, blastic plasmacytoid dendritic neoplasm (BPDCN), and hairy cell leukemia [63]. Its notable expression on AML blasts and leukemic stem cells (LSCs) positions it as an appealing target for CAR-T-cell therapy. A recent study involving CD123-based CAR-T cells in patients with r/r AML (NCT04318678) reported modest outcomes in an ongoing trial with 12 enrolled patients. The infused CAR-T-cell products exhibited potent antitumor activity in vitro against $CD123^+$ targets, and the infusions were well-tolerated, resulting in no adverse events. While transient cytopenias were observed, persistent marrow aplasia did not develop. At the

4–6-week follow-up, marginal effects were noted, emphasizing the ongoing need for advancements in CD123-based CAR-T cells. Another clinical trial (NCT03585517) evaluated the safety and efficacy of IM23 CAR-T cells in $CD123^+$ r/r AML patients, with results awaiting disclosure.

2.5 CAR-T-Cell Therapy for Chronic Lymphocytic Leukemia (CLL)

CLL is a hematological malignancy characterized by the gradual accumulation of abnormal B lymphocytes, or white blood cells, in the bone marrow, blood, and various tissues. It is the most prevalent form of adult leukemia in Western countries and predominantly affects older individuals. CLL typically progresses slowly, and in some cases, patients may not need immediate treatment. Nevertheless, it can exhibit an aggressive course under specific circumstances [64]. The pathogenesis of CLL involves the clonal expansion of mature B cells expressing CD5, a distinctive marker of CLL cells. These aberrant B cells accumulate due to defects in the apoptosis process, enabling their prolonged survival and proliferation. While the specific cause of CLL remains unclear, genetic and environmental factors may contribute to its development. Genetic abnormalities play a pivotal role in CLL, with chromosomal aberrations influencing the disease's clinical course and prognosis. The presence or absence of specific genetic markers, such as deletion 13q, trisomy 12, and mutations in TP53 and NOTCH1, can significantly influence disease progression and guide treatment decisions.

The clinical course of CLL varies widely, and some patients may remain asymptomatic for extended periods. Common clinical manifestations include lymphadenopathy (enlarged lymph nodes), splenomegaly (enlarged spleen), and cytopenias. Additionally, CLL cells can infiltrate other organs, leading to complications such as autoimmune hemolytic anemia or immune thrombocytopenia. The Rai and Binet staging systems are frequently employed to assess the extent of CLL and guide treatment decisions. These staging systems consider factors such as lymphocytosis, lymphadenopathy, and organomegaly. Patients with early-stage disease (Rai 0, Binet A) may not need immediate treatment and can be managed with a "watch and wait" approach, whereas those with advanced-stage disease may benefit from therapeutic intervention. In 2016, a novel prognostic model known as the CLL International Prognostic Index (CLL-IPI) was introduced, facilitating a more precise approach to the management of CLL [65].

In recent years, treatment options for CLL have undergone significant evolution. Chemoimmunotherapy, typified by the combination of fludarabine, cyclophosphamide, and rituximab (FCR), is a conventional first-line treatment for fit patients. However, the treatment landscape has shifted with the emergence of novel agents. B-cell receptor (BCR) pathway inhibitors, including ibrutinib, acalabrutinib, and zanubrutinib, have demonstrated efficacy in both treatment-naive and r/r CLL patients by targeting signaling pathways that promote CLL cell survival and proliferation.

Venetoclax, an inhibitor of the antiapoptotic protein BCL-2, has exhibited remarkable efficacy, particularly in CLL patients featuring a 17p deletion or TP53 mutation, which are associated with poorer outcomes. Combining venetoclax with other agents, such as rituximab or obinutuzumab, has further increased response rates [66].

The advent of targeted therapies has significantly influenced the landscape of stem cell transplantation in CLL. Allogeneic stem cell transplantation, once considered a potential curative option, is now reserved for high-risk patients because of the effectiveness of targeted therapies and the associated risks of transplantation. The management of CLL extends beyond disease-specific interventions. Supportive care, encompassing immunizations, antimicrobial prophylaxis, and monitoring for infectious complications, is crucial. Addressing comorbid conditions and optimizing overall health contribute to the comprehensive care of CLL patients. Research endeavors persist in exploring novel therapeutic strategies and understanding the underlying biology of CLL. The rise of precision medicine, guided by genetic and molecular profiling, holds promise for tailoring treatments to individual patients on the basis of the specific characteristics of their CLL cells.

CLL malignant cells express the CD19 receptor, rendering this disease a suitable candidate for treatment with CD19-based CAR-T cells. However, clinical studies in this context have yielded less than satisfactory results. In one phase 1 clinical trial (NCT00466531), CD19-targeted CAR-T cells were administered to 16 CLL patients. The outcomes revealed that three out of the 12 patients who underwent conditioning chemotherapy achieved CR and remained progression-free at a median follow-up of 53 months. While cytokine release syndrome was observed in all patients, grades 3 and 4 cytokine release syndrome and neurological events were identified in 10% of patients [67]. These findings indicated that CD19-targeted CAR-T-cell therapy was reasonably well-tolerated in r/r CLL patients. In another clinical study (NCT02640209), 38 r/r CLL patients received infusions of anti-CD19 CAR-T cells at doses ranging from 5×10^7 to 5×10^8 CAR-T cells, and long-term outcomes were assessed. With a median follow-up of 31.5 months, the ORR was 44% (90% CI, 29–60%), and CR was observed in 28% (90% CI, 16–44%) of the patients. The median overall survival for all patients was 64 months. These suboptimal results suggest the need for further studies on CD19-based CAR-T-cell therapies in patients with CLL [68].

Richter's syndrome is characterized by the transformation of CLL into highly aggressive lymphoma, typically DLBCL, which occurs in approximately 2–10% of CLL patients [69]. Presently, several ongoing clinical studies recruiting patients for CAR-T-cell therapy in conjunction with chemotherapeutic drugs exist. The NCT05873712 clinical study assessed the efficacy of Breyanzi in combination with Bruton's tyrosine kinase (BTK) inhibitors for Richter's syndrome. The ZUMA-25 clinical study (NCT05537766) investigated the impact of Tecartus in patients with r/r Richter's transformation. Additionally, in a phase 2 trial (NCT05672173), the synergistic effects of Breyanzi, immune checkpoint inhibitors, and Bruton's tyrosine kinase (BTK) inhibitors are under examination in individuals with Richter's transformation. These studies may yield valuable insights in the future, providing potential benefits for patients with Richter's syndrome.

3 Targets of CAR-T-Cell Therapy for Hematological Malignancies

The FDA has approved six CAR-T-cell therapies for hematological cancers since 2017. Among these, four are anti-CD19 CARs, and the two latest CAR-T-cell products are designed to target BCMA (Table 1). The human CD19 antigen, classified as a type I transmembrane protein within the immunoglobulin superfamily glycoprotein, is widely expressed throughout all stages of B-cell development until it reaches terminal differentiation into plasma cells [70]. In terms of functionality, CD19 is crucial in determining intrinsic B-cell signaling thresholds because it influences both BCR-dependent and BCR-independent signaling pathways. Its importance extends to antigen-independent B-cell development and immunoglobulin-induced activation of B cells [71]. Before their incorporation into CAR-T-cell therapy, CD19 monoclonal antibodies were studied in patients with progressive B-cell lymphoma, yielding suboptimal results [72]. However, in the current landscape, CAR-T-cell therapy relies heavily on CD19 and has expanded its exploration beyond B-cell malignancies.

BCMA, alternatively known as tumor necrosis factor (TNF) receptor superfamily member 17 (TNFRSF17), is a protein encoded by the TNFRSF17 gene in humans. BCMA recognizes B-cell activating factor (BAFF) and a proliferation-inducing ligand (APRIL) as ligands. This protein is predominantly expressed in mature B lymphocytes and plays a pivotal role in B-cell development and the autoimmune response. The interaction of BCMA with BAFF leads to the activation of NF-κB and MAPK8/JNK. Additionally, it binds to various Tumor necrosis factor (TNF) receptor–associated factor (TRAF) family members, potentially transducing signals for cell survival and proliferation. In addition to CAR-T-cell therapy in MM, BCMA has been explored for alternative treatment modalities, such as bispecific antibody constructs, including bispecific T-cell engager (BiTE) immuno-oncology therapies, and antibody–drug conjugates (ADCs) [48]. In addition to these two CAR-T-cell therapy targets, numerous other antigens currently under study are discussed here (Fig. 2).

3.1 Targets of CAR-T-Cell Therapy for Acute Lymphoblastic Leukemia

For B-cell malignancies, CD19, present throughout B-cell development, was selected as a viable target despite not being tumor specific. This choice is based on its elevated expression in the majority of malignant B cells, coupled with its absence on hematopoietic stem cells, thereby mitigating the risk of anemia. While CD19-based CAR-T-cell therapy has demonstrated success, challenges such as CD19 relapse and the limited persistence of CAR-T cells remain significant. Consequently, there is ongoing exploration of alternative targets beyond CD19.

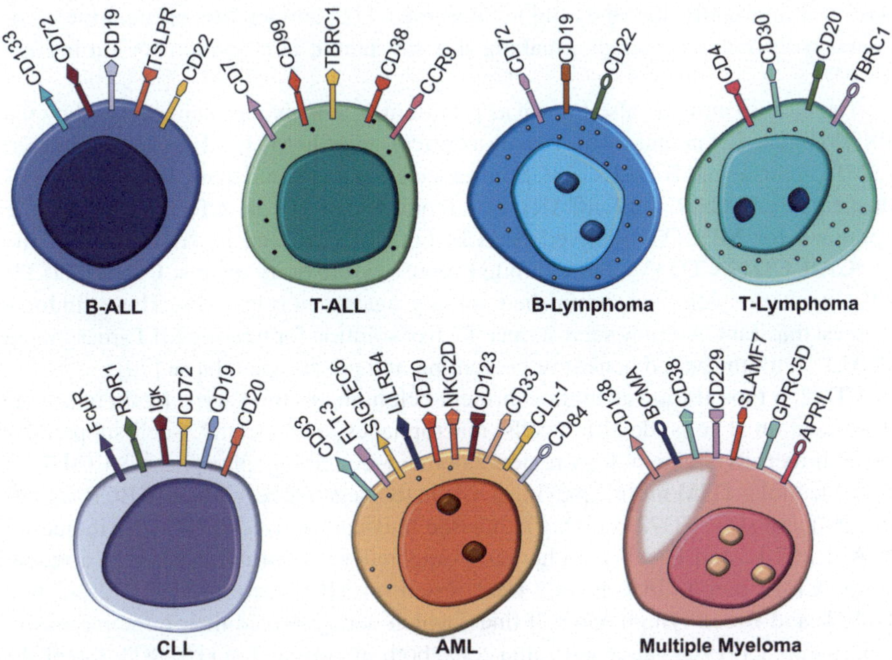

Fig. 2 The targets for CAR-T-cell therapy in hematological malignancies are illustrated, highlighting various identified targets for CAR-T-cell therapy in B- and T-cell malignancies, chronic lymphocytic leukemia, acute myeloid leukemia, and multiple myeloma

Thymic stromal lymphopoietin (TSLP), a cytokine derived from epithelial cells such as IL-7, plays a crucial role in hematopoietic cell maturation and the stimulation of pediatric ALL [73]. TSLP binds to its receptor, TSLPR, which is encoded by the cytokine receptor-like factor 2 (CRLF2) gene, and is overexpressed in 15% of B-ALL patients due to chromosomal rearrangements, increasing the risk of relapse [74]. The TSLPR signaling axis has been identified as crucial, making TSLPR an attractive immunotherapeutic target for B-ALL. Studies indicate that TSLPR CAR-T cells induce leukemia regression and exhibit long-term persistence of CAR-expressing T cells. In murine models, TSLPR CAR-T-cell therapy demonstrated efficacy comparable to that of CD19 CAR-T-cell therapy. Given their restricted surface expression on normal tissues, TSLPR-targeted CAR-T cells hold significant potential [75].

CD22 serves as an inhibitory co-receptor of the BCR, playing a crucial role in establishing signaling thresholds for B-cell activation and contributing to the regulation of B-cell survival and function. This protein is expressed by a majority (60–90%) of blasts in B-ALL, making it an attractive target for CAR-T-cell therapy [76]. In a recent human trial involving 34 r/r B-ALL patients, both pediatric and adult, who had previously failed CD19 CAR-T-cell therapy, CD22-based CAR-T-cell therapy was administered. The results demonstrated CR in 80% of patients by day 30. Importantly, the study did not observe CD22 antigen loss or mutation associated with relapsed patients, making it a substantial contribution for further trials [77].

The CD133 antigen, also known as prominin-1, is a glycoprotein encoded by the PROM1 gene in humans that serves as a critical target in MLL-AF4 and is expressed in MLL-rearranged B-ALLs [78]. Studies have demonstrated that CD19 and CD133 dual-targeting CAR-T (TanCAR) cell therapy can be an effective therapeutic approach for MLL-AF4-derived B-ALL, both in vitro and in vivo. Notably, the TanCAR-CD19/CD133 CAR exhibited robust cytotoxicity against $CD19^+CD133^+$ MLL leukemic cells but diminished activity against normal cells. These findings suggest that TanCAR may serve as an effective solution for treating MLL-rearranged B-ALL, with the added benefit of preventing antigen escape relapse [79].

CD72, a type II transmembrane protein belonging to the C-type lectin family, is expressed on B cells, dendritic cells, macrophages, and specific T-cell subpopulations. It recruits Src homology region 2 domain–containing phosphatase-1 (SHP-1) to intracellular ITIM motifs, playing a role in the negative regulation of BCR signaling [80]. Notably, CD72 is highly expressed in B-cell-derived neoplasms, including B-ALL, CLL, and B-cell lymphomas, compared with normal tissues. In a recent study, CD72-based CAR-T cells were tested on MLL-rearranged CD72-positive B-ALL and B-cell lymphoma cell lines, which demonstrated their effectiveness in eliminating CD19-negative leukemia cells, both in vitro and in vivo [81], establishing them as promising targets.

In contrast to B-ALL cells, T-ALL cells exhibit elevated levels of CD7, a transmembrane glycoprotein that plays a crucial role in T-cell and T-cell/B-cell interactions during early lymphoid development [82]. CAR-T cells engineered with CD7 have demonstrated cytotoxic effects against malignant T-cell lines and primary tumors, offering protective benefits in a mouse xenograft model of T-ALL [83]. In a recent phase 1 human trial (NCT04572308), 20 T-ALL patients were treated with CD7-based NS7CAR CAR-T cells, resulting in CR in the bone marrow for 19 patients within 28 days. Further exploration of CD7-based CAR-T cells through additional studies could provide valuable insights into their potential therapeutic applications in human therapy [84]. CD38 is a type II transmembrane glycoprotein with roles in cell adhesion, migration, and signal transduction, and it possesses NADase and cyclase activities [85]. Typically, in B cells, T cells, and MM cells, efforts have been made to induce CD38 expression in T-ALL patients via all-trans retinoic acid (ATRA) and subsequently treat them with CD38-based CAR-T cells [86]. Further investigations are needed to explore the potential of CD38 as a target in CAR-T-cell therapy for T-ALL.

CD99, a cell surface glycoprotein, plays crucial roles in leukocyte migration, T-cell adhesion, ganglioside GM1, and transmembrane protein transport, as well as T-cell death through a caspase-independent pathway. Given its strong expression in the majority of T-ALL and AML patients, CD99 has emerged as a promising antigen to target these diseases. In a recent study, a specific antibody was developed to target CD99 on leukemic cells over normal cells, and CAR-T cells were generated. The results demonstrated that CD99-based CAR-T cells exhibited robust cytotoxicity, specifically against CD99+ T-ALL cell lines and primary tumor cells, in vitro. Furthermore, they significantly prolong survival in cell line–derived xenografts or patient-derived xenograft models in vivo, highlighting CD99 as a promising target [87].

CCR9, a chemokine receptor that is responsive to CCL25 in T cells and highly expressed in CD8-positive lineage T cells, is a promising target for CAR-T-cell therapy. Moreover, CCR9 is expressed in more than 70% of T-ALL patients, further establishing its potential as an ideal target. In a recent investigation employing cell line models and patient-derived xenografts, CAR-T cells designed to target CCR9 displayed resistance to fratricide and demonstrated robust antileukemic activity, both in vitro and in vivo, even when the target antigen density was low. Importantly, this approach eliminates the need for T-cell aplasia and genome engineering, which complicates other methods, making it a promising avenue for further exploration [88].

3.2 *Targets of CAR-T-Cell Therapy for Lymphoma*

In addition to successful studies targeting CD19, several other potential targets are currently under investigation for lymphoma patients. CD20, a B-cell-specific antigen that is expressed in various lymphoid malignancies, has emerged as a promising target in CAR-T-cell therapy. The utilization of bispecific CAR-T cells has shown potential in eliminating lymphoma cells negative for either CD19 or CD20, thereby demonstrating dual antigen targeting of both CD19 and CD20 [89]. Currently, a diverse range of bispecific CAR-T cells are being investigated, with CD20 as a target [90].

CD4, a transmembrane glycoprotein, functions as a co-receptor for the TCR and is prominently present on immune cells such as helper T cells, monocytes, macrophages, and dendritic cells. A recent study introduced a novel CAR, the CD4 CAR, designed to redirect the antigen specificity of CD8+ cytotoxic T cells toward CD4-expressing cells. In coculture assays, CD4CAR-T cells efficiently eliminated CD4+ leukemic cell lines and primary samples from patients with peripheral T-cell lymphoma. Remarkably, in aggressive orthotropic T-cell lymphoma models, these CAR-T cells not only suppressed the growth of lymphoma cells but also significantly prolonged mouse survival. This innovative approach holds promise as a potential bridge to transplant or stand-alone therapy for treating patients with peripheral T-cell lymphoma [91].

TCRs are integral components on the surface of T cells that are responsible for recognizing antigens, subsequently triggering the T-cell immune response. Compared with an α-chain and a β-chain, the β-chain expresses either the T-cell receptor beta constant 1 (TRBC1) or the T-cell receptor beta constant 2 (TRBC2) genes in its constant domain. Malignant T cells typically express TRBC1, rendering it a viable target for CAR-T-cell therapy. In a recent study, a unique antibody with specificity for TRBC1 was identified, leading to the development of anti-TRBC1

CAR-T cells. The results demonstrated that these CAR-T cells effectively recognized and eliminated both normal and malignant TRBC1⁺ T cells while sparing TRBC2⁺ T cells, both in vitro and by ablating malignant T cells in vivo. These findings highlight the potential of TRBC1 as a promising target for CAR-T-cell therapy [92].

CD30, a member of the tumor necrosis factor receptor superfamily, is expressed at characteristic levels in specific hematopoietic malignancies, such as anaplastic large cell lymphoma and peripheral T-cell lymphoma. Notably, CD30 is restrictively expressed on T cells, making it an ideal target for CAR-T-cell therapy in lymphomas. In a recent study, researchers successfully developed CD30 CAR-T cells, demonstrating their effectiveness in treating recurrent and malignant CD30-positive peripheral T-cell lymphomas. These CAR-T cells demonstrated potent cytotoxicity against lymphoma cells, both in vitro and in a xenograft NCG mouse model, suggesting the promising potential of CD30-based CAR-T cells as therapeutic candidates for cancer treatment [93]. Similarly, CD30 is expressed in Hodgkin lymphoma, prompting numerous clinical trials to explore CD30-based CAR-T-cell therapy in patients with r/r Hodgkin lymphoma. In a phase 1/2 clinical trial (NCT02690545), 27 patients received treatment with anti-CD30 CAR-T cells, resulting in 67% of patients achieving CR within 6 weeks. However, 63% of patients experienced disease progression during the study, with a median follow-up of 9.5 months, indicating the need for further investigations [94].

3.3 *Targets of CAR-T-Cell Therapy for Multiple Myeloma*

CAR-T-cell therapy has emerged as a promising approach for the treatment of MM, a hematologic malignancy characterized by abnormal plasma cell proliferation. FDA-approved CAR-T-cell therapies, such as Abecma and Carvykti, have shown remarkable efficacy in clinical trials, with significant responses, including deep and durable remission, in heavily pretreated MM patients. While challenges remain, such as managing adverse effects and improving CAR-T-cell persistence, CAR-T-cell therapy represents a groundbreaking advancement in the pursuit of more effective treatments for MM. Several additional targets are currently under investigation for potential utilization in CAR-T-cell therapy for MM.

CD138, part of the syndecan family, is a type I transmembrane proteoglycan characterized by a core protein glycosylated with chondroitin and heparan sulfate moieties. Its diverse functions include roles in wound healing, cell adhesion, endocytosis, and potential nuclear translocation. CD138 interacts with numerous proteins, such as extracellular matrix components, integrins, growth factors, cytokines, and chemokines [95]. MM cells exhibit elevated levels of CD138, contributing to their survival and proliferation. Clinical trials have demonstrated the limited efficacy of anti-CD138 CAR-T-cell therapy for the treatment of chemotherapy-refractory MM, with 80% of patients experiencing stable disease for more than 3 months without notable toxicity. In another investigation, CD138 CAR-T-cell treatment successfully eradicated tumor cell lines and primary myeloma cells, both in vitro and in vivo, sparing normal cells. CD138 is also present in myeloma stem cells and can be effectively targeted and eliminated by CD138-specific CAR-T cell therapy. These findings underscore the potential of adoptively transferred CD138-specific CAR-T cells as a viable therapeutic approach for patients with r/r MM [96].

CD38 is another target highly expressed in MM cells. However, it is also expressed in some normal cells, and a study demonstrated the adverse effects of CD38-based CAR-T cells on normal hematopoietic cells in a mouse model, suggesting the need for further tuning [97]. Recent clinical studies have shown that dual-targeting CAR-T cells, where one target is CD38, can be useful for MM patients. For example, one study utilized bispecific BM38 CARs targeting BCMA and CD38 to assess the antimyeloma activity of BM38 CAR-T cells, in vitro and in vivo. In a phase 1 trial, 23 patients with r/r MM received infusions of BM38 CAR-T cells and displayed stronger in vitro cytotoxicity and potent antimyeloma activity in a xenograft mouse model. The ORR was 87%, with 52% achieving a stringent CR, highlighting CD38 as a potent target [98].

CD229 is an immune-modulating receptor expressed on the cell surfaces of B and T lymphocytes, with high expression observed in MM samples. Researchers have engineered CAR-T cells on the basis of CD229, which demonstrated high activity, both in vitro and in vivo, against MM plasma cells, memory B cells, and MM-propagating cells. However, a drawback was noted, as these CAR-T cells also targeted normal CD229-high T cells [99]. Another study addressed this issue by substituting a single amino acid in the CAR-binding domain to reduce CAR affinity, successfully achieving potent antileukemic activity while sparing normal cells [100]. Further studies are necessary for the development of CD229-based CAR-T cells for MM therapy.

SLAMF7 (CD319), a protein encoded by the SLAMF7 gene in humans, serves as a surface antigen that is expressed both in normal and malignant plasma cells in MM. As a self-ligand receptor, SLAMF7 features immunoreceptor tyrosine–based switch motifs in its cytoplasmic domain. Upon activation, these motifs undergo phosphorylation, recruiting SH2 domain–containing molecules to facilitate downstream signaling [101]. The application of CAR-T cells based on SLAMF7 resulted in swift cytolysis of untreated and r/r primary myeloma cells in vitro. Notably, this approach successfully resolved medullary and extramedullary myeloma manifestations in a murine xenograft model in vivo [102]. However, SLAMF7 is expressed on a subset of normal lymphocytes, including natural killer (NK) cells, T cells, and B cells. Researchers are actively addressing the challenge of potential off-target effects on normal cells associated with SLAMF7-based CAR-T cells. Further investigations are necessary to refine and enhance the therapeutic potential of this approach while minimizing its impact on normal cells.

The G protein–coupled receptor class C, group 5, member D (GPRC5D) is a surface protein expressed in MM cells that is expressed at low levels in other tissues. Recent clinical studies have revealed positive outcomes for GPRC5D-based CAR-T-cell therapy. In a phase 1, first-in-human, multicenter, open-label study (NCT04674813), GPRC5D-based CAR-T cells (BMS-986393) were administered to 33 r/r MM patients, some of whom had previously undergone BCMA-targeted therapies. With a median follow-up of 3.1 months, the overall response rate was 89% (17/19) in efficacy-evaluable patients, including responses in seven out of nine individuals treated with prior BCMA-directed therapies, including CAR-T cells (PMCID: PMC10428438). In another clinical study, patients with heavily pretreated MM, including those who relapsed after BCMA CAR-T-cell therapy (NCT04555551), received GPRC5D-targeted CAR-T-cell therapy (MCARH109). The results revealed a therapeutic response in 71% of the patients in the entire cohort, with manageable adverse effects associated with CAR-T-cell therapy. These findings underscore GPRC5D as a viable immunotherapeutic target in MM, confirming the potential effectiveness of GPRC5D-targeted CAR-T-cell therapy [103].

A proliferation-inducing ligand (APRIL) is a natural ligand that binds directly to B-cell maturation antigen (BCMA) and transmembrane activator and CAML interactor (TACI). It is hypothesized that APRIL-targeted CAR-T cells can recognize both BCMA and TACI expressed on MM cells, potentially reducing the risk of antigen escape and enhancing the effectiveness of myeloma therapy [104]. However, recent clinical studies have not yielded satisfactory results for APRIL-based CAR-T-cell therapy. In a phase 1 clinical trial (NCT03287804, AUTO2), the APRIL CAR was assessed in patients with r/r MM. Although the therapy was well-tolerated, only 45.5% of patients exhibited positive responses [105]. Further research is needed to develop more potent APRIL-based CAR-T-cell therapies for MM.

3.4 Targets of CAR-T-Cell Therapy for Acute Myeloid Leukemia

CAR-T-cell therapy in AML faces several challenges, as there are currently no FDA-approved CAR-T-cell therapies for this specific cancer type. AML is known for its heterogeneity, presenting a complex landscape that requires tailored therapeutic approaches. Despite the absence of approved treatments, the field is experiencing significant progress, with numerous clinical trials underway to explore the potential of CAR-T-cell therapy in AML. These trials are aimed to address the diverse molecular and genetic characteristics of AML, seeking innovative ways to target and eliminate leukemic cells. Ongoing research reflects a collective effort to advance treatment options for AML patients and underscores the urgency of unlocking effective CAR-T-cell therapies for this challenging hematologic malignancy.

In addition to targeting CD123 with CAR-T-cell therapy in AML, various other receptors are under investigation. CD38, which has been identified as a potential target in T-ALL and MM, is also being explored for AML therapy. CD33, a myeloid-specific sialic acid–binding receptor expressed on most AML blasts and LSCs, is present in approximately 90% of cases [106]. A clinical study (NCT01864902) involving AML patients treated with CD33-directed CAR-T-cell therapy demonstrated a significant reduction in the number of bone marrow blasts 2 weeks after therapy [107]. However, another phase 1 clinical trial (NCT03126864) did not yield favorable results with CD33-directed CAR-T-cell therapy in AML, indicating the need for further studies [108].

CLL-1, a myeloid cell marker that is overexpressed in AML samples but not expressed on normal hematopoietic cells, has robust antileukemic activity, both in vitro and in vivo, in CLL-1-based CAR-T-cell therapy of CLL-1^+ cell lines and primary AML patient samples [109]. CD70, a TNF-alpha family receptor expressed on AML bulk cells and LSCs but not on normal HSCs, exhibited superior proliferation and antitumor activity, in vitro and in vivo, in a recent study in which CD70-based CAR-T cells were developed. These findings suggest that CD70-targeted immunotherapy holds promise as a novel treatment strategy for patients with CD70-positive AML [110]. Leukocyte immunoglobulin-like receptor B4 (LILRB4) functions as an inhibitory receptor, playing a crucial role in immune checkpoint pathways. Its high expression in monocytic AML cells makes it an appealing target. In a recent study, CAR-T cells based on LILRB4 were developed and tested on AML samples. The results demonstrated efficient effector functions, both in vitro and in vivo, against LILRB4^+ AML cells. Importantly, there was no effect on normal CD34^+ umbilical cord blood cells in colony-forming unit assays or in a humanized hematopoietic-reconstituted mouse model. These findings suggest that LILRB4 could be a promising target for monocytic AML therapy [111].

CD84, a glycoprotein belonging to the signaling lymphocyte activation molecule (SLAM) family, is overexpressed on the surface of AML cells, both in de novo cases and at relapse, with lower expression detected in CD34^+ hematopoietic stem/progenitor cells (HSPCs). A recent study developed CAR-T cells based on CD84 and reported antileukemic activity in both in vitro and in vivo settings, albeit with potential myelotoxicity. Further studies are necessary to assess their suitability for clinical studies [112]. Sialic acid–binding Ig-like lectin 6 (Siglec-6) is a receptor that facilitates the uptake of extracellular vesicles through a noncanonical glycolipid binding pocket. It is expressed in approximately 60% of AML cases. Treatment with Siglec-6-based CAR-T cells demonstrated specific antileukemic activity against Siglec-6-expressing cells in preclinical models. In a xenograft AML model, this therapy induced CR. Notably, Siglec-6 is absent in normal HSCs, indicating that Siglec-6 CAR-T-cell therapy might effectively address AML without requiring subsequent allogeneic hematopoietic stem cell transplantation [113]. T-cell immunoglobulin mucin-3 (TIM-3), a surface molecule expressed on LSCs in almost all types of AML but not on HSCs, has become a specific target for AML. TIM-3-based CAR-T-cell therapy exhibited antileukemia activity against AML cell lines and primary AML blasts, suggesting that TIM-3-based CAR-T-cell therapy is a promising alternative target in AML therapy [114]. In contrast, targets such as FLT3 and CD93 have shown suboptimal results in AML because of their expression in other normal cells, emphasizing the need for AML-specific targets in CAR-T-cell therapy [115, 116].

3.5 Targets of CAR-T-Cell Therapy for Chronic Lymphocytic Leukemia

Currently, there is no approved CAR-T-cell therapy for patients with CLL. Although CD19-based CAR-T trials have been extensively explored for CLL patients, researchers have investigated several other potential targets, including CD20, which has also been explored in B-cell lymphoma. B lymphocytes express monoclonal immunoglobulins with either κ or λ light chains on their surface, and the clonal restriction in κ/λ expression means that malignant CLL cells within a specific patient express either the κ or λ light chain [117]. Consequently, CARs designed to target the light chain expressed by the tumor aim to selectively spare normal B cells that express the reciprocal light chain, thereby avoiding conditions such as hypogammaglobulinemia. This approach seeks to minimize the impact on normal B cells while precisely targeting malignant cells. In a recent development, researchers designed CAR-T cells on the basis of the κ light chain and tested them in CLL patients as part of the clinical trial NCT00881920. The results indicated that out of nine patients with relapsed NHL or CLL, two achieved CR after two or three infusions of CAR-T cells, and one achieved a partial response. These findings suggest that κ light chain–based CAR-T cells are feasible, safe, and capable of inducing complete clinical responses [118].

Receptor tyrosine kinase-like orphan receptor 1 (ROR1) plays a crucial role in modulating neurite growth in the central nervous system, particularly in the innervation of auditory hair cells by spiral ganglion neurons. While ROR1 is highly expressed in CLL, it is not present in normal B cells, making it an appealing target for therapeutic intervention. ROR1-based CAR-T cells have demonstrated cytotoxicity against primary CLL cells in vitro [119]. Furthermore, a clinical study involving CLL patients is underway to investigate the effectiveness of ROR1-based CAR-T-cell therapy (NCT02706392). This research aims to explore the potential of ROR1 as a target for CAR-T-cell therapy in CLL and assess its clinical impact. CD37, a tetraspanin prominently expressed on the surface of mature B cells, has been investigated as a potential target for CAR-T-cell therapy in CLL. CAR-T cells based on CD37 have demonstrated antileukemic activity in vitro. Additionally, bispecific anti-CD37/anti-CD19 CAR-T cells exhibited specificity, cytokine production, and cytotoxic activity in models of B- and T-cell lymphomas, both in vitro and in vivo, including patient-derived xenografts [120]. The immunoglobulin M Fc receptor (FcμR), which plays a role in the immune response by mediating the endocytosis of IgA and IgM during the immune response to microbes, is highly expressed in CLL samples and is present at low levels in normal B cells. CAR-T cells, which are based on FcμRs, respond to CLL cells, releasing pro-inflammatory cytokines and lytic factors, effectively eliminating leukemic cells while sparing normal B cells. Furthermore, these CAR-T cells inhibited Mec-1-induced leukemia in a xenograft mouse model, suggesting a superior therapeutic index for FcμR-based CAR-T cells in CLL treatment [121]. In addition to these targets, numerous studies are underway to identify novel targets aimed at mitigating the long-term B-cell aplasia commonly associated with current CAR-T-cell targets.

4 Conclusion

In conclusion, CAR-T-cell therapy has emerged as a groundbreaking and transformative approach in the treatment of hematological malignancies. The approval of multiple CAR-T-cell therapies, particularly those targeting CD19 and BCMA, signifies remarkable advancements in the field. While these therapies have demonstrated unprecedented success in inducing durable remission, challenges such as antigen escape and persistence issues necessitate ongoing research into alternative targets and enhanced engineering strategies. The expanding repertoire of potential targets, including CD20, CD22, and CCR9, holds promise for addressing the complexities and limitations associated with current CAR-T-cell therapies. As we navigate this evolving landscape, ongoing clinical trials and translational research efforts are crucial for unlocking the full potential of CAR-T-cell therapy, ultimately offering renewed hope for patients with hematological malignancies. These endeavors underscore the evolving landscape of CAR-T-cell therapy in acute lymphoblastic leukemia, demonstrating the commitment to advancing precision medicine in cancer treatment.

Declaration of Interest The author declares no competing interests.

References

1. Welniak LA, Blazar BR, Murphy WJ. Immunobiology of allogeneic hematopoietic stem cell transplantation. Annu Rev Immunol. 2007;25:139–70.
2. Quezada SA, Peggs KS, Simpson TR, Allison JP. Shifting the equilibrium in cancer immunoediting: from tumor tolerance to eradication. Immunol Rev. 2011;241(1):104–18.
3. Miliotou AN, Papadopoulou LC. CAR T-cell therapy: a new era in cancer immunotherapy. Curr Pharm Biotechnol. 2018;19(1):5–18.
4. Srivastava S, Riddell SR. Engineering CAR-T cells: design concepts. Trends Immunol. 2015;36(8):494–502.
5. Watanabe K, Nishikawa H. Engineering strategies for broad application of TCR-T- and CAR-T-cell therapies. Int Immunol. 2021;33(11):551–62.
6. Majzner RG, Mackall CL. Clinical lessons learned from the first leg of the CAR T cell journey. Nat Med. 2019;25(9):1341–55.
7. Gross G, Waks T, Eshhar Z. Expression of immunoglobulin-T-cell receptor chimeric molecules as functional receptors with antibody-type specificity. Proc Natl Acad Sci USA. 1989;86(24):10024–8.
8. Vairy S, Garcia JL, Teira P, Bittencourt H. CTL019 (tisagenlecleucel): CAR-T therapy for relapsed and refractory B-cell acute lymphoblastic leukemia. Drug Des Devel Ther. 2018;12:3885–98.

9. Mueller KT, Maude SL, Porter DL, Frey N, Wood P, Han X, et al. Cellular kinetics of CTL019 in relapsed/refractory B-cell acute lymphoblastic leukemia and chronic lymphocytic leukemia. Blood. 2017;130(21):2317–25.
10. Matsueda S, Chodon T, Koya RC. Cellular immunotherapy in the treatment of hematopoietic malignancies. Adv Exp Med Biol. 2019;1143:217–29.
11. Abramson JS. Anti-CD19 CAR T-cell therapy for B-cell non-Hodgkin lymphoma. Transfus Med Rev. 2020;34(1):29–33.
12. Denlinger N, Bond D, Jaglowski S. CAR T-cell therapy for B-cell lymphoma. Curr Probl Cancer. 2022;46(1):100826.
13. Mohty M, Dulery R, Gauthier J, Malard F, Brissot E, Aljurf M, et al. CAR T-cell therapy for the management of refractory/relapsed high-grade B-cell lymphoma: a practical overview. Bone Marrow Transplant. 2020;55(8):1525–32.
14. Grosser R, Cherkassky L, Chintala N, Adusumilli PS. Combination immunotherapy with CAR T cells and checkpoint blockade for the treatment of solid tumors. Cancer Cell. 2019;36(5):471–82.
15. Park JA, Cheung NV. Limitations and opportunities for immune checkpoint inhibitors in pediatric malignancies. Cancer Treat Rev. 2017;58:22–33.
16. Sterner RC, Sterner RM. CAR-T cell therapy: current limitations and potential strategies. Blood Cancer J. 2021;11(4):69.
17. Lulla P, Heslop HE. Checkpoint inhibition and cellular immunotherapy in lymphoma. Hematology Am Soc Hematol Educ Program. 2016;2016(1):390–6.
18. Fuca G, Reppel L, Landoni E, Savoldo B, Dotti G. Enhancing chimeric antigen receptor T-cell efficacy in solid tumors. Clin Cancer Res. 2020;26(11):2444–51.
19. Maude SL, Laetsch TW, Buechner J, Rives S, Boyer M, Bittencourt H, et al. Tisagenlecleucel in children and young adults with B-cell lymphoblastic leukemia. N Engl J Med. 2018;378(5):439–48.
20. Neelapu SS, Locke FL, Bartlett NL, Lekakis LJ, Miklos DB, Jacobson CA, et al. Axicabtagene ciloleucel CAR T-cell therapy in refractory large B-cell lymphoma. N Engl J Med. 2017;377(26):2531–44.
21. Schuster SJ, Tam CS, Borchmann P, Worel N, McGuirk JP, Holte H, et al. Long-term clinical outcomes of tisagenlecleucel in patients with relapsed or refractory aggressive B-cell lymphomas (JULIET): a multicentre, open-label, single-arm, phase 2 study. Lancet Oncol. 2021;22(10):1403–15.
22. Wang M, Munoz J, Goy A, Locke FL, Jacobson CA, Hill BT, et al. KTE-X19 CAR T-cell therapy in relapsed or refractory mantle-cell lymphoma. N Engl J Med. 2020;382(14):1331–42.
23. Abramson JS, Palomba ML, Gordon LI, Lunning MA, Wang M, Arnason J, et al. Lisocabtagene maraleucel for patients with relapsed or refractory large B-cell lymphomas (TRANSCEND NHL 001): a multicentre seamless design study. Lancet. 2020;396(10254):839–52.
24. Jacobson CA, Chavez JC, Sehgal AR, William BM, Munoz J, Salles G, et al. Axicabtagene ciloleucel in relapsed or refractory indolent non-Hodgkin lymphoma (ZUMA-5): a single-arm, multicentre, phase 2 trial. Lancet Oncol. 2022;23(1):91–103.
25. Munshi NC, Anderson LD Jr, Shah N, Madduri D, Berdeja J, Lonial S, et al. Idecabtagene vicleucel in relapsed and refractory multiple myeloma. N Engl J Med. 2021;384(8):705–16.
26. Shah BD, Ghobadi A, Oluwole OO, Logan AC, Boissel N, Cassaday RD, et al. KTE-X19 for relapsed or refractory adult B-cell acute lymphoblastic leukaemia: phase 2 results of the single-arm, open-label, multicentre ZUMA-3 study. Lancet. 2021;398(10299):491–502.
27. Berdeja JG, Madduri D, Usmani SZ, Jakubowiak A, Agha M, Cohen AD, et al. Ciltacabtagene autoleucel, a B-cell maturation antigen-directed chimeric antigen receptor T-cell therapy in patients with relapsed or refractory multiple myeloma (CARTITUDE-1): a phase 1b/2 open-label study. Lancet. 2021;398(10297):314–24.
28. Locke FL, Miklos DB, Jacobson CA, Perales MA, Kersten MJ, Oluwole OO, et al. Axicabtagene ciloleucel as second-line therapy for large B-cell lymphoma. N Engl J Med. 2022;386(7):640–54.

29. Fowler NH, Dickinson M, Dreyling M, Martinez-Lopez J, Kolstad A, Butler J, et al. Tisagenlecleucel in adult relapsed or refractory follicular lymphoma: the phase 2 ELARA trial. Nat Med. 2022;28(2):325–32.
30. Abramson JS, Solomon SR, Arnason J, Johnston PB, Glass B, Bachanova V, et al. Lisocabtagene maraleucel as second-line therapy for large B-cell lymphoma: primary analysis of the phase 3 TRANSFORM study. Blood. 2023;141(14):1675–84.
31. Hunger SP, Mullighan CG. Acute lymphoblastic leukemia in children. N Engl J Med. 2015;373(16):1541–52.
32. Terwilliger T, Abdul-Hay M. Acute lymphoblastic leukemia: a comprehensive review and 2017 update. Blood Cancer J. 2017;7(6):e577.
33. Kantarjian HM, DeAngelo DJ, Stelljes M, Martinelli G, Liedtke M, Stock W, et al. Inotuzumab ozogamicin versus standard therapy for acute lymphoblastic leukemia. N Engl J Med. 2016;375(8):740–53.
34. Kochenderfer JN, Wilson WH, Janik JE, Dudley ME, Stetler-Stevenson M, Feldman SA, et al. Eradication of B-lineage cells and regression of lymphoma in a patient treated with autologous T cells genetically engineered to recognize CD19. Blood. 2010;116(20):4099–102.
35. Swerdlow SH, Campo E, Pileri SA, Harris NL, Stein H, Siebert R, et al. The 2016 revision of the World Health Organization classification of lymphoid neoplasms. Blood. 2016;127(20):2375–90.
36. Coiffier B, Lepage E, Briere J, Herbrecht R, Tilly H, Bouabdallah R, et al. CHOP chemotherapy plus rituximab compared with CHOP alone in elderly patients with diffuse large-B-cell lymphoma. N Engl J Med. 2002;346(4):235–42.
37. Nowakowski GS, Blum KA, Kahl BS, Friedberg JW, Baizer L, Little RF, et al. Beyond RCHOP: a blueprint for diffuse large B cell lymphoma research. J Natl Cancer Inst. 2016;108(12):djw257.
38. Schuster SJ, Svoboda J, Chong EA, Nasta SD, Mato AR, Anak O, et al. Chimeric antigen receptor T cells in refractory B-cell lymphomas. N Engl J Med. 2017;377(26):2545–54.
39. Nygren L, Wasik AM, Baumgartner-Wennerholm S, Jeppsson-Ahlberg A, Klimkowska M, Andersson P, et al. T-cell levels are prognostic in mantle cell lymphoma. Clin Cancer Res. 2014;20(23):6096–104.
40. Mian A, Hill BT. Brexucabtagene autoleucel for the treatment of relapsed/refractory mantle cell lymphoma. Expert Opin Biol Ther. 2021;21(4):435–41.
41. Sehgal A, Hoda D, Riedell PA, Ghosh N, Hamadani M, Hildebrandt GC, et al. Lisocabtagene maraleucel as second-line therapy in adults with relapsed or refractory large B-cell lymphoma who were not intended for haematopoietic stem cell transplantation (PILOT): an open-label, phase 2 study. Lancet Oncol. 2022;23(8):1066–77.
42. Palumbo A, Anderson K. Multiple myeloma. N Engl J Med. 2011;364(11):1046–60.
43. Kumar SK, Rajkumar V, Kyle RA, van Duin M, Sonneveld P, Mateos MV, et al. Multiple myeloma. Nat Rev Dis Primers. 2017;3:17046.
44. Pratt G. Molecular aspects of multiple myeloma. Mol Pathol. 2002;55(5):273–83.
45. Rajkumar SV. Multiple myeloma: 2020 update on diagnosis, risk-stratification and management. Am J Hematol. 2020;95(5):548–67.
46. Facon T, Kumar S, Plesner T, Orlowski RZ, Moreau P, Bahlis N, et al. Daratumumab plus lenalidomide and dexamethasone for untreated myeloma. N Engl J Med. 2019;380(22):2104–15.
47. Atanackovic D, Radhakrishnan SV, Bhardwaj N, Luetkens T. Chimeric antigen receptor (CAR) therapy for multiple myeloma. Br J Haematol. 2016;172(5):685–98.
48. Shah N, Chari A, Scott E, Mezzi K, Usmani SZ. B-cell maturation antigen (BCMA) in multiple myeloma: rationale for targeting and current therapeutic approaches. Leukemia. 2020;34(4):985–1005.
49. Paul B, Rodriguez C, Usmani SZ. BCMA-targeted biologic therapies: the next standard of care in multiple myeloma therapy. Drugs. 2022;82(6):613–31.
50. Dohner H, Weisdorf DJ, Bloomfield CD. Acute myeloid leukemia. N Engl J Med. 2015;373(12):1136–52.

51. Papaemmanuil E, Gerstung M, Bullinger L, Gaidzik VI, Paschka P, Roberts ND, et al. Genomic classification and prognosis in acute myeloid leukemia. N Engl J Med. 2016;374(23):2209–21.
52. Kayser S, Levis MJ. The clinical impact of the molecular landscape of acute myeloid leukemia. Haematologica. 2023;108(2):308–20.
53. Aitken MJL, Ravandi F, Patel KP, Short NJ. Prognostic and therapeutic implications of measurable residual disease in acute myeloid leukemia. J Hematol Oncol. 2021;14(1):137.
54. Zhao JC, Agarwal S, Ahmad H, Amin K, Bewersdorf JP, Zeidan AM. A review of FLT3 inhibitors in acute myeloid leukemia. Blood Rev. 2022;52:100905.
55. Jonas BA, Pollyea DA. How we use venetoclax with hypomethylating agents for the treatment of newly diagnosed patients with acute myeloid leukemia. Leukemia. 2019;33(12):2795–804.
56. Martelli MP, Martino G, Cardinali V, Falini B, Martinelli G, Cerchione C. Enasidenib and ivosidenib in AML. Minerva Med. 2020;111(5):411–26.
57. Yang X, Wang J. Precision therapy for acute myeloid leukemia. J Hematol Oncol. 2018;11(1):3.
58. Fiorenza S, Turtle CJ. CAR-T cell therapy for acute myeloid leukemia: preclinical rationale, current clinical progress, and barriers to success. BioDrugs. 2021;35(3):281–302.
59. Sentman CL, Meehan KR. NKG2D CARs as cell therapy for cancer. Cancer J. 2014;20(2):156–9.
60. Barber A, Zhang T, Megli CJ, Wu J, Meehan KR, Sentman CL. Chimeric NKG2D receptor-expressing T cells as an immunotherapy for multiple myeloma. Exp Hematol. 2008;36(10):1318–28.
61. Baumeister SH, Murad J, Werner L, Daley H, Trebeden-Negre H, Gicobi JK, et al. Phase I trial of autologous CAR T cells targeting NKG2D ligands in patients with AML/MDS and multiple myeloma. Cancer Immunol Res. 2019;7(1):100–12.
62. Murad JM, Baumeister SH, Werner L, Daley H, Trebeden-Negre H, Reder J, et al. Manufacturing development and clinical production of NKG2D chimeric antigen receptor-expressing T cells for autologous adoptive cell therapy. Cytotherapy. 2018;20(7):952–63.
63. Testa U, Pelosi E, Frankel A. CD 123 is a membrane biomarker and a therapeutic target in hematologic malignancies. Biomark Res. 2014;2(1):4.
64. Hallek M, Cheson BD, Catovsky D, Caligaris-Cappio F, Dighiero G, Dohner H, et al. iwCLL guidelines for diagnosis, indications for treatment, response assessment, and supportive management of CLL. Blood. 2018;131(25):2745–60.
65. Hallek M, Al-Sawaf O. Chronic lymphocytic leukemia: 2022 update on diagnostic and therapeutic procedures. Am J Hematol. 2021;96(12):1679–705.
66. Fischer K, Al-Sawaf O, Bahlo J, Fink AM, Tandon M, Dixon M, et al. Venetoclax and obinutuzumab in patients with CLL and coexisting conditions. N Engl J Med. 2019;380(23):2225–36.
67. Geyer MB, Riviere I, Senechal B, Wang X, Wang Y, Purdon TJ, et al. Safety and tolerability of conditioning chemotherapy followed by CD19-targeted CAR T cells for relapsed/refractory CLL. JCI Insight. 2019;5(9):e122627.
68. Frey NV, Gill S, Hexner EO, Schuster S, Nasta S, Loren A, et al. Long-term outcomes from a randomized dose optimization study of chimeric antigen receptor modified T cells in relapsed chronic lymphocytic leukemia. J Clin Oncol. 2020;38(25):2862–71.
69. Thompson PA, Siddiqi T. Treatment of Richter's syndrome. Hematology Am Soc Hematol Educ Program. 2022;2022(1):329–36.
70. Scheuermann RH, Racila E. CD19 antigen in leukemia and lymphoma diagnosis and immunotherapy. Leuk Lymphoma. 1995;18(5–6):385–97.
71. Poe JC, Minard-Colin V, Kountikov EI, Haas KM, Tedder TF. A c-Myc and surface CD19 signaling amplification loop promotes B cell lymphoma development and progression in mice. J Immunol. 2012;189(5):2318–25.
72. Hekman A, Honselaar A, Vuist WM, Sein JJ, Rodenhuis S, ten Bokkel Huinink WW, et al. Initial experience with treatment of human B cell lymphoma with anti-CD19 monoclonal antibody. Cancer Immunol Immunother. 1991;32(6):364–72.

73. Cianferoni A, Spergel J. The importance of TSLP in allergic disease and its role as a potential therapeutic target. Expert Rev Clin Immunol. 2014;10(11):1463–74.
74. Tirado CA, Lin Y, Tang R, Bajpai A, Yeh W, Karamooz S, Rao A. A CRLF2 rearrangement in a pediatric patient with B-ALL detected by FISH within the context of a complex abnormal karyotype. J Assoc Genet Technol. 2022;48(4):164–7.
75. Erratum: Qin H, Cho M, Haso W, et al. Eradication of B-ALL using chimeric antigen receptor-expressing T cells targeting the TSLPR oncoprotein. Blood. 2015;126(5):629–39. Blood. 2016;128(10):1441.
76. Lanza F, Maffini E, Rondoni M, Massari E, Faini AC, Malavasi F. CD22 expression in B-cell acute lymphoblastic leukemia: biological significance and implications for inotuzumab therapy in adults. Cancers (Basel). 2020;12(2):303.
77. Pan J, Niu Q, Deng B, Liu S, Wu T, Gao Z, et al. CD22 CAR T-cell therapy in refractory or relapsed B acute lymphoblastic leukemia. Leukemia. 2019;33(12):2854–66.
78. Wuchter C, Ratei R, Spahn G, Schoch C, Harbott J, Schnittger S, et al. Impact of CD133 (AC133) and CD90 expression analysis for acute leukemia immunophenotyping. Haematologica. 2001;86(2):154–61.
79. Li D, Hu Y, Jin Z, Zhai Y, Tan Y, Sun Y, et al. TanCAR T cells targeting CD19 and CD133 efficiently eliminate MLL leukemic cells. Leukemia. 2018;32(9):2012–6.
80. Wu Y, Nadler MJ, Brennan LA, Gish GD, Timms JF, Fusaki N, et al. The B-cell transmembrane protein CD72 binds to and is an in vivo substrate of the protein tyrosine phosphatase SHP-1. Curr Biol. 1998;8(18):1009–17.
81. Nix MA, Mandal K, Geng H, Paranjape N, Lin YT, Rivera JM, et al. Surface proteomics reveals CD72 as a target for in vitro-evolved nanobody-based CAR-T cells in KMT2A/MLL1-rearranged B-ALL. Cancer Discov. 2021;11(8):2032–49.
82. Sempowski GD, Lee DM, Kaufman RE, Haynes BF. Structure and function of the CD7 molecule. Crit Rev Immunol. 1999;19(4):331–48.
83. Gomes-Silva D, Srinivasan M, Sharma S, Lee CM, Wagner DL, Davis TH, et al. CD7-edited T cells expressing a CD7-specific CAR for the therapy of T-cell malignancies. Blood. 2017;130(3):285–96.
84. Lu P, Liu Y, Yang J, Zhang X, Yang X, Wang H, et al. Naturally selected CD7 CAR-T therapy without genetic manipulations for T-ALL/LBL: first-in-human phase 1 clinical trial. Blood. 2022;140(4):321–34.
85. Li W, Liang L, Liao Q, Li Y, Zhou Y. CD38: an important regulator of T cell function. Biomed Pharmacother. 2022;153:113395.
86. Mihara K, Yoshida T, Ishida S, Takei Y, Kitanaka A, Shimoda K, et al. All-trans retinoic acid and interferon-alpha increase CD38 expression on adult T-cell leukemia cells and sensitize them to T cells bearing anti-CD38 chimeric antigen receptors. Blood Cancer J. 2016;6(5):e421.
87. Shi J, Zhang Z, Cen H, Wu H, Zhang S, Liu J, et al. CAR T cells targeting CD99 as an approach to eradicate T-cell acute lymphoblastic leukemia without normal blood cells toxicity. J Hematol Oncol. 2021;14(1):162.
88. Maciocia PM, Wawrzyniecka PA, Maciocia NC, Burley A, Karpanasamy T, Devereaux S, et al. Anti-CCR9 chimeric antigen receptor T cells for T-cell acute lymphoblastic leukemia. Blood. 2022;140(1):25–37.
89. Zah E, Lin MY, Silva-Benedict A, Jensen MC, Chen YY. T cells expressing CD19/CD20 bispecific chimeric antigen receptors prevent antigen escape by malignant B cells. Cancer Immunol Res. 2016;4(6):498–508.
90. Chen Z, Liu Y, Chen N, Xing H, Tian Z, Tang K, et al. Loop CD20/CD19 CAR-T cells eradicate B-cell malignancies efficiently. Sci China Life Sci. 2023;66(4):754–70.
91. Pinz K, Liu H, Golightly M, Jares A, Lan F, Zieve GW, et al. Preclinical targeting of human T-cell malignancies using CD4-specific chimeric antigen receptor (CAR)-engineered T cells. Leukemia. 2016;30(3):701–7.

92. Maciocia PM, Wawrzyniecka PA, Philip B, Ricciardelli I, Akarca AU, Onuoha SC, et al. Targeting the T cell receptor beta-chain constant region for immunotherapy of T cell malignancies. Nat Med. 2017;23(12):1416–23.
93. Wu Y, Chen D, Lu Y, Dong SC, Ma R, Tang WY, et al. A new immunotherapy strategy targeted CD30 in peripheral T-cell lymphomas: CAR-modified T-cell therapy based on CD30 mAb. Cancer Gene Ther. 2022;29(2):167–77.
94. Voorhees TJ, Zhao B, Oldan J, Hucks G, Khandani A, Dittus C, et al. Pretherapy metabolic tumor volume is associated with response to CD30 CAR T cells in Hodgkin lymphoma. Blood Adv. 2022;6(4):1255–63.
95. Hayashida K, Bartlett AH, Chen Y, Park PW. Molecular and cellular mechanisms of ectodomain shedding. Anat Rec (Hoboken). 2010;293(6):925–37.
96. Sun C, Mahendravada A, Ballard B, Kale B, Ramos C, West J, et al. Safety and efficacy of targeting CD138 with a chimeric antigen receptor for the treatment of multiple myeloma. Oncotarget. 2019;10(24):2369–83.
97. Drent E, Groen RW, Noort WA, Themeli M, Lammerts van Bueren JJ, Parren PW, et al. Preclinical evaluation of CD38 chimeric antigen receptor engineered T cells for the treatment of multiple myeloma. Haematologica. 2016;101(5):616–25.
98. Mei H, Li C, Jiang H, Zhao X, Huang Z, Jin D, et al. A bispecific CAR-T cell therapy targeting BCMA and CD38 in relapsed or refractory multiple myeloma. J Hematol Oncol. 2021;14(1):161.
99. Radhakrishnan SV, Luetkens T, Scherer SD, Davis P, Vander Mause ER, Olson ML, et al. CD229 CAR T cells eliminate multiple myeloma and tumor propagating cells without fratricide. Nat Commun. 2020;11(1):798.
100. Vander Mause ER, Baker JM, Dietze KA, Radhakrishnan SV, Iraguha T, Omili D, et al. Systematic single amino acid affinity tuning of CD229 CAR T cells retains efficacy against multiple myeloma and eliminates on-target off-tumor toxicity. Sci Transl Med. 2023;15(705):eadd7900.
101. Guo H, Cruz-Munoz ME, Wu N, Robbins M, Veillette A. Immune cell inhibition by SLAMF7 is mediated by a mechanism requiring src kinases, CD45, and SHIP-1 that is defective in multiple myeloma cells. Mol Cell Biol. 2015;35(1):41–51.
102. Gogishvili T, Danhof S, Prommersberger S, Rydzek J, Schreder M, Brede C, et al. SLAMF7-CAR T cells eliminate myeloma and confer selective fratricide of SLAMF7(+) normal lymphocytes. Blood. 2017;130(26):2838–47.
103. Mailankody S, Devlin SM, Landa J, Nath K, Diamonte C, Carstens EJ, et al. GPRC5D-targeted CAR T cells for myeloma. N Engl J Med. 2022;387(13):1196–206.
104. Schmidts A, Ormhoj M, Choi BD, Taylor AO, Bouffard AA, Scarfo I, et al. Rational design of a trimeric APRIL-based CAR-binding domain enables efficient targeting of multiple myeloma. Blood Adv. 2019;3(21):3248–60.
105. Lee L, Lim WC, Galas-Filipowicz D, et al (2023) Limited efficacy of APRIL CAR in patients with multiple myeloma indicate challenges in the use of natural ligands for CAR T-cell therapy [published correction appears in J Immunother Cancer. 11(7):e006699corr1]. J Immunother Cancer. 11(6):e006699. https://doi.org/10.1136/jitc-2023-006699. PMID: 37399355. PMCID: PMC10314698
106. Drach J, Lopez-Berestein G, McQueen T, Andreeff M, Mehta K. Induction of differentiation in myeloid leukemia cell lines and acute promyelocytic leukemia cells by liposomal all-trans-retinoic acid. Cancer Res. 1993;53(9):2100–4.
107. Wang QS, Wang Y, Lv HY, Han QW, Fan H, Guo B, et al. Treatment of CD33-directed chimeric antigen receptor-modified T cells in one patient with relapsed and refractory acute myeloid leukemia. Mol Ther. 2015;23(1):184–91.
108. Tambaro FP, Singh H, Jones E, Rytting M, Mahadeo KM, Thompson P, et al. Autologous CD33-CAR-T cells for treatment of relapsed/refractory acute myelogenous leukemia. Leukemia. 2021;35(11):3282–6.

109. Wang J, Chen S, Xiao W, Li W, Wang L, Yang S, et al. CAR-T cells targeting CLL-1 as an approach to treat acute myeloid leukemia. J Hematol Oncol. 2018;11(1):7.
110. Sauer T, Parikh K, Sharma S, Omer B, Sedloev D, Chen Q, et al. CD70-specific CAR T cells have potent activity against acute myeloid leukemia without HSC toxicity. Blood. 2021;138(4):318–30.
111. John S, Chen H, Deng M, Gui X, Wu G, Chen W, et al. A novel anti-LILRB4 CAR-T cell for the treatment of monocytic AML. Mol Ther. 2018;26(10):2487–95.
112. Pérez Amill L, Armand Ugon M, Peña Pérez S, Val Casals M, Santos C, Frigola G, et al. P1356: CD84-targeted cart-cell therapy: a novel strategy for the treatment of haematological malignancies. Hemasphere. 2023;7(Suppl):e29611a0.
113. Jetani H, Navarro-Bailon A, Maucher M, Frenz S, Verbruggen C, Yeguas A, et al. Siglec-6 is a novel target for CAR T-cell therapy in acute myeloid leukemia. Blood. 2021;138(19):1830–42.
114. Lee WS, Ye Z, Cheung AMS, Goh YPS, Oh HLJ, Rajarethinam R, et al. Effective killing of acute myeloid leukemia by TIM-3 targeted chimeric antigen receptor T cells. Mol Cancer Ther. 2021;20(9):1702–12.
115. Pedersen MG, Moller BK, Bak RO. Recent advances in the development of anti-FLT3 CAR T-cell therapies for treatment of AML. Biomedicines. 2022;10(10):2441.
116. Richards RM, Zhao F, Freitas KA, Parker KR, Xu P, Fan A, et al. NOT-gated CD93 CAR T cells effectively target AML with minimized endothelial cross-reactivity. Blood Cancer Discov. 2021;2(6):648–65.
117. Fialkow PJ, Najfeld V, Reddy AL, Singer J, Steinmann L. Chronic lymphocytic leukaemia: clonal origin in a committed B-lymphocyte progenitor. Lancet. 1978;2(8087):444–6.
118. Ramos CA, Savoldo B, Torrano V, Ballard B, Zhang H, Dakhova O, et al. Clinical responses with T lymphocytes targeting malignancy-associated kappa light chains. J Clin Invest. 2016;126(7):2588–96.
119. Hudecek M, Schmitt TM, Baskar S, Lupo-Stanghellini MT, Nishida T, Yamamoto TN, et al. The B-cell tumor-associated antigen ROR1 can be targeted with T cells modified to express a ROR1-specific chimeric antigen receptor. Blood. 2010;116(22):4532–41.
120. Scarfo I, Ormhoj M, Frigault MJ, et al. Anti-CD37 chimeric antigen receptor T cells are active against B- and T-cell lymphomas. Blood. 2018;132(14):1495–506. Blood. 2018;132(23):2527.
121. Faitschuk E, Hombach AA, Frenzel LP, Wendtner CM, Abken H. Chimeric antigen receptor T cells targeting Fc mu receptor selectively eliminate CLL cells while sparing healthy B cells. Blood. 2016;128(13):1711–22.

CAR-T-Cell Therapy for Solid Tumors

Sridevi Duggirala

Abstract Chimeric antigen receptor T-cell (CAR-T) therapy has transformed cancer treatment, particularly for hematologic malignancies. However, its application in solid tumors presents a complex and challenging landscape. The CAR-T-cell method is a personalized treatment approach in which immune cells from donor cells express recombinant/chimeric antigen receptors on their surface to recognize and target tumor-specific antigens (TSAs). This review provides a comprehensive exploration of CAR-T-cell therapy in solid tumors, with a focus on concepts, current clinical outcomes, challenges faced, and innovative strategies being employed to increase its effectiveness. Through a synthesis of fundamental concepts, clinical advancements, and ongoing research, this chapter aims to offer valuable insights into the evolving field of CAR-T-cell therapy for solid malignancies. This chapter addresses in depth the fundamental concept of CAR-T cells, the current antigens explored, and the challenges for CAR-T cells to traverse the tumor microenvironment (TME).

Keywords Personalized cancer therapy · Targeted immunotherapy · T-cell therapy · Chimeric antigen receptor · CAR-T therapy · Solid tumors · Tumor-associated antigens · Tumor microenvironment · Clinical trials

Abbreviations

CAR-T-cell therapy	Chimeric antigen receptor T-cell therapy
CXCR	C-X-C chemokine receptor
EGFR	Epidermal growth factor receptor
HER2	Human epidermal growth factor receptor 2
MHC	Major histocompatibility complex

S. Duggirala (✉)
Department of Molecular Oncology, Cancer Institute (W.I.A.), Adyar, Chennai, India
e-mail: s.duggirala@cancerinstitutewia.org

A. K. Mishra, S. K. Malonia (eds.), *Cell-based Immunotherapies for Cancer*,
https://doi.org/10.1007/978-3-031-88695-9_6

MSLN	Mesothelin
PD-L1	Programmed death ligand 1
scFv	Single-chain variable fragment
TAA	Tumor-associated antigen
TME	Tumor microenvironment

1 Introduction

Chimeric antigen receptor T-cell (CAR-T) therapy represents an innovation in cancer treatment, with unparalleled success in treating hematologic malignancies by harnessing the body's immune system to combat cancer cells [1, 2]. However, the translation of this revolutionary therapeutic approach to solid tumors introduces a complex and multifaceted challenge, posing significant hurdles owing to the distinct intricacies of the tumor microenvironment (TME) and the diverse heterogeneity inherent in solid cancers [3–5].

The landscape of solid tumors, which spans a spectrum of cancers, such as breast, lung, pancreatic, and colorectal cancers, presents formidable obstacles to the efficacy of CAR-T-cell therapy (Figure 1) [6, 7]. Unlike hematologic malignancies characterized by clearly defined tumor-associated antigens (TAAs), the identification of suitable targets in solid tumors remains an intricate puzzle [8]. Moreover, the interaction between the immune system, tumor cells, and complex tumor microenvironment creates a hostile milieu that often impedes the infiltration, proliferation, and persistence of CAR-T cells within solid tumors, challenging the success of this promising therapy in this domain [9–11].

In this chapter, we delve into the fundamental concepts underpinning CAR-T-cell therapy, exploring its mechanisms of action, the different targets used and target recognition, and the intricate challenges encountered in applying this innovative treatment modality to solid tumors [12]. By synthesizing current clinical outcomes with specific examples of various solid tumors, ongoing research efforts, and innovative strategies, we aim to offer comprehensive insights into the evolving landscape of CAR-T-cell therapy in the context of solid malignancies [13]. This chapter not only highlights the barriers hindering CAR-T-cell efficacy in solid tumors but also provides a glimpse of innovative solutions and emerging strategies that hold promise in transforming the paradigm of cancer treatment.

1.1 Fundamental Concept of CAR-T Cells

Chimeric antigen receptor T-cell (CAR-T) therapy represents a transformative approach in cancer treatment [14, 15], designed to empower a patient's immune system to specifically recognize and eliminate cancer cells. At the heart of

CAR-T-cell therapy lies the engineering of T cells to express synthetic receptors known as chimeric antigen receptors (CARs) [5]. These CARs are composed of distinct domains: an extracellular antigen-recognition domain derived from a single-chain variable fragment (scFv) of an antibody, a transmembrane domain, and an intracellular signaling domain [16, 17]. The scFv segment plays a pivotal role, as it serves as the antigen-binding domain, conferring specificity to CAR-T cells to recognize and engage tumor-specific antigens (TSAs) independently of major histocompatibility complex (MHC) presentation [13, 18]. The effectiveness of CAR-T-cell therapy relies heavily on this scFv, as it determines the CAR's ability to recognize and bind to tumor antigens with high affinity, initiating downstream signaling pathways essential for T-cell activation and the subsequent destruction of cancer cells [7, 19–21]. Several generations of CARs evolved over the period. The challenge for each CAR is to maintain accessibility to the tumor and induce effective lysis.

1.2 Current Clinical Status of CAR-T-Cell Therapy for Solid Tumors

More than 200 interventional trials, many of which are phase 1/phase 2, open-label, single-group assignment trials, have been conducted to assess the effectiveness of CAR-T-cell therapy in solid tumors [7, 22]. Many patients with solid tumor malignancies, such as melanoma, colon, colorectal, cervical, liver, peritoneal, pancreatic, gastric, renal, lung, breast, sarcoma, prostate, thyroid, and retinoblastoma, are part of the studies (Fig. 1) [3, 9, 23]. Clinical trials involving CAR-T cells for solid tumors are emerging vigorously, and Table 1 lists the important antigens employed for successful CAR-T-cell therapy and ongoing clinical trials. Figure 1 lists the important TAAs/TSAs explored for an effective CAR-T-cell therapy pertaining to solid tumors and Fig. 1b describes the complex TME of the solid tumors and the TAAs explored for the effective cancer cell lysis.

2 Potential Target Antigens for CAR-T-Cell Therapy in Lung Cancer

The TAAs that are under evaluation for CAR-T-cell therapy for lung cancer include the epidermal growth factor receptor (EGFR), human epidermal growth factor receptor 2 (HER2), mesothelin (MSLN), CEA (Carcino embryonic Antigen), PSCA (Prostate Stem Cell Antigen) , ROR1(Receptor Tyrosine Kinase-like Orphan Receptor 1), programmed death ligand (PD-L1), and mucin 1 (MUC1) [6, 34].

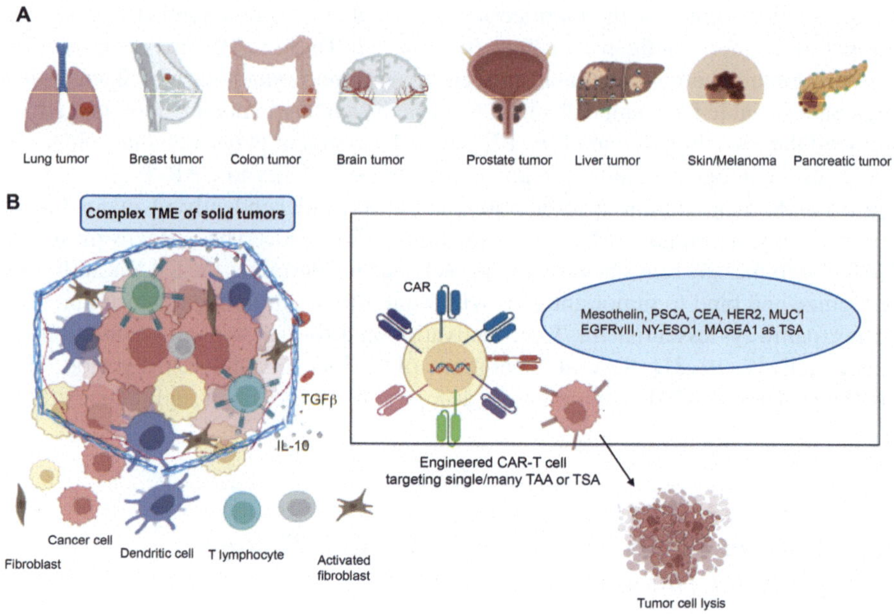

Fig. 1 CAR-T-cell therapy for solid tumors. (**a**) Schematic representation of various organ sites affected by solid tumors. (**b**) Illustration of the complex tumor microenvironment (TME), comprising cancer cells, fibroblasts, dendritic cells, T lymphocytes, and other key components that influence tumor progression and immune response. Engineered CAR-T cells targeting single or multiple TAAs are being explored as a therapeutic strategy to enhance tumor cell lysis and improve treatment efficacy

2.1 EGFR

EGFR is the driver mutation for lung cancer and is expressed at a greatly increased level in lung malignancies, making it a suitable target for CAR-T-cell therapy. A phase 1 clinical trial (NCT01869166) recruited advanced non-small cell lung cancer (NSCLC) patients with more than 50% EGFR-positive expression on tumor cells, and they received CAR-T-cell therapy involving second-generation lentivirus-transduced T cells (EGFR CAR) [4]. Peripheral blood was used to generate CAR-T cells, which were stimulated *in vitro* for 10–13 days before treatment. It was proven to be efficient and safe for the treatment of EGFR-positive relapsed NSCLC. None of the patients displayed toxic side effects in this clinical trial; partial remission was observed for two patients, and five patients were stable for 8 months. The biopsies revealed that EGFR-specific cytotoxicity was achieved by the infiltration of CAR-T cells specifically targeting EGFR. Zhang et al. reported that EGFR variant III CAR-T cells can effectively prevent lung cancer metastasis and recurrence by inhibiting the growth of tumors in vivo and causing tumor cytotoxicity through the release of cytokines such as IFN-γ and TNF-α [1].

Table 1 List of selected CAR-T-cell clinical trials for solid tumors

Clinical trial (NCT number)	CAR developed	TAA chosen	Target tumors	Phase 1	Ongoing status	Reference
NCT04976218	EGFR-TGFβR-KO CAR-T cells	EGFR	EGFR-positive solid tumors	Phase 1	Recruiting	[7]
NCT03182816	Anti-CTLA-4/PD-1-expressing EGFR-CAR-T	EGFR	EGFR-positive advanced solid tumors	Phase 1	Unknown	[24]
NCT02873390	PD-1 antibody–expressing EGFR-CAR-T cells	EGFR	EGFR-positive advanced solid tumors	Phase 1	Unknown	[7]
NCT05341492	EGFR/B7-H3 CAR-T cells	EGFR and B7-H3	Advanced TNBC	Early phase 1	Recruiting	[25]
NCT04489862	MSLN CAR-T cells	MSLN	Non-small cell lung cancer	Early phase 1	Recruiting	[26]
NCT04348643	CEA CAR-T cells	CEA	Solid tumor (metastatic colorectal cancer) Lung cancer Colorectal cancer Liver cancer Pancreatic cancer Gastric cancer Breast cancer	Phase 1 Phase 2	Recruiting	[27]
NCT01109095	HER2 CAR-T cells	HER2	Glioblastoma multiforme	Phase 1	Completed	[28]
NCT03500991	HER2 CAR-T cells	HER2	HER2-positive recurrent/refractory pediatric CNS tumors	Phase 1	Recruiting	[28]
	HER2 CAR-T cells	HER2	HER2-positive gastric cancer	Phase 1 phase 2	Recruiting	
NCT04650451			HER2-positive breast cancer HER2 protein overexpression Solid tumor			[29] [30]
NCT04511871	HER2 CAR-T cells	HER2	Solid tumor Gastric cancer	Phase 1	Recruiting	[31] [31]

(continued)

Table 1 (continued)

Clinical trial (NCT number)	CAR developed	TAA chosen	Target tumors	Phase 1	Ongoing status	Reference
NCT02580747	4-1BB-CD3ζ	Mesothelin	Pancreatic cancer, ovarian tumor, triple-negative breast cancer, malignant mesothelioma, endometrial cancer		Unknown	[32]
NCT03356795	CAR-T cells	GD2, PSMA, MUC1, MSLN, or other markers	Cervical cancer	Phase 1 phase 2	Unknown	[33]

CAR-T (EGFR) cells function as immune killers by modulating the TME in cold tumors such as NSCLC [4]. This finding was confirmed by another clinical trial, NCT04153799, in which C-X-C chemokine receptor (CXCR) type 5–modified anti-EGFR CAR-T cells were used to assess the safety of these CARs in EGFR-positive NSCLC patients [35].

2.2 *HER2*

While HER2 is a highly promising CAR-targeted antigen for lung cancer and has shown promising therapeutic advantages in sarcoma patients, however respiratory distress has been noted 15 minutes after the infusion of 1×10^{10} HER2-targeted CAR-T cells in a few cases. The autoimmune response is associated with low expression of HER2 in lung tumors, potentially jeopardizing the safety and effectiveness of CAR-T cells [36].

2.3 *MSLN*

In NSCLC patients, high tumor aggressiveness and a poor prognosis have been linked to the overexpression of the cell adhesion glycoprotein mesothelin (MSLN) [37]. It is a prime target for investigation during the creation of CARs since it facilitates cancer invasion and metastasis, and low levels of MSLN are found in normal tissues. In preclinical mouse models, MSLN CAR-T-cell therapy decreased the tumor burden. Sixty-nine percent of patients with lung adenocarcinoma express

MSLN [38]. Many clinical trials, such as NCT02580747, are being conducted to treat non-small cell lung cancer (NSCLC) with anti-MSLN CAR-T cells. Both the growth of xenografts and the formation of tumors are suppressed by MLSN-CAR-T cells, and these effects are sustained for an extended period in the internal organs of the mice [39].

2.4 *Programmed Death Ligand 1 (PD-L1)*

The role of PD-1 and PD-L1 during immunotherapy in all tumors has been well-elucidated. By suppressing the cytotoxic T-cell response, T-cell activation, and proliferation, PD-1 and PD-L1 enable cancer cells to evade the immune system [40]. The use of monoclonal antibodies to block the PD-1–PD-L1 interaction has obtained FDA approval. Autologous CAR-T cells that target PD-L1 and CD80/CD86 were investigated in an early phase 1 clinical trial (NCT03060343) to treat recurrent or refractory non-small cell lung cancer [35]. The safety and efficacy of anti-PD-L1 CAR-T-cell therapy in advanced PD-L1-positive NSCLC patients were analyzed in phase 1 clinical research (NCT03330834). The activation of T cells involves the use of scFv from the variable regions of a PD-L1 monoclonal antibody and CD28/4-1BB/CD3ζ [35].

3 Targets of Triple-Negative Breast Cancer (TNBC) for CAR-T-Cell Therapy

TAAs used for CAR-T-cell therapy in breast cancer belong to the RTK (Receptor Tyrosine Kinases) family of surface proteins or serum tumor markers. These genes include HER2 (20–30% of patients present with HER2 overexpression), ERBB1 (15–30% of breast cancer patients present with this positive marker, which is related to poor clinical outcomes and larger tumor sizes at diagnosis), mesothelin (cell surface glycoprotein), CEA, fibroblast activation protein (FAP), and VEGF - R2 (Vascular Endothelial Growth Factor Receptor 2) [30].

3.1 *HER2*

Although the monoclonal antibody targeting HER, trastuzumab, the first FDA-approved targeted therapy, has improved survival and clinical outcomes for patients, CAR-T-cell therapy has also been explored in the context of HER2 (clinical trials: NCT04650451, NCT03740256, NCT03696030, and NCT04430595) [41]. HER2–CAR-T cells target $HER2^{+}$ tumors to reduce growth and regression. The potential of

HER2–CAR-T cells as therapeutic agents was evaluated in mouse xenograft models, in which trastuzumab-resistant cell line–derived xenograft mouse models (JIMT-1) could penetrate the tumor matrix and reduce the size of established solid tumors, thereby resulting in improved long-term survival [42].

3.1.1 Cell Surface Proteins

There are approximately eleven surface proteins whose expression is altered during breast cancer. The expression of cell surface proteins, such as MUC1 and MSN (Moesin), is altered, and their overexpression is associated with unfavorable prognoses and outcomes. A CAR that targets the TnMUC antigen was used in a phase 1 clinical trial of TnMUC1-CAR-T cells (NCT04025216). This CAR is composed of dual CD2 and CD3ζ intracellular signaling domains. This prolonged the CAR-T-cell survival and allowed them to recognize the epitope consisting of the Tn glycan of MUC1. It also includes a mouse anti-human scFv with a CD8a transmembrane region [43]. A variety of MUC1 domains, including the additional domain of the cleaved form of MUC1 (NCT04020575), the aberrantly glycated form of MUC1 (NCT04025216), and the whole MUC1 (NCT02587689), have been used in clinical trials for CAR-T-cell treatment [44]. In the TNBC cell line and xenograft models, third-generation MSLN-CAR-T cells had strong inhibitory effects. The co-administration of PD-1 and MSLN further enhanced T-cell cytotoxicity and persistence. Based on the release of high doses of cytokines, including TNF-α and INF-γ, the third generation of EpCAM-CAR-T cells was shown to be cytotoxic. NCT02915445 is a phase 1 clinical trial that aims to determine the safety and effectiveness of EpCAM-CAR-T cells, specifically the third generation [45].

4 Targets of Glioblastoma (GBM) for CAR-T-Cell Therapy

T-cell therapies for GBM have the limitations of low levels of tumor-specific antigen expression and a low mutational burden. Several clinical trials are in place with potential targets for CAR-T in GBM, and one of the key immunotherapeutic strategies is to employ the target IL13Ra2. It is expressed in more than 75% of GBMs and is a membrane-bound protein that is absent in normal brain tissue and activates the mTOR pathway, which favors tumor growth [46]. A clinical trial (NCT02208362) investigated the efficacy and feasibility of this molecule in patients with recurrent or refractory GBM. When CAR-T cells are given via the intraventricular route, a transient complete response for 7.5 months and an improved quality of life are observed [47].

Patients with recurrent GBM were given a second generation of 4–1BB, CD3ζ anti-EGFRvIII CAR-T cells in the clinical trial NCT02209376, which induced the upregulation of PD-L1, TGF-β (Transforming growth factor-beta), and

IL-10 (Interleukin-10) (which are immunosuppressive), thereby making the therapy less efficient. This trial demonstrated that this CAR could traverse the blood-brain barrier and that there was no cross-reaction with wild-type EGFR [48].

5 CAR-T-Cell Therapy for Pancreatic/Colorectal Cancer

CAR-T-cell therapy for PDAC requires further insights into the TME, improvements in off-target effects, and the incorporation of CAR-T-cell therapy with novel agents, which will improve treatment outcomes for patients with PDAC (Pancreatic ductal adenocarcinoma). Among the many preclinical studies with CAR-T cells for PDAC in mice and cell line models, CAR-T cells targeting mesothelin is the only option explored (NCT03323944) [49]. The key observation is that there is no dose-limiting toxicity, cytokine release syndrome, or neurological issues, and the patients had stable disease. Many clinical trials have been initiated using CEA, HER2, and MUC1, but toxicity and respiratory distress issues have been observed [50].

6 Challenges to Successful CAR-T-Cell Therapy in Solid Tumors

CAR-T-cell application to solid tumors poses several challenges that need to be addressed for successful outcomes. Some of these challenges include the following:

6.1 Tumor Heterogeneity

Solid tumors exhibit significant heterogeneity in terms of antigen expression. Identifying a single, universally expressed target antigen for CAR-T cells in solid tumors is challenging. The inability of CAR-T cells to traverse the surface of the tumor to facilitate binding to the target protein and the presence of subpopulations with different antigen profiles can lead to incomplete elimination of the tumor [14, 49, 51]. This trafficking has been a challenge for glioblastomas known for limited T-cell infiltration, but improvements in T-cell infiltration were observed when CAR-T cells, which express chemokine receptors, were used [35, 47].

In the presence of the extracellular matrix and stromal cells, CAR-T cells must combat the immunosuppressive tumor microenvironment to restrict access to the target tumor cells [51]. TILs and immune cells produce a TME that is antagonistic to the autocrine release of inhibitory cytokines by CAR-T cells. Additionally, the production of cytokines such as CXCL1 (C-X-C motif chemokine ligand), CXCL2, and other CXCL cytokines inhibit CAR-T cells from reaching their target [6, 52].

T-cell exhaustion is the key challenge in CAR-T-cell therapy. The inhibition of T-cell effector function and T-cell proliferation leads to resistance, poor clinical outcomes, and relapse [53, 54]. Exhaustion is a hyporesponsive state of T cells in which elevated expression of inhibitory immune checkpoint receptors and decreased expression of effector cytokines are observed, ultimately leading to the loss of T-cell cytotoxicity [14]. Continuous exposure of T cells to TSA/TAA allows them to differentiate into dysfunctional states with poor function [55].

6.2 *Limited Trafficking and Penetration*

CAR-T-cell therapy has major limitations, with its inability to traffic and infiltrate solid tumors because of the immunosuppressive tumor microenvironment and the tumor stroma, which provide a solid barrier that limits the mobility and penetration of CAR-T cells [5]. One adaptable strategy to combat this is to adapt to the local delivery route rather than the systemic delivery route, which limits the travel of CAR-T cells and minimizes on-target, off-tumor toxicity. Superior treatment results have been achieved for glioblastoma via intraventricular injection of CAR-T cells in clinical trials NCT02208362 and NCT03389230. However, this approach can be used for single-tumor lesions/oligometastatic diseases.

The extracellular matrix of the stroma is primarily composed of heparin sulfate proteoglycans (HSPGs), thereby allowing the CAR-T cells to pass into the tumors. CAR-T cells were engineered to express heparinase to degrade HSPG, and they demonstrated improved antitumor activity and tumor infiltration [5, 56].

6.3 *Antigen Escape and Relapse*

Tumor cells can undergo antigen loss or downregulation, leading to CAR-T-cell resistance and relapse. Solid tumors are highly heterogeneous, and the emergence of antigen-negative tumor variants poses a significant challenge in maintaining the long-term effectiveness of CAR-T-cell therapy. CAR-T cells interact with an adequate number of epitopes of the target surface antigen on cells and are targeted to kill the cells. However, tumor cells bypass this mechanism by expressing different forms of antigen (without the epitope), and this process is referred to as antigen escape [57]. Lineage switching and antigen downregulation are mechanisms of antigen escape, where the former refers to the antigen expression of phenotypically different but genetically related diseases, and the latter is the mechanism in which surface target antigen expression is very low such that CAR-T-cell activation escapes [58]. Single-antigen-targeting CARs were initially used as good deliverables, but the tumors of the patients treated with these CARs displayed the loss of this antigen. This antigen escape is a challenge for acute lymphoid leukemia patients who are administered a CD19-targeted CAR. Approximately 30–60% of relapsed patients exhibit downregulation of CD19 [59]. Antigen escape is also an issue for glioblastoma, where a CAR targeting IL13Ra2 is used. Tumor recurrence is associated with decreased IL13Ra2 expression [47].

6.4 Cytokine Release Syndrome (CRS) and Neurotoxicity

Systemic activation of CAR-T cells can lead to severe side effects, such as cytokine release syndrome (CRS) and neurotoxicity. These side effects can be particularly challenging to manage in solid tumor settings, where localized effects might be preferred. The major toxicities that arise from CAR-T cells in humans and have not been revealed in mouse models include CRS and neurotoxicity syndrome [60]. The interaction between CAR-T cells and tumor cells activates host macrophages (bystander cells). The pyroptotic products of tumor cells, such as damage-associated molecular patterns (DAMPs), ATP, and dsDNA, activate macrophages. This overactivation of immune effector cells leads to abnormal levels of IL-1, IL-6, IFN-γ and GM -CSF (granulocyte-macrophage colony-stimulating factor) production, thereby leading to constitutional symptoms of CRS [61]. Tumor pyroptosis, unlike apoptosis, is a form of programmed cell death that is characterized by cell swelling, lysis, and pro-inflammatory factor release and is a prelude to the onset of CRS. This massive release of cytokines leads to clinical manifestations such as myalgia, arthralgia, rash, fever, diarrhea, cardiac arrest, hypotension, and renal/multi-organ failure and may lead to death [62].

6.5 Optimal CAR-T-Cell Design

Designing CAR-T cells with the right co-stimulatory signals and cytokine profiles is critical. Determining the optimal combination of CAR design elements to achieve sustained antitumor activity without inducing excessive toxicity is an ongoing challenge. The optimal design of CAR-T cells is a challenge for solid tumors because of the thick stroma. The "Warburg effect" of tumor cells promotes the immunosuppressive function of cytotoxic T cells. As mentioned previously, the high frequency of immunosuppressive signals such as PD-1/PD-L1 impairs the proliferation of CAR-T cells [63]. CAR-T-cell antigen–binding is key for an effective therapeutic response, and it must bind to its target epitope for CAR-T-cell activation and cytokine secretion. A minimum threshold level of this activation is essential because, if surpassed, it results in toxic levels of cytokines and the consequences of CRS. It is important for optimum activation during the therapeutic window of a thoroughly engineered CAR whose activation kinetics depend on the antigen on the malignant cells, its binding affinity to the epitope, the tumor burden, and the co-stimulatory elements on the CAR [64]. The hinge and the transmembrane are possible modulation sites for the activation of CAR-T cells. Alterations in the amino acid region of the CD8-α-derived hinge and transmembrane in the CD-19-targeted CAR resulted in low levels of cytokine release and reduced CAR-T-cell proliferation [65]. This optimization resulted in decreased toxicity during the phase 1 clinical trials, with complete remission in 54.5% of the B-cell lymphoma patients who presented with reduced/no CRS events.

7 Emerging Solutions

Several emerging solutions are being explored to address the challenges of applying CAR-T-cell therapy to solid tumors (Fig. 2).

7.1 Multi-antigen Targeting

The development of CAR-T cells that target multiple antigens expressed on solid tumors can mitigate the risk of antigen escape. The clinical efficacy, durability, and survival of dual-targeting CAR-T-cell therapy have been well developed to reduce

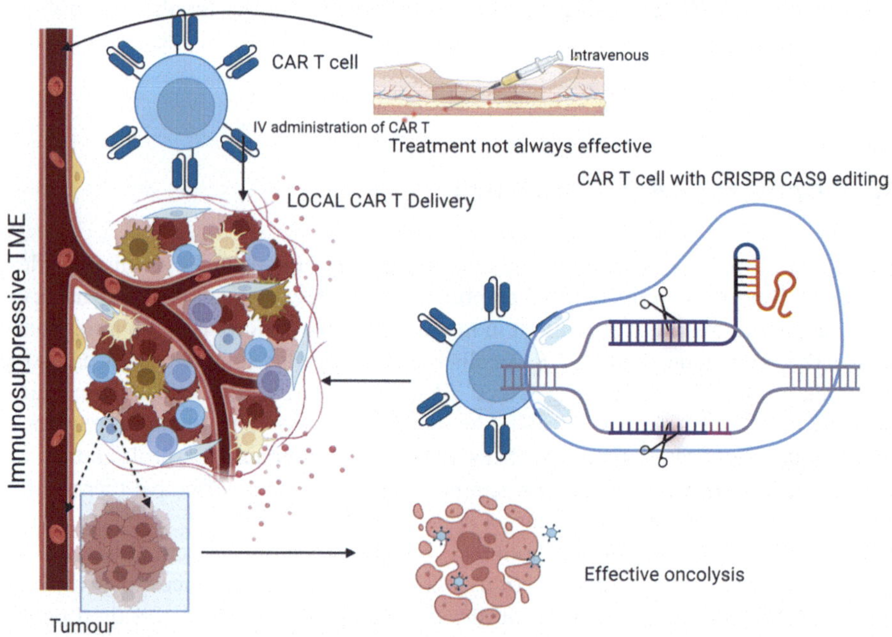

Fig. 2 Strategies to overcome immunosuppressive tumor microenvironment. This figure illustrates various approaches to enhance CAR-T-cell efficacy in immunosuppressive tumors, including multi-antigen targeting to improve tumor recognition, localized CAR-T-cell administration to minimize off-target effects and enhance infiltration, and CRISPR/Cas9 gene editing to optimize CAR-T-cell function and resistance to inhibitory signals

the risk of relapse. Dual-targeting CAR-T-cell therapy refers to two pooled single-CAR-T-cell products with different antigen specificities or a single-CAR-T-cell product that can target two TAAs (Bi-CAR-Ts). The main structures used are the Si-CAR, bivalent tandem CAR, bivalent loop CAR, and bicistronic CAR. The dual CAR-T cells are those for which the two individual CARs target different antigens in a single T cell and the tandem CAR-T cells wherein the two antigen-binding domains are connected in tandem to a single CAR.

By targeting several antigens simultaneously, these engineered cells can increase efficacy and reduce the chances of tumor recurrence due to antigen loss [66]. As mentioned before, antigen escape is the key challenge in CAR-T-cell immunotherapy, and multi-antigen-targeted CAR-T cells are being explored for their ability to reduce relapse effectively. Some of the major multi-antigen-targeted therapies include pooled CAR-T cells, in which two engineered T-cell lines (each expressed as a distinct antigen-specific CAR) are explored [67]. Pooled CAR-T-cell engineering involves a combination of T-cell lines, each expressing a distinct antigen-specific CAR. Boolean logic gates constitute the strategy included here. It comprises "AND," "OR," and NOT gates [68]. CAR-T cells with the "AND" gate require both predetermined antigens for activation, and those with the "OR" gate require either one of the antigens for activation. CAR-T cells with NOT gates are useful for identifying normal cells by deactivation if a particular antigen is detected [68, 69].

7.2 Combination Therapies

Combining CAR-T-cell therapy with other treatments, such as checkpoint inhibitors, targeted therapies, or traditional chemotherapy/radiotherapy, can increase the effectiveness of CAR-T-cell therapy in solid tumors [70]. These combinations can help modulate the immunosuppressive TME and improve CAR-T-cell persistence and infiltration. Inhibited auto-immunity is observed after many cycles of chemotherapy, increasing the efficacy of adoptive T-cell transfer in patients [71]. Chemotherapy prior to CAR-T-cell administration improved their sustenance by inhibiting autoimmunity and removing suppressive cells. This conditioning chemotherapy can reduce the immunogenicity of CAR-T cells, thereby allowing them to secrete only cytokines that are homeostatic to CAR-T cells to reduce the effects of CRS [72].

Radiotherapy has been proven to increase the immune response by reversing T-cell exhaustion, releasing tumor antigens that can augment CAR-T-cell function, and recruiting cytotoxic T cells. Radiotherapy sensitizes antigen-negative tumor cells to CAR-T-cell-mediated apoptosis via TRAIL (Tumor Necrosis Factor-related Apoptosis-Inducing Ligand) [73]. In addition to chemotherapy/radiotherapy, standard treatment regimens combined with CAR-T-cell therapy include oncolytic viruses, cytokines, cancer vaccines, checkpoint inhibitors, and metabolic inhibitors [74]. The standard dosage of carboplatin had a synergistic effect on CAR-T-cell therapy, which targets Erb-B, and the tumor cells are sensitive to degradation [75].

Oncolytic virotherapy is a novel cancer immunotherapy in which cancer killer viruses infect tumor cells selectively so that TAAs are present in the immune system as vaccines and can be genetically engineered [76].

Combinations of cancer vaccines with CAR-T-cell therapy are used to address the downregulation of target antigens, and these vaccines can promote CAR-T cells through stimulation via antigen-presenting cells (APCs). An engineered helper-dependent oncolytic adenovirus that expresses a PD-L1 blocking antibody (CAd-VECPD-L1 administered intramurally), when evaluated for its antitumor efficacy in combination with (HER2)-specific CAR-T cells in a prostate cancer mouse xenograft model, resulted in better outcomes and prolonged survival than individual monotherapies [75].

7.3 PD-1 Blockade Improves CAR-T-Cell Therapy Outcomes

To evade T-cell recognition and destruction, cancer cells utilize a vital signaling complex, the PD-1/PDL-1 complex [77, 78]. PD-1 is a receptor protein present in T cells and functions as a checkpoint molecule that works during the immune surveillance response for the self-recognition of normal cells [79]. PD-L1 is expressed mostly under inflammatory conditions by macrophages and activated T and B cells. PD-1 and its ligand PD-L1 inhibit T survival, activation, proliferation, and cytotoxic secretion in cancer cells. Blockage of PD-1 along with CAR-T-cell administration is promising, and improvements in the treatment outcomes of solid tumors have been reported [80]. The PD-1 gene alteration disrupted CAR-T-cell function, improved antitumor activity, and had fewer toxicity effects on T-cell proliferation [81]. CRISPR/Cas9 technology was used to disrupt PD-1s, and trials are underway to generate PD-1 knockout CAR-T cells to target various solid malignancies. In addition to increased production of cytokines such as IFN-γ and IL-2, there is increased antitumor activity in knockouts [82].

CAR-T cells were engineered to release PD-1 scFv and then combined with PD-antibody therapy. The combined treatment exhibited superior antitumor effectiveness. To improve the delivery of CAR-T cells and improve their effectiveness, the precise localized delivery of PD-1 scFv works in both a paracrine and autocrine manner [83].

7.4 Engineering CAR-T Cells for TME Adaptation

Strategies to alter the TME to improve CAR-T-cell activity are being explored (Fig. 2). This includes the use of agents to normalize the tumor vasculature, reduce the number of immunosuppressive cells, or alter the extracellular matrix to improve CAR-T-cell infiltration [84]. Modifying CAR-T cells to resist the inhibitory signals present in the TME can enhance their functionality within solid tumors. This includes incorporating genetic modifications that increase resistance to immunosuppressive factors or providing them with receptors that recognize and counteract these signals. Enabling T cells with specific cytokines improved the efficacy of the therapy [85]. T cells with tumor-specific chemokine receptors that target Hodgkin's tumor were engineered such that high levels of thymus- and activation-regulated chemokine/CC chemokine ligand 17 (TARC/CCL17) and macrophage-derived chemokines are produced by the engineered cells [86].

7.5 Local Delivery Methods

The technology of administering CAR-T cells through locoregional delivery mechanisms is gaining momentum and being used to treat solid tumors effectively. Directly administering CAR-T cells into the tumor site or using localized delivery approaches can improve their penetration into solid tumors (Fig. 2) [87]. Strategies such as intra-tumoral injection or regional perfusion can help overcome barriers to infiltration and reduce systemic toxicity. The use of different modalities, such as imaging/specialized equipment, cost, and technology, is a challenge for delivering CAR-T cells [88]. Access to the tumor site is a challenge for locoregional delivery, but it has been well-addressed for glioma and hepatic carcinoma, thereby reducing the degree of peripheral toxicity induced by chemo- or CAR-T-cell therapy. A delivery system using the Rickham reservoir/catheter system for gliomas, where catheters can be placed into the tumor bed/resection cavity/ventricular regions (intraventricular device) to target local recurrence, was adopted for adult and pediatric cancer patients. Measuring the persistence of CAR-T cells and the levels of inflammatory cytokines through these devices is feasible, although the installation of these devices is a greater risk than a simple IV [88].

7.6 Genetic Engineering Advances

Continued advancements in genetic engineering techniques have enabled the development of more sophisticated CAR designs. The incorporation of inducible safety switches, the optimization of CAR signaling domains, and the use of gene-editing tools such as CRISPR/Cas9 to increase CAR-T-cell persistence and function are actively being researched [89]. Editing of cellular gene profiles is achieved via CRISPR/CAS9 technology, where the gene loci involved in the expression of immune checkpoint inhibitors can be deleted. Improved gene editing in primary T cells, especially with CAR-T cells, has enabled the development of next-generation CAR-T cells [90]. An improved inhibition of the tumor response and enhanced cytotoxicity were observed when CAR-T cells were used to target mesothelin, and the PDCD1 gene locus was knocked out via CRISPR/CAS9 [91]. Elimination of many genes enhances the persistence of CAR-T cells and tackles T-cell exhaustion

[92]. Studies by Alishah et al. revealed that knockout of the TGFβRII gene in CAR-T cells improved resistance to exhaustion, and the durability of these cells improved, as TGF-β is the key inhibitory cytokine in the tumor environment, and knockout of receptors of this gene resulted in improved outcomes [93].

7.7 *Next-Generation CAR Constructs*

Advancements in CAR design, such as incorporating synthetic biology elements, developing universal CAR-T cells, or utilizing different immune cell types, such as natural killer cells or gamma delta T cells, hold promise for improving the effectiveness of CAR-T-cell therapy in solid tumors [94]. Improved gene editing in primary T cells, especially with CAR-T cells, has enabled progress in the next generation of CAR-T cells [95].

The objective of CAR-T-cell therapy is to transition the TME from an immunosuppressive nature to an immunopermissive nature for an efficient immune response [95]. The fourth generation of CARs involves the modification of the second generation of CARs by the constitutive expression of IL-12. Those T cells that are engineered with fourth-generation CARs are termed T cells redirected for universal cytokine-mediated killing (TRUCKS). Here, cytokines are used as chemoattractants for T-cell proliferation and survival [96]. These CAR-T cells are engineered to carry cassettes one for the nuclear factor of the activated T-cell (NFAT)–responsive cassette and another for the transgenic cytokines such as IL-12. Transgene expression is induced when CD3ζ-containing CARs encounter their specific target. Tumor killing occurs via the activation of these CARs, which triggers the production and secretion of particular cytokines to promote tumor cytotoxicity. Mechanisms such as exocytosis or ligand death receptors facilitate tumor killing [97]. The fifth generation of CARs is also in the pipeline and is being actively explored. It is also based on second-generation CARs, which contain a truncated cytoplasmic IL-2 receptor and a binding site for the transcription factor STAT3 [98].

8 Conclusion

CAR-T-cell therapy represents a groundbreaking advancement in cancer treatment, demonstrating remarkable success in hematologic malignancies. However, its efficacy in solid tumors remains limited due to challenges such as tumor heterogeneity, immunosuppressive microenvironments, and physical barriers that hinder cell infiltration. Addressing these obstacles through strategies that enhance CAR-T-cell persistence, expand antigen specificity, and modulate the tumor microenvironment is essential for improving therapeutic outcomes. Additionally, identifying reliable biomarkers for patient selection could further refine treatment strategies, leading to more precise and effective applications of CAR-T therapy in solid tumors. Continued research and innovation in this field will be crucial for unlocking the full potential of CAR-T-cell therapy across a broader range of cancers.

Acknowledgments The author would like to acknowledge SERB TARE (Science and Engineering Research Board, Ministry of Science and Technology) grant number, TAR/2022/000213, Ministry of Science and Technology, Government of India, for funding the ongoing projects.

References

1. Zhang Z, Jiang J, Wu X, Zhang M, Luo D, Zhang R, et al. Chimeric antigen receptor T cell targeting EGFRvIII for metastatic lung cancer therapy. Front Med. 2019;13(1):57–68.
2. Albelda SM. CAR T cell therapy for patients with solid tumors: key lessons to learn and unlearn. Nat Rev Clin Oncol. 2024;21(1):47–66.
3. Albelda SM. CAR T cell therapy for patients with solid tumors: key lessons to learn and unlearn. Nat Rev Clin Oncol. 2023;21:1–20.
4. Feng K, Guo Y, Dai H, Wang Y, Li X, Jia H, et al. Chimeric antigen receptor-modified T cells for the immunotherapy of patients with EGFR-expressing advanced relapsed/refractory non-small cell lung cancer. Sci China Life Sci. 2016;59(5):468–79.
5. Sterner RC, Sterner RM. CAR-T cell therapy: current limitations and potential strategies. Blood Cancer J. 2021;11(4):69.
6. Kandra P, Nandigama R, Eul B, Huber M, Kobold S, Seeger W, et al. Utility and drawbacks of Chimeric Antigen Receptor T Cell (CAR-T) therapy in lung cancer. Front Immunol. 2022;13:903562.
7. Maalej KM, Merhi M, Inchakalody VP, Mestiri S, Alam M, Maccalli C, et al. CAR-cell therapy in the era of solid tumor treatment: current challenges and emerging therapeutic advances. Mol Cancer. 2023;22(1):20.
8. Braun DA, Wu CJ. Antigen discovery and therapeutic targeting in hematologic malignancies. Cancer J. 2017;23(2):115–24.
9. Lindo L, Wilkinson LH, Hay KA. Befriending the hostile tumor microenvironment in CAR T-cell therapy. Front Immunol. 2020;11:618387.
10. Guzman G, Reed MR, Bielamowicz K, Koss B, Rodriguez A. CAR-T therapies in solid tumors: opportunities and challenges. Curr Oncol Rep. 2023;25(5):479–89.
11. Liu G, Rui W, Zhao X, Lin X. Enhancing CAR-T cell efficacy in solid tumors by targeting the tumor microenvironment. Cell Mol Immunol. 2021;18(5):1085–95.
12. Hou AJ, Laurence LC, Chen YY. Navigating CAR-T Cells through the solid-tumor microenvironment. Nat Rev Drug Discov. 2021;20(7):531–50.
13. Brown CE, Mackall CL. CAR T cell therapy: inroads to response and resistance. Nat Rev Immunol. 2019;19(2):73–4.
14. Gumber D, Wang LD. Improving CAR-T immunotherapy: overcoming the challenges of T cell exhaustion. EBioMedicine. 2022;77:103941.
15. Borch TH, Andersen R, Ellebaek E, Met Ö, Donia M, Svane IM. Future role for adoptive T-cell therapy in checkpoint inhibitor-resistant metastatic melanoma. J Immunother Cancer. 2020;8(2):e000668.
16. Hwang J, Byeon Y, Kim D, Park S. Recent insights of T cell receptor-mediated signaling pathways for T cell activation and development. Exp Mol Med. 2020;52(5):750–61.
17. Sadelain M, Rivière I, Riddell S. Therapeutic T cell engineering. Nature. 2017;545(7655):423–31.
18. Verdegaal EME. Adoptive cell therapy: a highly successful individualized therapy for melanoma with great potential for other malignancies – ScienceDirect. Curr Opin Immunol. 2016;39:90–5.

19. Glover M, Avraamides S, Maher J. How can we engineer CAR T cells to overcome resistance? Biologics. 2021;15:175–98.
20. Muñoz-López P, Ribas-Aparicio RM, Becerra-Báez EI, Fraga-Pérez K, Flores-Martínez LF, Mateos-Chávez AA, et al. Single-chain fragment variable: recent progress in cancer diagnosis and therapy. Cancers (Basel). 2022;14(17):4206.
21. Mao R, Kong W, He Y. The affinity of antigen-binding domain on the antitumor efficacy of CAR T cells: moderate is better. Front Immunol. 2022;13:1032403.
22. Moghimi B, Barrett D. CAR T cells for solid tumors. Curr Stem Cell Rep. 2017;3:269–78.
23. Holzinger A, Abken H. Advances and challenges of CAR T cells in clinical trials. Recent Results Cancer Res. 2020;214:93–128.
24. Zhang Y, Zhang Z, Ding Y, Fang Y, Wang P, Chu W, et al. Phase I clinical trial of EGFR-specific CAR-T cells generated by the piggyBac transposon system in advanced relapsed/refractory non-small cell lung cancer patients. J Cancer Res Clin Oncol. 2021;147(12):3725–34.
25. Geng P, Chi Y, Yuan Y, Yang M, Zhao X, Liu Z, et al. Novel chimeric antigen receptor T cell-based immunotherapy: a perspective for triple-negative breast cancer. Front Cell Dev Biol. 2023;11:1158539.
26. Wang C, Yuan X, Xue J. Targeted therapy for rare lung cancers: status, challenges, and prospects. Mol Ther. 2023;31(7):1960–78.
27. Zhang C, Wang Z, Yang Z, Wang M, Li S, Li Y, et al. Phase I escalating-dose trial of CAR-T therapy targeting CEA+ metastatic colorectal cancers. Mol Ther. 2017;25(5):1248–58.
28. Ahmed N, Brawley V, Hegde M, Bielamowicz K, Wakefield A, Ghazi A, et al. Autologous HER2 CMV bispecific CAR T cells are safe and demonstrate clinical benefit for glioblastoma in a Phase I trial. J Immunother Cancer. 2015;3(Suppl 2):O11.
29. Mercogliano MF, Bruni S, Mauro FL, Schillaci R. Emerging targeted therapies for HER2-positive breast cancer. Cancers (Basel). 2023;15(7):1987.
30. Yang Y, Liu J, Lu C, Wei J. CAR-T cell therapy for breast cancer: from basic research to clinical application. Int J Biol Sci. 2022;18(6):2609–26.
31. Yeo D, Giardina C, Saxena P, Rasko JEJ. The next wave of cellular immunotherapies in pancreatic cancer. Mol Ther Oncolytics. 2022;24:561–76.
32. Zhai X, Mao L, Wu M, Liu J, Yu S. Challenges of anti-mesothelin CAR-T-cell therapy. Cancers (Basel). 2023;15(5):1357.
33. Ferrall L, Lin KY, Roden RBS, Hung C, Wu T-. Cervical cancer immunotherapy: facts and hopes. Clin Cancer Res. 2021;27(18):4953–73.
34. Lu Y, Zeng T, Zhang H, Li Y, Zhu X, Liu H, et al. Nano-immunotherapy for lung cancer. Nano TransMed. 2023;2(1):e9130018.
35. Qu J, Mei Q, Chen L, Zhou J. Chimeric antigen receptor (CAR)-T-cell therapy in non-small-cell lung cancer (NSCLC): current status and future perspectives. Cancer Immunol Immunother. 2021;70(3):619–31.
36. Morgan RA, Yang JC, Kitano M, Dudley ME, Laurencot CM, Rosenberg SA. Case report of a serious adverse event following the administration of T cells transduced with a chimeric antigen receptor recognizing ERBB2. Mol Ther. 2010;18(4):843–51.
37. Thomas A, Chen Y, Steinberg SM, Luo J, Pack S, Raffeld M, et al. High mesothelin expression in advanced lung adenocarcinoma is associated with KRAS mutations and a poor prognosis. Oncotarget. 2015;6(13):11694–703.
38. Zhong S, Cui Y, Liu Q, Chen S. CAR-T cell therapy for lung cancer: a promising but challenging future. J Thorac Dis. 2020;12(8):4516–21.
39. Chen L, Chen F, Li J, Pu Y, Yang C, Wang Y, et al. CAR-T cell therapy for lung cancer: potential and perspective. Thorac Cancer. 2022;13(7):889–99.
40. Sacher AG, Gandhi L. Biomarkers for the clinical use of PD-1/PD-L1 inhibitors in non-small-cell lung cancer: a review. JAMA Oncol. 2016;2(9):1217–22.
41. He Q, Hu H, Yang F, Song D, Zhang X, Dai X. Advances in chimeric antigen receptor T cells therapy in the treatment of breast cancer. Biomed Pharmacother. 2023;162:114609.

42. Tóth G, Szöllősi J, Abken H, Vereb G, Szöőr Á. A small number of HER2 redirected CAR T cells significantly improves immune response of adoptively transferred mouse lymphocytes against human breast cancer xenografts. Int J Mol Sci. 2020;21(3):1039.
43. Yin L, Chen G, Xiang Z, Liu Y, Li X, Bi J, et al. Current progress in chimeric antigen receptor-modified T cells for the treatment of metastatic breast cancer. Biomed Pharmacother. 2023;162:114648.
44. Zhang H, Zhu S, Deng W, Li R, Zhou H, Xiong H. The landscape of chimeric antigen receptor T cell therapy in breast cancer: perspectives and outlook. Front Immunol. 2022;13:887471.
45. Landeros N, Castillo I, Pérez-Castro R. Preclinical and clinical trials of new treatment strategies targeting cancer stem cells in subtypes of breast cancer. Cells. 2023;12(5):720.
46. Zeng J, Zhang J, Yang Y, Wang F, Jiang H, Chen H, et al. IL13RA2 is overexpressed in malignant gliomas and related to clinical outcome of patients. Am J Transl Res. 2020;12(8):4702–14.
47. Brown CE, Alizadeh D, Starr R, Weng L, Wagner JR, Naranjo A, et al. Regression of glioblastoma after chimeric antigen receptor T-cell therapy. N Engl J Med. 2016;375(26):2561–9.
48. O'Rourke DM, Nasrallah MP, Desai A, Melenhorst JJ, Mansfield K, Morrissette JJD, et al. A single dose of peripherally infused EGFRvIII-directed CAR T cells mediates antigen loss and induces adaptive resistance in patients with recurrent glioblastoma. Sci Transl Med. 2017;9(399):eaaa0984.
49. Akce M, Zaidi MY, Waller EK, El-Rayes BF, Lesinski GB. The potential of CAR T cell therapy in pancreatic cancer. Front Immunol. 2018;9:2166.
50. Tran E, Longo DL, Urba WJ. A milestone for CAR T cells. N Engl J Med. 2017;377(26):2593–6.
51. Johnson A, Townsend M, O'Neill K. Tumor microenvironment immunosuppression: a roadblock to CAR T-cell advancement in solid tumors. Cells. 2022;11(22):3626.
52. Ahrends T, Borst J. The opposing roles of CD4+ T cells in anti-tumor immunity. Immunology. 2018;154(4):582–92.
53. Andrea AE, Chiron A, Mallah S, Bessoles S, Sarrabayrouse G, Hacein-Bey-Abina S. Advances in CAR-T cell genetic engineering strategies to overcome hurdles in solid tumors treatment. Front Immunol. 2022;13:830292.
54. Qin Y, Xu G. Enhancing CAR T-cell therapies against solid tumors: mechanisms and reversion of resistance. Front Immunol. 2022;13:1053120.
55. Majzner RG, Mackall CL. Tumor antigen escape from CAR T-cell therapy. Cancer Discov. 2018;8(10):1219–26.
56. Zhang B, Qin D, Mo Z, Li Y, Wei W, Wang Y, et al. Hurdles of CAR-T cell-based cancer immunotherapy directed against solid tumors. Sci China Life Sci. 2016;59(4):340–8.
57. Hegde M, Mukherjee M, Grada Z, Pignata A, Landi D, Navai SA, et al. Tandem CAR T cells targeting HER2 and IL13Rα2 mitigate tumor antigen escape. J Clin Invest. 2016;126(8):3036–52.
58. Shah NN, Fry TJ. Mechanisms of resistance to CAR T cell therapy. Nat Rev Clin Oncol. 2019;16(6):372–85.
59. Fousek K, Watanabe J, Joseph SK, George A, An X, Byrd TT, et al. CAR T-cells that target acute B-lineage leukemia irrespective of CD19 expression. Leukemia. 2021;35(1):75–89.
60. Zhang Y, Qin D, Shou AC, Liu Y, Wang Y, Zhou L. Exploring CAR-T cell therapy side effects: mechanisms and management strategies. J Clin Med. 2023;12(19):6124.
61. Xiao X, Huang S, Chen S, Wang Y, Sun Q, Xu X, et al. Mechanisms of cytokine release syndrome and neurotoxicity of CAR T-cell therapy and associated prevention and management strategies. J Exp Clin Cancer Res. 2021;40:367.
62. Chohan KL, Siegler EL, Kenderian SS. CAR-T cell therapy: the efficacy and toxicity balance. Curr Hematol Malig Rep. 2023;18(2):9–18.
63. Liu Z, Zhou Z, Dang Q, Xu H, Lv J, Li H, et al. Immunosuppression in tumor immune microenvironment and its optimization from CAR-T cell therapy. Theranostics. 2022;12(14):6273–90.
64. Shafer P, Kelly LM, Hoyos V. Cancer therapy with TCR-engineered T cells: current strategies, challenges, and prospects. Front Immunol. 2022;13:835762.
65. Stoiber S, Cadilha BL, Benmebarek M, Lesch S, Endres S, Kobold S. Limitations in the design of chimeric antigen receptors for cancer therapy. Cells. 2019;8(5):472.

66. Rafiq S, Hackett CS, Brentjens RJ. Engineering strategies to overcome the current roadblocks in CAR T cell therapy. Nat Rev Clin Oncol. 2020;17(3):147–67.
67. Han X, Wang Y, Wei J, Han W. Multi-antigen-targeted chimeric antigen receptor T cells for cancer therapy. J Hematol Oncol. 2019;12:128.
68. Jo Y, Ali LA, Shim JA, Lee BH, Hong C. Innovative CAR-T cell therapy for solid tumor; current duel between CAR-T spear and tumor shield. Cancers (Basel). 2020;12(8):2087.
69. Jensen MC, Riddell SR. Design and implementation of adoptive therapy with chimeric antigen receptor-modified T cells. Immunol Rev. 2014;257(1):127–44.
70. Ando M, Ito M, Srirat T, Kondo T, Yoshimura A. Memory T cell, exhaustion, and tumor immunity. Immunol Med. 2020;43(1):1–9.
71. Korell F, Berger TR, Maus MV. Understanding CAR T cell-tumor interactions: paving the way for successful clinical outcomes. Med. 2022;3(8):538–64.
72. Wang AX, Ong XJ, D'Souza C, Neeson PJ, Zhu JJ. Combining chemotherapy with CAR-T cell therapy in treating solid tumors. Front Immunol. 2023;14:1140541.
73. Qin VM, Haynes NM, D'Souza C, Neeson PJ, Zhu JJ. CAR-T plus radiotherapy: a promising combination for immunosuppressive tumors. Front Immunol. 2022;12:813832.
74. Al-Haideri M, Tondok SB, Safa SH, Maleki AH, Rostami S, Jalil AT, et al. CAR-T cell combination therapy: the next revolution in cancer treatment. Cancer Cell Int. 2022;22:365.
75. Zhang X, Wu Y, Xu T, Cui M. CAR-T cells in the treatment of ovarian cancer: a promising cell therapy. Biomol Ther. 2023;13(3):465.
76. Muthukutty P, Yoo SY. Oncolytic virus engineering and utilizations: cancer immunotherapy perspective. Viruses. 2023;15(8):1645.
77. Francisco LM, Sage PT, Sharpe AH. The PD-1 pathway in tolerance and autoimmunity. Immunol Rev. 2010;236:219–42.
78. Huang MY, Jiang XM, Wang BL, Sun Y, Lu JJ. Combination therapy with PD-1/PD-L1 blockade in non-small cell lung cancer: strategies and mechanisms. Pharmacol Ther. 2021;219:107694.
79. Song W, Zhang M. Use of CAR-T cell therapy, PD-1 blockade, and their combination for the treatment of hematological malignancies. Clin Immunol. 2020;214:108382.
80. Fedorov VD, Themeli M, Sadelain M. PD-1– and CTLA-4–based inhibitory chimeric antigen receptors (iCARs) divert off-target immunotherapy responses. Sci Transl Med. 2013;5(215):215ra172.
81. McGowan E, Lin Q, Ma G, Yin H, Chen S, Lin Y. PD-1 disrupted CAR-T cells in the treatment of solid tumors: promises and challenges. Biomed Pharmacother. 2020;121:109625.
82. Wei W, Chen Z, Wang K. CRISPR/Cas9: a powerful strategy to improve CAR-T cell persistence. Int J Mol Sci. 2023;24(15):12317.
83. Grosser R, Cherkassky L, Chintala N, Adusumilli PS. Combination immunotherapy with CAR T cells and checkpoint blockade for the treatment of solid tumors. Cancer Cell. 2019;36(5):471–82.
84. Kankeu Fonkoua LA, Sirpilla O, Sakemura R, Siegler EL, Kenderian SS. CAR T cell therapy and the tumor microenvironment: current challenges and opportunities. Mol Ther Oncolytics. 2022;25:69–77.
85. Hou AJ, Chang ZL, Lorenzini MH, Zah E, Chen YY. TGF-β-responsive CAR-T cells promote anti-tumor immune function. Bioeng Transl Med. 2018;3(2):75–86.
86. Di Stasi A, De Angelis B, Rooney CM, Zhang L, Mahendravada A, Foster AE, et al. T lymphocytes coexpressing CCR4 and a chimeric antigen receptor targeting CD30 have improved homing and antitumor activity in a Hodgkin tumor model. Blood. 2009;113(25):6392–402.
87. Sagnella SM, White AL, Yeo D, Saxena P, van Zandwijk N, Rasko JEJ. Locoregional delivery of CAR-T cells in the clinic. Pharmacol Res. 2022;182:106329.
88. Khawar MB, Ge F, Afzal A, Sun H. From barriers to novel strategies: smarter CAR T therapy hits hard to tumors. Front Immunol. 2023;14:1203230.
89. Li C, Mei H, Hu Y. Applications and explorations of CRISPR/Cas9 in CAR T-cell therapy. Brief Funct Genomics. 2020;19(3):175–82.

90. Chen X, Zhong S, Zhan Y, Zhang X. CRISPR–Cas9 applications in T cells and adoptive T cell therapies. Cell Mol Biol Lett. 2024;29:52.
91. Cherkassky L, Morello A, Villena-Vargas J, Feng Y, Dimitrov DS, Jones DR, et al. Human CAR T cells with cell-intrinsic PD-1 checkpoint blockade resist tumor-mediated inhibition. J Clin Invest. 2016;126(8):3130–44.
92. Tao R, Han X, Bai X, Yu J, Ma Y, Chen W, et al. Revolutionizing cancer treatment: enhancing CAR-T cell therapy with CRISPR/Cas9 gene editing technology. Front Immunol. 2024;15:1354825.
93. Alishah K, Birtel M, Masoumi E, Jafarzadeh L, Mirzaee HR, Hadjati J, et al. CRISPR/Cas9-mediated TGFβRII disruption enhances anti-tumor efficacy of human chimeric antigen receptor T cells in vitro. J Transl Med. 2021;19:482.
94. Celichowski P, Turi M, Charvátová S, Radhakrishnan D, Feizi N, Chyra Z, et al. Tuning CARs: recent advances in modulating chimeric antigen receptor (CAR) T cell activity for improved safety, efficacy, and flexibility. J Transl Med. 2023;21:197.
95. Dimitri A, Herbst F, Fraietta JA. Engineering the next-generation of CAR T-cells with CRISPR-Cas9 gene editing. Mol Cancer. 2022;21:78.
96. Tokarew N, Ogonek J, Endres S, von Bergwelt-Baildon M, Kobold S. Teaching an old dog new tricks: next-generation CAR T cells. Br J Cancer. 2019;120(1):26–37.
97. Hong M, Clubb JD, Chen YY. Engineering CAR-T cells for next-generation cancer therapy. Cancer Cell. 2020;38(4):473–88.
98. Wang J, Wang L. CAR-T cell therapy: where are we now, and where are we heading? Blood Sci. 2023;5(4):237–48.

Genetic Engineering and Synthetic Biology for Next-Generation CAR-T Cells

Sovon Acharya, Deepesh Kumar Gupta, Neetu Bhardwaj, Madhu Malinee, Munindra Ruwali, and Alok Kumar

Abstract Rapidly developing bioengineer cells have made excellent contributions to the development of cell-based immunotherapies, offering several genetic tools to improve performance, novel functions, and diverse applications. These synthetic cells have emerged as potential therapeutic approaches for various diseases, including cancer. Advances in synthetic biology and genetic engineering have paved the way for innovative strategies in next-generation cell-based cancer therapies. These approaches aim to increase the efficacy, specificity, and safety of cancer treatments by harnessing the power of engineered cells. Synthetic biology enables the design and construction of novel genetic circuits, whereas genetic engineering allows for precise manipulation of cellular functions. This chapter overviews recent advancements in the field, highlighting critical strategies for developing cell-based cancer therapies. These include engineering chimeric antigen receptor (CAR)- T cells, developing synthetic gene networks for improved control and optimized performance of engineered therapeutic cells, and utilizing gene-editing technologies such

S. Acharya
Vapiano Laboratories Pvt. Ltd., Bhubaneswar, Odisha, India

Amity Institute of Biotechnology, Amity University, Gurgaon, Haryana, India

D. K. Gupta
Division of Oncology, Section of Cellular Therapy, Washington University, School of Medicine, St. Louis, MO, USA

Center of Gene and Cellular Immunotherapy, Washington University School of Medicine, St. Louis, MO, USA

N. Bhardwaj
Department of Medical Oncology, All India Institute of Medical Sciences, New Delhi, India

M. Malinee · A. Kumar (✉)
Department of Immunology, University of Pittsburgh, Pittsburgh, PA, USA
e-mail: kumaral@pitt.edu

M. Ruwali
Department of Education in Science and Mathematics (DESM), National Council of Educational Research and Training (NCERT), New Delhi, India

A. K. Mishra, S. K. Malonia (eds.), *Cell-based Immunotherapies for Cancer*,
https://doi.org/10.1007/978-3-031-88695-9_7

as CRISPR for targeted cancer interventions. We also discuss the challenges associated with these strategies and highlight future directions in the field.

Keywords Cancer · Immunotherapy · Cell-based therapy · Immunity · CAR-T · Synthetic biology · Genetic engineering

Abbreviations

TCR	T-cell receptor
CAR	chimeric antigen receptor
TRUCKs	T cells redirected for antigen-unrestricted cytokine-initiated killing
CRS	cytokine release syndrome
CRES	CAR-T encephalopathy syndrome
TIL	tumor-infiltrating lymphocyte
ImmTAC	immune-mobilizing monoclonal T-cell receptors
MHC	major histocompatibility complex
HLA	human leukocyte antigen
scFv	single-chain fragment of antibody
ADR	allogeneic immune defense receptor
GM-CSF	granulocyte-macrophage colony-stimulating factor
GVHD	graft-versus-host disease
OXPHOS	oxidative phosphorylation
FAO	fatty acid oxidation
TNFRSF9	tumor necrosis factor receptor superfamily member 9
GEMS	generalized extracellular molecule sensor
MESA	modular extracellular sensor architecture
SUPRA-CAR	split, universal and programmable CAR

1 Introduction

Cancer is the second most common cause of death worldwide, and the available treatment options are not efficient enough to reduce the global burden of this disease [1, 2]. This critical situation prompts us to better consider cancer as "the emperor of all maladies of this century" [3]. Chemotherapy, radiotherapy, and surgery have been the three significant cancer treatment modalities for decades. Although chemotherapeutic drugs can potentially kill fast-dividing cancer cells, they cannot distinguish between normal and malignant cells. As a result, normal cells with high cell division rates (especially bone marrow and hair follicle cells) are also killed by chemotherapeutic drugs as collateral damage to patients [4].

Another relatively new type of treatment, called "cancer immunotherapy," has been developed over the past decade. In the cancer immunotherapy approach, the immune system is activated to screen, fight, and destroy cancer cells from the body. Various forms of immunotherapy are currently being used in clinical settings to combat cancer, e.g., immune system modulators, which encompass agonistic and antagonistic monoclonal antibodies that target both costimulatory and immune inhibitory receptors present on the surface of T cells; engineered T-cell transfer therapy; and cancer vaccines [5]. Cancer immunotherapy emerged after successful clinical trials of antagonistic antibodies against two important immune checkpoint inhibitors, namely, programmed cell death 1 (PD-1) and cytotoxic T-lymphocyte antigen type 4 (CTLA-4), which were discovered by Tasuku Honjo and James P Alison [6–10]. These two great scientists were awarded the Nobel Prize jointly in Physiology or Medicine in 2018 for these fundamental discoveries [11]. Immune cell therapy involves T cells, natural killer (NK) cells, dendritic cells (DCs), and tumor-infiltrating lymphocytes (TILs), which, after being collected from patients, are genetically modified and expanded ex vivo before being retransplanted [5].

Another form of cell-based immunotherapy is "chimeric antigen receptor (CAR)-T cell" (CAR-T) therapy. CARs are synthetic proteins that are simplified functional versions of TCRs that enable antigen recognition on cells that have downregulated MHC, as most cancers do (Fig. 1) [12]. CAR-T cells are engineered to express synthetic receptors that allow antigen-specific recognition, activation, and elimination of the target cell. CAR is incorporated into T cells, and additional receptors are expressed on the surface of T cells. CAR-T cells are considered living drugs and are engineered to target cancer cells (Fig. 2) [13]. While CAR-T-cell therapy has shown considerable success in treating hematological malignancies (anti-CD19-based therapies), its efficacy is not promising in the case of solid tumors. The therapeutic efficacy is compromised by different contributing factors, including the immune-suppressive tumor microenvironment (TME) [14]. In the direction of CAR-T-cell improvements, numerous research groups worldwide have studied how to design better CAR-T-cell-based cell products. Here, we describe in detail the

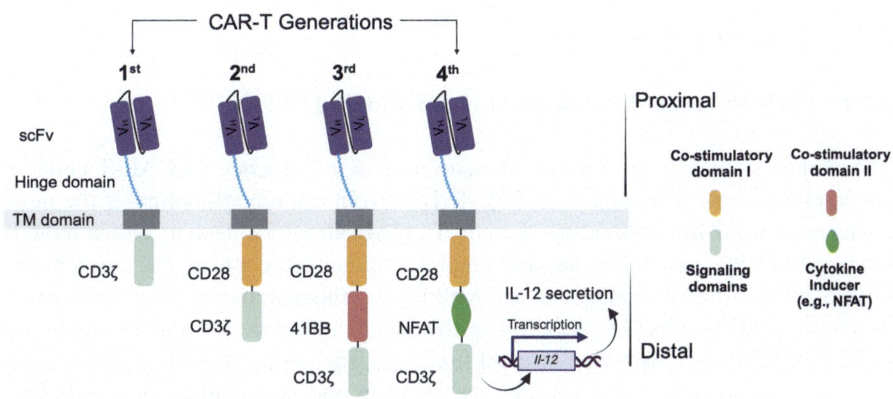

Fig. 1 Schematics of TCR and CAR-T-cell signaling. *TCR* T-cell receptor, *CAR* chimeric antigen receptor, *APC* antigen-presenting cell, single-chain variable fragment. (Figure created with Biorender.com)

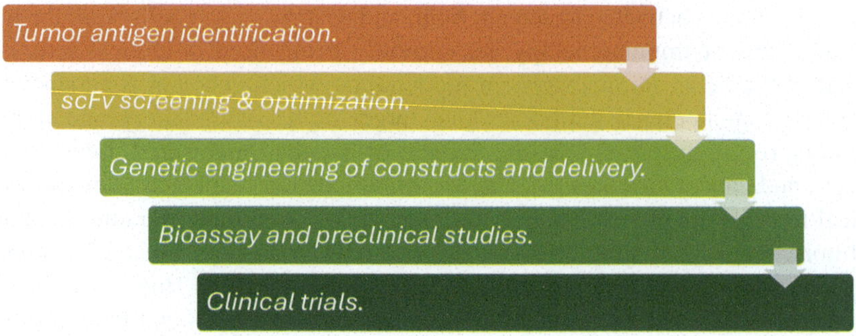

Fig. 2 Schematic representation of the CAR-T-cell therapy workflow. The entire process involves [1] T-cell collection, [2] T-cell selection and activation, [3] genetic engineering, [4] cell expansion, and [5] quality control and administration of CAR-T cells

synthetic receptor iterations developed throughout CAR-T-cell research and the genetic engineering approaches used in cancer immunotherapy treatment.

2 Chimeric Antigen Receptor (CAR)

CAR-T cells have made great strides in treating malignant hematological cancers. In 1993, the first CAR was reported by Eshhar et al., where the authors combined the killing potential of T cells with the antibody specificity in binding with their target [12]. Typically, the CAR is a recombinant single transmembrane receptor composed of an extracellular antigen-recognizing domain, a flexible hinge, a transmembrane domain, and an intracellular T-cell-activating domain (Fig. 1).

2.1 CAR Structure

2.1.1 CAR Proximal Domain (or Ligand-Binding Domain)

The affinity of the ligand-binding domain affects the efficacy of CAR-T cells. A single-chain variable fragment (scFv), derived from immunoglobulins, is the most abundantly used ligand-binding domain in CAR structure design. Other ligand-binding domains are nanobodies and ligands to cognate receptors. The antigen epitope location, affinity, avidity, and accessibility are the parameters to consider when designing CAR-T cells [15]. Modifying affinity can influence not only the signaling process of CARs but also other effector functions, such as proliferation, persistence, and cytokine secretion. Antigen epitope location and accessibility also influence CAR function. The flexibility of the modular structure of CARs allows them to target difficult epitopes, such as large bulky cell surface receptors [16]. The location of the epitope is crucial for modulating the distance of the "immune synapse" to determine the delivery of cytotoxic granules [16].

2.1.1.1 Nanobodies in CAR-T Cells

Many tumor-associated intracellular antigens cannot be targeted by classical CAR-T cells. Nanobodies can be used to solve this problem. It could be a suitable alternative to scFv for the generation of CAR constructs [17]. The antigen recognition region of CAR-T cells typically consists of a single-chain variable fragment (scFv), comprising heavy-chain variable fragments connected to light-chain variable fragments via a flexible linker. However, scFvs do not always fold efficiently and tend to aggregate. To overcome the problems of complex folding and assembly steps, the stability of scFv and traditional antibodies, nanobodies could be used to generate CAR-T cells for efficient function [18].

Nanobodies, also referred to as variable regions of heavy-chain antibodies (VHHs), are notably smaller than conventional monoclonal antibodies [19]. Their exceptionally stable physicochemical properties enable them to withstand harsh conditions such as acidity or high pressure. They maintain their high affinity for antigens. These characteristics facilitate the rapid distribution of nanobodies throughout the body following administration. Additionally, nanobodies exhibit favorable tissue distribution and low immunogenicity in humans, attributed to their high sequence homology with the human VH3 gene family. Recent research has demonstrated the successful integration of nanobodies as an optimal antigen recognition domain for generating tumor antigen-specific CAR-T cells [20].

2.1.2 Spacer or Hinge Domain

The hinge domain links the antigen-binding region "scFV" to the transmembrane domain, which offers flexibility and enhanced effectiveness. Sterically inaccessible epitopes can be targeted by specific engineering of the spacer domain. This approach could control synaptic cleft distances [16]. By altering the spacer domain length, the diffusion of lytic granules by granzymes and pore formation by perforin can also occur. As a result, the cytolytic activity is increased [21].

2.1.3 CAR Transmembrane Domain

The transmembrane membrane domain is an integral part of the signal transduction process. The transmembrane area functions like the fulcrum to transduce alerts from the ligand reputation domain (LRD) to the cytoplasmic/intracellular site. Under normal physiological conditions, the T-cell-CD3 complex is formed during immune activation [22, 23].

2.1.4 CAR Distal Domain

In T cells, the intracellular signaling domain of the TCR-CD3 complex is responsible for transmitting the crucial signal that initiates the signaling cascade. Similarly, the distal part of CAR-T cells increases the activity of T cells and has the potential to kill them by releasing cytotoxic granules [23].

2.2 Synthetic Cell Bridging Approaches for CAR-T Cells

Synthetic biology plays a very important role in designing cell-based therapies such as CAR-T-cell therapy. Initially, the integration of exogenous T-cell receptors (TCRs) into the autologous T cells of cancer patients was performed. As a result, the possibility and affinity of binding with the target cell improved. Later, the addition of small peptides and antibody fragments to the TCR for enhanced interaction between T cells and tumors was explored via synthetic biology approaches. There are different approaches to enhance cell bridging events, as discussed earlier, such as the fusion of neopeptide tags to antibody fragments [24]. The primary aim of artificial cell bridging is to improve the efficiency of a) targeted migration of cells and b) the degree of activation of cells for their functions. Genetically engineered CAR-T cells (axicabtagene & tisagenlecleucel) were approved by the U.S. FDA & EMA to treat large B-cell lymphoma & B-cell acute lymphoblastic leukemia (B-ALL) [25]. Rodgers et al. reported that the construction of peptide-specific switchable CAR-T cells could improve tumor-specific activation in a reversible manner. Genetically modified antibody fragments specific for tumor antigens could improve the strength of immunological synapses between cancer cells and CAR-T cells [26].

The adoptive transfer of genetically engineered T cells (e.g., CAR-T cells, NK-T cells, and TILs) that target tumor-associated antigens (TAAs) presents a distinctive opportunity to integrate advanced systems and synthetic biology methodologies for developing cancer therapeutics with innovative functionalities [27].

2.3 Different Generations of CAR-T Cells

Approximately three decades ago, scientists developed a chimeric molecule by fusing an antibody-like variable region with TCR-derived constant regions, marking the inception of CAR technology. These constructs, known as "first-generation CARs," heralded the beginning of the CAR era [12] (Fig. 3). They contain a single-chain variable fragment (scFv) for the recognition of antigens. The hinge (H) and transmembrane (TD) domains are required for intracellular signal transduction. Moreover, an intracellular CD3ζ domain is incorporated for activation [28]. First-generation CAR-T cells exhibited antigen-specific activation and cytotoxicity but

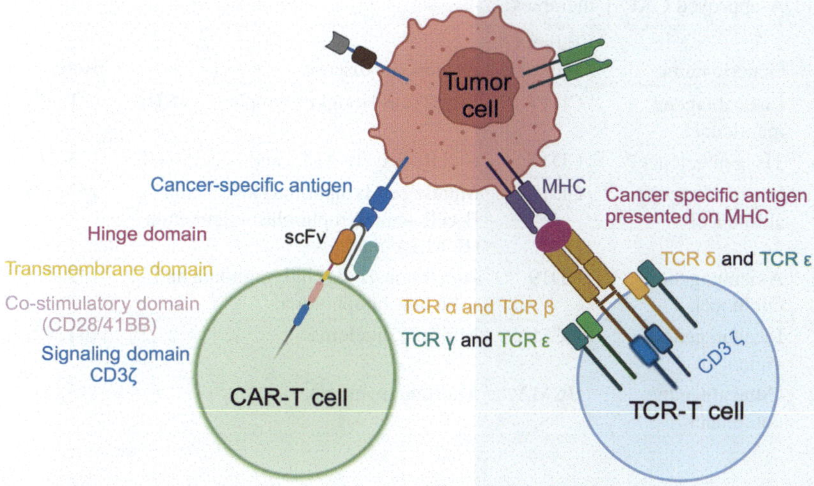

Fig. 3 Bioengineered CAR-T generation. First-generation CAR-T cells—intact with the extracellular domain scFv and the intracellular signaling domain CD3ζ. second-generation CAR-T cells—incorporated with a scFv domain combined with two intracellular domains: CD3ζ and an additional costimulatory domain (e.g., CD28 or 4-1BB). Third-generation CAR-T cells are equipped with a scFv domain combined with three intracellular domains: CD3ζ and two additional costimulatory domains. Fourth-/next-generation CAR-T cells—CAR-T cells complemented with cytokine-expressing CAR-T cells (as illustrated above). (Figure created with Biorender.com)

displayed inadequate proliferation rates in vivo. Consequently, "second-generation CAR-T cells" were developed that incorporate a costimulatory domain between the transmembrane domain and the CD3ζ domain to address this limitation. This addition resolved the limitations of first-generation CAR-T cells. second-generation CAR-T cells have demonstrated promising clinical outcomes, with instances of complete remission observed in clinical trials involving B-cell acute lymphoblastic leukemia (B-ALL) patients and chronic lymphocytic leukemia patients as early as 2011. The advent of second-generation CARs heralded a transformative shift in the treatment landscape of hematological malignancies [28] (Fig. 3).

The "third-generation CAR" is characterized by the incorporation of two costimulatory domains, OX40 and CD28, between the transmembrane and CD3z domains. This modification further improved the performance of CAR-T cells [29] (Fig. 3). Furthermore, modulating immune regulation could be pivotal for allowing the CAR to effectively fight solid tumors. CAR-T cells were further engineered to release inflammatory cytokines (e.g., IL-12, IL-7, IL-15, IL-18, and IL-23) upon CAR signaling. These modified CAR-T cells are generally called fourth-generation CAR-T cells (Fig. 3) [30]. T cells redirected for antigen-unrestricted cytokine-initiated killing (TRUCKs), often referred to as fourth-generation CAR-T cells, utilize both the

Table 1 FDA-approved CAR-T therapies

Brand name	Generic name	Target antigen	Against the disease	Reference
Breyanji	Lisocabtagene maraleucel	CD19	B-cell non-Hodgkin lymphoma (NHL)	[31, 32]
Kymriah	Tisagenlecleucel	CD19	B-cell ALL (B-ALL) and B-cell NHL	[33–36]
Tecartus	Brexucabtagene autoleucel	CD19	Mantle cell lymphoma (MCL) and B-cell acute lymphoblastic leukemia (B-ALL) in adults	[37, 38]
Yescarta	Axicabtagene ciloleucel	CD19	B-cell non-Hodgkin lymphoma and follicular lymphonma	[39, 40]
Abecma	Idecabtagene vicleucel	BCMA	Multiple myeloma	[41, 42]
Carvykti	Ciltacabtagene autoleucel	BCMA	Multiple myeloma	[43, 44]

direct antitumor capabilities of CAR-T cells and the immune-modulating effects of the administered cytokine following CAR signaling.

Since 2017, regulatory agencies (Food and Drug Administration (FDA), USA) have approved six CAR-T-cell-based therapies for treating different blood lymphomas and, most recently, for treating multiple myeloma (Table 1).

2.4 *Limitations of CAR-T-Cell Therapy*

The primary obstacles faced by CAR-T cells at present include a) cytokine release syndrome (CRS), which stems from T-cell overactivation. b) Off-target effects resulting from the erroneous targeting and killing of healthy cells due to the recognition of tumor-associated antigens on nontumor cells in vivo [45, 46]. Clinical trials have indicated that CAR-T-cell therapy can induce damage to the central nervous system, leading to CAR-T-cell encephalopathy syndrome (CRES), a form of neurological toxicity [47]. These clinical side effects frequently prevent patients with compromised physical conditions from continuing CAR-T-cell therapy. Changes in cell distribution and decreased CAR-T-cell numbers in the body contribute to tumor recurrence in some patients after treatment [48]. Hence, there is an urgent need to develop next-generation CAR-T-cell therapies that specifically target tumor cells without harming the host's nontumor cells.

2.5 *Novel CAR Designs*

CAR-T-cell modifications have developed rapidly to increase efficiency and reduce unwanted responses. In this section, we explore the major groups of new types of CARs, which include dual antigen-targeting CARs (BiCARs), gated CARs, and armed CARs.

2.5.1 Dual/Multi-antigen Targeting

While single antigen targeting fails in some patients for several reasons, the concept of multitarget CARs has been developed. Sometimes, the expression of antigens on the tumor fades away from the tumor cell. A new generation of CAR-T cells has been developed using gene-editing approaches that target two or more antigens. These enhanced iterations of CAR-T cells demonstrate heightened functionality, extended survival, and improved targeting capabilities. TandemCAR and trivalent-CAR designs aim to target multiple tumor-associated antigens, thereby increasing the precision of target identification [49]. Multitarget CAR-T cells express high-density CARs on the surface of T cells. In a recent study, Yan and colleagues engineered CAR-T cells targeting CD19 and CD123, which resulted in reduced tumor growth and a lower relapse rate in B-ALL patients [50]. Compared to CAR-T cells with HER2 alone, BiCAR-T cells expressing HER2 and IL13Ra2 effectively cleared glioblastoma tumors with reduced antigen escape variants in a mouse xenograft model system [51]. Another study revealed that, compared with single CAR-T cells, BiCAR-T cells are inferior to single CAR-T cells in terms of the ability to secrete IL2 due to steric hindrance when both antigens are simultaneously targeted to T cells. This hinders positive feedback on IL2 [52]. Further, TandemCAR shows benefits over BiCAR's structures and has improved functionality by reducing steric interactions with tumor antigens [53].

2.5.2 Gated CARs

To overcome the problem of imbalanced antigens affecting the efficacy of CAR-T cells, further modifications in CAR design are needed to discriminate between tumor cells and normal host cells (Fig. 4). To target tumor cells lacking tumor-specific antigens, employing logic gating in the CAR design is a promising approach to kill cancer cells while preventing on-target/off-tumor toxicity [54]. The most common logic gates employed in the CAR structure are AND, OR, and NOT.

2.5.3 Armored CAR

For hematological malignancies, CAR-T cells are designed to target individual malignant cells. Conversely, when dealing with solid tumors, T cells must overcome multiple layers of barriers to effectively localize to the tumor site and eliminate malignant cells. These barriers include a) antigen escape, b) restricted trafficking and limited tumor infiltration, and c) an immunosuppressive TME [55]. Severe life-threatening toxicity is also one of the limitations that needs to be overcome, as most patients receive immunotherapy because of immune-related adverse events (irAEs) [56].

Within the solid tumor mass, various cell types and their interactions with each other through different chemical messengers (e.g., cytokines and chemokines) form

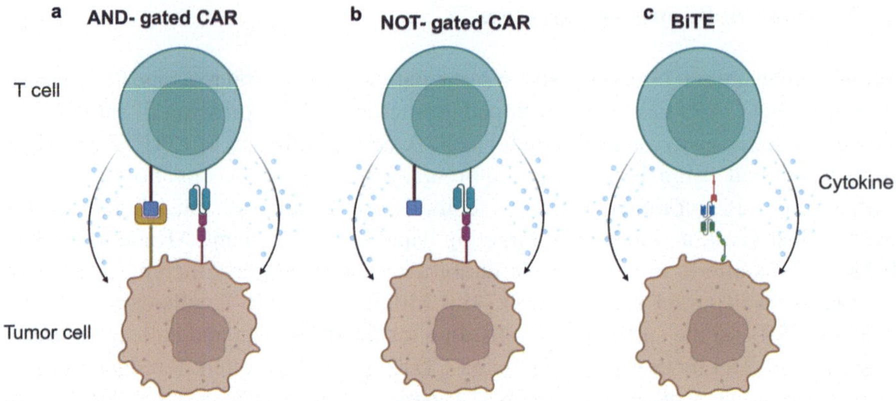

Fig. 4 Novel CAR-T-cell designs. (**a**) AND gate CAR-T design showing the CD3ζ signaling domain (in light green) in the CAR receptor and the CD28/4-1BB costimulatory domains in another receptor (in light blue). (**b**) NOT gates allow engineered T cells to distinguish normal tissues by preventing T-cell activation when double antigen recognition is present. (**c**) BiTE is a bispecific antibody designed to target CD3 and tumor-specific antigens simultaneously. (Figure created with Biorender.com)

the tumor microenvironment (TME), which supports the growth and survival of the tumor. They also create an immune-suppressive environment to escape immune surveillance [57]. Different types of cells present within solid tumors include cancer-associated fibroblasts (CAFs), immune cells such as natural killer (NK) cells, macrophages, cytotoxic ($CD8^+$), helper T cells ($CD4^+$), B cells, and endothelial cells. Certain immune cells create an immunosuppressive environment by releasing transforming growth factor-β (TGF-β), IL-10, and IL-4. Conversely, specific regulatory T (Treg) cells directly impede the function of effector T cells or CAR-T cells [58].

To circumvent this issue, researchers have modified CAR-T cells to generate a fourth-generation CAR for solid tumors (Fig. 3). These cells are referred to as "Armored CAR-T cells" and are engineered to express surface proteins and a third-generation CAR, aiming to diminish immunosuppression and enhance antitumor activity [59].

One particular subset of these cells is known as T cells Redirected toward Universal Cytokine Killing (TRUCK), which is capable of generating cytokines with the potential to disrupt the immunosuppressive effects of cytokines present in solid tumors [60]. By engineering CAR-T cells, the cytokine profile of the TME could be modified, potentially bolstering immune activity and enhancing the anticancer efficacy of both the immune system and CAR-T cells (Fig. 5).

Following infection, immune cells such as dendritic cells (DCs), monocytes, and macrophages secrete the cytokine IL-12. This cytokine leads to activation of the immune system by stimulating the differentiation of naïve T cells into Th1 cells. IL-12 release creates a positive loop of immune activation. As a result, the efficacy of CAR-T cells could increase in solid tumors. A study suggested that CD19-targeting CAR-T cells, which constitutively secrete IL-12, exhibit greater antitumor efficacy than CD19-targeting CAR-T cells alone, both in vitro and in vivo [61].

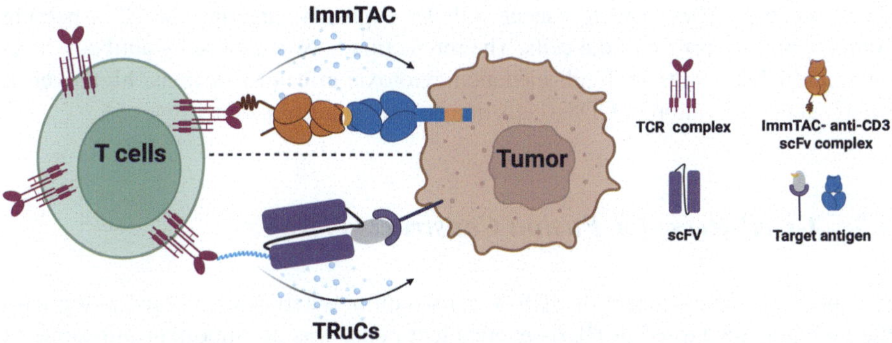

Fig. 5 Recombinant CAR-T-cell designs. The immune-mobilizing monoclonal T-cell receptor against cancer (ImmTAC) molecule, which has a T-cell receptor (TCR)-targeting domain, interacts with the target antigen on tumor cells and the anti-CD3 scFv intact with the TCR complex (upper panel) on the other hand. TRUC CAR-T cells showing an intact TCR complex can be effectively reprogrammed for cancer therapy by recombinantly fusing a scFv domain to its TCRα, TCRβ, CD3γ, CD3δ, or CD3ε subunit (lower panel). (Figure created with Biorender.com)

3 Next-Generation CARs

3.1 Recombinant TCRs

The TCR, present on the surface of T cells, is a complex composed of six receptor subunits that help them perform several important signaling events. The α and β chains of the TCR bind to the antigen via the peptide–MHC complex as well as signaling subunits of the CD3 complex (ϵ, γ, and δ) and the CD3ζ homodimer [62]. The TCR of tumor-infiltrating lymphocytes (TILs) undergoes a dramatic change, influencing tumor-dominant T cells. TCR alteration determines the fate of T cells, such as whether they proliferate or differentiate. TCR diversity and increased TILs are associated with antitumor effects. Additionally, it affects TCR engineering in tumor-infiltrating lymphocytes [63].

3.2 Immune-Mobilizing Monoclonal T-Cell Receptors (ImmTCs)

ImmTACs are a class of receptors synthesized from soluble, engineered, and affinity-enhanced monoclonal TCRs (mTCRs). ImmTACs are synthesized as hybrid proteins with an engineered TCR-based targeting system along with a single-chain fragment of antibody (scFv) effector function, and TCRs function as antibodies that recognize antigens within the immune system [63] (Fig. 5). While antibodies are limited to targeting the cell surface or secreted proteins, TCRs have the ability to recognize peptides derived from intracellular targets presented by human leukocyte antigen (HLA) or MHC. ImmTACs offer a unique ability to inhibit tumor growth by redirecting

T-cell function. They bind to cancer cells by explicitly targeting the HLA-peptide complex on the surface of the cells. The interaction between the scFv antibody fragment and CD3 results in T-cell-mediated cytotoxic effector functions. Moreover, it triggered the activation of CD8+ cells in a dose-dependent manner [64, 65].

3.3 T-Cell-Receptor Fusion Constructs (TRuCs)

Recently, the development of TCR-T cells with new target specificity, along with the potential for targeting HLA-independent cells, was an important milestone in CAR-T-cell research. TRuCs feature an antibody-based binding domain fused to the T-cell receptor (TCR) subunit engineered specifically for efficient recognition of tumor surface antigens [66]. TRuCs are composed of a ligand-specific antibody fused to the extracellular N-termini of five TCR subunits (TCRα, TCRβ, CD3ϵ, CD3γ, and CD3δ). This configuration endows engineered T cells with novel target specificity and the ability to eliminate target cells independently of HLA, thereby activating upon encountering corresponding target cells [67]. The T-cell antigen coupler (TAC) is an emerging platform that utilizes endogenous TCRs independently, offering increased antitumor efficacy while minimizing toxicity. The chimeric protein is conjugated to recognize antigens via CD3 domain binding. As a result, more TCR-CD3 complex formation takes place, and a prominent T-cell response is observed [30, 67].

3.4 Bites-CAR

Bispecific antibodies are compact antibodies with the ability to target two distinct antigens. There are two categories of bispecific antibodies: (1) IgG-like and (2) non-IgG-like. BiTEs belong to the non-IgG-like subtype of antibodies, featuring two antigen recognition domains linked together through a linker [68].

Bispecific T-cell engagers (BiTEs) target specific antigens present on the surface of cancer cells. BiTE-CAR is a combination of bispecific antibodies and CAR-T cells. Bispecific antibodies are secreted. Later, they could activate CAR-T cells and endogenous T cells to obtain a robust antitumor effect. This strategy fails to have a termination system, which could lead to adverse effects during treatment [69].

3.5 SynNotch Receptor

The SynNotch receptor is an engineered receptor that maintains the inherent transmembrane domain of the Notch receptor. It substitutes the intracellular and extracellular domains with customized domains, allowing it to identify specific ligands

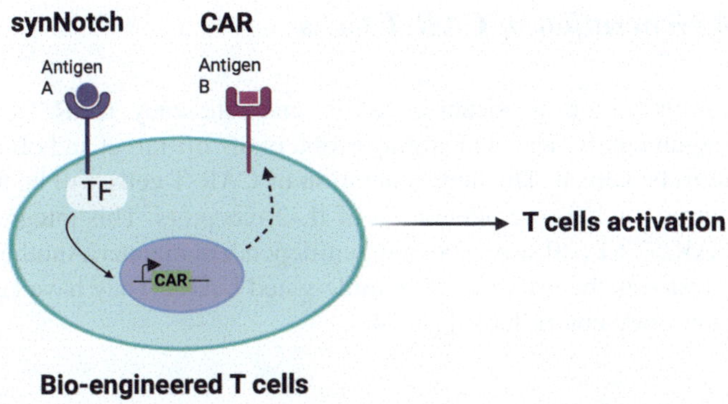

Fig. 6 SynNotch CAR-T design. An AND gate approach where a SynNotch receptor combines with a CAR to increase antigen specificity and safety. Under the recognition of an "antigen A" by the SynNotch receptor, an intracellular transcription factor (TF) is released and induces the activation of the CAR receptor specific to a second antigen, "antigen B", allowing the activation of T cells. (Figure created with Biorender.com)

and activate particular genes simultaneously [70]. Cho et al. introduced a method to coexpress SynNotch and CAR on the surface of T cells, with each receptor targeting distinct antigens [71] (Fig. 6). This dual-receptor system enhances T-cell targeting to tumor cells, and SynNotch also has the potential to regulate cytokine expression [71].

3.6 Allogeneic Immune Defense Receptor

Allogeneic immune defense receptor (ADR) is a tailored receptor designed to facilitate the elimination of autoreactive T cells and effectively prevent rejection mediated by allogeneic T cells [72]. Activated lymphocytes can transiently modulate the expression of tumor necrosis factor receptor superfamily member 9 (TNFRSF9/4-1BB) and the ADR receptor, which bind to 4-1BB on activated lymphocytes, thereby preventing lymphocyte attack on ADR-T cells. Allogeneic T lymphocytes engineered to express both ADR and CAR can potentially prevent host rejection without compromising CAR functionality. These cells exhibit robust survival in mouse models and demonstrate sustained tumor eradication in hematopoietic malignancies and solid cancers. This approach holds promise as a successful protocol for delivering universal CAR-T-cell products. Other artificial receptors, such as Tango, modular extracellular sensor architecture (MESA), and generalized extracellular molecule sensor (GEMS), are also accessible for T-cell engineering.

3.7 *Fifth Generation of CAR-T Cells*

To enhance proliferation, persistence, safety, and efficiency, CAR-T-cell therapy has evolved significantly. The challenging problem of "off-target and off-tumor toxicity" has yet to be solved. The new generation of CAR-T cells will be armed with different features, such as the integration of IL-2 receptors. This integration could activate the JAK/STAT pathway in an antigen-dependent manner. Another approach that is being tested is the use of lenalidomide-gated CARs. They have better safety profiles and are more controllable [73, 74].

3.7.1 SUPRA-CAR

SUPRA-CAR stands for the split, universal, and programmable (SUPRA) CAR system, which has improved specificity and controllability. They have a universal signal transduction receptor, zipCAR, zipFv, a soluble antigen-binding component expressed on T cells. The zipFv consists of a leucine zipper and a single-chain variable fragment (scFv). This type of zipCAR has an extracellular cognate zipper that binds specifically with the zipper on the zipFv. The intracellular part of zipCAR has a signaling domain. To elicit a T-cell response, these zippers help create cell bridges. The SUPRA-CAR system has an ON/OFF system to fine-tune T-cell activation [75, 76].

SUPRA-CARs use the biotin-avidin system and leucine zippers to segregate the single-chain variable fragment (scFv) from the receptor domain. This segregation helps control T-cell activity within the host immune system [77]. After the course of treatment, when the patient is cured or cannot tolerate the treatment, it can be terminated by stopping the injection of the scFv. However, the drawback of this system is that it is not cost-effective. The in vitro synthesis and delivery of scFv increases the cost of treatment [77].

4 Gene-Editing Approach for Cancer Immunotherapy

The main idea behind genome editing is to modify cell function by introducing new gene inserts or eliminating genes to improve cell functions. In CAR-T cells, removing or adding these genes could facilitate or eliminate adverse clinical conditions. We discuss the role of granulocyte–macrophage colony–stimulating factor (GM-CSF). Its primary function is to increase the proliferation and activation of macrophages and blood monocytes. This event promotes proinflammatory conditions at the tissue level. As a result of gene knockout through different methods, such as CRISPR, GM-CSF production can be explicitly suppressed. This is a much more effective method than neutralizing antibodies against GM-CSF [78].

A study conducted by Sachdeva et al. revealed that the use of transcription activator-like effector nuclease (TALEN) led to the successful knockout of GM-CSF and did not impact the regular proliferation and activation of CAR-T cells. By downregulating the expression of GM-CSF, the secretion of CRS biomarkers such as IL-6, monocyte chemoattractant protein 1 (MCP-1), and IL-8 could be abolished. Treatment with GM-CSF-knockout CAR-T cells typically does not induce cytokine release syndrome (CRS) in patients [79].

4.1 Gene Engineering Revolutionizes "Off-The-Shelf" Cell Strategies

At present, autologous cell therapy is the predominant approach in adoptive cell therapy. Throughout treatment, the immune system may reject specific types of stem or immune cells, leading to decreased efficacy. Furthermore, the emergence of graft-versus-host disease (GVHD) can impact the effectiveness of cell therapy [80]. Some medications, cytokines, or monoclonal antibodies may mitigate the adverse effects of the immune response and induce tolerance to graft-versus-host disease (GVHD). With the current advancements in gene-editing technology, allogenic "off-the-shelf" CAR-T-cell production has become a reality. Compared with autologous cells, it offers advantages such as cost-effectiveness, personalization, and customization [81].

5 Conclusion

Research into CAR-T cells is rapidly advancing, with hundreds of ongoing clinical trials fueling this expansion. A significant factor driving this growth is the identification of numerous antigens on tumor cells that represent promising targets for CAR-T cells. While FDA-approved CAR-T-cell therapies currently target only 'CD19' and 'B-cell maturation antigen (BCMA)' antigens, efforts have led to the development of CAR-T-cell therapies targeting other antigens commonly found in blood cancers. Moreover, there are endeavors to create therapies that target multiple antigens simultaneously.

Although CAR-T-cell therapy is very efficacious in treating lymphomas, its efficacy for treating solid cancers is limited. The reasons include immunologically cold tumors, a lack of identification of unique antigens, and a suppressive tumor microenvironment. We need to address these issues to make solid cancer treatment with CARs effective. Next-generation sequencing is needed to identify a specific antigen present in the cancer of a specific patient. Several research groups are focusing on identifying unique antigens present in solid cancers that can be targeted by CAR-T cells.

Efficient and representative models of CAR-T-cell therapy are needed. In preclinical models of human CAR-T cells, NSG mice bearing tumors, which are usually derived from cell lines, are commonly employed and often involve the transfer of non-MHC-matched CAR-T cells. We developed a murine immunocompetent model for CAR-T-cell therapy that targets CD105. CD105 is a cell surface protein that is typically expressed by activated endothelium but is highly overexpressed in various tumor types, including melanoma and acute myelogenous leukemia [82, 83]. The development of murine CD105 CAR-T cells has laid the groundwork for further investigations into CD105 as a potential target for human CAR-T-cell therapy [84].

The immunosuppressive tumor microenvironment is another factor that limits the ability of CAR-T cells to target solid cancers. Nutrient deficiency and hypoxic and acidic environments led to exhaustion of T cells. Suppression of PGC-1α serves as the primary driver of this altered metabolic fate. PGC-1α coordinates mitochondrial biogenesis, mitochondrial fusion, OXPHOS/FAO, and antioxidant activities that promote mitochondrial health and capacity. We incorporated engineered PGC-1α molecules that are resistant to posttranslational regulation [85].

Furthermore, the next generation of CARs should be able to resist acidic/hypoxic and immunosuppressive tumor microenvironments at solid tumor sites. These should be implemented in the fifth and further generations of CARs.

Competing Financial Interest The authors declare that they have no competing financial interests.

References

1. Weiss C. One in four dies of cancer. Questions about the epidemiology of malignant tumours. Recent Results Cancer Res. 2021;218:15–29.
2. Collaborators GBDCoD. Global, regional, and national age-sex specific mortality for 264 causes of death, 1980–2016: a systematic analysis for the Global Burden of Disease Study 2016. Lancet. 2017;390(10100):1151–210.
3. Mukherjee S. The emperor of all maladies: a biography of cancer. Can Fam Physician. 2011;57(4):460.
4. Chan J, Adderley H, Alameddine M, Armstrong A, Arundell D, Fox R, et al. Permanent hair loss associated with taxane chemotherapy use in breast cancer: A retrospective survey at two tertiary UK cancer centres. Eur J Cancer Care (Engl). 2021;30(3):e13395.
5. Fan T, Zhang M, Yang J, Zhu Z, Cao W, Dong C. Therapeutic cancer vaccines: advancements, challenges, and prospects. Signal Transduct Target Ther. 2023;8(1):450.
6. Hodi FS, O'Day SJ, McDermott DF, Weber RW, Sosman JA, Haanen JB, et al. Improved survival with ipilimumab in patients with metastatic melanoma. N Engl J Med. 2010;363(8):711–23.
7. Robert C, Long GV, Brady B, Dutriaux C, Maio M, Mortier L, et al. Nivolumab in previously untreated melanoma without BRAF mutation. N Engl J Med. 2015;372(4):320–30.
8. Rotte A. Combination of CTLA-4 and PD-1 blockers for treatment of cancer. J Exp Clin Cancer Res. 2019;38(1):255.
9. Ishida Y, Agata Y, Shibahara K, Honjo T. Induced expression of PD-1, a novel member of the immunoglobulin gene superfamily, upon programmed cell death. EMBO J. 1992;11(11):3887–95.

10. Leach DR, Krummel MF, Allison JP. Enhancement of antitumor immunity by CTLA-4 blockade. Science. 1996;271(5256):1734–6.
11. Zang X. 2018 Nobel Prize in medicine awarded to cancer immunotherapy: immune checkpoint blockade – a personal account. Genes Dis. 2018;5(4):302–3.
12. Eshhar Z, Waks T, Gross G, Schindler DG. Specific activation and targeting of cytotoxic lymphocytes through chimeric single chains consisting of antibody-binding domains and the gamma or zeta subunits of the immunoglobulin and T-cell receptors. Proc Natl Acad Sci USA. 1993;90(2):720–4.
13. De Marco RC, Monzo HJ, Ojala PM. CAR T cell therapy: a versatile living drug. Int J Mol Sci. 2023;24(7):6300.
14. Martinez M, Moon EK. CAR T cells for solid tumors: new strategies for finding, infiltrating, and surviving in the tumor microenvironment. Front Immunol. 2019;10:128.
15. Chmielewski M, Hombach A, Heuser C, Adams GP, Abken H. T cell activation by antibody-like immunoreceptors: increase in affinity of the single-chain fragment domain above threshold does not increase T cell activation against antigen-positive target cells but decreases selectivity. J Immunol. 2004;173(12):7647–53.
16. Jayaraman J, Mellody MP, Hou AJ, Desai RP, Fung AW, Pham AHT, et al. CAR-T design: elements and their synergistic function. EBioMedicine. 2020;58:102931.
17. Bao C, Gao Q, Li LL, Han L, Zhang B, Ding Y, et al. The application of nanobody in CAR-T therapy. Biomol Ther. 2021;11(2):238.
18. Xia B, Lin K, Wang X, Chen F, Zhou M, Li Y, et al. Nanobody-derived bispecific CAR-T cell therapy enhances the anti-tumor efficacy of T cell lymphoma treatment. Mol Ther Oncolytics. 2023;30:86–102.
19. Jin BK, Odongo S, Radwanska M, Magez S. Nanobodies: a review of generation, diagnostics and therapeutics. Int J Mol Sci. 2023;24(6):5994.
20. Mo F, Duan S, Jiang X, Yang X, Hou X, Shi W, et al. Nanobody-based chimeric antigen receptor T cells designed by CRISPR/Cas9 technology for solid tumor immunotherapy. Signal Transduct Target Ther. 2021;6(1):80.
21. Guest RD, Hawkins RE, Kirillova N, Cheadle EJ, Arnold J, O'Neill A, et al. The role of extracellular spacer regions in the optimal design of chimeric immune receptors: evaluation of four different scFvs and antigens. J Immunother. 2005;28(3):203–11.
22. Liu J, Zhong JF, Zhang X, Zhang C. Allogeneic CD19-CAR-T cell infusion after allogeneic hematopoietic stem cell transplantation in B cell malignancies. J Hematol Oncol. 2017;10(1):35.
23. Tasian SK, Gardner RA. CD19-redirected chimeric antigen receptor-modified T cells: a promising immunotherapy for children and adults with B-cell acute lymphoblastic leukemia (ALL). Ther Adv Hematol. 2015;6(5):228–41.
24. Li XC, Kloc M, Ghobrial RM. Memory T cells in transplantation – progress and challenges. Curr Opin Organ Transplant. 2013;18(4):387–92.
25. Neelapu SS, Locke FL, Bartlett NL, Lekakis LJ, Miklos DB, Jacobson CA, et al. Axicabtagene ciloleucel CAR T-cell therapy in refractory large B-cell lymphoma. N Engl J Med. 2017;377(26):2531–44.
26. Rodgers DT, Mazagova M, Hampton EN, Cao Y, Ramadoss NS, Hardy IR, et al. Switch-mediated activation and retargeting of CAR-T cells for B-cell malignancies. Proc Natl Acad Sci USA. 2016;113(4):E459–68.
27. Chakravarti D, Wong WW. Synthetic biology in cell-based cancer immunotherapy. Trends Biotechnol. 2015;33(8):449–61.
28. Wu Y, Huang Z, Harrison R, Liu L, Zhu L, Situ Y, et al. Engineering CAR T cells for enhanced efficacy and safety. APL Bioeng. 2022;6(1):011502.
29. Hombach AA, Abken H. Costimulation by chimeric antigen receptors revisited the T cell antitumor response benefits from combined CD28-OX40 signalling. Int J Cancer. 2011;129(12):2935–44.

30. Chmielewski M, Abken H. TRUCKs: the fourth generation of CARs. Expert Opin Biol Ther. 2015;15(8):1145–54.
31. Iragavarapu C, Hildebrandt G. Lisocabtagene Maraleucel for the treatment of B-cell lymphoma. Expert Opin Biol Ther. 2021;21(9):1151–6.
32. Kamdar M, Solomon SR, Arnason J, Johnston PB, Glass B, Bachanova V, et al. Lisocabtagene maraleucel versus standard of care with salvage chemotherapy followed by autologous stem cell transplantation as second-line treatment in patients with relapsed or refractory large B-cell lymphoma (TRANSFORM): results from an interim analysis of an open-label, randomised, phase 3 trial. Lancet. 2022;399(10343):2294–308.
33. Schuster SJ, Bishop MR, Tam CS, Waller EK, Borchmann P, McGuirk JP, et al. Tisagenlecleucel in adult relapsed or refractory diffuse large B-cell lymphoma. N Engl J Med. 2019;380(1):45–56.
34. Wang XJ, Wang YH, Li SCT, Gkitzia C, Lim ST, Koh LP, et al. Cost-effectiveness and budget impact analyses of tisagenlecleucel in adult patients with relapsed or refractory diffuse large B-cell lymphoma from Singapore's private insurance payer's perspective. J Med Econ. 2021;24(1):637–53.
35. Wakase S, Teshima T, Zhang J, Ma Q, Fujita T, Yang H, et al. Cost effectiveness analysis of tisagenlecleucel for the treatment of adult patients with relapsed or refractory diffuse large B Cell lymphoma in Japan. Transplant Cell Ther. 2021;27(6):506 e1–e10.
36. Goto H, Makita S, Kato K, Tokushige K, Fujita T, Akashi K, et al. Efficacy and safety of tisagenlecleucel in Japanese adult patients with relapsed/refractory diffuse large B-cell lymphoma. Int J Clin Oncol. 2020;25(9):1736–43.
37. Ball G, Lemieux C, Cameron D, Seftel MD. Cost-effectiveness of brexucabtagene autoleucel versus best supportive care for the treatment of relapsed/refractory mantle cell lymphoma following treatment with a bruton's tyrosine kinase inhibitor in Canada. Curr Oncol. 2022;29(3):2021–45.
38. Frey NV. Approval of brexucabtagene autoleucel for adults with relapsed and refractory acute lymphocytic leukemia. Blood. 2022;140(1):11–5.
39. Locke FL, Ghobadi A, Jacobson CA, Miklos DB, Lekakis LJ, Oluwole OO, et al. Long-term safety and activity of axicabtagene ciloleucel in refractory large B-cell lymphoma (ZUMA-1): a single-arm, multicentre, phase 1-2 trial. Lancet Oncol. 2019;20(1):31–42.
40. Neelapu SS, Dickinson M, Munoz J, Ulrickson ML, Thieblemont C, Oluwole OO, et al. Axicabtagene ciloleucel as first-line therapy in high-risk large B-cell lymphoma: the phase 2 ZUMA-12 trial. Nat Med. 2022;28(4):735–42.
41. Mann H, Comenzo RL. Evaluating the therapeutic potential of idecabtagene vicleucel in the treatment of multiple myeloma: evidence to date. Onco Targets Ther. 2022;15:799–813.
42. Munshi NC, Anderson LD Jr, Shah N, Madduri D, Berdeja J, Lonial S, et al. Idecabtagene vicleucel in relapsed and refractory multiple myeloma. N Engl J Med. 2021;384(8):705–16.
43. Berdeja JG, Madduri D, Usmani SZ, Jakubowiak A, Agha M, Cohen AD, et al. Ciltacabtagene autoleucel, a B-cell maturation antigen-directed chimeric antigen receptor T-cell therapy in patients with relapsed or refractory multiple myeloma (CARTITUDE-1): a phase 1b/2 open-label study. Lancet. 2021;398(10297):314–24.
44. Martin T, Usmani SZ, Berdeja JG, Agha M, Cohen AD, Hari P, et al. Ciltacabtagene autoleucel, an anti–B-cell maturation antigen chimeric antigen receptor T-Cell therapy, for relapsed/refractory multiple myeloma: CARTITUDE-1 2-year follow-up. J Clin Oncol. 2022;41(6):1265–74.
45. Brudno JN, Kochenderfer JN. Recent advances in CAR T-cell toxicity: mechanisms, manifestations and management. Blood Rev. 2019;34:45–55.
46. Neelapu SS. Managing the toxicities of CAR T-cell therapy. Hematol Oncol. 2019;37(Suppl 1):48–52.
47. Belin C, Devic P, Ayrignac X, Dos Santos A, Paix A, Sirven-Villaros L, et al. Description of neurotoxicity in a series of patients treated with CAR T-cell therapy. Sci Rep. 2020;10(1):18997.
48. Li S, Tang H, Li C, Ma J, Ali M, Dong Q, et al. Synthetic biology technologies and genetically engineering strategies for enhanced cell therapeutics. Stem Cell Rev Rep. 2023;19(2):309–21.

49. Hegde M, Mukherjee M, Grada Z, Pignata A, Landi D, Navai SA, et al. Tandem CAR T cells targeting HER2 and IL13Ralpha2 mitigate tumor antigen escape. J Clin Invest. 2019;129(8):3464.
50. Yan LE, Zhang H, Wada M, Fang L, Feng J, Zhang W, et al. Targeting two antigens associated with B-ALL with CD19-CD123 compound Car T cell therapy. Stem Cell Rev Rep. 2020;16(2):385–96.
51. Hegde M, Corder A, Chow KK, Mukherjee M, Ashoori A, Kew Y, et al. Combinational targeting offsets antigen escape and enhances effector functions of adoptively transferred T cells in glioblastoma. Mol Ther. 2013;21(11):2087–101.
52. Wilkie S, van Schalkwyk MC, Hobbs S, Davies DM, van der Stegen SJ, Pereira AC, et al. Dual targeting of ErbB2 and MUC1 in breast cancer using chimeric antigen receptors engineered to provide complementary signaling. J Clin Immunol. 2012;32(5):1059–70.
53. Navai SA, Ahmed N. Targeting the tumour profile using broad spectrum chimaeric antigen receptor T-cells. Biochem Soc Trans. 2016;44(2):391–6.
54. Abbott RC, Hughes-Parry HE, Jenkins MR. To go or not to go? Biological logic gating engineered T cells. J Immunother Cancer. 2022;10(4):e004185.
55. Rodriguez-Garcia A, Palazon A, Noguera-Ortega E, Powell DJ Jr, Guedan S. CAR-T cells hit the tumor microenvironment: strategies to overcome tumor escape. Front Immunol. 2020;11:1109.
56. Vaddepally R, Doddamani R, Sodavarapu S, Madam NR, Katkar R, Kutadi AP, et al. Review of immune-related adverse events (irAEs) in non-small-cell lung cancer (NSCLC)-their incidence, management, multiorgan irAEs, and rechallenge. Biomedicines. 2022;10(4):790.
57. Kim SJ, Khadka D, Seo JH. Interplay between solid tumors and tumor microenvironment. Front Immunol. 2022;13:882718.
58. Hosseinalizadeh H, Rabiee F, Eghbalifard N, Rajabi H, Klionsky DJ, Rezaee A. Regulating the regulatory T cells as cell therapies in autoimmunity and cancer. Front Med (Lausanne). 2023;10:1244298.
59. Hawkins ER, D'Souza RR, Klampatsa A. Armored CAR T-cells: the next chapter in T-cell cancer immunotherapy. Biologics. 2021;15:95–105.
60. Dai H, Wang Y, Lu X, Han W. Chimeric antigen receptors modified T-cells for cancer therapy. J Natl Cancer Inst. 2016;108(7):djv439.
61. Pegram HJ, Purdon TJ, van Leeuwen DG, Curran KJ, Giralt SA, Barker JN, et al. IL-12-secreting CD19-targeted cord blood-derived T cells for the immunotherapy of B-cell acute lymphoblastic leukemia. Leukemia. 2015;29(2):415–22.
62. Schamel WW, Arechaga I, Risueno RM, van Santen HM, Cabezas P, Risco C, et al. Coexistence of multivalent and monovalent TCRs explains high sensitivity and wide range of response. J Exp Med. 2005;202(4):493–503.
63. Zhao Q, Jiang Y, Xiang S, Kaboli PJ, Shen J, Zhao Y, et al. Engineered TCR-T cell Immunotherapy in anticancer precision medicine: pros and cons. Front Immunol. 2021;12:658753.
64. Liddy N, Bossi G, Adams KJ, Lissina A, Mahon TM, Hassan NJ, et al. Monoclonal TCR-redirected tumor cell killing. Nat Med. 2012;18(6):980–7.
65. He Q, Jiang X, Zhou X, Weng J. Targeting cancers through TCR-peptide/MHC interactions. J Hematol Oncol. 2019;12(1):139.
66. Getts D, Hofmeister R, Quintas-Cardama A. Synthetic T cell receptor-based lymphocytes for cancer therapy. Adv Drug Deliv Rev. 2019;141:47–54.
67. Alcover A, Alarcon B, Di Bartolo V. Cell biology of T cell receptor expression and regulation. Annu Rev Immunol. 2018;36:103–25.
68. Lejeune M, Kose MC, Duray E, Einsele H, Beguin Y, Caers J. Bispecific, T-cell-recruiting antibodies in B-cell malignancies. Front Immunol. 2020;11:762.
69. Kegyes D, Constantinescu C, Vrancken L, Rasche L, Gregoire C, Tigu B, et al. Patient selection for CAR T or BiTE therapy in multiple myeloma: which treatment for each patient? J Hematol Oncol. 2022;15(1):78.

70. Hyrenius-Wittsten A, Su Y, Park M, Garcia JM, Alavi J, Perry N, et al. SynNotch CAR circuits enhance solid tumor recognition and promote persistent antitumor activity in mouse models. Sci Transl Med. 2021;13(591):eabd8836.
71. Cho JH, Okuma A, Al-Rubaye D, Intisar E, Junghans RP, Wong WW. Engineering Axl specific CAR and SynNotch receptor for cancer therapy. Sci Rep. 2018;8(1):3846.
72. Mo F, Watanabe N, McKenna MK, Hicks MJ, Srinivasan M, Gomes-Silva D, et al. Engineered off-the-shelf therapeutic T cells resist host immune rejection. Nat Biotechnol. 2021;39(1):56–63.
73. Jan M, Scarfo I, Larson RC, Walker A, Schmidts A, Guirguis AA, et al. Reversible ON- and OFF-switch chimeric antigen receptors controlled by lenalidomide. Sci Transl Med. 2021;13(575):eabb6295.
74. Zhang C, Liu J, Zhong JF, Zhang X. Engineering CAR-T cells. Biomark Res. 2017;5:22.
75. Cho JH, Collins JJ, Wong WW. Universal chimeric antigen receptors for multiplexed and logical control of T cell responses. Cell. 2018;173(6):1426–38 e11.
76. Cho JH, Okuma A, Sofjan K, Lee S, Collins JJ, Wong WW. Engineering advanced logic and distributed computing in human CAR immune cells. Nat Commun. 2021;12(1):792.
77. Woodsworth DJ, Holt RA. Cell-based therapeutics: making a faustian pact with biology. Trends Mol Med. 2017;23(2):104–15.
78. Cox MJ, Manriquez Roman C, Tapper EE, Siegler EL, Chappell D, Durrant C, et al. GM-CSF disruption in CART cells modulates T cell activation and enhances CART cell anti-tumor activity. Leukemia. 2022;36(6):1635–45.
79. Sachdeva M, Duchateau P, Depil S, Poirot L, Valton J. Granulocyte-macrophage colony-stimulating factor inactivation in CAR T-cells prevents monocyte-dependent release of key cytokine release syndrome mediators. J Biol Chem. 2019;294(14):5430–7.
80. Cooke KR, Luznik L, Sarantopoulos S, Hakim FT, Jagasia M, Fowler DH, et al. The biology of chronic graft-versus-host disease: a task force report from the national institutes of health consensus development project on criteria for clinical trials in chronic graft-versus-host disease. Biol Blood Marrow Transplant. 2017;23(2):211–34.
81. Zhao J, Lin Q, Song Y, Liu D. Universal CARs, universal T cells, and universal CAR T cells. J Hematol Oncol. 2018;11(1):132.
82. Altomonte M, Montagner R, Fonsatti E, Colizzi F, Cattarossi I, Brasoveanu LI, et al. Expression and structural features of endoglin (CD105), a transforming growth factor beta1 and beta3 binding protein, in human melanoma. Br J Cancer. 1996;74(10):1586–91.
83. Cosimato V, Scalia G, Raia M, Gentile L, Cerbone V, Visconte F, et al. Surface endoglin (CD105) expression on acute leukemia blast cells: an extensive flow cytometry study of 1002 patients. Leuk Lymphoma. 2018;59(9):2242–5.
84. Lontos K, Wang Y, Colbert M, Kumar A, Joshi S, Philbin M, et al. Fully murine CD105-targeted CAR-T cells provide an immunocompetent model for CAR-T cell biology. Onco Targets Ther. 2022;11(1):2131229.
85. Lontos K, Wang Y, Joshi SK, Frisch AT, Watson MJ, Kumar A, et al. Metabolic reprogramming via an engineered PGC-1alpha improves human chimeric antigen receptor T-cell therapy against solid tumors. J Immunother Cancer. 2023;11(3):e006522.

CAR T-Cell Therapy-Associated Toxicities

Bela G. Santani, Alok K. Mishra, and Ritesh Thakare

Abstract Chimeric antigen receptor T-cell (CAR-T) therapies have transformed the treatment of several hematologic malignancies, offering a powerful new option for patients with otherwise refractory cancers. Despite their groundbreaking success, these therapies are associated with a range of toxicities, including cytokine-release syndrome (CRS), neurological toxicities (ICANS), hematologic, and organ-specific adverse effects, along with immune-related adverse events (IRAEs). Understanding and managing these toxicities is critical for optimizing patient outcomes and expanding the use of CAR T-cell therapies. This chapter offers a broad overview of the mechanisms underlying CAR T-cell-related toxicity, highlighting the importance of a proactive and structured approach to toxicity management. The discussion includes an in-depth analysis of CRS and neurological toxicities, detailing their pathophysiology, clinical presentation, risk factors, and current management strategies. Additionally, the chapter covers hematologic, hepatic, pulmonary, renal, and immunogenic toxicities, emphasizing the need for vigilant monitoring and timely intervention. To mitigate these toxicities, this chapter outlines key strategies, such as pretreatment risk assessment, patient education, and innovations in CAR T-cell manufacturing. Furthermore, novel approaches and adjunct therapies designed to reduce the severity of adverse events without compromising therapeutic

B. G. Santani
Department of Microbiology, Sant Gadge Baba Amravati University (SGBAU), Amravati, Maharashtra, India

Foundation Medicine, Inc., Boston, MA, USA

A. K. Mishra
Department of Hematopoietic Biology & Malignancy, Division of Cancer Medicine, The University of Texas MD Anderson Cancer Center, Houston, TX, USA

Department of Molecular, Cell and Cancer Biology, UMass Chan Medical School, Worcester, MA, USA

R. Thakare (✉)
Department of Molecular, Cell and Cancer Biology, UMass Chan Medical School, Worcester, MA, USA
e-mail: ritesh.thakare@umassmed.edu

A. K. Mishra, S. K. Malonia (eds.), *Cell-based Immunotherapies for Cancer*,
https://doi.org/10.1007/978-3-031-88695-9_8

efficacy have been explored. Integrating these advancements into clinical practices can increase the safety and effectiveness of CAR T-cell therapies, ultimately improving patient care and expanding the potential applications of this transformative treatment.

Keywords Chimeric antigen receptor T-cell (CAR-T) · Cytokine-release syndrome (CRS) · Immune effector cell-associated neurotoxicity syndrome (ICANS) · Immune-related adverse events (IRAEs)

1 Introduction

1.1 Definition and Overview of CAR T-Cell Therapy

Chimeric antigen receptor T-cell (CAR-T) therapies have emerged as a revolutionary frontier in cancer treatment, harnessing the patient's immune system to combat human malignancies. CAR T-cell therapies include genetically modifying patients T-cells, endowing them with chimeric antigen receptors (CARs) that augment their ability to target and kill cancer cells. This individualized and focused approach marks a departure from conventional cancer treatments, providing a flexible and adaptive method customized to the distinct genetic profile of every patient. This innovative therapeutic approach has garnered substantial attention and optimism because of its remarkable success in treating hematologic malignancies, including certain types of B-cell lymphomas, such as diffuse large B-cell lymphoma, B-cell acute lymphoblastic leukemia (B-ALL), follicular lymphoma, and mantle cell lymphoma [1, 2]. Ongoing research is exploring the extension of CAR T-cell therapies to solid tumors, broadening the potential scope of their impact, considering their tremendous success in hematologic malignancies [3, 4].

1.2 Mechanism of Action and Clinical Application

The mechanism of action underlying CAR T-cell therapies is intricate and potent. CAR T-cells are autologous T-cells that are isolated from patients and genetically modified ex vivo to express CAR receptors that recognize specific antigens expressed on cancer cells [5]. CAR receptors typically consist of an extracellular antigen-binding domain, a transmembrane domain, and intracellular signaling domains that activate T-cell function upon antigen recognition. When these reprogrammed CAR T-cells are infused back into the patient, they effectively seek and engage with cancer cells bearing the targeted antigen. The subsequent activation of T-cells triggers a cascade of immune responses, leading to the destruction of malignant T-cells. The activation process involves cytokine release, proliferation, and

Table 1 FDA-approved CAR-T therapies for the treatment of hematologic malignancies

CAR-T therapies	Brand name (FDA approval year)	Target antigen	Hematologic malignancies	Trial population	Clinicaltrial.gov identifier (trial name)	References
Tisagenlecleucel (tisa-cel)	Kymriah (2017)	CD19	R/R B-ALL	Children and young	NCT02435849 (ELIANA)	[7]
			R/R B-NHL	Adult	NCT02445248 (JULIET)	[8]
			R/R follicular lymphoma	Adult	NCT03568461 (ELARA)	[9]
Axicabtagene ciloleucel (axi-cel)	Yescarta (2017)		R/R B-NHL	Adult	NCT02348216 (ZUMA-1)	[10]
			R/R follicular lymphoma	Adult	NCT03105336 (ZUMA-5)	[11]
Brexucabtagene autoleucel (brexu-cel)	Tecartus (2021)		R/R B-ALL	Adult	NCT02614066 (ZUMA-3)	[12]
			R/R MCL	Adult	NCT02601313 (ZUMA-2)	[13]
Lisocabtagene maraleucel (liso-cel)	Breyanzi (2022)		R/R B-NHL	Adult	NCT02631044 (TRANSCEND NHL 001)	[14]
Idecabtagene vicleucel (ide-cel)	Abecma (2021)	BCMA	R/R MM	Adult	NCT03361748 (KARMMA)	[15]
Ciltacabtagene autoleucel (cilta-cel)	Carvykti (2022)		R/R MM	Adult	NCT03548207 (CARTITUDE-1)	[16]

Abbreviations: *RR* Relapsed/refractory, *B-ALL* B-cell acute lymphoblastic leukemia, *NHL* Non-Hodgkin lymphoma, *MCL* Mantle cell lymphoma, *MM* Multiple myeloma

infiltration into tumor tissues [6]. Clinical applications of CAR T-cell therapies have demonstrated significant success in the treatment of hematologic malignancies, particularly B-cell malignancies. To date, the Food and Drug Administration (FDA) has approved six CAR T-cell therapies for treating hematologic cancers, including Kymriah, Yescarta, Tecartus, Breyanzi, Abecma, and Carvykti. The information on these CAR T-cell therapies and associated clinical trials is tabulated in Table 1.

CAR T-cell therapy, which targets clusters of differentiation-19 (CD-19), a surface marker on B-cells, has demonstrated promising results in various hematologic malignancies. These malignancies include pediatric and B-cell acute lymphocytic leukemia (B-ALL), chronic lymphocytic leukemia (CLL), and non-Hodgkin lymphoma (NHL). In 2017, the Food and Drug Administration (FDA) approved the first CAR T-cell therapy, tisagenlecleucel, directed at CD-19. This approval was specific to patients with B-ALL, high-grade B-cell lymphoma, diffuse large B-cell lymphoma (DLBCL), and DLBCL arising from follicular lymphoma [17, 18]. The FDA subsequently approved axicabtagene ciloleucel for the treatment of relapsed/refractory DLBCL, follicular lymphoma [18, 19]; lisocabtagene maraleucel [20] for large B-cell lymphoma; brexucabtagene autoleucel [21] for mantle cell lymphoma; anti-B-cell maturation antigen (BCMA) [22] for multiple myeloma; idecabtagene vicleucel [2] for large B-cell lymphoma; and ciltacabtagene autoleucel [22] for mantle cell lymphoma and multiple myeloma, as indicated in Table 1.

1.3 Importance of Understanding Toxicities in CAR-T Therapies

Despite their significant success in treating hematological malignancies, CAR T-cell therapies are also associated with unique toxicities and off-target activities that warrant careful consideration and management. CAR T-cells tend to attack normal cells expressing similar surface proteins. This off-target activity results in adverse effects, leading to substantial therapy-associated mortality and morbidity and thus preventing it from becoming the first-line treatment option for cancer. Although complications arising from CAR T-cell therapy are generally reversible, there are multiple instances where they can result in fatal outcomes or irreversible damage to organs [23]. In the long-term aftermath of CAR T-cell therapy, individuals may experience neuropsychiatric adverse events such as anxiety, depression, or cognitive difficulties. A study reported that up to 37.5% of long-term survivors of CAR T-cell therapy exhibited neuropsychiatric symptoms [24]. These findings emphasize the importance of vigilance in monitoring immediate complications and the potential long-term consequences of CAR T-cell therapy.

CAR T-cell therapy-associated toxicity is associated mainly with toxic levels of systemic cytokine release and hyperimmune cell cross-reactivity, which results in cytokine-release syndrome (CRS) [25], immune effector cell-associated neurotoxicity syndrome (ICANS), macrophage-activation syndrome (MAS), and hemophagocytic lymphohistiocytosis [26, 27]. Additionally, CAR T-cell therapy-associated toxicities include cytopenias, infections, tumor lysis syndrome (TLS), and acute anaphylaxis [28]. The typical toxicities associated with CAR T-cell therapies involve constitutional symptoms such as fever, fatigue, anorexia, and malaise. However, these toxicities affect various organ systems and lead to neurological, hematological, hepatic, gastrointestinal, cardiovascular, pulmonary, renal, and musculoskeletal complications. Figure 1 shows a snapshot of the diverse toxicities linked to CAR T-cell therapy.

Understanding and managing CAR T-cell therapy toxicity is essential for patient safety and effective treatment. CAR T-cell activation and proliferation can lead to systemic responses such as CRS, neurological toxicity, and hematologic complications, which require vigilant monitoring and tailored interventions. Effective management helps balance therapeutic efficacy with patient safety, allowing for adjustments in treatment strategies and informed consent processes. As CAR T-cell

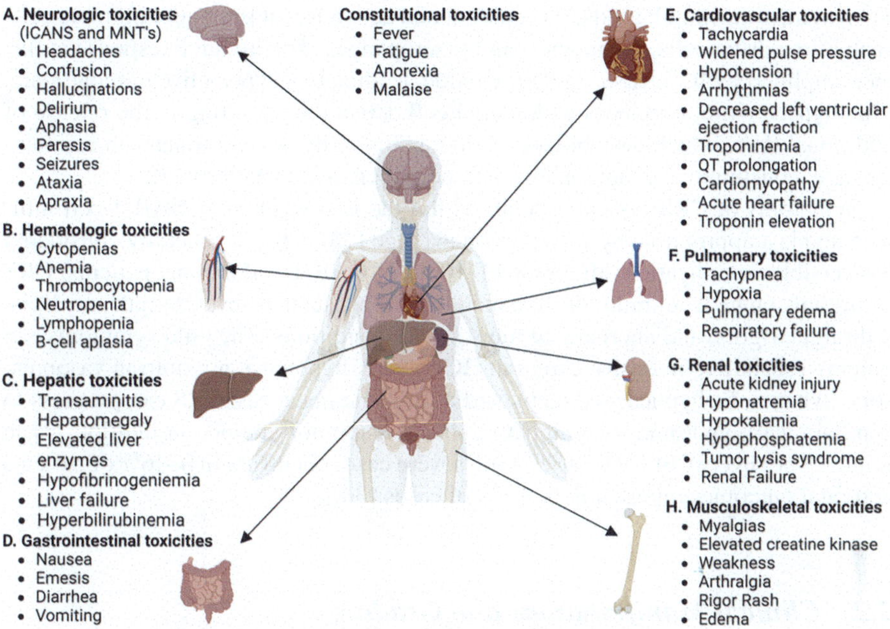

Fig. 1 Clinical presentation of CAR T-cell therapy-associated toxicities

therapies have advanced, comprehending the factors influencing toxicity development and severity has become crucial [29]. Research into the mechanisms of adverse events is necessary for refining CAR T-cell engineering and identifying predictive biomarkers for high-risk patients. This approach aligns with precision medicine trends and supports regulatory approvals and clinical practice guidelines [30].

This chapter covers common CAR T-cell therapy toxicities, their mechanisms, clinical manifestations, and evolving management strategies.

2 Cytokine-Release Syndrome (CRS)

2.1 *Definition and Pathophysiology*

Cytokine-release syndrome (CRS) is the most common toxicity associated with CAR T-cell therapy. It is triggered primarily by the massive release of cytokines following the infusion of CAR T-cells. This cytokine surge is initiated when CAR T-cells are activated and expanded through interactions between tumor cells and immune effector cells [31]. The precise mechanism of CRS remains elusive, but it is thought to involve the release of various inflammatory cytokines, including IL-2, IL-8, IL-10, IFN-γ, tumor necrosis factor (TNF)-α, and GM-cerebrospinal fluid (CSF), following the interaction between CAR T-cells and their target tumor cells.

This interaction stimulates the recruitment and activation of bystander immune cells such as dendritic cells, monocytes, and macrophages. The immune response is further amplified by IL-1, IL-6, and nitric oxide, leading to a hyperactive immune state. This hyperactivation induces endothelial cell activation, resulting in the release of additional IL-6, which destabilizes membrane integrity and culminates in capillary leakage, consumptive coagulopathy, and hemodynamic instability [32].

Symptoms of CRS typically begin within the first week after CAR T-cell infusion and last approximately 7–8 days on average [33]. CRS is generally considered a reversible complication, affecting 42–100% of CAR T-cell therapy patients. CRS symptoms range from mild, flu-like manifestations, such as fever, fatigue, myalgia, arthralgia, rigors, and anorexia, to more severe conditions. The mild symptoms can quickly escalate to intensive care unit (ICU) care, hypotension requiring vasopressors, hypoxia, tachypnea, and tachycardia. In some instances, CRS can progress to capillary leak syndrome, coagulopathy, shock, respiratory failure, or multiple organ failure. The severity of CRS varies, with severe cases occurring in 0–46% of patients and fatal outcomes reported in 0–9.1% of cases [34].

2.2 Clinical Manifestations and Grading

The evaluation of CRS and its severity relies on clinical symptoms, signs, laboratory biomarkers, radiological findings, and inflammatory markers. The presence of varying degrees of fever, hypotension, tachycardia, nausea, hypoxia, and vomiting are considered critical indicators for the clinical diagnosis of CRS. Specific laboratory biomarkers and inflammatory markers are sequentially measured to assess and monitor the severity of CRS. These include a complete blood count (CBC), renal function test, liver function tests, coagulation profile, serum ferritin level, lactate dehydrogenase (LDH) level, IL-6 level, and C-reactive protein (CRP) level [35]. Additionally, the levels of cytokines such as TNF-α, IFN-γ, IL-5, and IL-13 are also utilized for both CRS diagnosis and severity assessment [36]. Vital signs are frequently assessed on the basis of institutional policies, typically for 4 hours for patients with grade 1 or 2 CRS and every 1–2 hours for those with grade 3 or 4 CRS [33, 37].

Several grading systems have been designed to evaluate CRS induced by CAR T-cell therapy. A consensus grading scale developed in 2014 was adopted by the National Cancer Institute (NCI) to grade CRS in CD19-directed CAR T-cell clinical trials [36]. The clinical trial programs at the University of Pennsylvania and the Children's Hospital of Philadelphia have their own modified grading scales [38]. Furthermore, the NCI Common Terminology Criteria for Adverse Events (CTCAE) version 4.0 has also been used for clinical trials of blinatumomab and other therapeutic antibodies [39]. Owing to the use of multiple grading systems and the need for standardization, the American Society for Blood and Marrow Transplantation (ASBMT) introduced a consensus grading system in 2018. The severity of CRS is categorized into four grades by the American Society for Transplantation and Cellular Therapy (ASTCT) on the basis of clinical signs and symptoms (Table 2) [38, 40, 41].

Table 2 CRS grades and clinical manifestation

CRS grades	Clinical manifestation	Interventions
CRS grade 1	Fever, fatigue, nausea, myalgia, headache, malaise, etc., but no hypoxia or hypotension.	Symptomatic treatment given with IV hydration, antipyretics, and infectious disease workflow with blood culture and imaging.
CRS grade 2	Fever (temperature ≥38 °C), hypoxia that require low-flow oxygen (O_2) <40% FiO_2. Hypotension without need for vasopressors.	Moderate intervention is required with supportive measures, similar to grade 1 CRS. Tocilizumab (8 mg/kg) given q8h, limit dosage to maximum of 3 doses in 24 h and total of 4 doses. Corticosteroids given if no improvement seen with Tocilizumab. Dexamethasone 10 gm q6h or methylprednisolone 1 mg/kg q12h.
CRS grade 3	Fever, hypoxia that require high-flow O_2 (high-flow nasal cannula, mask, non-invasive ventilation) ≥40% FiO_2. Hypotension with need for one vasopressor.	Aggressive intervention is required with grade 1 supportive care. ICU support with vasopressor and supplemental oxygen. Tocilizumab 8 mg/kg q24h (maximum dose: 800 mg) Dexamethasone 10–20 mg q6h or Methylprednisolone 1 mg/kg q12h.
CRS grade 4	Fever, hypoxia that require positive pressure (e.g. intubation, mechanical ventilation CPAP (Continuous Positive Airway Pressure), and and BiPAP (Bilevel Positive Airway Pressure). Hypotension with need for multiple vasopressors.	Aggressive intervention is required with grade 1 supportive care in ICU, vasopressor support, and supplemental O_2 with positive pressure. Tocilizumab 8 mg/kg q24h (maximum dose: 800 mg) Methylprednisolone 1000 mg q24h for 3 days.

2.3 *Risk Factors for CRSs*

Multiple factors influence the risk of developing CRS in response to CAR T-cell therapy, each of which plays a distinct role in the onset and severity of this systemic inflammatory response. One critical determinant is the quantity of CAR T-cells administered. Higher doses of CAR T-cells increase the likelihood of an exaggerated immune response, leading to CRS. Numerous studies have shown that the infusion dose of CAR T-cells impacts both safety and efficacy; higher doses are associated with more severe cases of CRS [42, 43]. Another contributing factor is patient's disease burden. Patients with a high tumor load are more prone to intense activation of CAR T-cells and subsequent cytokine release. The peak expansion of CAR T-cells also plays a role in increasing the risk of CRS. The rapid and robust proliferation of CAR T-cells can amplify the cytokine surge, potentially overwhelming the body's regulatory mechanisms. Among others, thrombocytopenia, characterized by low platelet counts, is another risk factor for CRS [44]. A diminished platelet count compromises the ability of the body to regulate clotting, contributing to the overall systemic response observed in CRS.

Additionally, preexisting endothelial activation before CAR T-cell infusion heightens the risk of CRS, as an already sensitized endothelium is more susceptible to the inflammatory signals triggered by CAR T-cells. Lymphodepletion therapy, which commonly involves drugs such as fludarabine and cyclophosphamide, is also a key risk factor [42]. This preparatory regimen aims to create favorable conditions for CAR T-cell engraftment by suppressing the patient's existing immune system. However, it also contributes to the risk of CRS by further amplifying the systemic inflammatory response upon CAR T-cell infusion.

In summary, the risk of developing CRS with CAR T-cell therapy is determined by a combination of factors, including the number of CAR T-cells administered, peak CAR T-cell expansion, disease burden, thrombocytopenia, preexisting endothelial activation, and the choice of lymphodepletion drugs.

2.4 Management and Treatment Strategies

Aggressive supportive care is the cornerstone of managing CRS in patients receiving CAR T-cell therapy. This includes daily assessment and grading of symptoms and monitoring inflammatory markers. Several studies have indicated that tumor burden, levels of ferritin, LDH, and CRP and Eastern Cooperative Oncology Group (ECOG) performance status before CAR T-cell therapy can predict the onset of CRS [40]. From a treatment perspective, blocking the IL-6 receptor with tocilizumab is the primary pharmacologic intervention for CRS. Tocilizumab, an anti-IL-6 receptor antibody, is often used as the first line of defense, and corticosteroids are employed when CRS patients do not respond to tocilizumab. The US FDA has established the risk evaluation and mitigation strategy (REMS) to ensure patient safety when undergoing CAR T-cell therapy. According to the REMS, two doses of tocilizumab must be available for administration within 2 hours of CAR T-cell infusion [45]. While corticosteroids are crucial in managing CRS, their use necessitates careful consideration due to the risk of immunosuppression, which can hinder the proliferation and expansion of CAR T-cells. As a result, corticosteroids are not permitted from the first day of lymphodepletion chemotherapy [46].

3 Neurological Toxicity

3.1 Introduction to the Neurological Toxicities Associated with CAR T-Therapies

Neurotoxicity and neurological complications in cancer treatment are not uncommon. Traditional anticancer therapies have long been associated with well-documented neurotoxicity [47, 48]. Both systemic and intrathecal chemotherapy, as well as radiotherapy, are known to induce neurological side effects in the central nervous system (CNS) and peripheral nervous system (PNS). Some CNS complications may manifest during treatment, whereas others might not appear months or years later. However, most cases of neurotoxicity during CAR T-cell therapy typically resolve within 3–8 weeks [49, 50].

The advent of CAR T-cell therapy introduced a new spectrum of neurological toxicities. It was initially termed CAR T-cell-related encephalopathy syndrome (CRES) and is now more commonly referred to as immune effector cell-associated neurotoxicity syndrome (ICANS). ICANS represents a severe complication that occurs in CAR T-cell therapy, with the potential for life-threatening consequences. Although ICANS is generally a reversible complication, it occurs less frequently and typically later than CRS does. The incidence of ICANS ranges from 3% to 64%, with severe cases reported in 0–54% of patients [23].

3.2 Mechanism and Pathophysiology of ICANS

The mechanism and pathophysiology of ICANS remain inadequately understood. Several theories have been proposed, including CAR T-cell trafficking to the CNS and disruption of the blood–brain barrier. Additionally, the activation of myeloid cells within the CNS, IL-1/IL-6 secretion by monocytes, and GM-CSF-mediated stimulation of monocytes following CAR T-cell administration have been suggested as potential contributors to the development of ICANS [51, 52]. Moreover, elevated levels of IL-15 and N-methyl-D-aspartate receptor agonists (glutamate and quinolinic acid) have also been implicated in the development of ICANS. CAR T-cells may break the blood–brain barrier and induce neurotoxicity, resulting in mild confusion and severe encephalopathy.

ICANS is associated with the activation of myeloid cells, mainly through the stimulation of monocytes by GM-CSF after CAR T-cell therapy, leading to IL-1/IL-6 secretion by monocytes [51, 52]. Higher serum levels of several cytokines, including IL-2, IL-2 receptor-α, IL-5, IL-6, IL-10, IL-15, CRP, ferritin, GM-CSF, granzyme B, and TNF-α, have been reported in several studies [25, 52]. However, none of these cytokines are specific for ICANS, as severe cases of CRS have also been reported because of their elevated status.

Neurotoxicity in CAR T-cell therapy is characterized by elevated cerebrospinal fluid (CSF) protein and white blood cell counts, indicating increased trafficking across the blood–CSF barrier. High levels of IL-6, tumor necrosis factor (TNF)-α, TNFRp55/p75, and interferon-γ in the CSF suggest failure of the blood–brain barrier to prevent the transit of circulating cytokines into the CSF [53, 54]. Interestingly, the detection of CAR T-cells in the CSF, even in patients without neurotoxicity, suggests that the mere presence of CAR T-cells in the CNS is insufficient to induce neurological dysfunction [52, 55]. However, severe neurotoxicity is associated with an enrichment of CD4+ CAR T-cells in the CSF, indicating potential differences in CNS migration between CAR T-cell subsets.

In a nonhuman primate model of CD20-directed CAR T-cell-induced neurotoxicity, both CAR T-cells and non-CAR T-cells accumulated in the CSF and brain parenchyma, further highlighting the complex mechanisms underlying neurotoxicity in CAR T-cell therapy [51].

3.3 Clinical Manifestations of ICANS

ICANS presents with a range of symptoms, including headache, aphasia, seizures, and altered mental status. These symptoms can vary from aphasia and mild confusion to severe encephalopathy and delirium. Common manifestations include transient impairments in attention and cognitive processing, which may present as delirium or confusion in 50–66% of patients receiving CAR T-cell therapy [7, 52, 56, 57]. Seizures are reported with varying incidences across clinical trials and are more common in pediatric patients and adults receiving CD28-directed CAR products. However, they are infrequent in adults without a prior history of seizures or severe neurotoxicity [7, 25, 58]. Language disturbances, such as difficulties with word finding, are likely manifestations of neurotoxicity.

Acute cerebral edema is a particularly severe complication associated exclusively with CD19-directed CAR T-cell therapy. This condition was notably severe in a clinical trial of adults with acute lymphoblastic leukemia (ALL), where five patients succumbed to fatal cerebral edema, leading to the trial's termination [59]. Fatalities related to cerebral edema have also been reported in trials involving CAR constructs with either 4–1BB or CD28 costimulatory domains across various patient populations, including non-Hodgkin lymphoma (NHL), children and adults with B-ALL, and chronic lymphocytic leukemia (CLL) [52, 56, 60]. Cerebral edema can develop rapidly, often while cytokine-release syndrome (CRS) is resolving, and may progress from mild confusion to severe outcomes such as coma and brain death. In addition to cerebral edema, other severe neurologic events following CD19 CAR T-cell therapy include cortical necrosis, acute cerebral hemorrhage, brainstem and deep brain structure edema and hemorrhage, multifocal thrombotic microangiopathy, and subacute encephalomalacia [7, 56]. Postmortem studies have revealed widespread neuronal and white matter injury with perivascular CD8+ T-cell infiltration, microglial activation, macrophage infiltration, and CAR T-cell infiltration in brain tissue and CSF [52].

In the diagnosis of ICANS, clinical symptoms such as headaches, agitation, impaired consciousness, hallucinations, lethargy, aphasia, tremors, encephalopathy, and seizures are critical [61]. Diagnostic approaches include biomarkers indicating inflammatory responses and neurological involvement, lumbar puncture for cerebrospinal fluid analysis, neuroimaging (MRI) to detect structural abnormalities or inflammation, and electroencephalography (EEG) to assess brain activity. These measures help confirm ICANS and differentiate it from other conditions, such as infections, stroke, chemical toxicity, or CNS malignancies [25]. Misdiagnosis of ICANS can occur due to overlapping symptoms with stroke, chemotherapy-related toxicity, CNS involvement in malignancies, and viral infections [62]. Accurate differentiation is crucial for appropriate treatment, as each condition requires different management strategies. A comprehensive evaluation, including clinical assessments, imaging studies, and laboratory analyses, is essential for precise diagnosis and effective treatment planning.

Patients with severe neurotoxicity (grade ≥4) have elevated serum levels of IL-2, IL-10, IL-15, and MCP-1 within the first 36 hours, along with an initial peak of IL-6, compared with those with milder neurotoxicity (grade ≤3). A high angiopoietin-2 ratio before lymphodepletion chemotherapy indicates preexisting endothelial activation dysregulation, increasing the risk of severe neurotoxicity. While cytokine-based algorithms are sensitive and specific for predicting neurotoxicity, practical limitations exist in real-time clinical settings. Fever ≥38.9 °C within 36 hours of CAR T-cell infusion alone shows 82% specificity and 100% sensitivity for life-threatening neurotoxicity, and the addition of a serum MCP-1 assay improved the specificity to 94% [52]. This highlights the potential for the use of these clinical markers for timely intervention.

3.4 ICANS Grading

The evaluation and grading of neurotoxicity have evolved with the implementation of the immune effector cell-associated encephalopathy (ICE) criteria, a refinement replacing the CARTOX-10 criteria that were previously used. This change aims to introduce greater objectivity in the assessment of neurotoxicity severity (Table 3) [35, 40, 63].

Table 3 Encephalopathy assessment tools for grading of ICANS

	ICE [40]		CARTOX-10 [25]	
Parameters	Task	Scoring point	Task	Scoring point
Orientation	Orientation to the month, year, hospital, and city.	4	Orientation to month, year, hospital, city, prime minister/president of country.	5
Naming	Ability to name three objects ability (e.g., point to clock, button, pen)	3	Ability to name 3 objects (e.g., point to clock, button, pen)	3
Follow commands	Ability to follow simple commands (e.g., "Close your eyes or "Show me 2 fingers")	1	–	–
Writing ability	Ability to write simple standard sentences (e.g., "Our national bird is the bald eagle"):	1	Ability to write simple standard sentences (e.g., "Our national bird is the bald eagle")	1
Attention ability	Ability to count backward from 100 to 10	1	Ability to count backward from 100 to 10	1

Table 4 ASTCT-ICANS consensus grading for adults [40]

Grade	Clinical manifestation	Intervention/management
1	*ICE score*: 7–9 *Consciousness*: Awakens spontaneously *Seizure*: NA *Motor findings*: NA *Elevated ICP/cerebral edema*: NA	Supportive management with EEG, lumbar puncture, and neuroimaging. Tocilizumab given when concurrent CRS is observed.
2	*ICE score*: 3–6 *Consciousness*: Awakens to a voice *Seizure*: NA *Motor findings*: NA *Elevated ICP/cerebral edema*: NA	Grade 1 supportive management. Dexamethasone 10 mg, q6h, or methylprednisolone equivalent.
3	*ICE score*: 0–2 *Consciousness*: Awakens only to tactile stimuli *Seizure*: Any seizure that resolves rapidly, with/without interventions. *Elevated ICP/cerebral edema*: Focal or local edema on imaging.	Grade 1 supportive management. ICU admission/airway protection. Dexamethasone 10–20 mg q6h or methylprednisolone equivalent. If deterioration in health observed, grade 4 steroids are given. For seizures control, antiepileptics and benzodiazepines given.
4	*ICE score*: 0 *Consciousness*: Unarousable, or coma, or vigorous stimuli requires to wake patient. *Seizure*: Repetitive life-threatening seizures (>5 min) or Clinical or Electrical seizures with no baseline. Deep focal weakness such as hemiparesis or paraparesis. *Elevated ICP/cerebral edema*: Diffuse cerebral edema on imaging. Papilledema, cranial nerve VI palsy, decerebrate/decorticate posture, or Cushing's triad (hypertension, bradycardia, and abnormal breathing).	Grade 1 supportive management. ICU admission/airway protection, Intubation, and mechanical ventilation. Lower elevated ICP with hyperventilation, hyperosmolar therapy, and/or neurosurgical intervention. Antiepileptics given for Seizers. If refractory, siltuximab or anakinra and high-dose methylprednisolone (1000 mg/day) is given.

Abbreviations: *NA* Not applicable, *ICE* Immune effector cell-associated encephalopathy, *CNS* Central nervous system, *EEG* Electroencephalogram, *ICP* Intracranial pressure

CARTOX-10 is updated to the ICE tool. ICE adds a command-following assessment without changing the scoring system. The scoring system used is as follows: 10: no impairment; 7–9: Grade 1 ICANS; 3–6: Grade 2 ICANS; 0–2: Grade 3 ICANS; and 0: Grade 4 ICANS (unable patient or inability to perform ICE assessment) [40] Table 4.

3.5 *Risk Factors for Neurological Toxicity*

The risk factors for ICANS associated with CAR T-cell therapy are multifaceted and contribute to both the development and severity of this neurotoxic complication. CRS, which frequently occurs alongside ICANS, is a significant factor that often correlates with an increased risk of ICANS [64]. Additionally, preexisting neurologic dysfunction increases susceptibility to neurotoxic effects. A high disease burden, indicated by a greater tumor load, increases the number of target T-cells for CAR T-cell therapy, which may intensify the immune response and subsequent neurotoxicity. Elevated LDH levels and thrombocytopenia are markers of disease activity that can be correlated with the severity of ICANS [65]. Furthermore, increased ferritin levels within 72 hours after CAR T-cell infusion are reflective of an inflammatory response and may signal impending neurotoxicity. The use of lymphodepletion therapy with agents such as fludarabine and cyclophosphamide can modify the patient's immune environment, affecting the risk and severity of neurotoxicity [66]. Other patient-specific factors, such as age and overall performance status, play a role, with older age and poorer performance status being linked with a greater risk of neurotoxicity [67]. The type and stage of the malignancy being treated can influence ICANS risk, as certain malignancies may inherently pose a higher risk owing to their biological characteristics. The number of CAR T-cells administered can impact the risk of ICANS, as higher doses may lead to a more intense immune response and neurotoxic effects. Additionally, elevated levels of inflammatory biomarkers, such as CRP and procalcitonin, have been associated with increased risk and severity of ICANS [68]. Genetic factors may contribute to the variability in patient responses to CAR T-cell therapy and susceptibility to ICANS. A history of infections, particularly those affecting the central nervous system, and baseline immune status, including prior immunosuppressive treatments, may further influence the risk of neurotoxicity. Understanding these risk factors is crucial for predicting and managing the potential for neurological toxicity in patients receiving CAR T-cell therapy.

3.6 Management and Treatment Strategies

Timely identification of high-risk patients for neurotoxicity is crucial for enabling timely intervention. Evaluating cytokine profiles and clinical markers can help determine whether a patient is a candidate for neurotoxicity. Management approaches for neurotoxicity in the context of CAR T-cell therapy vary among researchers. Some advocate a conservative approach, emphasizing supportive care alone since most neurological symptoms resolve within 21 days after CAR T-cell infusion. In contrast, others recommend a more proactive strategy involving early and aggressive interventions with immunomodulators [69]. Many patients with neurotoxicity also experience CRS. Standard treatments for both CRS and ICANS include tocilizumab, an IL-6 receptor antibody, and/or corticosteroids to modulate T-cell activity. However, the impact of these interventions on the course of neurotoxicity still needs to be fully understood. The use of tocilizumab is inconsistent across clinical trials, with some centers using it as a first-line therapy for ICANS and others not employing it at all [25, 70].

Dexamethasone is commonly used as a first-line therapy for CAR T-cell-associated neurotoxicity because of its excellent ability to penetrate the CNS and its established role in treating cerebral edema and inflammation in conditions such as brain tumors and neurotrauma [71]. Low to moderate doses of dexamethasone (10 mg q12h or q8h) were used to treat ICANS. The high-dose methylprednisolone regime is reserved for severe neurotoxicity cases, given its proven efficacy and safety in neuroinflammatory disorders [72]. However, the impact of high-dose corticosteroids (e.g., 1 g/day methylprednisolone) on antitumor efficacy remains to be formally assessed, as they may cause significant lymphopenia and reduce the number of circulating CAR T-cells.

Other biologics include siltuximab, a chimeric monoclonal antibody that blocks IL-6, masking its interaction with both soluble and membrane-bound IL-6 receptors. Siltuximab has been used to manage neurologic adverse events in patients receiving CAR T-cell therapy for glioblastoma [73]. Anakinra, an FDA-approved IL-1 receptor antagonist for adult rheumatoid arthritis and neonatal-onset multisystem inflammatory disease, has been demonstrated to be effective in treating life-threatening macrophage-activation syndrome and systemic juvenile idiopathic arthritis. In a murine model, anakinra administered at CAR T-cell infusion effectively abrogated CRS and neurotoxicity [74]. Other potential therapies for CRS and neurotoxicity include platelet hypertransfusion, plasma exchange, angiopoietin-1 augmentation, or small-molecule inhibition of cytokine signaling pathways [75]. Levetiracetam is commonly used to treat seizures in patients receiving CAR T-cell therapy.

For life-threatening neurotoxicity, such as rapidly progressive neurologic symptoms or cerebral edema, aggressive supportive care is needed. Interventions for severe neurologic symptoms include sedation, hyperosmolar therapy (e.g., hypertonic saline or mannitol), high-dose corticosteroids, and aggressive seizure management. The evidence does not support invasive neuromonitoring, which poses significant risks, particularly in coagulopathic patients. Early detection of deterioration risk remains the most effective strategy to prevent fulminant cerebral edema, as its progression can be rapid and potentially irreversible once it is established. Apart from ICANS, a distinct neurologic toxicity has been identified as movement and neurocognitive treatment-emergent adverse events. This phenomenon was recently observed in patients receiving BCMA CAR T-cell therapy for multiple myeloma [76]. Movement- and neurocognitive treatment-emergent adverse events are similar to those of Parkinsonian syndrome, despite the absence of substantia nigra degeneration in neuropathology and the lack of BCMA expression in the brain parenchyma, eliminating the possibility of on-target, off-tumor toxicity. Currently, there is no established treatment for this particular entity.

4 Hematologic Toxicities in CAR-T Therapies

4.1 Overview of Hematologic Toxicities

Hematological toxicities in CAR T-cell therapy represent a spectrum of adverse effects on the blood and bone marrow, reflecting the profound impact of this treatment on the hematopoietic system. As CAR T-cell therapy becomes more prominent in treating certain hematologic malignancies, it also presents significant hematological challenges. After CRS and ICANS, hematological toxicity is one of the most frequently observed adverse events, with an incidence exceeding 90%, and it is associated with poor outcomes [77–79]. These toxicities involve mainly cytopenias, which are reductions in essential blood cells, including red (anemia), white (neutropenia), and platelets (thrombocytopenia). These cytopenias can result from the intense immune response triggered by CAR T-cells, which affects both malignant and healthy cells in the bone marrow. Additionally, hematologic toxicity includes immune constituent alterations, poor CD4+ T-cell or CD19+ T-cell recovery, and hypogammaglobulinemia [80, 81]. These factors heighten the risk of infection and are linked to prolonged hospitalizations [82].

4.2 Cytopenias: Thrombocytopenia, Anemia, and Neutropenia

Cytopenias refers to a deficiency or decrease in one or more cell types in the peripheral blood, specifically affecting cells of the hematopoietic system. The criteria for defining cytopenia and its recovery have been established by the reporting guidelines of the Center for International Blood and Marrow Transplant Research (CIBMTR) [83].

Anemia, characterized by decreased red blood cell production or increased destruction, requires careful monitoring and management to alleviate symptoms such as fatigue and shortness of breath. Anemia is identified by a hemoglobin concentration of less than 120 g/L in men and 110 g/L in women, with levels less than 60 g/L considered severe. Anemia increases susceptibility to infections, contributes to fatigue and shortness of breath, and impairs blood clotting. Thrombocytopenia is a decrease in the platelet count to less than 100×10^9/L, which increases the risk of bleeding and bruising. Severe thrombocytopenia (platelet count $<20 \times 10^9$/L) in patients receiving CAR T-cell therapy has led to fatalities due to intracranial hemorrhage or other life-threatening bleeding events during clinical trials [10, 84]. Neutropenia, a reduction in the neutrophil count to 1.5×10^9/L, significantly increases susceptibility to infections, posing a serious concern for patients receiving CAR T-cell therapy. Fatalities have been reported in patients with severe neutropenia (neutrophils; $<0.5 \times 10^9$/L) due to severe infections [13, 85].

A recent meta-analysis of hematological toxicity following CAR T-cell therapy, which included 52 studies involving 2004 patients, highlighted the incidence of neutropenia (80%), thrombocytopenia (61%), and anemia (68%) of any grade. The incidence rates of grade ≥3 neutropenia, thrombocytopenia, and anemia were 60%, 33%, and 32%, respectively. Among the 52 studies, 46 reported the incidence of hematological toxicity involving one or more cytopenias. Specifically, neutropenia was reported in 40 studies (n = 1652), thrombocytopenia in 41 studies (n = 1619), and anemia in 40 studies (n = 1638). CAR T-cell therapies targeting non-CD19 antigens have shown higher rates of hematological toxicity than do CD19-targeted therapies, with CD28-targeted CAR T-cell therapies exhibiting a greater tendency to

Table 5 Hematological toxicities reported in CAR-T therapy for leukemia, lymphoma, and multiple myeloma [25, 84, 86, 87]

Cytopenia	Neutropenia		Thrombocytopenia		Anemia	
Hematological malignancies	Any grade AE (%)	≥3 grade AE (%)	Any grade AE (%)	≥3 grade AE (%)	Any grade AE (%)	≥3 grade AE (%)
Leukemia	244 (62)	390 (48)	254 (60)	450 (28)	176 (69)	350 (41)
Lymphoma	737 (83)	985 (60)	742 (60)	825 (32)	721 (68)	979 (24)
Multiple myeloma	132 (88)	215 (58)	182 (57)	261 (40)	136 (53)	350 (31)

Lymphoma is collectively referred for DLBCL, MCL, and HL; Leukemia is referred for all types of leukemia

induce hematological toxic effects than 4-1BB-targeted therapies [86]. Age also plays a role in the incidence of hematological toxicity, particularly in lymphoma patients, where those under 60 years of age are more likely to suffer from hematological toxicity. The incidence of thrombocytopenia is greater in older patients [87].

The pooled results for hematological toxicities from all these clinical trials encompassing three hematological malignancies (leukemia, lymphoma, and multiple myeloma) are tabulated in Table 5 [25, 84, 86, 87].

4.3 *Risk Factors and Monitoring for Hematologic Toxicities*

Identifying risk factors is crucial for the preemptive management of hematological toxicities with CAR T-cell therapy. These risk factors include patient-specific variables, disease characteristics, and treatment-related factors. Rigorous monitoring through regular blood tests, particularly complete blood counts, is essential for the timely detection and intervention of patients with cytopenias. Paying close attention to trends in hematologic parameters helps predict and prevent severe complications. In cases of persistent or severe cytopenias, further investigations, such as a bone marrow biopsy, are recommended to assess for secondary bone marrow malignancies [88].

4.4 *Management and Treatment Approaches*

The management of hematological toxicities requires a multifaceted approach, with treatment algorithms tailored to the specific type and severity of cytopenia. This includes platelet transfusions, erythropoiesis-stimulating agents, and red blood cell transfusions to address thrombocytopenia and anemia. For persistent cytopenias, the infusion of autologous or allogeneic stem cells has also been employed [89]. CD34+ selected stem cell boosts have been explored as a potential strategy [90, 91]. The management of anemia and thrombocytopenia involves the administration of erythrocytes and platelets, respectively, while prolonged neutropenia is managed with GM-CSF. GM-CSF has also been suggested as a potential therapy to manage CAR T-cell toxicity [92, 93]. The use of erythropoietin or thrombopoietin analogs has been reported in cases of prolonged cytopenias [94, 95].

Typically, cytopenias tend to resolve spontaneously over time. However, in cases where cytopenias persist or recovery is delayed, a bone marrow biopsy may be warranted to assess the possibility of secondary bone marrow malignancy [96]. Hematologic toxicity can lead to additional complications, such as infections or hemorrhage, which are often associated with cytokine release following CAR T-cell infusion. Improved management of CRS may enhance hematopoietic recovery following CD19 CAR T-cell therapy, with extended growth factors and transfusional support playing a critical role [97]. Hematologic toxicity following lymphodepletion chemotherapy may require antiviral prophylaxis, such as acyclovir, and antimicrobial and antifungal prophylaxis, especially in patients with severe or persistent neutropenia [98].

Taken together, these findings suggest that CAR T-cell therapy associated with hematological toxicity poses a significant hurdle. Predominant grade ≥3 side effects involve hematological issues, with some cases leading to fatalities due to infections or severe hemorrhage in the initial stages. However, during long-term follow-up, hematological toxicity is typically less life-threatening, and most affected patients recover to adequate levels within 3 months. Clinicians must proactively address hematological toxicity early in CAR T-cell therapy to prevent life-threatening infections.

5 Organ System-Specific Toxicities

The impact of CAR T-cell therapy extends across multiple organ systems owing to its off-target activity. The emergence of CRS, ICANS, and other significant toxicities has underscored the occurrence of multiorgan toxicity following CAR T-cell therapy. While CRS and ICANS manifest systemically, it is essential to recognize that CAR T-cell therapies can also influence specific organs, presenting distinct clinical challenges. A snapshot of organ system-specific toxicity is shown in Fig. 1.

5.1 Hepatic Toxicity

Hepatic toxicity associated with CAR T-cell therapies has emerged as a significant challenge. These toxicities are characterized by abnormal liver function tests, including elevated alanine aminotransferase (ALT) and aspartate aminotransferase (AST) in the liver. Serum transaminases and bilirubin levels are reported to increase following the onset of CRS and ICANS during CAR T-cell therapy [43, 61]. The interaction between CAR T-cells and the liver involves several mechanisms contributing to hepatic damage. Patients may present with a range of symptoms, from mild hepatomegaly to severe hepatic dysfunction. The activation of CAR T-cells triggers a cascade of cytokine release, leading to immune-mediated hepatocellular damage. Risk factors for hepatic toxicity include the intensity of the lymphodepletion regimen, tumor burden, and preexisting liver conditions [99]. The management of hepatic toxicity requires a multidisciplinary approach that regularly involves monitoring liver function tests and early detection. In mild cases, symptomatic relief and temporary discontinuation of CAR T-cell therapy may suffice, whereas severe cases may require immunosuppressive agents, such as corticosteroids, to mitigate the inflammatory response.

5.2 *Pulmonary Toxicities*

The pathophysiology of pulmonary toxicity involves a complex interaction between CAR T-cells and lung tissue. CRS and ICANS contribute to the release of proinflammatory cytokines, leading to pulmonary vascular leakage and edema. Pulmonary toxicities associated with CAR T-cell therapies can present a spectrum of clinical symptoms, ranging from mild cough and dyspnea to life-threatening acute respiratory distress syndrome (ARDS) [39]. These toxicities can lead to hypoxia, pulmonary edema, dyspnea, and pneumonitis. Hypoxia has been reported in 6–15% of patients, with some cases being severe enough to require mechanical ventilation [61]. Early recognition of these symptoms is crucial for prompt intervention. Continuous monitoring of respiratory status and early identification of pulmonary toxicity are vital. Imaging, such as chest X-rays and CT scans, plays a crucial role in assessing the extent of lung involvement. In severe cases, management strategies, such as oxygen supplementation and mechanical ventilation, require supportive care.

5.3 *Renal Toxicity*

Renal toxicity is emerging as a significant concern with CAR T-cell therapies. The primary cause of renal injury during CAR T-cell therapy is often decreased renal perfusion, resulting from cytokine-induced vasodilation and intravascular dehydration caused by insensible fluid loss during high fevers [100]. Acute renal toxicity following CAR T-cell infusion is typically reversible and has multiple contributing factors. Tumor lysis syndrome and the effects of certain drugs, including antibiotics, may also contribute to renal injury. Additionally, electrolyte imbalances, such as hypokalemia (Low K^+ in blood), hyponatremia (Low Na^+ in blood), and hypophosphatemia, are frequently observed. The clinical presentation can range from asymptomatic proteinuria to acute kidney injury [101]. Understanding the spectrum of renal toxicity is crucial for timely intervention. Risk factors for renal toxicity include preexisting renal dysfunction, the intensity of the conditioning regimens, and the cumulative dose of CAR T-cells. Management involves supportive care and targeted interventions, with adequate hydration essential to prevent acute kidney injury. Diuretics may be used judiciously, and in severe cases, renal replacement therapy may be needed [102]. The development of nephroprotective strategies and personalized dosing regimens is necessary to optimize the safety profile of CAR T-cell therapies.

5.4 Cardiovascular Toxicity

CAR T-cell therapies are associated with cardiovascular toxicity, adding complexity to the clinical landscape. One notable concern is the development of cardiomyopathy and myocarditis, conditions characterized by heart muscle inflammation, which can lead to impaired cardiac function [103]. Monitoring cardiac health becomes paramount, with vigilance for signs such as chest pain, dyspnea, and changes in cardiac biomarkers. Additionally, CAR T-cell therapies may contribute to the onset of arrhythmias, disrupting the normal rhythm of the heart. Patients receiving CAR T-cell therapy require close monitoring for any irregularities in heart rhythm, including a thorough cardiac evaluation. Addressing arrhythmias promptly prevents complications and ensures the patient's overall cardiovascular well-being. In clinical trials, cases of cardiac arrest have been reported as early as 7 days after CAR T-cell infusion. For example, a patient with B-ALL experienced a significant decrease in the left ventricular ejection fraction to less than 25% of the normal baseline value [61]. Similarly, reversible reductions in the cardiac ejection fraction have been documented in multiple other patients [100, 104].

5.5 Musculoskeletal Toxicity

Musculoskeletal toxicity can manifest as a range of adverse events affecting bones, joints, and surrounding tissues. Creatine kinase has been observed in patients receiving anti-CD19 CAR T-cell therapy for B-ALL, indicating musculoskeletal toxicity [43]. Patients receiving CAR T-cell therapy may experience arthralgia (joint pain) and myalgia (muscle pain), which contribute to discomfort and impaired mobility, impacting their quality of life [100]. Long-term considerations include the potential impact of CAR T-cell therapies on bone health, with osteoporosis characterized by reduced bone density and increased fragility being a concern. Monitoring bone health through appropriate assessments and considering interventions to support bone density should be integral to the comprehensive care plan for patients receiving CAR T-cell therapy.

5.6 *Infectious Diseases*

CAR T-cell treatments can compromise the immune system's ability to defend against pathogens, increasing vulnerability to infections owing to their immunosuppressive nature. Addressing infectious diseases is critical to patient care during CAR T-cell therapy. Patients receiving CAR T-cell therapy often experience neutropenia and lymphopenia, which predisposes them to opportunistic infections. This compromised immune system is a characteristic state induced by CAR T-cell therapies and faces an elevated risk of opportunistic bacterial, viral, fungal, or parasitic infections. Furthermore, distinguishing the symptoms of CRS, such as fever, tachycardia, and hypotension, from sepsis resulting from opportunistic infection can be challenging. A patient with chronic lymphocytic leukemia died from fever, renal failure, and hypotension after CD19-directed CAR T-cell therapy. Subsequent investigations revealed elevated levels of inflammatory cytokines in the serum before CAR T-cell infusion, indicating a preexisting infection [105]. Bacterial infections, including bacteremia, urinary tract infections, and Salmonella, have been reported after CAR T-cell treatment [106, 107]. Additionally, viral infections such as influenza virus, herpes zoster virus, and respiratory syncytial virus have been reported after CAR T-cell infusion [104, 108].

In summary, mitigating organ system-specific toxicity following CAR T-cell therapy demands a nuanced and vigilant approach. Research exploring the role of anti-inflammatory agents, hepatoprotective agents, and the development of nephroprotective strategies, along with comprehensive care, in mitigating pulmonary toxicity is warranted.

6 Immunogenicity and Immune-Related Adverse Events During CAR T-Cell Therapy

Cytokine release is a vital component of the therapeutic mechanism of CAR T-cell therapy, as it directs the immune response against cancer cells. However, this process also introduces immune-related adverse events (IRAEs). IRAEs are adverse events that result from the immune system's response to CAR T-cell therapy and manifest as side effects and complications ranging from mild to severe. The principal IRAEs linked with CAR T-cells include CRS, neurotoxicity, immunosuppression, and on-target/off-tumor recognition [109]. Increased immune activation and cytokine release can lead to systemic inflammatory responses, affecting various organs and tissues. Common IRAEs with CAR T-cell therapy include gastroenteritis and dermatitis, whereas more severe IRAEs may involve adrenal failure, diabetic ketoacidosis, pneumonitis, and myocarditis. Other IRAEs include hypophysitis, hepatitis, thyroiditis, nephritis, and pancreatitis [109]. Severe but rare IRAEs, such as hemophagocytic lymphohistiocytosis (HLH) and macrophage-activation syndrome (MAS), can result in serious complications. HLH/MAS is characterized by excessive immune activation, infiltration of lymphohistiocytic tissues, and the onset of multiorgan failure. Treatment typically involves tocilizumab and corticosteroids, with etoposide, a conventional chemotherapeutic agent, administered in cases resistant to these interventions [25]. Effective management of IRAEs is critical for ensuring the efficacy and safety of CAR T-cell therapy.

6.1 Endocrinopathies

Immune-related adverse events, particularly endocrinopathies such as thyroiditis and thyroid dysfunction, are well documented in patients receiving immune checkpoint inhibitors. Most commonly, cytotoxic T-lymphocyte antigen 4 (CTLA-4), programmed death ligand 1 (PDL-1), and programmed death 1 (PD-1) antibodies have been linked to endocrinopathies [110, 111]. However, such endocrinopathies are very uncommon in patients receiving CAR T-cell therapy. Only three case studies have reported endocrine system dysregulation following CAR T-cell therapy. These disruptions may manifest as thyroiditis, adrenal insufficiency, or other hormonal imbalances, leading to clinical symptoms such as fatigue, weight changes, and alterations in metabolic function. One rare case involved an acute lymphoblastic leukemia patient who developed CRS and significant arthritis after CAR T-cell therapy. The patient experienced relief of joint symptoms following glucocorticoid therapy [112].

Additionally, Hashimoto's thyroiditis has been reported in two patients who underwent CAR T-cell therapy for relapsed refractory diffuse large B-cell lymphoma. In these patients, elevated peroxidase and globulin antibodies develop after CAR T-cell therapy, leading to the onset of Hashimoto's thyroiditis. Notably, both patients achieved complete remission at 1 and 3 months posttherapy [113].

6.2 Dermatologic Toxicity

CAR T-cell therapy can induce a range of dermatologic toxicities, ranging from mild rashes to severe skin conditions such as Stevens–Johnson syndrome. However, these types of CAR T-cell therapy are relatively rare. Dermatologic toxicities are more commonly associated with other forms of immunotherapy, with reported incidence rates varying from 4% to 36% [114]. Among the adverse cutaneous events, the most prevalent manifestation is a maculopapular eruption, followed by erythematous rashes, petechiae, purpura, and bullous eruptions. The skin reactions are mild to moderate in severity and typically resolve independently without intervention. The latency period for onset ranges from 1 day to 19 months following the initial CAR T-cell infusion, highlighting the importance of ongoing clinical vigilance [114].

The mechanisms underlying cutaneous toxicity in CAR T-cell therapy are not well understood. One possible cause is off-target activity, where CAR T-cells may cross-react with normal tissues expressing antigens similar to those expressed by malignant

T-cells. Additionally, the cytokine storm associated with CRS, a well-described adverse event in CAR T-cell therapy, may contribute to cutaneous toxicity. Elevated levels of proinflammatory cytokines linked to CRS may play a role in the formation of bullous and other cutaneous reactions. Furthermore, on-target off-tumor effects, where CAR T-cells attack normal tissue cells expressing a similar target antigen, represent another potential mechanism for these dermatologic toxicities [115].

Few case reports have been published on CAR T-cell-associated dermatologic toxicity. For example, a study by Hu et al. reported a 40-year-old male subject with relapsed/refractory B-cell lymphoma who underwent dual-targeted CD19/CD22-CAR T-cell therapy. The subject developed a painful maculopapular rash and tense bullae on his fingers within 5 days of CAR T-cell infusion. Flow cytometry confirmed the presence of CAR T-cells in the bullae fluid, which linked these symptoms directly to the therapy [116]. Another clinical trial involving 41 patients with relapsed/refractory Hodgkin's lymphoma treated with CD30-directed CAR T-cell therapy. Among the 41 patients, 19 developed a maculopapular rash, which spontaneously resolved within 10 days. A quantitative polymerase chain reaction of a biopsy sample confirmed the presence of the CD30 CAR transgene [117].

Given the potential impact of dermatologic toxicity, prompt intervention is often necessary. Management may involve topical treatments, systemic corticosteroids, or, in severe cases, temporary discontinuation of CAR T-cell therapy.

6.3 Gastrointestinal Toxicity

Gastrointestinal toxicity represents another aspect of IRAEs linked with CAR T-cell therapy. A range of symptoms include nausea, diarrhea, vomiting, and abdominal pain. These toxicities can escalate to gastrointestinal perforation or colitis in severe cases [100, 105]. In a clinical trial with 130 subjects receiving CAR T-cell therapy for large B-cell lymphoma, 24 subjects had involvement of the gastrointestinal tract. Among these 24 patients, a small subset (3/24) experienced gastrointestinal perforation, leading to poor treatment outcomes and complicated post-CAR T-cell treatment courses [118]. Despite these risks, some patients with gastrointestinal involvement appear to achieve durable remission following CAR T-cell therapy. Effective monitoring and management of these toxicities require a combination of supportive care, the use of antidiarrheal medications, and the administration of immunosuppressive agents when necessary.

6.4 Management of IRAEs

Understanding the risk factors associated with IRAEs is essential for personalized patient care. Preexisting comorbidities, disease characteristics, and patient-specific factors may contribute to the likelihood and severity of IRAEs. Monitoring laboratory parameters, imaging studies, and clinical assessments is crucial for early detection and intervention. Managing IRAEs often involves a multidisciplinary approach, including collaboration between oncologists, endocrinologists, dermatologists, and gastroenterologists.

7 Strategies for Minimizing Toxicities and Future Perspective

7.1 Pretreatment Risk Assessment and Patient Selection

Effective management of CAR T-cell therapy begins with a thorough pretreatment risk assessment and careful patient selection. This process involves evaluating the patient's overall health, tumor burden, and potential for adverse reactions to the therapy. Health assessments should consider comorbidities such as cardiovascular disease or renal impairment, as these conditions may increase the risk of severe toxicity. Additionally, a high tumor burden has been associated with a greater risk of CRS and neurotoxicity. Imaging and biomarker assessments can better stratify patients according to their risk. These tools help in creating more personalized treatment plans. Furthermore, evaluating a patient's genetic predisposition, including polymorphisms related to the immune response, may offer additional insights into their risk profile for CAR T-cell therapy-related toxicity.

7.2 Patient Education and Informed Consent

It is crucial to educate patients and their families about the potential risks and benefits of CAR T-cell therapy. Informed consent should be thorough, addressing both short-term and long-term risks associated with CAR T-cell therapy. Clinicians must ensure that patients fully understand possible side effects, such as CRS, neurotoxicity, and the potential need for intensive care. Patients should have ample time to consider the risks and benefits, with the consent process being a continuous dialog rather than a one-time event. Offering resources and support, including psychosocial care and clear guidelines on what to expect during treatment, can help reduce anxiety and better prepare patients for their treatment journey.

7.3 CAR T-Manufacturing and Product Modifications

Modifications in CAR T-cell manufacturing processes can also play a role in reducing toxicity. Innovations in how these cells are engineered and expanded have led to safer and more effective therapies. The choice of the costimulatory domain and the affinity of the CAR for its target antigen can influence the activation and persistence of CAR T-cells, thereby affecting their toxicity profiles. Improved techniques for T-cell selection, expansion, and cryopreservation have enhanced the consistency and quality of CAR T-cell products, reducing variability in patient responses. Careful titration of CAR T-cell doses, based on patient-specific factors such as body surface area and tumor burden, can help minimize adverse effects [119].

7.4 Novel Approaches and Adjunct Therapies for Toxicity Management

CAR T-cell-related toxicity management has advanced with the development of novel approaches and adjunct therapies designed to mitigate severity while preserving therapeutic efficacy. Cytokine inhibitors, such as tocilizumab (an IL-6 receptor antagonist), have become the standard for managing CRS [120]. Corticosteroids are effective for treating neurotoxicity, especially in severe cases, but their use must be balanced to avoid diminishing CAR T-cell activity. Early intervention with these agents can prevent progression to severe, life-threatening stages. A notable challenge in CAR T-cell therapy is the limited persistence of CAR T-cells following infusion, often due to heightened expression of PD-1, leading to immune exhaustion and potential disease relapse. To address this, immune checkpoint inhibitors targeting PD-1 have been investigated [121]. Preclinical studies suggest that combining PD-1 checkpoint blockade with CAR T-cell therapy can reverse CAR T-cell exhaustion and enhance tumor cell death, indicating promising clinical outcomes. Additionally, research into next-generation CAR T-cells includes the development of safety switches or suicide genes to reduce toxicity while maintaining efficacy. For example, the inducible caspase-9 safety switch, initially designed to mitigate graft-versus-host disease (GVHD) in allogeneic stem cell transplants, can also be applied to manage significant toxicities in CAR T-cell therapies [121].

8 Conclusion

CAR T-cell therapy has transformed the landscape of cancer treatment, offering hope to patients with otherwise refractory malignancies. However, the success of this innovative therapy is tempered by significant toxicity, ranging from CRS to ICANS, hematologic, and organ-specific complications. This chapter explores these toxicity mechanisms, risk factors, and clinical management, underscoring the importance of a proactive and comprehensive approach to patient care. The strategies outlined, including thorough pretreatment risk assessments, patient education, and advancements in CAR T-cell manufacturing, are critical for minimizing toxicity while preserving therapeutic efficacy. Novel approaches and adjunct therapies pave the way for more refined and personalized CAR-T-related adverse event management. As CAR T-cell therapies evolve, ongoing research into predictive biomarkers and personalized toxicity management will be essential. The integration of these strategies into clinical practices not only improves patient outcomes but also expands the applicability of CAR T-cell therapies to a broader patient population. By overcoming the curve in toxicity management, the field can maximize the therapeutic potential of CAR T-cell therapies, ultimately leading to better, safer, and more effective cancer treatments.

Acknowledgments The authors acknowledge UMass Chan Medical School, Foundation Medicine Inc., MD Anderson Cancer Center, and SGBAU for their support. Authors also acknowledge the assistance of AI tools in refining the language and enhancing the clarity of the chapter.

Author Disclosure The authors have no conflicts of interest to disclose.

References

1. Braendstrup P, Levine BL, Ruella M. The long road to the first FDA-approved gene therapy: chimeric antigen receptor T-cells targeting CD19. Cytotherapy. 2020;22:57–69.
2. Sharma P, Kanapuru B, George B, Lin X, Xu Z, Bryan WW, et al. FDA approval summary: idecabtagene vicleucel for relapsed or refractory multiple myeloma. Clin Cancer Res. 2022;28(9):1759–64.
3. D'Aloia MM, Zizzari IG, Sacchetti B, Pierelli L, Alimandi M. CAR T-cells: the long and winding road to solid tumors review-article. Cell Death Dis. 2018;9:282.
4. Safarzadeh Kozani P, Safarzadeh Kozani P, Ahmadi Najafabadi M, Yousefi F, Mirarefin SMJ, Rahbarizadeh F. Recent advances in solid tumor CAR T-cell therapy: driving tumor cells from hero to zero? Front Immunol. 2022;13:795164.
5. Sermer D, Brentjens R. CAR T-cell therapy: full speed ahead. Hematol Oncol. 2019;37(S1):95–100.
6. Alnefaie A, Albogami S, Asiri Y, Ahmad T, Alotaibi SS, Al-Sanea MM, et al. Chimeric antigen receptor T-cells: an overview of concepts, applications, limitations, and proposed solutions. Front Bioeng Biotechnol. 2022;10:797440.
7. Maude SL, Laetsch TW, Buechner J, Rives S, Boyer M, Bittencourt H, et al. Tisagenlecleucel in children and young adults with B-cell lymphoblastic leukemia. N Engl J Med. 2018;378(5):439–48.
8. Schuster SJ, Bishop MR, Tam CS, Waller EK, Borchmann P, McGuirk JP, et al. Tisagenlecleucel in adult relapsed or refractory diffuse large B-cell lymphoma. N Engl J Med. 2019;380(1):45–56.
9. Fowler NH, Dickinson M, Dreyling M, Martinez-Lopez J, Kolstad A, Butler J, et al. Tisagenlecleucel in adult relapsed or refractory follicular lymphoma: the phase 2 ELARA trial. Nat Med. 2022;28(2):325–32.
10. Neelapu SS, Locke FL, Bartlett NL, Lekakis LJ, Miklos DB, Jacobson CA, et al. Axicabtagene ciloleucel CAR T-cell therapy in refractory large B-cell lymphoma. N Engl J Med. 2017;377(26):2531–44.
11. Jacobson CA, Chavez JC, Sehgal AR, William BM, Munoz J, Salles G, et al. Axicabtagene ciloleucel in relapsed or refractory indolent non-Hodgkin lymphoma (ZUMA-5): a single-arm, multicentre, phase 2 trial. Lancet Oncol. 2022;23(1):91–103.

12. Shah BD, Ghobadi A, Oluwole OO, Logan AC, Boissel N, Cassaday RD, et al. KTE-X19 for relapsed or refractory adult B-cell acute lymphoblastic leukemia: phase 2 results of the single-arm, open-label, multicenter ZUMA-3 study. Lancet. 2021;398(10299):491–502.
13. Wang M, Munoz J, Goy A, Locke FL, Jacobson CA, Hill BT, et al. KTE-X19 CAR T-cell therapy in relapsed or refractory mantle-cell lymphoma. N Engl J Med. 2020;382(14):1331–42.
14. Abramson JS, Palomba ML, Gordon LI, Lunning MA, Wang M, Arnason J, et al. Lisocabtagene maraleucel for patients with relapsed or refractory large B-cell lymphomas (TRANSCEND NHL 001): a multicenter seamless design study. Lancet. 2020;396(10254):839–52.
15. Munshi NC, Anderson LD, Shah N, Madduri D, Berdeja J, Lonial S, et al. Idecabtagene vicleucel in relapsed and refractory multiple myeloma. N Engl J Med. 2021;384(8):705–16.
16. Berdeja JG, Madduri D, Usmani SZ, Jakubowiak A, Agha M, Cohen AD, et al. Ciltacabtagene autoleucel, a B-cell maturation antigen-directed chimeric antigen receptor T-cell therapy in patients with relapsed or refractory multiple myeloma (CARTITUDE-1): a phase 1b/2 open-label study. Lancet. 2021;398(10297):314–24.
17. Si Lim SJ, Grupp SA, DiNofia AM. Tisagenlecleucel for treatment of children and young adults with relapsed/refractory B-cell acute lymphoblastic leukemia. Pediatr Blood Cancer. 2021;68:e29123.
18. Bouchkouj N, Kasamon YL, de Claro RA, George B, Lin X, Lee S, et al. FDA approval summary: axicabtagene ciloleucel for relapsed or refractory large B-cell lymphoma. Clin Cancer Res. 2019;25(6):1702–8.
19. Bouchkouj N, Zimmerman M, Kasamon YL, Wang C, Dai T, Xu Z, et al. FDA approval summary: axicabtagene ciloleucel for relapsed or refractory follicular lymphoma. Oncologist. 2022;27:587–94.
20. St-Pierre F, Gordon LI. Lisocabtagene maraleucel in the treatment of relapsed/refractory large B-cell lymphoma. Future Oncol. 2023;19:19–28.
21. Deshpande A, Wang Y, Munoz J, Jain P. Brexucabtagene autoleucel: a breakthrough in the treatment of mantle cell lymphoma. Drugs Today. 2022;58:283–98.
22. Chekol Abebe E, Yibeltal Shiferaw M, Tadele Admasu F, Asmamaw Dejenie T. Ciltacabtagene autoleucel: the second anti-BCMA CAR T-cell therapeutic armamentarium of relapsed or refractory multiple myeloma. Front Immunol. 2022;13:991092.
23. Xiao X, Huang S, Chen S, Wang Y, Sun Q, Xu X, et al. Mechanisms of cytokine release syndrome and neurotoxicity of CAR T-cell therapy and associated prevention and management strategies. J Exp Clin Cancer Res. 2021;40:367.
24. Ruark J, Mullane E, Cleary N, Cordeiro A, Bezerra ED, Wu V, et al. Patient-reported neuropsychiatric outcomes of long-term survivors after chimeric antigen receptor T-cell therapy. Biol Blood Marrow Transplant. 2020;26(1):34–43.
25. Neelapu SS, Tummala S, Kebriaei P, Wierda W, Gutierrez C, Locke FL, et al. Chimeric antigen receptor T-cell therapy-assessment and management of toxicities. Nat Rev Clin Oncol. 2018;15:47–62.
26. Roex G, Timmers M, Wouters K, Campillo-Davo D, Flumens D, Schroyens W, et al. Safety and clinical efficacy of BCMA CAR T-cell therapy in multiple myeloma. J Hematol Oncol. 2020;13(1):164.
27. Adkins S. CAR T-cell therapy: adverse events and management. J Adv Pract Oncol. 2019;10(4):21–8.
28. Logue JM, Peres LC, Hashmi H, Colin-Leitzinger CM, Shrewsbury AM, Hosoya H, et al. Early cytopenias and infections after standard of care idecabtagene vicleucel in relapsed or refractory multiple myeloma. Blood Adv. 2022;6(24):6109–19.
29. Sterner RC, Sterner RM. CAR T-cell therapy: current limitations and potential strategies. Blood Cancer J. 2021;11:69.
30. Thompson JA. New NCCN guidelines: recognition and management of immunotherapy-related toxicity. JNCCN: J Natl Compr Cancer Netw. 2018;16:594–6.
31. Teachey DT, Lacey SF, Shaw PA, Melenhorst JJ, Maude SL, Frey N, et al. Identification of predictive biomarkers for cytokine release syndrome after chimeric antigen receptor T-cell therapy for acute lymphoblastic leukemia. Cancer Discov. 2016;6(6):664–79.

32. Gauthier J, Turtle CJ. Insights into cytokine release syndrome and neurotoxicity after CD19-specific CAR T-cell therapy. Curr Res Transl Med. 2018;66:50–2.
33. Owusu KA, Schiffer M, Perreault S. Chimeric antigen receptor T-cells: toxicity and management considerations. AACN Adv Crit Care. 2022;33(4):301–7.
34. Abramson JS, Solomon SR, Arnason J, Johnston PB, Glass B, Bachanova V, et al. Lisocabtagene maraleucel as second-line therapy for large B-cell lymphoma: primary analysis of the phase 3 TRANSFORM study. Blood. 2023;141(14):1675–84.
35. Chou CK, Turtle CJ. Assessment and management of cytokine release syndrome and neurotoxicity following CD19 CAR T-cell therapy. Expert Opin Biol Ther. 2020;20:653–64.
36. Lee DW, Gardner R, Porter DL, Louis CU, Ahmed N, Jensen M, et al. Current concepts in the diagnosis and management of cytokine release syndrome. Blood. 2014;124(2):188–95.
37. Yáñez L, Sánchez-Escamilla M, Perales MA. CAR T-cell toxicity: current management and future directions. HemaSphere. 2019;3:e186.
38. Porter D, Frey N, Wood PA, Weng Y, Grupp SA. Grading of cytokine release syndrome associated with the CAR T-cell therapy tisagenlecleucel. J Hematol Oncol. 2018;11:35.
39. Frey N, Porter D. Cytokine release syndrome with chimeric antigen receptor T-cell therapy. Biol Blood Marrow Transplant. 2019;25:e123–7.
40. Lee DW, Santomasso BD, Locke FL, Ghobadi A, Turtle CJ, Brudno JN, et al. ASTCT consensus grading for cytokine release syndrome and neurologic toxicity associated with immune effector cells. Biol Blood Marrow Transplant. 2019;25:625–38.
41. Acharya UH, Dhawale T, Yun S, Jacobson CA, Chavez JC, Ramos JD, et al. Management of cytokine release syndrome and neurotoxicity in chimeric antigen receptor (CAR) T-cell therapy. Expert Rev Hematol. 2019;12:195–205.
42. Turtle CJ, Hanafi LA, Berger C, Gooley TA, Cherian S, Hudecek M, et al. CD19 CAR T-cells of defined CD4+:CD8+ composition in adult B-cell ALL patients. J Clin Invest. 2016;126(6):2123–38.
43. Lee DW, Kochenderfer JN, Stetler-Stevenson M, Cui YK, Delbrook C, Feldman SA, et al. T-cells expressing CD19 chimeric antigen receptors for acute lymphoblastic leukemia in children and young adults: a phase 1 dose-escalation trial. Lancet. 2015;385(9967):517–28.
44. Du M, Huang L, Kou H, Li C, Hu Y, Mei H. Case report: ITP treatment after CAR T-cell therapy in patients with multiple myeloma. Front Immunol. 2022;13:898341.
45. Booth JP, Kusoski CL, Kennerly Shah JM. The pharmacist's role in chimeric antigen receptor T-cell therapy. J Oncol Pharm Pract. 2020;26:1725–31.
46. Alexander M, Culos K, Roddy J, Shaw JR, Bachmeier C, Shigle TL, et al. Chimeric antigen receptor T-cell therapy: a comprehensive review of clinical efficacy, toxicity, and best practices for outpatient administration. Transplant Cell Ther. 2021;27:558–70.
47. Stone JB, DeAngelis LM. Cancer-treatment-induced neurotoxicity-focus on newer treatments. Nat Rev Clin Oncol. 2016;13:92–105.
48. Wang AB, Housley SN, Flores AM, Kircher SM, Perreault EJ, Cope TC. A review of movement disorders in chemotherapy-induced neurotoxicity. J Neuroeng Rehabil. 2021;18:16.
49. Tallantyre EC, Evans NA, Parry-Jones J, Morgan MPG, Jones CH, Ingram W. Neurological updates: neurological complications of CAR-T therapy. J Neurol. 2021;268(4):1544–54.
50. Hunter BD, Jacobson CA. CAR T-cell associated neurotoxicity: mechanisms, clinicopathologic correlates, and future directions. J Natl Cancer Inst. 2019;111(7):646–54.
51. Taraseviciute A, Tkachev V, Ponce R, Turtle CJ, Snyder JM, Liggitt HD, et al. Chimeric antigen receptor T-cell–mediated neurotoxicity in nonhuman primates. Cancer Discov. 2018;8(6):750–63.
52. Gust J, Hay KA, Hanafi LA, Li D, Myerson D, Gonzalez-Cuyar LF, et al. Endothelial activation and blood–brain barrier disruption in neurotoxicity after adoptive immunotherapy with CD19 CAR T-cells. Cancer Discov. 2017;7(12):1404–19.
53. Gust J, Taraseviciute A, Turtle CJ. Neurotoxicity associated with CD19-targeted CAR T-cell therapies. CNS Drugs. 2018;32(12):1091–101.

54. Santomasso BD, Park JH, Salloum D, Riviere I, Flynn J, Mead E, et al. Clinical and biological correlates of neurotoxicity associated with car t-cell therapy in patients with B-cell acute lymphoblastic leukemia. Cancer Discov. 2018;8(8):958–71.
55. Santomasso B, Park JH, Riviere I, Mead E, Halton E, Diamonte C, et al. Biomarkers associated with neurotoxicity in adult patients with relapsed or refractory B-ALL (R/R B-ALL) treated with CD19 CAR T-cells. J Clin Oncol. 2017;35(15_suppl):3019.
56. Schuster SJ, Svoboda J, Chong EA, Nasta SD, Mato AR, Anak Ö, et al. Chimeric antigen receptor T-cells in refractory B-cell lymphomas. N Engl J Med. 2017;377(26):2545–54.
57. Fry TJ, Shah NN, Orentas RJ, Stetler-Stevenson M, Yuan CM, Ramakrishna S, et al. CD22-targeted CAR T-cells induce remission in B-ALL that is naive or resistant to CD19-targeted CAR immunotherapy. Nat Med. 2018;24(1):20–8.
58. Gardner RA, Finney O, Annesley C, Brakke H, Summers C, Leger K, et al. Intent-to-treat leukemia remission by CD19 CAR T-cells of defined formulation and dose in children and young adults. Blood. 2017;129(25):3322–31.
59. Puzanov I, Diab A, Abdallah K, Bingham CO, Brogdon C, Dadu R, et al. Managing toxicities associated with immune checkpoint inhibitors: consensus recommendations from the Society for Immunotherapy of Cancer (SITC) Toxicity Management Working Group. J Immunother Cancer. 2017;5(1):1–28.
60. Locke FL, Neelapu SS, Bartlett NL, Lekakis LJ, Jacobson CA, Braunschweig I, et al. Preliminary results of prophylactic tocilizumab after axicabtageneciloleucel (axi-cel; KTE-C19) treatment for patients with refractory, aggressive non-Hodgkin lymphoma (NHL). Blood. 2017;130(Suppl_1):1547.
61. Brudno JN, Kochenderfer JN. Toxicities of chimeric antigen receptor T-cells: recognition and management. Blood. 2016;127:3321–30.
62. Messmer AS, Que YA, Schankin C, Banz Y, Bacher U, Novak U, et al. CAR T-cell therapy and critical care: a survival guide for medical emergency teams. Wien Klin Wochenschr. 2021;133:1318–25.
63. Murthy HS, Yassine F, Iqbal M, Alotaibi S, Moustafa MA, Kharfan-Dabaja MA. Management of CAR T-cell related toxicities: what did the learning curve teach us so far?, Hematol Oncol Stem Cell Ther. 2022;15:100–11.
64. Morris EC, Neelapu SS, Giavridis T, Sadelain M. Cytokine release syndrome and associated neurotoxicity in cancer immunotherapy. Nat Rev Immunol. 2022;22:85–96.
65. Castaneda-Puglianini O, Chavez JC. Assessing and management of neurotoxicity after CAR-T therapy in diffuse large B-cell lymphoma. J Blood Med. 2021;12:775–83.
66. Holtzman NG, Xie H, Bentzen S, Kesari V, Bukhari A, El Chaer F, et al. Immune effector cell-associated neurotoxicity syndrome after chimeric antigen receptor T-cell therapy for lymphoma: predictive biomarkers and clinical outcomes. Neuro-Oncology. 2021;23(1):112–21.
67. Santomasso BD, Gust J, Perna F. How I treat unique and difficult-to-manage cases of CAR T-cell therapy–associated neurotoxicity. Blood. 2023;141:2443–51.
68. Amidi Y, Eckhardt CA, Quadri SA, Malik P, Firme MS, Jones DK, et al. Forecasting immune effector cell-associated neurotoxicity syndrome after chimeric antigen receptor t-cell therapy. J Immunother Cancer. 2022;10(11):e005459.
69. Teachey DT, Bishop MR, Maloney DG, Grupp SA. Toxicity management after chimeric antigen receptor T-cell therapy: one size does not fit "ALL". Nat Rev Clin Oncol. 2018;15:218.
70. Grupp S. Beginning the CAR T-cell therapy revolution in the US and EU. Curr Res Transl Med. 2018;66:62–4.
71. Michinaga S, Koyama Y. Pathogenesis of brain edema and investigation into anti-edema drugs. Int J Mol Sci. 2015;16:9949–75.
72. Berkovich RR. Acute multiple sclerosis relapse. CONTINUUM Lifelong Learn Neurol. 2016;22:799–814.
73. Chen F, Teachey DT, Pequignot E, Frey N, Porter D, Maude SL, et al. Measuring IL-6 and sIL-6R in serum from patients treated with tocilizumab and/or siltuximab following CAR T-cell therapy. J Immunol Methods. 2016;434:1–8.

74. Norelli M, Camisa B, Barbiera G, Falcone L, Purevdorj A, Genua M, et al. Monocyte-derived IL-1 and IL-6 are differentially required for cytokine-release syndrome and neurotoxicity due to CAR T-cells. Nat Med. 2018;24(6):739–48.
75. Saharinen P, Eklund L, Alitalo K. Therapeutic targeting of the angiopoietin-TIE pathway. Nat Rev Drug Discov. 2017;16:635–61.
76. Cohen AD, Parekh S, Santomasso BD, Pérez-Larraya JG, van de Donk NWCJ, Arnulf B, et al. Incidence and management of CAR-T neurotoxicity in patients with multiple myeloma treated with ciltacabtagene autoleucel in CARTITUDE studies. Blood Cancer J. 2022;12(2):32.
77. Nahas GR, Komanduri KV, Pereira D, Goodman M, Jimenez AM, Beitinjaneh A, et al. Incidence and risk factors associated with a syndrome of persistent cytopenias after CAR T-cell therapy (PCTT). Leuk Lymphoma. 2020;61(4):940–3.
78. Jain T, Knezevic A, Pennisi M, Chen Y, Ruiz JD, Purdon TJ, et al. Hematopoietic recovery in patients receiving chimeric antigen receptor T-cell therapy for hematologic malignancies. Blood Adv. 2020;4(15):3776–87.
79. Fried S, Avigdor A, Bielorai B, Meir A, Besser MJ, Schachter J, et al. Early and late hematologic toxicity following CD19 CAR T-cells. Bone Marrow Transplant. 2019;54(10):1643–50.
80. Baird JH, Epstein DJ, Tamaresis JS, Ehlinger Z, Spiegel JY, Craig J, et al. Immune reconstitution and infectious complications following axicabtagene ciloleucel therapy for large B-cell lymphoma. Blood Adv. 2021;5(1):143–55.
81. Logue JM, Zucchetti E, Bachmeier CA, Krivenko GS, Larson V, Ninh D, et al. Immune reconstitution and associated infections following axicabtagene ciloleucel in relapsed or refractory large B-cell lymphoma. Hematologica. 2021;106(4):978.
82. Rejeski K, Perez A, Iacoboni G, Penack O, Bücklein V, Jentzsch L, et al. The CAR-HEMATOTOX risk-stratifies patients for severe infections and disease progression after CD19 CAR-T in R/R LBCL. J Immunother Cancer. 2022;10(5):e004475.
83. Li H, Zhao L, Sun Z, Yao Y, Li L, Wang J, et al. Prolonged hematological toxicity in patients receiving BCMA/CD19 CAR T-cell therapy for relapsed or refractory multiple myeloma. Front Immunol. 2022;13:1019548.
84. Yan Z, Cao J, Cheng H, Qiao J, Zhang H, Wang Y, et al. A combination of humanized anti-CD19 and anti-BCMA CAR T-cells in patients with relapsed or refractory multiple myeloma: a single-arm, phase 2 trial. Lancet Hematol. 2019;6(10):e521–9.
85. Tong C, Zhang Y, Liu Y, Ji X, Zhang W, Guo Y, et al. Optimized tandem CD19/CD20 CAR-engineered T-cells in refractory/relapsed B-cell lymphoma. Blood. 2020;136(14):1632–44.
86. Zhao X, Yang J, Zhang X, Lu XA, Xiong M, Zhang J, et al. Efficacy and safety of CD28- or 4-1BB-based CD19 CAR T-cells in B-cell acute lymphoblastic leukemia. Mol Ther Oncolytics. 2020;18:272–81.
87. Luo W, Li C, Zhang Y, Du M, Kou H, Lu C, et al. Adverse effects in hematologic malignancies treated with chimeric antigen receptor (CAR) T-cell therapy: a systematic review and meta-analysis. BMC Cancer. 2022;22(1):98.
88. Sharma N, Reagan PM, Liesveld JL. Cytopenia after CAR T-cell therapy—a brief review of a complex problem. Cancers. 2022;14:1501.
89. Xia Y, Zhang J, Li J, Zhang L, Li J, Fan L, et al. Cytopenias following anti-CD19 chimeric antigen receptor (CAR) T-cell therapy: a systematic analysis for contributing factors. Ann Med. 2022;54:2951–65.
90. Mullanfiroze K, Lazareva A, Chu J, Williams L, Burridge S, Silva J, et al. CD34 selected stem cell boost can safely improve cytopaenias following car-t therapy. Bone Marrow Transplant. 2022;57(10):4715–8.
91. Rejeski K, Burchert A, Iacoboni G, Sesques P, Fransecky L, Bücklein V, et al. Safety and feasibility of stem cell boost as a salvage therapy for severe hematotoxicity after CD19 CAR T-cell therapy. Blood Adv. 2022;6:4719–25.
92. Sterner RM, Sakemura R, Cox MJ, Yang N, Khadka RH, Forsman CL, et al. GM-CSF inhibition reduces cytokine release syndrome and neuroinflammation but enhances CAR T-cell function in xenografts. Blood. 2019;133(7):697–709.

93. Miller KC, Johnson PC, Abramson JS, Soumerai JD, Yee AJ, Branagan AR, et al. Effect of granulocyte colony-stimulating factor on toxicities after CAR T-cell therapy for lymphoma and myeloma. Blood Cancer J. 2022;12(10):146.
94. Cordeiro A, Bezerra ED, Hirayama AV, Hill JA, Wu QV, Voutsinas J, et al. Late events after treatment with CD19-targeted chimeric antigen receptor modified T-cells. Biol Blood Marrow Transplant. 2020;26(1):26–33.
95. Baur R, Jitschin R, Kharboutli S, Stoll A, Völkl S, Büttner-Herold M, et al. Thrombopoietin receptor agonists for acquired thrombocytopenia following anti-CD19 CAR T-cell therapy: a case report. J Immunother Cancer. 2021;9(7):e002721.
96. Jain T, Olson TS, Locke FL. How I treat cytopenias after CAR T-cell therapy. Blood. 2023;141:2460–9.
97. Juluri KR, Wu QV, Voutsinas J, Hou J, Hirayama AV, Mullane E, et al. Severe cytokine release syndrome is associated with hematologic toxicity following CD19 CAR T-cell therapy. Blood Adv. 2022;6(7):2055–68.
98. Buie LW. Balancing the CAR T: perspectives on efficacy and safety of CAR T-cell therapy in hematologic malignancies. Am J Manag Care. 2021;27:S243–52.
99. Garcia-Recio M, Wudhikarn K, Pennisi M, Alonso-Trillo R, Flynn J, Shouval R, et al. The international prognostic index is associated with outcomes in diffuse large B-cell lymphoma after chimeric antigen receptor T-cell therapy. Transplant Cell Ther. 2021;27(3):233–40.
100. Brudno JN, Somerville RPT, Shi V, Rose JJ, Halverson DC, Fowler DH, et al. Allogeneic T-cells that express an anti-CD19 chimeric antigen receptor induce remissions of B-cell malignancies that progress after allogeneic hematopoietic stem-cell transplantation without causing graft-versus-host disease. J Clin Oncol. 2016;34(10):1112–21.
101. Gutgarts V, Jain T, Zheng J, Maloy MA, Ruiz JD, Pennisi M, et al. Acute kidney injury after CAR T-cell therapy: low incidence and rapid recovery. Biol Blood Marrow Transplant. 2020;26(6):1071–6.
102. Jhaveri KD, Rosner MH. Chimeric antigen receptor T-cell therapy and the kidney: what the nephrologist needs to know. Clin J Am Soc Nephrol. 2018;13:796–8.
103. Rao A, Stewart A, Eljalby M, Ramakrishnan P, Anderson LD, Awan FT, et al. Cardiovascular disease and chimeric antigen receptor cellular therapy. Front Cardiovasc Med. 2022;9:932347.
104. Kochenderfer JN, Dudley ME, Carpenter RO, Kassim SH, Rose JJ, Telford WG, et al. Donor-derived CD19-targeted T-cells cause regression of malignancy persisting after allogeneic hematopoietic stem cell transplantation. Blood. 2013;122(25):4129–39.
105. Brentjens R, Yeh R, Bernal Y, Riviere I, Sadelain M. Treatment of chronic lymphocytic leukemia with genetically targeted autologous T-cells: case report of an unforeseen adverse event in a phase I clinical trial. Mol Ther. 2010;18:666–8.
106. Kochenderfer JN, Dudley ME, Kassim SH, Somerville RPT, Carpenter RO, Maryalice SS, et al. Chemotherapy-refractory diffuse large B-cell lymphoma and indolent B-cell malignancies can be effectively treated with autologous T-cells expressing an anti-CD19 chimeric antigen receptor. J Clin Oncol. 2015;33(6):540–9.
107. Maude SL, Frey N, Shaw PA, Aplenc R, Barrett DM, Bunin NJ, et al. Chimeric antigen receptor T-cells for sustained remissions in leukemia. N Engl J Med. 2014;371(16):1507–17.
108. Kochenderfer JN, Dudley ME, Feldman SA, Wilson WH, Spaner DE, Maric I, et al. B-cell depletion and remissions of malignancy along with cytokine-associated toxicity in a clinical trial of anti-CD19 chimeric-antigen-receptor-transduced T-cells. Blood. 2012;119(12):2709–20.
109. Bonifant CL, Jackson HJ, Brentjens RJ, Curran KJ. Toxicity and management in CAR T-cell therapy. Mol Ther Oncolytics. 2016;3:16011.
110. Chong EA, Melenhorst JJ, Lacey SF, Ambrose DE, Gonzalez V, Levine BL, et al. PD-1 blockade modulates chimeric antigen receptor (CAR)-modified T-cells: refueling the CAR. Blood. 2017;129:1039–41.

111. Kambhampati S, Gray L, Fakhri B, Lo M, Vu K, Arora S, et al. Immune-related adverse events associated with checkpoint inhibition in the setting of CAR T-cell therapy: a case series. Clin Lymphoma Myeloma Leuk. 2020;20(3):e118–23.
112. Wang LX, Chen X, Jia M, Wang S, Shen J. Arthritis of large joints shown as a rare clinical feature of cytokine release syndrome after chimeric antigen receptor T-cell therapy: a case report. Medicine (United States). 2018;97(16):e0455.
113. Chen P, Xia Y, Lei W, Zhong S, Jiang H, Ren L, et al. Case report: Hashimoto's thyroiditis after CD19 chimeric antigen receptor T-cell therapy. Front Immunol. 2022;13:995496.
114. Nusbaum KB, Dulmage B, Choi JN, Jaglowski SM, Korman AM. Cutaneous manifestations of chimeric antigen receptor T-cell therapy: an introduction for dermatologists. J Am Acad Dermatol. 2022;87:597–604.
115. Feng KC, Guo YL, Liu Y, Dai HR, Wang Y, Lv HY, et al. Cocktail treatment with EGFR-specific and CD133-specific chimeric antigen receptor-modified T-cells in a patient with advanced cholangiocarcinoma. J Hematol Oncol. 2017;10(1):1–11.
116. Hu Y, Zheng W, Qiao J, Cui Q, Zhu Y, Chang AH, et al. Bullous and exanthematous lesions associated with chimeric antigen receptor T-cell therapy in a patient with diffuse large B-cell lymphoma. JAMA Dermatol. 2020;156:1026–8.
117. Ramos CA, Grover NS, Beaven AW, Lulla PD, Wu MF, Ivanova A, et al. Anti-CD30 CAR T-cell therapy in relapsed and refractory Hodgkin lymphoma. J Clin Oncol. 2020;38(32):3794–804.
118. Cortes-Bullich A, Perez A, Bachmeier C, Mhaskar R, Chavez JC, Shah B, et al. Outcomes of CD19 chimeric antigen receptor T-cell therapy in patients with gastrointestinal tract involvement of large B-cell lymphoma. Transplant Cell Ther. 2021;27(9):768-e1.
119. Watanabe N, Mo F, McKenna MK. Impact of manufacturing procedures on CAR T-cell functionality. Front Immunol. 2022;13:876339.
120. Si S, Teachey DT. Spotlight on tocilizumab in the treatment of CAR T-cell-induced cytokine release syndrome: clinical evidence to date. Ther Clin Risk Manag. 2020;16:705–14.
121. Straathof KC, Pulè MA, Yotnda P, Dotti G, Vanin EF, Brenner MK, et al. An inducible caspase 9 safety switch for T-cell therapy. Blood. 2005;105(11):4247–54.

T-Cell Receptor-Engineered T (TCR-T) Cell Therapy: Advancements, Challenges, and Future Perspectives

Pratirodh Koirala, Dibya Ghimire, Alok K. Mishra, and Samyabrata Bhaduri

Abstract T-cell receptor-engineered T (TCR-T) cell therapy is a promising adoptive cell therapy (ACT) for solid tumor treatment. TCR-T-cell therapy involves technology to engineer T cells that express tumor-specific TCRs, increasing their affinity for cancer antigens, which may improve their cytotoxicity. Unlike chimeric antigen receptor T (CAR-T) cells, which target only membrane antigens, TCR-T cells exhibit a broader antigen repertoire, enabling the recognition of membrane and intracellular protein antigens presented by MHC-I and expanding the potential application of TCR-T-cell therapy. Given the importance of TCR-T-cell cancer treatment, this chapter discusses the methods and platforms used to discover cancer-specific TCRs, preclinical strategies for TCR-T-cell therapy development, and an overview of the current clinical landscape of TCR-T-cell therapy. This chapter summarizes the cancer antigen classes recognized by TCR-T cells, strategies for selecting antigens, indications, and patients for clinical studies. It also addresses challenges associated with TCR-T-cell therapy, including manufacturing bottlenecks, associated toxicity, and potential resistance mechanisms. Finally, this chapter offers perspective on the future development of improved TCR-T-cell therapy.

Keywords Adoptive cell therapy · TCR · TCR screening and optimization · Cancer antigen · MHC-I · Off-target toxicity · TCR-T resistance · Exhaustion

P. Koirala (✉)
Takeda Pharmaceuticals, Cambridge, MA, USA

D. Ghimire
Massachusetts General Hospital, Boston, MA, USA

A. K. Mishra
Department of Molecular, Cell and Cancer Biology, UMass Chan Medical, Worcester, MA, USA

Department of Hematopoietic Biology & Malignancy, Division of Cancer Medicine, The University of Texas MD Anderson Cancer Center, Houston, TX, USA

S. Bhaduri
Stylus Medicine, Cambridge, MA, USA

A. K. Mishra, S. K. Malonia (eds.), *Cell-based Immunotherapies for Cancer*,
https://doi.org/10.1007/978-3-031-88695-9_9

Abbreviations

AAV	Adeno-associated virus
ACT	Adoptive cell therapy
APC	Antigen-presenting cell
BCMA	B-cell maturation antigen
CAR	Chimeric antigen receptor
CD28	Cluster of differentiation 28
CD3	Cluster of differentiation 3
CDR	Complementary determining region
CRISPR	Clustered regularly interspaced short palindromic repeats
D	Diverse (TCR domain)
DNA	Deoxyribonucleic acid
ERAP	Endoplasmic reticulum aminopeptidase
ERp57	Protein disulfide isomerase A3
FDA	US Food and Drug Administration
gp100	Glycoprotein 100
HLA	Human leukocyte antigen
ICIs	Immune checkpoint inhibitors
IFN-γ	Interferon-gamma
IL15	Interleukin-15
IL2	Interleukin-2
IL21	Interleukin-21
KI	Knockin
KO	Knockout
MAGE	Melanoma-associated antigen
MART-1	Melanoma antigen recognized by T cells-1
MGC	Minigene cassette
MHC	Major histocompatibility complex
MHC-I	Major histocompatibility complex class I
PBMCs	Peripheral blood mononuclear cells
pHLA	Peptide-human leukocyte antigen
PLC	Peptide loading complex
RNA	Ribonucleic acid
scRNA	Single-cell RNA
SMH	Somatic hypermutation

1 Introduction

Immunotherapy is a revolutionary method for enhancing the ability of the human immune system to fight cancer or other diseases. It is remarkably successful in cancer treatment. Various immunotherapeutic options have been developed for cancer treatment and are broadly classified into two groups.

1.1 Immunomodulators

Immunomodulators are small or large molecules designed to increase the ability of the human immune system to eliminate or control tumors. Various immune modulators, including adjuvants, agonists, cytokines, and immune checkpoint inhibitors, are available for cancer treatment. The US Food and Drug Administration (FD)A has approved sixteen immune modulators comprising four cytokines, two adjuvants, nine immune checkpoint inhibitors (ICIs), and a small molecule [1, 2]. Among these options, ICIs are first-line options for treating non-small cell lung carcinoma and melanoma. However, only a subset of patients respond positively to ICIs. Consequently, ongoing efforts are focused on developing improved ICIs and adoptive cell therapy (ACT) to improve treatment outcomes [1–4].

1.2 Adoptive Cell Therapy (ACT)

Adoptive cell therapy utilizes live lymphocytes or T-cell drug products for cancer treatment. Drug products can include natural T cells, such as tumor-infiltrating lymphocytes (TILs), or engineered T-cell therapy, such as chimeric antigen receptor T (CAR) and T-cell receptor-engineered T (TCR-T) cells [5] (Table 1 and Fig. 1). In TIL therapy, T cells are isolated from resected tumors, expanded with the help of interleukin 2 (IL2), and finally transfused into patients [6]. While TILs were the first

Table 1 Comparative overview of T lymphocyte-based adoptive cell therapies

Aspect	TILs	CAR-T cells	TCR-T cells
Therapeutic approach	Isolation and expansion of naturally occurring TILs	Genetic modification of T cells to express chimeric antigen receptors (CARs)	Engineering T cells to express tumor antigen-specific TCRs
Target recognition	Broad recognition, often undefined antigens	Targets specific antigens defined by the CAR	Recognizes tumor antigens presented by MHC-I
Antigen specificity	Variable specificity depending on TILs from patient	Specificity determined by the design of the CAR	High specificity due to engineered TCRs
Versatility of antigen targets	Variable, limited to antigens present in the tumor microenvironment	Primarily targets membrane-associated antigens	Recognizes both membrane and intracellular protein antigens presented by MHC-I
Source	Isolated from the patient's tumor tissue	Autologous/Allogeneic T cells obtained after apheresis	Autologous/Allogeneic T cells obtained after apheresis
Clinical development stage	FDA-approved for melanoma treatment	FDA-approved and used in clinical settings for blood cancer treatment	Approved for treatment of MAZE-A4 positive metastatic synovial sarcoma

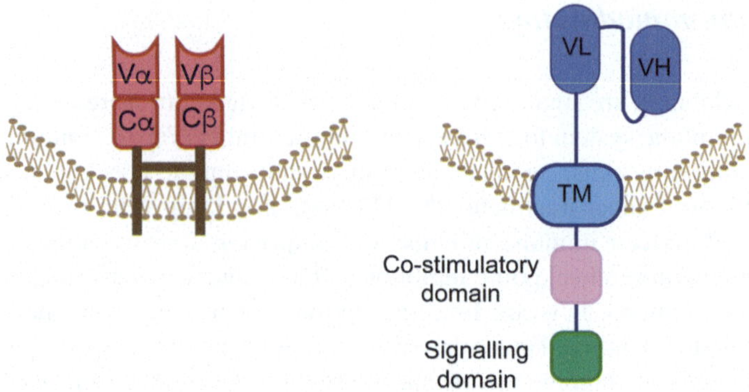

Fig. 1 Schematic representation of the TCR and CAR constructs. Left: synthesized TCR construct; right: third generation CAR construct with a hinge, costimulatory, and signaling domain

adoptive cell therapy developed to treat melanoma, challenges arise in sourcing T cells from resected tumors and expanding them to manufacture T-cell drug products at the required dose. Moreover, T cells isolated from tumors usually lack fitness and have limited expansion potential. Despite their recent success in melanoma, TIL therapy has shown limited efficacy in other tumor types [7]. Therefore, developing more precise and potent adoptive cell therapies with broader applications will help more patients. TCR-T-cell therapy is an enhanced alternative to TIL therapy, offering a robust platform for engineering T cells. By expressing tumor-specific TCRs capable of interacting with tumor antigens presented by MHC-I, TCR-T-cell therapy has demonstrated potent cytotoxicity against tumor cells [5] (*refer to* Fig. 2).

Moreover, the versatility of TCR-T cells extends their applicability beyond CAR-T-cell therapy. While CAR-T cells rely on chimeric antigen receptors (CARs) to target tumor-associated receptors and eliminate receptor-expressing cells, TCR-T cells exhibit a broader ability to recognize membrane antigens and extend their reach to intracellular protein antigens presented by major histocompatibility complex class I (MHC-I). CAR-T-cell therapy has shown remarkable success in blood cancer treatment. Currently, six FDA-approved CAR-T-cell therapies targeting CD19 or BCMA antigens are available [8]. In contrast to blood cancers, CAR-T-cell therapy faces several challenges when it is used to target solid tumors. Solid tumors have a dense and compact architectural structure that conceals the target antigens, making it difficult for CAR-T cells to access them [9]. Moreover, the irregular vasculature and hostile tumor microenvironment hinder T-cell adherence and function within these tumors [10]. TCR-T-cell therapy can achieve improved efficacy in solid tumors because of its increased antigen availability. The enhanced avidity enables TCR-T cells to be activated by only a few receptors per cell [11]. Consequently, it can address the challenges of limited antigen accessibility within solid tumors and antigen escape in blood cancers and works as a complementary therapy to CAR-T-cell therapy.

The TCR is a multimeric protein composed of an antigen-binding pocket formed by a TCRαβ heterodimer and a CD3 signal transduction domain. The TCR binds peptides present in the MHC-I groove, which initiates the phosphorylation of the CD3 subunit and downstream proteins, leading to the upregulation of transcriptional activity and ultimately resulting in T-cell activation, differentiation,

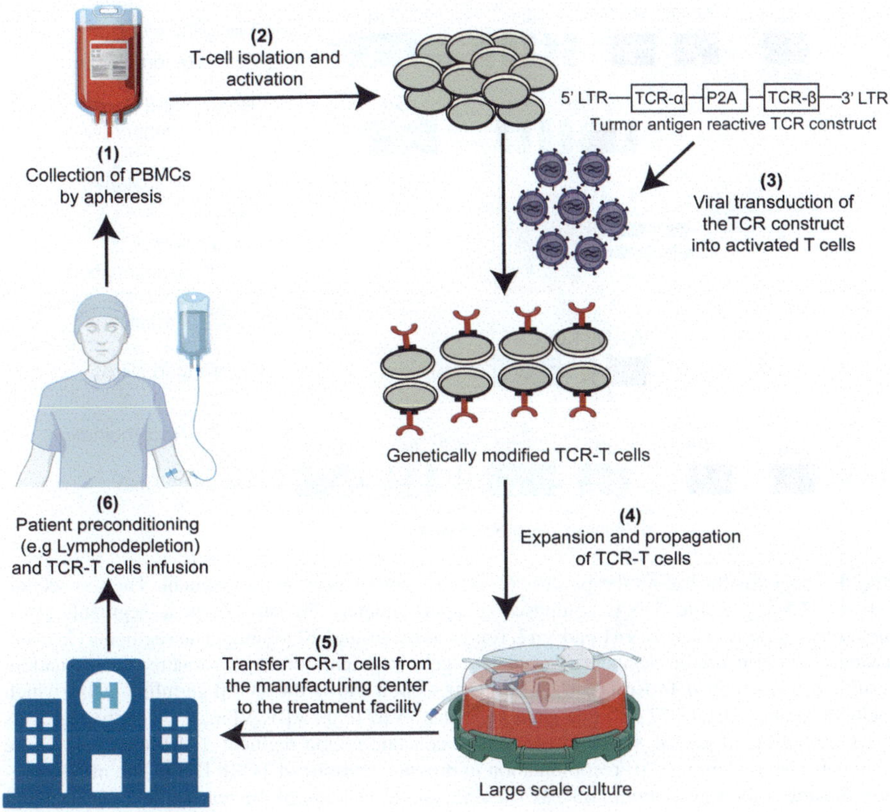

Fig. 2 Schematic representation of the T-cell therapy process. This figure outlines the step-by-step process of T-cell receptor (TCR)–engineered T-cell therapy for cancer treatment. Step 1: Peripheral blood mononuclear cells (PBMCs) are collected from the patient via apheresis. Step 2: T cells are isolated and activated from the collected PBMCs. Step 3: Activated T cells are genetically modified through viral transduction to express a tumor-reactive and optimized T-cell receptor (TCR). Step 4: These engineered T cells are expanded in large-scale culture. Step 5: The modified T cells are transported from the manufacturing facility to the treatment center. Step 6: After patient preconditioning (chemotherapy, radiation, or lymphodepletion), the TCR-T cells are infused to target tumor cells

maturation, and clonal expansion [10]. The TCRαβ protein shows similar homology and sequence diversity to those of immunoglobulins [11]. Like immunoglobulin, TCRαβ also possesses variable (V), diverse (D), and joining (J) domains, which can undergo somatic hypermutation, which is key to the generation of TCR repertoire diversity (Fig. 3). Such diversity is reduced in viral infection, aging, and tumors [12, 13]. In melanoma and pancreatic ductal carcinoma, clonally expanded tumor-infiltrating T cells with less diverse TCRs than peripheral blood mononuclear cells (PBMCs) have been observed [14]. However, not all TCRs identified in TILs are equally reactive to tumor antigens [15]. Identifying tumor antigens and reactive TCRs is critical for successfully engineering TCR-T cells. A comparison of the three T-cell-based therapies is summarized in Table 1.

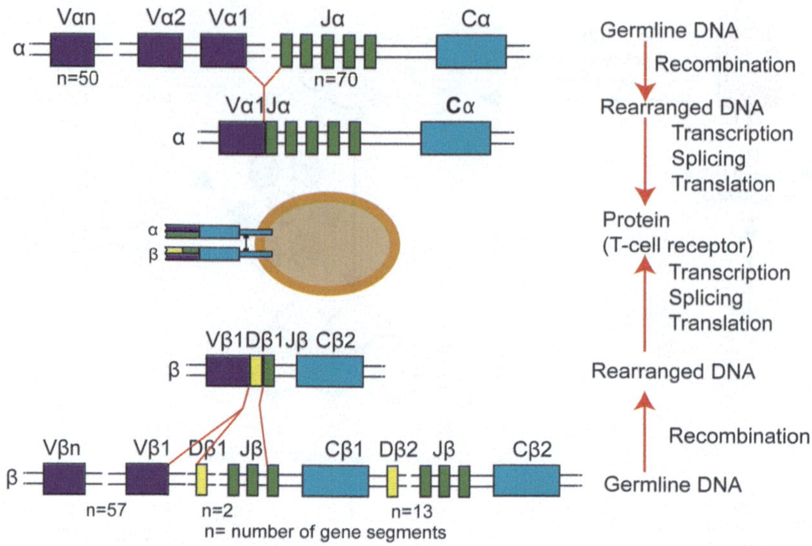

Fig. 3 Diagram illustrating the process of T-cell receptor gene rearrangement. The top section depicts TCRα germline DNA, composed of approximately 50 variable gene segments (Vα), including examples such as Vα1 and Vα2, followed by around 70 joining gene segments (Jα), and a single constant region segment (Cα). These gene segments undergo somatic recombination, resulting in rearranged TCRα DNA. The bottom section shows the TCRβ germline DNA, which includes approximately 57 variable gene segments (Vβ) (e.g., Vβ1, Vβn), 2 diversity segments (Dβ1 and Dβ2), 13 joining segments (Jβ), and 2 constant region segments (Cβ1 and Cβ2). These segments also undergo V(D)J recombination to produce rearranged TCRβ DNA. The middle section illustrates the transcription, mRNA splicing, and translation of the rearranged DNA, leading to the production of the T-cell receptor (TCR-αβ protein). Arrow indicates the flow of information from germline DNA, through recombination and RNA processing, to the formation of functional TCR proteins

2 Tumor Antigens, Types, and Their Uses in TCR-T Development

Genetic instability is a pervasive trait in cancer cells, resulting in gene mutation, deletion, and amplification that leads to aberrant gene, protein, and antigen expression or the creation of novel antigens [13], which elicit an immune response. An ideal antigen for TCR-T-cell therapy is expressed uniformly in cancer cells, has a foreign structure, and is effectively presented by MHC-I [15]. Antigens can be categorized on the basis of their origin and distribution within tissue. Currently, two primary categories of tumor antigens are under consideration for TCR-T-cell therapy (Fig. 4).

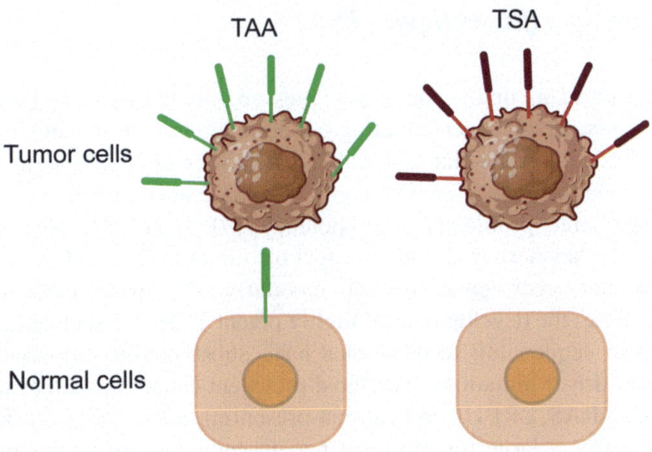

Fig. 4 Types of tumor antigens and their distribution in different tissues. Tumor-associated antigens (TAAs) are overexpressed in tumor cells. Tumor-specific antigens are exclusively expressed in tumor cells and not expressed in normal cells

2.1 Tumor-Associated Antigens (TAAs)

Tumor-associated antigens (TAAs) are overexpressed in tumors and exhibit limited expression in normal cells [16]. TAAs are shared across patients, presenting them as attractive targets for TCR-T-cell therapy and holding potential for application across multiple indications and patients. However, TAA targeting also poses a greater risk of toxicity because it can cross-react with normal cells expressing such antigens [17]. One of the reasons for such toxicity is the class of TAAs called tissue differentiation antigens (TDAs), which are expressed not only in tumors but also in tissues of tumor origin. For example, melanoma antigens recognized by T cells (MART-1) and glycoprotein 100 (gp100) are expressed at low levels in melanoma and melanocytes [18]. Clinical trials targeting these two antigens with TCR-T cells have shown low efficacy and high cutaneous, ocular, or auditory toxicity [5, 19]. Another example where the expression of TAA is not restricted to the tissue of tumor origin is a cancer-germline antigen (CGA), such as New York esophageal squamous cell carcinoma 1 (NY-ESO-1) [17]. NY-ESO-1 is expressed in ovarian cancer, cervical cancer, melanoma, and other tissues, such as spermatocytes and placenta. However, TCR-T-cell therapy targeting NY-ESO-1 in melanoma has a nearly 50% overall response rate (ORR) with no major toxicity [16]. Another CGA in melanoma targeted by TCR-T cells is MAG-3. Initial clinical data show that this could be an exciting target. However, MAG3-TCR-T cells exhibit cross-reactivity with MAGE-12 (which is expressed in brain tissues), leading to severe neurotoxicity and fatality [20].

2.2 Tumor-Specific Antigen (TSA)

Unlike TAAs, TSAs are unique antigens expressed only in tumors and not in normal tissues. TAAs arise from cancer-initiating gene mutations, cancer-inducing viral antigens, and additional genetic changes such as alternation of open reading frames and posttranscriptional or translational changes. TSAs are ideal targets for TCR-T cells because of their safety profile and T-cell potency [17]. Since TSAs are not expressed in normal tissue, they have minimal on-target off-tumor toxicity. Moreover, negative selection does not occur against neoantigens during the thymic development of T cells, which allows the development of highly potent TCRs for such antigens [18].

Advances in sequencing technologies have substantially supported works to identify cancer driver mutations. The most prevalent cancer mutations are those in TP53, PIK3C, KRAS, LRP1B, and antigen-presenting cell (APC) [13]. These driver mutations generate a large repertoire of neoantigens that are interestingly shared among patients and presented to T cells by the same HLA class. TCR-T-cell therapy targeting TP53, PIK3C, and KRAS mutations is currently being tested in clinical trials, and the initial results show low efficacy with minimal safety concerns [21]. Approximately 15% of cancer cases are associated with viruses. Epstein–Barr virus, human papillomavirus (HPV), hepatitis B virus, human herpes virus-8, human lymphotropic virus 1, and hepatitis C virus can cause cancer [22–24]. Among all cancer-causing viruses, the pathogenesis of HPV has been well studied. Two HPV proteins, E6 and E7, are known to enhance tumorigenesis by deregulating tumor suppressors. TCR-T cells targeting the E6 and E7 antigens have been evaluated in clinical trials and have shown partial efficacy of approximately 50% in the case of E7+ epithelial cancers with minimal toxicity [25]. Several other clinical trials have evaluated TCR-T cells that target other viral antigens; such TCR-T cells are applicable in only specific sets of cancers, and most of these trials exhibit minimal toxicity, although their efficacy is still limited [26, 27].

2.3 Antigen Processing and Presentation in Tumor Cells

Even if a cancer cell generates neoantigens, a significant challenge encountered by any immunotherapy is presenting such antigens to immune cells; MHC-I plays a pivotal role in TCR-T function, serving as a platform to present the cancer antigen to the TCR. MHC-I is a membrane glycoprotein composed of heavy chains made of $\alpha1$,$\alpha2$, and $\alpha3$ domains and a soluble protein $\beta2$ microglobulin ($\beta2$ M). The $\alpha1$ and $\alpha2$ domains fold to form a groove that works as an antigen-binding pocket and eventually interacts with the TCR [28]. In addition to the antigen presentation domains involved in the formation of the MHC-I heterodimer, the MHC-I pathway consists of several other intracellular proteins that are involved in peptide processing, MHC-I loading, folding, and transportation to the membrane from the endoplasmic reticulum [29]. Misfolded or defective proteins in tumor cells undergo proteasomal degradation, leading to the formation of peptide antigens of 9–16 amino acid residues. The transporter associated with the antigen processing (TAP) complex transports newly synthesized peptides into the endoplasmic reticulum, where the peptides are processed by endoplasmic reticulum aminopeptidase (ERAP1) and ERAP2. After processing peptide antigens, they are loaded into the peptide loading complex (PLC) formed by MHC class I and B2M. Three additional proteins, ERp57, tapasin, and calreticulin, are also associated with PLC. ERp57 is required for further antigen peptide trimming, and the other two proteins help in MHC-I folding. The Golgi apparatus and transporter vehicle transport the stable MHC-peptide complex to the membrane, where it can be recognized by T cells [30] (Fig. 5).

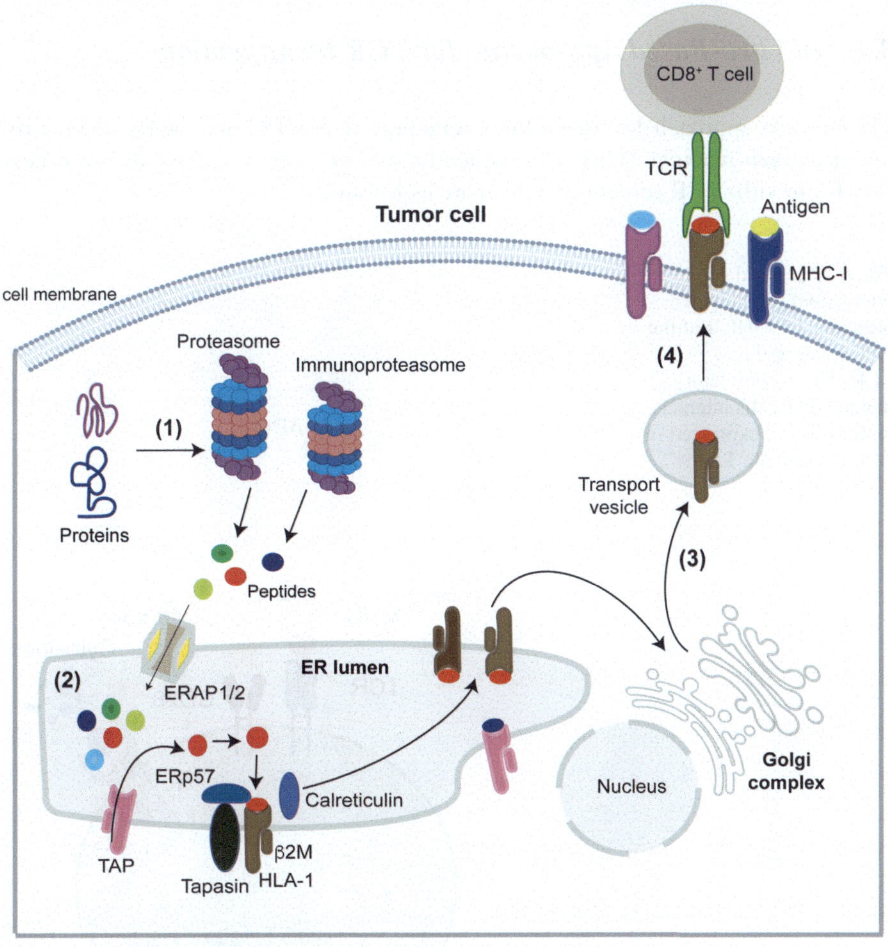

Fig. 5 Antigen processing and expression into MHC-I: Misfolded proteins are degraded by ubiquitination, further processed into the endoplasmic reticulum, and uploaded into HLA-1. The antigen HLA and B2M complex is transported to the membrane with the help of the Golgi vesicles. The antigen MHC-I complex in the membrane can interact with the TCR

3 TCR Discovery for TCR-T-Cell Development and Optimization

After T cells recognize the peptide-MHC-I complex, they are activated, proliferate, secrete cytokines, and kill the target [25] (Fig. 6). A reporter assay based on T-cell phenotype and function accompanied by single-cell RNA sequencing enables unbiased identification of cancer antigen-responsive TCRs for TCR-T-cell development [31]. Among the rapidly evolving landscape of TCR identification methods, two major methods for screening novel TCRs are discussed here.

3.1 *In Vitro-Based Approaches for TCR Identification*

The in vitro approach leverages the interaction of p-MHC and TCRs to identify tumor antigen-reactive TCRs. Consequently, two essential components are necessary for in vitro TCR screening, which are as follows:

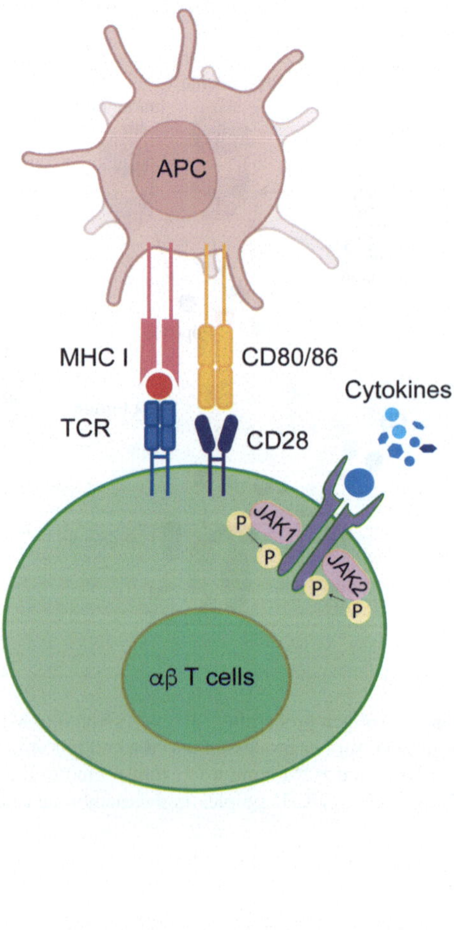

Fig. 6 T-cell and APC interactions. The antigen presented by MHC-I of the APC binds to the TCR. The costimulatory domain of T cells interacts with its ligand expressed in APCs to activate T cells

A. *Purified T Cells*: HLA-specific T cells can be isolated from resected tumors, patients, or healthy donors (of a specific HLA type) [32]. T cells isolated from resected tumors may express the memory and chemokine receptors necessary for tumor infiltration [33]. While infiltrated T cells are overactivated or exhausted and lack fitness and persistence, T cells obtained from PBMCs are easy to source and exhibit a better phenotype and function [34].
B. *Tumor Antigen-Presenting Cells*: systems that facilitate the presentation of tumor antigens are critical for identifying TCRs. Such systems can encompass approaches such as phage displays, antigen-presenting cells, or cell lines engineered to express antigen libraries. Among these methods, phage display has substantial potential for TCR identification but lacks the innate capacity to process antigens [35]. In contrast, engineered antigen-presenting cells retain their capacity to process antigens naturally, which makes them the preferred choice over phage display for TCR identification. One such system was created in the early 1990s to identify TCRs specific to the melanoma antigen MART-1. A cDNA library from the HLA-A2+ 501mel melanoma cell line was transfected into MART1-negative and HLA-A2+ MDA-MB231 breast cancer cells, which were subsequently used as antigen-presenting cells [36]. Once the T-cell and antigen-presenting systems are established, the cells are cocultured, and antigen-responsive T cells can be identified by assessing their activation phenotype or intracellular cytokines. Typically, activated cells are sorted and expanded, and TCRα or/and TCRβ sequencing determines the TCR sequences of activated cells. Similarly, V(D)J sequencing has also been utilized for functional TCR sequence identification [37].

Recently, the combination of single-cell RNA (scRNA) sequencing with deep sequencing has been utilized to sequence the TCRs of activated T cells. This approach is more efficient because it eliminates the need for cell sorting or single-cell culture. Preparing a full-length cDNA library for ScRNA could enable TCR and TCR sequencing in pairs [38], a capability that bulk RNA sequencing cannot achieve. While scRNA sequencing is still evolving, its advantages and unbiased analysis make it the favored TCR screening method in the future. Lu et al. utilized single-cell RNA (scRNA) sequencing to identify TCRs that are reactive to tumor antigens. This approach engineers the autologous antigen-presenting cells (APCs) to express tandem minigene cassettes (MGC) of antigens identified through nonsynonymous mutation analysis via exome sequencing. The antigen-reactive T cells were identified by their ability to produce IFN-γ when autologous T cells were cocultured with APC-expressing MCG. Subsequently, antigen-reactive TCRs are synthesized, cloned, and inserted into retroviruses, which are subsequently transduced into T cells to generate novel TCR-T cells (Fig. 7) [39].

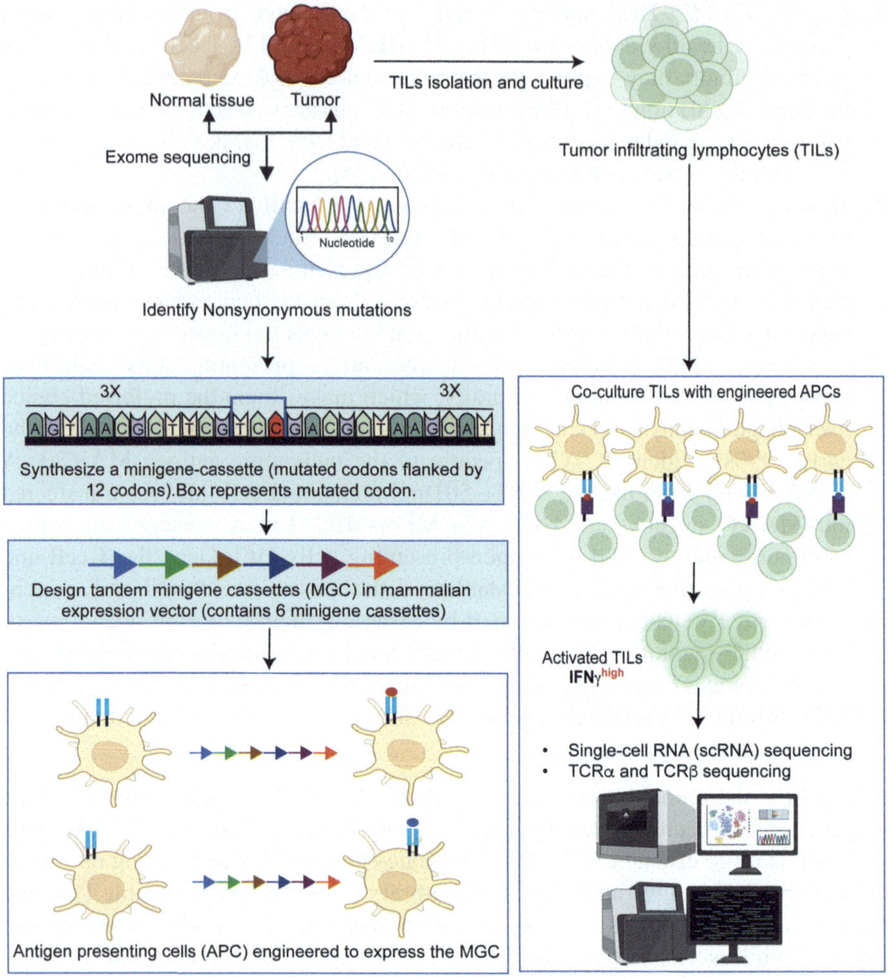

Fig. 7 Schematic representation of the identification of tumor antigen-specific TCRs of T cells isolated from resected tumors

3.2 In Vivo Approaches for TCR Identification

While the in vitro system is flexible, convenient, and effective in identifying TCRs, in vivo systems present more physiologically relevant models regarding cell–cell interactions, communication, and dimensionality. Humanized "VelociT mice" that express human TCRαβ, CD4, CD8 extracellular domain, MHC-I (HLA-A2), and MHC-II (HLA-DR2) with normal innate and adaptive immune functions serve as models for TCR screening [40]. The successful identification of NY-ESO-1-specific TCRs was achieved via this model. VelociT mice were immunized with the HLA-A2-restricted 9mer NY-ESO1$_{157–165}$. After 14 days, the spleen and lymph nodes were harvested, pHLA-tetramer+CD8+ cells were sorted, and TCRA and TCRB sequences were analyzed via bulk RNA sequencing. Once the sequence was identified, TCRs were synthesized and knocked into the TRAC locus of primary human T cells via CRISPR technology. These cells demonstrated robust cytotoxicity when cocultured with A375 (NY-ESO-1+) cells and in vivo models, indicating that VelociT or similar mouse models could be used for tumor antigen-specific TCR identification [40].

3.3 TCR Optimization

The overall success of TCR-T-cell therapy is intricately influenced by the level of the exogenous TCR, antigen specificity, and sensitivity of target binding. Therefore, optimization is indispensable for ensuring therapeutic efficacy, persistence, and safety (Table 2).

3.4 TCR Expression Optimization

The proper assembly of TCR domains facilitates correct folding and interaction, enhancing TCR expression and functions. Two prominent methods for TCR optimization include codon optimization or amino acid substitution in the Vα or Vβ sequence [41]. TCR optimization enhances the stability and assembly of the variable alpha–constant alpha (Vα–Cα) interface. This results in improved assembly, elevated TCR expression, effector cytokine release, and in vivo cytotoxicity [42]. The erroneous pairing between the TCRα and TCRβ chains of transgenes and endogenous TCRs generates nonspecific and cross-reactive TCRs. To mitigate this risk, swapping human Cα and Cβ with the corresponding mouse Cα and Cβ transgenes reduces the chances of mispairing, making such TCRs safe for clinical use [43]. Replacing endogenous TCRs via the use of gene editing CRISPR technology to knock out (KO) TRAC and TRBC loci and subsequently knocking in (KI) transgene TCRs via homologous recombination via lentivirus or adeno-associated virus (AAV)

Table 2 TCR optimization strategy

Potential pitfall	Optimization strategy
Minimizing wild type and synthesized TCRαβ mispairing	Substitute human TCR with murine TCR Knockout wild-type TCR and knock in the synthetic TCR
Enhancing expression of transduced TCR	Codon optimization
Enhancing stability of TCR	Single Amino acid substitution in TCR domains, improving conformation
Improving TCR sensitivity	Identify high-affinity TCR

at those loci minimizes mispairing and generates cells with single-copy TCRs [44]. Given that both the CRISPR and viral delivery techniques are safe for clinical use, TCR-T engineering using KOs and KIs is currently being tested in clinical trials.

3.5 TCR Affinity Optimization

A single TCR can interact with either a single or multiple pMHCs [45]. The strength of such interactions, measured by the dissociation constant (KD), defines the TCR affinity and gauges the TCR sensitivity to a peptide antigen [46]. The strength of multiple TCRs interacting with pMHCs is termed TCR avidity and is a critical factor for T-cell fitness and functions [47]. While a robust affinity is desirable for increased TCR sensitivity, it can reduce TCR specificity and promote cross-reactivity. TCRs that exhibit a strong affinity for TAAs are typically eliminated during thymic selection [48]. However, yeast or mammalian display systems can identify high-affinity TCRs (in the pico-molar range). Similarly, alanine scanning can identify critical amino acids for substitution in different domains of the TCR to increase affinity [49]. Additionally, somatic hypermutation (SMH), a mechanism akin to optimizing B-cell antibody optimization, can induce point and tandem mutations in the TCRα complementary region (CDR). This sequence alteration in the CDR enhances TCR affinity, improving TCR activity in vitro and in vivo [50].

4 Challenges Associated with TCR-T-Cell Therapy

Resistance to TCR and mitigation: MHC-I loss is a primary immune escape mechanism propelled by deregulation through mutations, transcriptional silencing due to promoter methylation, and loss of heterozygosity of proteins associated with MHC-I [51]. TCR function is indispensable for its interaction with pMHCs. Hence, the loss or downregulation of MHC-I presents a formidable challenge to TCR-T-cell therapy. The therapeutic outcome of TCR-T cells depends on T-cell functional activity and fitness. The autologous T cells derived from patients to engineer TCR-T cells are more differentiated than those from healthy donors [52]. Moreover, during manufacturing, T cells are stimulated with anti-CD3/CD28, promoting overactivation, which compromises their fitness. These cells may lack persistence, exhibit exhaustion phenotypes, and become unresponsive to antigens, leading to minimal therapeutic efficacy [53].

The selection and stratification of the right patient population are crucial in addressing challenges such as MHC-I mismatch, heterogeneous antigen expression, and antigen loss during TCR-T-cell therapy [54]. Multiplex TCR-T-cell designs targeting different HLA haplotypes and multiple targets can overcome MHC-I deregulation [54, 55].

The early memory phenotype is the key to TCR-T-cell persistence and durable effector function. Activation-free transduction and preconditioning of culture media with components such as IL21, IL15, and GSK-3 inhibitors significantly improve the early memory population of drug products [48]. Additionally, enhancing T-cell metabolic fitness by expressing metabolism-enhancing genes or knocking down checkpoint inhibitors such as PD1 could also increase the persistence of TCR-T drug products [56]. TCR-T-associated toxicity: TCR-T-associated toxicity can result from on-target off-tumor effects or cross-reactivity [54]. On-target off-tumor toxicity has been observed in TCR-T-cell therapy, particularly when targeting tumor-associated antigens (TAAs). TAAs are overexpressed in tumors but also expressed in normal cells, leading TCR-T cells to target both tumor and normal cells indiscriminately. For example, clinical toxicity was observed in a clinical trial of TCR-T cells targeting the melanoma antigens gp100 and MART-1, which are also expressed in melanocytes. The cross-reactivity of TCR-T cells also presents significant safety concerns. Specifically, MAGE-A3 TCR-T cells, initially intended to target melanoma cancer cells, cross-react with MAGE-A12, which is typically expressed in heart and brain tissues, leading to fatal cardiotoxicity and neurotoxicity [57]. High-affinity TCRs may have reduced specificity for the target antigen, increasing the likelihood of cross-reactivity [35]. The models available to study the in vivo toxicity of immune therapy are limited, and this limitation is particularly notable in the context of adoptive cell therapy. As a result, in silico and in vitro methods, along with clinical chemistry, are more frequently utilized to identify and mitigate the potential risks associated with TCR-T-cell therapy. While in vitro cancer cell cytotoxicity can be used to evaluate the efficacy of TCR-T cells, a normal cell-sparing assay is used to evaluate the off-target effects and cross-reactivity of TCR-T cells. In a normal cell-sparing assay, panels of primary cells are cocultured with TCR-T cells to assess TCR-T cytotoxicity [58]. Ideally, TCR-T cells with minimal cytotoxicity toward normal cells would be preferred for clinical use. Additionally, various computational in silico methods are available to identify the affinity of TCR-T cells for antigens found in cancer cells or normal cells. These algorithms can also scan amino acid sequences present in target antigens and modify the TCR to increase its specificity for cancer cells [52].

5 Future Perspectives on TCR-T-Cell Therapy Development and Conclusions

TCR-T cells constitute a potentially groundbreaking adoptive cell therapy approach currently under evaluation in numerous clinical trials to treat solid tumors. Initial clinical efficacy results show promise, with a favorable safety profile when tumor-specific antigens are targeted [17, 54, 59]. TCR-T cells demonstrate outstanding potential for recognizing intracellular antigens and a high affinity for binding tumor antigens, which makes them ideal for advanced solid tumor treatment. Further development and optimization efforts are essential to advance TCR-T-cell therapy.

Identifying cancer antigens/neoantigens, establishing a robust platform for TCR screening, and fine-tuning selected TCRs are essential steps to increase the coverage and effectiveness of solid tumor treatment [17, 55, 60].

Additionally, the adoption of genome editing and engineering technologies to replace endogenous TCRs with synthesized TCRs to eliminate TCR mismatch and site-directed insertion of synthesized TCRs will likely increase both efficacy and safety. Indeed, strategies to improve TCR-T-cell persistence and reduce exhaustion are critical for enhancing TCR-T-cell treatment outcomes. TCR-T-cell persistence can be augmented through the engineering of cytokines, such as IL21 or IL15, which enrich the T_{SCM} population or a gene that improves metabolic fitness. When the checkpoint inhibitors PD1, TIM3, and LAG3 are knocked out, TIGIT results in favorable preclinical outcomes and holds promise for translation in the clinical setting [17, 54, 59, 61]. Combination therapy is a cornerstone of cancer treatment, and the synergistic potential of combining a cancer vaccine with TCR-T-cell therapy could revolutionize cancer treatment paradigms. The incorporation of a cancer vaccine alongside TCR-T-cell therapy can augment the TCR-T-cell phenotype, enhance infiltration capabilities, and increase functionality in vivo. This combined approach holds promise in combating challenges such as tumor heterogeneity, optimizing antigen presentation, and overcoming the suppressive tumor microenvironment. By leveraging the complementary mechanisms of action offered by cancer vaccines and TCR-T-cell therapy, this combination strategy may improve therapeutic outcomes and pave the way for more effective cancer treatments [62].

In conclusion, TCR-T-cell therapy represents a promising therapeutic model for solid tumors that can potentially combat adverse tumor microenvironments and address the challenges associated with low-antigen-expressing solid tumors. However, the success of TCR-T cells relies on the immune escape mechanism, affinity, and avidity optimization. Moreover, the adoption of methods to increase T-cell persistence and antigen presentation is crucial for achieving positive outcomes with TCR-T-cell therapy.

Declaration of Interests The authors declare that they have no competing interests.

References

1. Wang G. Cancer immunotherapy and immunomodulation. Curr Med Chem. 2019;26(17):2988.
2. Khalil DN, Smith EL, Brentjens RJ, Wolchok JD. The future of cancer treatment: immunomodulation, CARs and combination immunotherapy. Nat Rev Clin Oncol. 2016;13:394.
3. Vaddepally RK, Kharel P, Pandey R, Garje R, Chandra AB. Review of indications of FDA-approved immune checkpoint inhibitors per NCCN guidelines with the level of evidence. Cancers. 2020;12:738.
4. Wolchok J. Putting the immunologic brakes on cancer. Cell. 2018;175:1452.
5. Rohaan MW, Gomez-Eerland R, van den Berg JH, Geukes Foppen MH, van Zon M, Raud B, et al. MART-1 TCR gene-modified peripheral blood T cells for the treatment of metastatic melanoma: a phase I/IIa clinical trial. Immuno-Oncol Technol. 2022;15:100089.

6. Rosenberg SA, Spiess P, Lafreniere R. A new approach to the adoptive immunotherapy of cancer with tumor-infiltrating lymphocytes. Science. 1986;233(4770):1318.
7. Manfredi F, Cianciotti BC, Potenza A, Tassi E, Noviello M, Biondi A, et al. TCR redirected T cells for cancer treatment: achievements, hurdles, and goals. Front Immunol. 2020;11:1689.
8. Cappell KM, Kochenderfer JN. Long-term outcomes following CAR T-cell therapy: what we know thus far. Nat Rev Clin Oncol. 2023;20:359.
9. Simsek H, Klotzsch E. The solid tumor microenvironment—breaking the barrier for T cells: how the solid tumor microenvironment influences T cells. BioEssays. 2022;44:e2100285.
10. Kankeu Fonkoua LA, Sirpilla O, Sakemura R, Siegler EL, Kenderian SS. CAR T-cell therapy and the tumor microenvironment: current challenges and opportunities. Mol Ther Oncolytics. 2022;25:69–77.
11. Cole DK, Pumphrey NJ, Boulter JM, Sami M, Bell JI, Gostick E, et al. Human TCR-binding affinity is governed by MHC class restriction. J Immunol. 2007;178(9):5727.
12. Lee S, Margolin K. Tumor-infiltrating lymphocytes in melanoma. Curr Oncol Rep. 2012;14(5):468.
13. Mendiratta G, Jones MK, Stites EC. How often is each gene mutated within the cancer patient population? Mol Cell Oncol. 2022;9(1):2065176.
14. Lee S, Zhao L, Little LD, Westin SN, Jazarei AA, Fleming ND, et al. Distinct T-cell receptor repertoire diversity of clinically defined high-grade serous ovarian cancer treatment subgroups. iScience. 2021;24(2):102053.
15. Waarts MR, Stonestrom AJ, Park YC, Levine RL. Targeting mutations in cancer. J Clin Invest. 2022;132(8):e154943.
16. Robbins PF, Kassim SH, Tran TLN, Crystal JS, Morgan RA, Feldman SA, et al. A pilot trial using lymphocytes genetically engineered with an NY-ESO-1-reactive T-cell receptor: long-term follow-up and correlates with response. Clin Cancer Res. 2015;21(5):1019.
17. Shafer P, Kelly LM, Hoyos V. Cancer therapy with TCR-engineered T cells: current strategies, challenges, and prospects. Front Immunol. 2022;13:835762.
18. Moran AE, Hogquist KA. T-cell receptor affinity in thymic development. Immunology. 2012;135:261.
19. Johnson LA, Morgan RA, Dudley ME, Cassard L, Yang JC, Hughes MS, et al. Gene therapy with human and mouse T-cell receptors mediates cancer regression and targets normal tissues expressing cognate antigen. Blood. 2009;114(3):535.
20. Cameron BJ, Gerry AB, Dukes J, Harper JV, Kannan V, Bianchi FC, et al. Identification of a titin-derived HLA-A1-presented peptide as a cross-reactive target for engineered MAGE A3-directed T cells. Sci Transl Med. 2013;5(197):197ra103.
21. Wang QJ, Yu Z, Griffith K, Hanada KI, Restifo NP, Yang JC. Identification of T-cell receptors targeting KRAS-mutated human tumors. Cancer Immunol Res. 2016;4(3):204.
22. de Martel C, Plummer M, Vignat J, Franceschi S. Worldwide burden of cancer attributable to HPV by site, country and HPV type. Int J Cancer. 2017;141(4):664.
23. Perz JF, Armstrong GL, Farrington LA, Hutin YJF, Bell BP. The contributions of hepatitis B virus and hepatitis C virus infections to cirrhosis and primary liver cancer worldwide. J Hepatol. 2006;45(4):529.
24. Farrell PJ. Epstein–Barr virus and cancer. Annu Rev Pathol. 2019;14:29–53.
25. Hennecke J, Wiley DC. T-cell receptor-MHC interactions up close. Cell. 2001;104:1.
26. Miller NJ, Church CD, Dong L, Crispin D, Fitzgibbon MP, Lachance K, et al. Tumor-infiltrating Merkel cell polyomavirus-specific T cells are diverse and associated with improved patient survival. Cancer Immunol Res. 2017;5(2):137.
27. Meng F, Zhao J, Tan AT, Hu W, Wang SY, Jin J, et al. Immunotherapy of HBV-related advanced hepatocellular carcinoma with short-term HBV-specific TCR expressed T cells: results of dose escalation, phase I trial. Hepatol Int. 2021;15(6):1402.
28. Ferm MT, Grönberg A. Human MHC class I antigens are associated with a 90-kDa cell surface protein. Scand J Immunol. 1991;34(2):243.
29. Hewitt EW. The MHC class I antigen presentation pathway: strategies for viral immune evasion. Immunology. 2003;110:163.

30. Yang K, Halima A, Chan TA. Antigen presentation in cancer — mechanisms and clinical implications for immunotherapy. Nat Rev Clin Oncol. 2023;20:604.
31. Zong S, Mi T, Flores LG, Alpert A, Olivares S, Patel K, et al. Very rapid cloning, expression and identifying specificity of T-cell receptors for T-cell engineering. PLoS One. 2020;15(2):e0228112.
32. Li Q, Ding ZY. The ways of isolating neoantigen-specific T cells. Front Oncol. 2020;10:1347.
33. Satooka H, Ishigaki H, Todo K, Terada K, Agata Y, Itoh Y, et al. Characterization of tumor-infiltrating lymphocytes in a tumor rejection cynomolgus macaque model. Sci Rep. 2020;10(1):8414.
34. Zhang DKY, Adu-Berchie K, Iyer S, Liu Y, Cieri N, Brockman JM, et al. Enhancing CAR-T-cell functionality in a patient-specific manner. Nat Commun. 2023;14(1):506.
35. Campillo-Davo D, Flumens D, Lion E. The quest for the best: how TCR affinity, avidity, and functional avidity affect TCR-engineered T-cell antitumor responses. Cells. 2020;9:1720.
36. Kawakami Y, Eliyahu S, Delgado CH, Robbins PF, Rivoltini L, Topalian SL, et al. Cloning of the gene coding for a shared human melanoma antigen recognized by autologous T cells infiltrating into tumor. Proc Natl Acad Sci USA. 1994;91(9):3515.
37. Li N, Yuan J, Tian W, Meng L, Liu Y. T-cell receptor repertoire analysis for the diagnosis and treatment of solid tumor: a methodology and clinical applications. Cancer Commun. 2020;40:473.
38. Lu YC, Zheng Z, Lowery FJ, Gartner JJ, Prickett TD, Robbins PF, et al. Direct identification of neoantigen-specific TCRs from tumor specimens by high-throughput single-cell sequencing. J Immunother Cancer. 2021;9(7):e002595.
39. Lu YC, Zheng Z, Robbins PF, Tran E, Prickett TD, Gartner JJ, et al. An efficient single-cell RNA-Seq approach to identify neoantigen-specific T-cell receptors. Mol Ther. 2018;26(2):379.
40. Moore MJ, Zhong M, Hansen J, Gartner H, Grant C, Huang M, et al. Humanization of T-cell–mediated immunity in mice. Sci Immunol. 2021;6(66):eabj4026.
41. Jorritsma A, Gomez-Eerland R, Dokter M, Van De Kasteele W, Zoet YM, Doxiadis IIN, et al. Selecting highly affine and well-expressed TCRs for gene therapy of melanoma. Blood. 2007;110(10):3564.
42. Degirmencay A, Thomas S, Mohammed F, Willcox BE, Stauss HJ. Modifications outside CDR1, 2 and 3 of the TCR variable β domain increase TCR expression and antigen-specific function. Front Immunol. 2023;14:14.
43. Cohen CJ, Zhao Y, Zheng Z, Rosenberg SA, Morgan RA. Enhanced antitumor activity of murine-human hybrid T-cell receptor (TCR) in human lymphocytes is associated with improved pairing and TCR/CD3 stability. Cancer Res. 2006;66(17):8878.
44. Legut M, Dolton G, Mian AA, Ottmann OG, Sewell AK. CRISPR-mediated TCR replacement generates superior anticancer transgenic T cells. Blood. 2018;131(3):311.
45. Wooldridge L, Ekeruche-Makinde J, Van Den Berg HA, Skowera A, Miles JJ, Tan MP, et al. A single autoimmune T-cell receptor recognizes more than a million different peptides. J Biol Chem. 2012;287(2):1168.
46. Tan MP, Gerry AB, Brewer JE, Melchiori L, Bridgeman JS, Bennett AD, et al. T-cell receptor binding affinity governs the functional profile of cancer-specific CD8+ T cells. Clin Exp Immunol. 2015;180(2):255.
47. Margulies DH. TCR avidity: it's not how strong you make it, it's how you make it strong. Nat Immunol. 2001;2(8):669.
48. Sabatino M, Hu J, Sommariva M, Gautam S, Fellowes V, Hocker JD, et al. Generation of clinical-grade CD19-specific CAR-modified CD81 memory stem cells for the treatment of human B-cell malignancies. Blood. 2016;128(4):519.
49. Malecek K, Zhong S, McGary K, Yu C, Huang K, Johnson LA, et al. Engineering improved T-cell receptors using an alanine-scan guided T-cell display selection system. J Immunol Methods. 2013;392(1–2):1–11.
50. Bassan D, Gozlan YM, Sharbi-Yunger A, Tzehoval E, Eisenbach L. Optimizing T-cell receptor avidity with somatic hypermutation. Int J Cancer. 2019;145(10):2816.

51. Taylor BC, Balko JM. Mechanisms of MHC-I downregulation and role in immunotherapy response. Front Immunol. 2022;13:844866.
52. Jaravine V, Mösch A, Raffegerst S, Schendel DJ, Frishman D. Expitope 2.0: a tool to assess immunotherapeutic antigens for their potential cross-reactivity against naturally expressed proteins in human tissues. BMC Cancer. 2017;17(1):892.
53. Ren H, Cao K, Wang M. A correlation between differentiation phenotypes of infused T cells and anti-cancer immunotherapy. Front Immunol. 2021;12:745109.
54. Liu Y, Yan X, Zhang F, Zhang X, Tang F, Han Z, et al. TCR-T immunotherapy: the challenges and solutions. Front Oncol. 2022;11:794183.
55. Zah E, Lin MY, Anne SB, Jensen MC, Chen YY. T cells expressing CD19/CD20 bispecific chimeric antigen receptors prevent antigen escape by malignant B cells. Cancer Immunol Res. 2016;4(6):498.
56. Wang Z, Li N, Feng K, Chen M, Zhang Y, Liu Y, et al. Phase I study of CAR-T cells with PD-1 and TCR disruption in mesothelin-positive solid tumors. Cell Mol Immunol. 2021;18(9):2188.
57. Morgan RA, Chinnasamy N, Abate-Daga D, Gros A, Robbins PF, Zheng Z, et al. Cancer regression and neurological toxicity following anti-MAGE-A3 TCR gene therapy. J Immunother. 2013;36(2):133.
58. Donnadieu E, Luu M, Alb M, Anliker B, Arcangeli S, Bonini C, et al. Time to evolve: predicting engineered T-cell-associated toxicity with next-generation models. J Immunother Cancer. 2022;10:e003486.
59. Rath JA, Arber C. Engineering strategies to enhance TCR-based adoptive T-cell therapy. Cells. 2020;9:1485.
60. Baulu E, Gardet C, Chuvin N, Depil S. TCR-engineered T-cell therapy in solid tumors: state of the art and perspectives. Sci Adv. 2023;9:eadf3700.
61. Ishihara M, Nishida Y, Kitano S, Kawai A, Muraoka D, Momose F, et al. A phase 1 trial of NY-ESO-1-specific TCR-engineered T-cell therapy combined with a lymph node-targeting nanoparticulate peptide vaccine for the treatment of advanced soft tissue sarcoma. Int J Cancer. 2023;152(12):2554.
62. Gumber D, Wang LD. Improving CAR-T immunotherapy: overcoming the challenges of T-cell exhaustion. eBioMedicine. 2022;77:103941.

Unconventional T Cells: Biology, Functions, and Their Implications in Cancer

Surojit Karmakar, Reshmi Suresh, Sourabh Yadav, and Girdhari Lal

Abstract T cells are the backbone of adaptive immune responses against various diseases, including cancer. Conventional αβ T cells recognize neoantigens from cancer cells through antigen-presenting cells and generate antigen-specific responses against cancer. They are also quite efficient at generating immunological memory and protecting against the reappearance of the same disease. Despite these strong checkpoints, cancer still progresses via immune escape through different mechanisms. Conventional T cells have been studied for a long time to generate effective therapeutic approaches with variable success rates against many forms of cancer. In addition to conventional αβ T cells, several unconventional T lymphocytes, such as γδ T cells, invariant natural killer T (iNKT) cells, and mucosal-associated invariant T (MAIT) cells, exist in the body and perform widespread and unique functions. γδ T cells can recognize some unique antigens from different metabolic pathways, invariant T-cell receptors (TCRs) on iNKT cells can recognize lipid antigens restricted to the major histocompatibility complex (MHC) I-related molecule CD1d, and MAIT cells can recognize many unique microbial antigens shared by both microbes and cancer cells but not expressed by normal mammalian cells. These cells are also potent effector cells of the immune system that act on cancer cells to various degrees. In this chapter, we highlight the biology, functions, and therapeutic potential of unconventional T cells, such as γδ T cells, iNKT, and MAIT cells, and their importance in the tumor microenvironment and cancer progression.

Keywords Cellular therapy · Gamma-delta T cells · Mucosal-associated invariant T (MAIT) cells · Natural killer (NK) cells · CAR-T cells · NKT cells

S. Karmakar · R. Suresh · S. Yadav · G. Lal (✉)
BRIC-National Centre for Cell Science (NCCS), Pune, Maharashtra, India
e-mail: glal@nccs.res.in

A. K. Mishra, S. K. Malonia (eds.), *Cell-based Immunotherapies for Cancer*,
https://doi.org/10.1007/978-3-031-88695-9_10

1 Introduction

In the past decade, there have been unprecedented advances in cancer immunotherapy. Immunotherapy via immune checkpoint blockade or adoptive cellular therapy (ACT) represents a significant landmark in cancer therapeutics [1, 2]. Among them, ACTs have successfully treated blood malignancies and provided new hope for curing many advanced and solid tumors [3–5]. Although the majority of ACT strategies use αβ T cells isolated from the peripheral blood of cancer patients, the use of other immune cells, such as natural killer (NK) cells, invariant NKT (iNKT) cells, and γδ T cells, is also gaining importance recently [6–9]. In humans, PBMCs represent approximately 0.095% of iNKT cells, 2.175% of γδ T cells, and 2.99% of the total T-cell population of mucosal-associated invariant T (MAIT) cells [10]. MAIT cells recognize microbial metabolites presented through the non-classical major histocompatibility complex (MHC) 1b molecule known as MHC-I-related protein 1 (MR1). Younger individuals had higher γδ T and mucosal-associated invariant T (MAIT) cells than older individuals [10]. ACT is performed by isolating autogenic or allogeneic immune cells from the patient's body [11, 12]. These purified cells are activated and propagated *in vitro* or genetically modified to express chimeric antigen receptors (CARs) and then adoptively transferred into the patient's body [12, 13]. Like conventional αβ T cells and NK cells, unconventional immune cells such as iNKT, γδ, and MAIT cells also hold immense potential in cancer therapeutics [14–16]. These cells develop from common lymphoid progenitor cells in the thymus but are selected differently because they are exposed to different types of antigens (iNKT and MAIT cells) or are based on the type of TCR expressed (γδ T cells). The origin and development pathways of these cells are summarized in Fig. 1. Among the different functions performed by iNKTs, γδ T cells, and MAIT cells, they strongly regulate tumor growth [17–19]. Despite their low frequency, they can control tumor growth directly or indirectly by regulating the functions of other immune cells [20, 21]. Owing to their cytotoxic and immune modulatory potential, scientists are trying to utilize various means to weaponize them to fight against different cancer types [12, 21–23]. In contrast, these cells also have many limitations, such as their significantly lower frequency and short life span within the body and in the tumor microenvironment [24, 25]. This chapter discusses various properties of iNKT, MAIT, and γδ T cells and their roles in the immune response to cancer.

2 Unconventional T Lymphocytes

T cells are important players in the complex world of the immune system. Conventional T lymphocytes, represented by $CD4^+$ helper and $CD8^+$ cytotoxic T cells, are the traditional architects of adaptive immunity. They recognize antigens presented by MHC molecules, coordinating immune responses by activating other

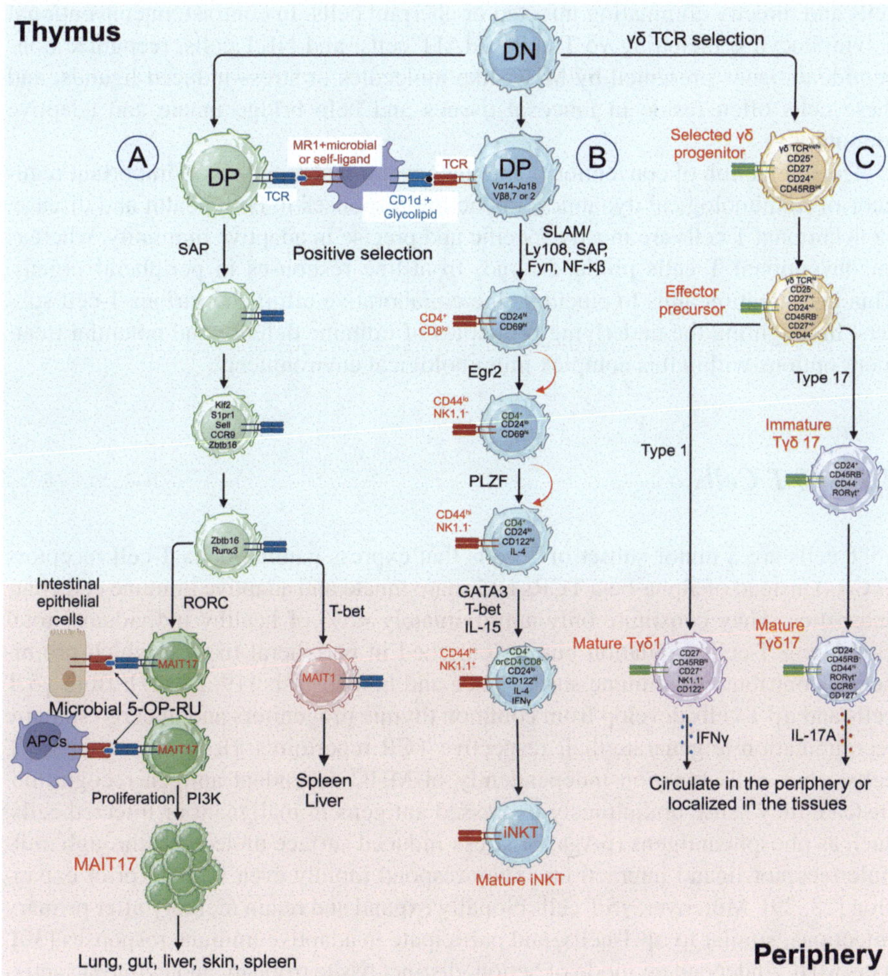

Fig. 1 Origin and development of unconventional T lymphocytes. Unconventional T lymphocytes develop from common lymphoid progenitors in the thymus. These cells differentiate into γδ T cells, MAIT cells, or iNKT cells from double-negative ($CD4^-CD8^-$) populations. (**a**) Double-negative (DN) cells are differentiated into MAIT cells if they experience microbial antigens presented to them by APCs (antigen-presenting cells) through MR-1. These cells then undergo different stages of differentiation until they develop into MAIT1 or MAIT17 cells in the periphery, which are then localized into different tissue spaces. (**b**) iNKTs develop from DN cells when glycolipids are presented to them through CD1d molecules by APCs (antigen-presenting cells). Then, they undergo stages of differentiation. (**c**) γδ T cells also develop from DN cells when γδTCR is expressed instead of αβTCR, and γδTCR-expressing cells are selected. They then undergo a series of differentiation steps and ultimately develop γδ1 or γδ17 in the periphery and localize in the tissues. (Image created with Biorender.com). DP Double positive ($CD4^+CD8^+$), DN Double-negative ($CD4^-CD8^-$), MR1 MHC-related gene protein 1, SAP SLAM-associated protein, SLAM Signaling lymphocyte activation molecule, NF-κβ Nuclear factor- κβ, Egr2 Early growth response gene, PLZF Promyelocytic leukemia zinc finger, T-bet T-box transcription factor 21, GATA3 GATA binding protein 3, IL Interleukin, RORC RAR-related orphan receptor C, CCR Chemokine receptor, PI3K Phosphatidylinositol 3-kinase, Klf2 Kruppel-like factor 2, IFN Interferon

cells and directly eliminating infected or aberrant cells. In contrast, unconventional T lymphocytes, including γδ T cells, MAIT cells, and NKT cells, recognize non-peptide antigens presented by MHC-like molecules or stress-induced ligands, and these cells often reside in mucosal tissues and help bridge innate and adaptive immunity.

The interaction of conventional and unconventional T cells is an important regulator of immunological dynamics, impacting responses in both health and disease. Conventional T cells are mostly specific and precise in adaptive immunity, whereas unconventional T cells produce rapid, front-line responses in peripheral organs. This investigation aims to elucidate the collaborative efforts of various T-cell subsets, highlighting the underlying principles of immune defense and potential treatment options within this complex immunological environment.

2.1 γδ T Cells

γδ T cells are a minor subset of T cells that express gamma-delta T-cell receptors (TCRs) instead of alpha-beta TCRs and share innate and adaptive immune cell characteristics. They constitute only approximately <5% of healthy individuals' total circulating T-cell population but are enriched in peripheral tissues, which prominently contribute to immune surveillance and homeostasis [19, 26, 27]. Both γδ T cells and αβ T cells develop from common thymic progenitors and undergo somatic recombination to generate their respective TCR repertoires. However, unlike αβ T cells, γδ T cells function independently of MHC-dependent antigen recognition. Instead, they sense ubiquitously expressed antigens in malignant or infected cells, such as phosphoantigens (pAgs) or stress-induced surface molecules, through multiple receptor-ligand interactions. They respond rapidly even without prior expansion [28, 29]. Moreover, γδ T cells clonally expand and retain memory after primary infections, similar to αβ T cells, and participate in adaptive immune responses [30]. This MHC-independent mode of action, distinct tissue tropism, and cytotoxic activity make γδ T cells an ideal alternative therapy against MHC-deficient tumors where αβ T-cell-based therapies fail or are less effective [31]. In addition. γδ T cells pose a low risk of alloreactivity and developing graft vs. host disease (GvHD) and are therefore, suitable for adoptive transfer [32, 33].

2.1.1 γδ T-Cell Subsets and Their Key Features

γδ T-cell subsets are defined on the basis of their TCR chain sequences generated by somatic recombination. Human γδ T-cell subtypes are defined by their δ chain, whereas in mice, they are defined on the basis of their γ chain. Among the eight δ-chain segments (Vδ1–8), subsets with Vδ1, Vδ2, and Vδ3 are the most prominent in humans [25]. Vδ1 T and Vδ2 T cells are the most studied subtypes because of their distinct physiological locations and potent innate-like antitumor activity [19,

34, 35]. While Vδ2 T cells constitute circulating γδ T cells in the blood, Vδ1 T cells reside in peripheral tissues and have distinct functional mechanisms. Details of their origin, distinct features, and functionality in antitumor immunity are discussed below.

2.1.2 Vδ2 T Cells

Although the Vδ1 and Vδ2 subsets develop in the fetal liver at 5–7 weeks, Vδ2 T cells are the first to develop, and after they develop in the fetal liver, from 8–15 weeks, Vδ2 T cells further develop in the thymus [36]. During V(D)J rearrangement, the Vδ2 TCR chain selectively pairs with a Vγ9 chain to generate Vγ9Vδ2 T cells, which dominate in the blood and spleen. The Vγ9Vδ2 T-cell population expands following frequent infections and human age and becomes the major type of circulating γδ T cell in adults (60–95%) [37–39]. Vγ9Vδ2 T cells recognize intracellular metabolites such as phosphoantigens (pAgs) present on transformed or infected cells due to dysregulated cholesterol synthesis [40]. Isopentyl pyrophosphate (IPP) is also a pAg that is overproduced in cancer cells due to an upregulated mevalonate pathway, which can elicit chemotaxis and activation of Vγ9Vδ2 T cells [41]. (E)-4-hydroxy-3-methyl-but-2-enyl pyrophosphate (HMB-PP) is produced by microbes through the non-mevalonate pathway and is another pAg recognized by Vγ9Vδ2 T cells [42]. Butyrophilin-3A (BTN3A) subfamily molecules are a group of immunoglobulin families present on various cells, such as innate immune cells and cancer cells. These pAgs bind to the intracellular domain of butyrophilin subfamily 3 member A1 (BTN3A1), causing conformational changes in their transmembrane domain and forming a complex with BTN2A1 [43–46]. The BTN3A1-BTN2A1 complex is directly recognized by the Vγ9 chain of Vγ9Vδ2 T cells, which is germline-encoded to bind BTN proteins [47, 48]. However, the δ chain of Vδ2 cells shows variations in their Jδ segments, suggesting the selection and clonal expansion of these cells upon microbial exposure and antigen stimulation. As a result, a semi-invariant TCR repertoire of Vγ9Vδ2 T cells have developed in humans, making them transferrable between individuals, with minimal autoreactivity [49–52]. Selective expansion of Vγ9Vδ2 T cells ex vivo was possible upon stimulation with phosphoantigens in the presence of IL-2 [53, 54], and these T cells exhibited antitumor activity in vitro [55]. The accumulation of synthetic analogs of pAg, such as bromohydrin pyrophosphates (BrHPP) and 2-methyl-3-butenyl-1-pyrophosphate (2M3B1PP), can directly stimulate Vγ9Vδ2 T-cell expansion [56, 57]. A class of drugs called amino bisphosphonates, which include pamidronate and zoledronate, has been shown to induce the activation and expansion of Vγ9Vδ2 T cells in vivo [57–59]. Amino bisphosphonates inhibit farnesyl pyrophosphatase synthase of the mevalonate pathway in tumor cells, leading to IPP accumulation and subsequent activation of Vγ9Vδ2 T cells. Most clinical trials have utilized pAgs or bisphosphonates for in vivo activation of Vδ2 T cells or ex vivo activation of patient-derived Vδ2 T cells. However, in vitro, the tumor-killing capacity of Vδ2 T cells has not been recapitulated in clinical settings. This inconsistency has been partly attributed to discrepancies in the pharmacokinetic profiles of amino bisphosphonates in cancer immunotherapy, as they are rapidly cleared from circulation by bone absorption or excretion [60]. Moreover, pAg-induced activation has been attributed to the anergy and exhaustion of Vδ2 T cells in preclinical models and clinical trials [61–63]. Monoclonal antibodies against BTN3A1 have recently been shown to promote Vγ9Vδ2 T-cell activation and cytotoxicity toward tumors [64, 65]. In a phase I trial (NCT04243499), the agonistic BTN3A antibody ICT01 caused a reduction in the number of circulating Vδ2 T cells, which was attributed to their ability to home to tumors and was marked by an increase in Vγ9^{+} T cells within the tumor [66]. With BTN3A1 interacting with BTN2A, which binds to the Vγ9 chain, the diversity of their antigens and binding mechanisms remains unclear. A plausible explanation implies that various ligands bind to different regions of a single γδTCR, mediating adaptive and innate responses [67]. Vγ9Vδ2 T cells can also cross-present antigens like dendritic cells [68–70]. Vδ2 T cells are also activated and elicit cytotoxicity independent of their TCR through different receptor-ligand interactions. Several natural killer (NK) cell receptors (NKRs), which bind to stress-induced ligands that are overexpressed on tumor cells, are expressed on Vδ2T cells. This includes natural killer group 2 member D (NKG2D), which binds to MHC class I chain-related proteins A and B (MICA and MICB), as well as multiple UL16-binding proteins (ULBPs) and DNAX accessory molecule-1 (DNAM-1), which binds to nectin-2 and nectin-like 5 expressed on tumor cells [71–73]. In addition, Vγ9Vδ2 T cells express FcγRIII or CD16 and, therefore, can mediate antibody-dependent cellular cytotoxicity (ADCC), which can be harnessed for antitumor activity [74–77].

2.1.3 Vδ1 T Cells

Vδ1 T cells constitute the major γδT cell subset within peripheral tissues, including the dermis, gut epithelium, liver, and spleen, and constitute approximately one-third of circulating γδT cells [78]. They are predominant mediators of tissue homeostasis and are involved in the first line of defense against viral infections. Vδ1 T cells have highly variable antigen recognition and effector mechanisms. Compared with Vγ9Vδ2 T cells, their TCR repertoire is more diverse and individualized, with specific clones dominating in adults [79, 80]. Most of these studies focused on the abundant Vγ9Vδ2 T cells and their application, and the mechanisms underlying the antigen-specific clonal expansion of Vδ1 T cells are still unknown. However, despite the individualized TCR repertoire of Vδ1 T cells, they can recognize diverse sets of ligands that are conserved among humans. Some Vδ1 T cells can recognize lipid antigens presented by the non-classical MHC molecules CD1c and CD1d [81–84]. Like Vδ2 T cells, Vδ1 T cells also express natural killer cell receptors (NKRs), such as natural killer group 2 member D (NKG2D) and DNAX accessory molecule 1 (DNAM-1, also known as CD226), which recognize their ligands on tumor cells [85–87]. A subset of Vδ1 T cells in intestinal epithelial tissues express another NKR called NKp46. The expressions of these molecules are associated with increased cytotoxic activity against colorectal cancer cells in vitro, and their presence is directly correlated with a lower risk of metastatic progression [88]. The expression of NKp30 and NKp44 can be induced upon TCR activation of Vδ1 T cells in the presence of the cytokines IL-2 or IL-15 [89, 90]. Given their tissue-specific localization, Vδ1 T cells also express high levels of receptors for tissue homing and retention [91–93]. This makes Vδ1 T cells suitable for adoptive cell therapies for solid cancers, as they can effectively infiltrate the tumor bed [61]. Owing to their tissue-resident characteristics and localization in tumors, Vδ1 T-cell subsets are easily accessible for therapeutic interventions in situ via the use of cell monoclonal antibodies and different cell engagers [25]. Moreover, their role in tissue homeostasis and rapid surveillance against disturbances associated with their localization makes them intrinsically regulated to their microenvironments. This approach becomes especially suitable for immunotherapeutic settings to achieve on-tumor cytotoxicity without off-target toxic effects.

2.2 *Invariant Natural Killer (iNKT) Cells*

iNKT cells belong to a heterogeneous group of NKT cells, which have the properties of both NK and T cells. iNKT cells constitute a subset of αβ T cells and are composed of Vα14-Jα18 (mouse) or Vα24-Jα18 (human) chains paired with Vβ7, β8, and β2 in mice and Vα24 and Vβ11 in humans, which are restricted to antigen presentation by non-classical MHC class I molecules, such as the CD1d molecule [94]. When iNKT cells recognize glycolipid antigens presented by CD1d, they quickly generate high amounts of cytokines such as IL-4, TNF-α, and IFN-γ. The sphingolipid alpha-galactosylceramide (α-GalCer), a microbial glycolipid from various species, such as *Borrelia burgdorferi* and *Sphingomonas* spp., and endogenous mammalian-linked glycolipids, such as α-GlcCer, can trigger a strong response by iNKT cells [94]. iNKT cells can modulate the immune system and play a significant role in the pathology or resolution of microbial and viral infections, cancers, malignancies, and autoimmune illnesses [94–99]. iNKT cells can be classified into three functional groups on the basis of the expression of transcription factors, such as T-box transcription factor 21 (T-bet), promyelocytic leukemia zinc finger (PLZF), and retinoic acid-related orphan receptor gamma t (RORγt), which contribute to their complexity and are classified into iNKT1, iNKT2, and iNKT17. iNKT2 and iNKT17 cells are found mainly in developmental stage 2, and iNKT1 cells are found mainly in developmental stage 3. Various transcription factors are involved in iNKT cell progression through developmental stages and their effector functions. T-bet is essential for the development, homeostasis, and function of iNKT cells [100]. RORγt, along with Runx1, also regulates the iNKT17 pathway and the growth of iNKT cells (Fig. 1). PLZF is involved in directing the development of effector iNKT cells [101, 102].

2.2.1 Th1-Like iNKT Cells

Th1-like iNKT cells express the transcription factor T-bet, a characteristic transcription factor of Th1 cells. It also expresses interleukin 2 receptor beta (CD122, IL-2Rβ) and requires IL-15 for development. These cells are present mostly in the mouse spleen and liver and produce IFN-γ. They lack IL-17RB and can be either $CD4^+$ or $CD4^-$ [103]. These iNKT cells produce mainly Th1-type cytokines in response to IL-12, but they can also produce some Th2-type cytokines with IFN-γ in response to the TCR signaling [104].

2.2.2 Th2-Like iNKT Cells

Upon activation, Th2-like iNKT cells can produce Th2-like cytokines such as IL-4, IL-9, IL-10, and IL-13. These cells are characterized mainly by the expression of IL-17RB and CD4. These cells are abundant in the lungs and are reportedly involved in Th2-type airway inflammation [103–105]. Although the expression of T-bet in Th1-like iNKT cells has been confirmed, the major transcription factor for Th2-like iNKT cells is still unknown. Although GATA3 is involved in the production of Th2-type cytokines in $CD4^+$ T cells, its role in iNKT cells is still obscure [104]. However, scientists have reported that the production of Th2-type cytokines is severely impaired in the few iNKT cells that develop in GATA3-deficient mice. GATA3 expression in the absence of T-bet or RORγt may lead to the development of Th2-like iNKT cells [106]. In addition to GATA3, another transcription factor, E4 promoter-binding protein 4 (E4BP4; also known as NFIL3), may also play a role in the production of Th2-type cytokines by iNKT cells. However, E4BP4 is also necessary for the production of Th17-type cytokines, so its role in Th2-like iNKT cells is still not well defined [104].

2.2.3 Th17-Like iNKT Cells

Th17-like iNKT cells are primarily present in peripheral lymph nodes, lungs, and skin. Most iNKT cells that resemble Th17 cells are present within the $CD4^-NK1.1^-$ subpopulation [107, 108]. Other phenotypes of IL-17A-producing iNKT cells belong to the $CD4^-IL\text{-}17RB^-$phenotype, overlapping with Th17-like $CD4^-NK1.1^-$ cells [109]. These cells also express cytokines such as IL-21 and IL-22. Like MHC-restricted Th17 cells, Th17-iNKT cells express the chemokine receptor CCR6 and require the transcription factor RORγt for development [104].

2.2.4 Other iNKT Cell Subsets

iNKT cell subsets, similar to the $CD4^+$ T-cell population and dependent on MHC, include Tfh and Treg cells. Recently, a $PD1^+CXCR5^+$ IL-21-producing B cell lymphoma 6 (BCL-6)-dependent 'follicular helper' iNKT cell subgroup has also been described [110, 111]. iNKT cells that express Forkhead-box P3 (FoxP3) have been demonstrated in vitro by artificial induction [112]. IL-10-producing iNKT cells are also found in pathological and physiological conditions, including in adipose tissues [113, 114] and within the peripheral blood of systemic lupus patients [115].

Although NKT cells are great killers, their number is relatively low in the body. However, it shows good potential in regulating different diseases, including cancer.

2.3 Mucosal-Associated Invariant T (MAIT) Cells

MAIT cells are a separate subset of unconventional T lymphocytes that have received intense interest due to their particular characteristics and activities in immune surveillance [116, 117]. Unlike ordinary T cells, MAIT cells have a T-cell receptor (TCR) that recognizes microbial-derived vitamin B metabolites presented by the non-classical MHC-Ib molecule MHC-related protein 1 (MR-1) [118–120]. Due to this atypical antigen recognition, MAIT cells have the specialized ability to recognize a wide range of bacteria and fungi, establishing them as essential participants in mucosal immunity [118, 119].

2.3.1 Phenotypic Characteristics and Distribution

MAIT cells are distinguished by the expression of a semi-invariant TCR composed of an invariant TCR chain (V7.2-J33 in humans) combined with a restricted set of TCR chains [121]. Because MAIT cells are semi-invariant, they can recognize microbial ligands while maintaining specificity [121]. Furthermore, MAIT cells frequently express high levels of the C-type lectin receptor CD161 and the IL-18 receptor, indicating responsiveness to mucosal cytokines [122]. MAIT cells are abundant in mucosal tissues such as the lungs, intestines, liver, and peripheral blood [123]. Because of their abundance in mucosal regions, MAIT cells serve as frontline defenders against microbial invaders that frequently breach these barriers [124].

2.3.2 Antigen Recognition Activation and Functions

MAIT cells recognize small compounds derived from the riboflavin (vitamin B2) metabolic pathway, which is extensively conserved among microorganisms but is missing from mammalian cells [116, 125, 126]. Recently, it has been shown that host tissue-derived antigens such as sulfated bile acid and cholic acid 7-sulfate (CA7S) bind to MR-1 and promote MAIT cell development and function [127]. This distinct identification enables MAIT cells to respond quickly to various bacterial and fungal diseases. MAIT cells release a variety of cytokines upon activation, including IFN-γ and TNF-α, which aid in the recruitment and activation of other immune cells [128, 129]. MAIT cells play a variety of roles in immunological defense. They impact the activation of other immune cells, modulate the integrity of mucosal barriers, and participate in tissue healing processes in addition to their direct antimicrobial capabilities. Their capacity to connect the innate and adaptive branches of the immune system emphasizes their importance in preserving mucosal homeostasis [121, 130].

3 Unconventional T Lymphocytes in the Cancer Immune Response

Attention in cancer immunology has increasingly shifted to unconventional T cells, which are distinct from conventional T cells in many ways [131–133]. γδ T cells, MAIT cells, and NKT cells are examples of unconventional players who bring unique skills to the forefront of the cancer immune response [134]. Unlike their conventional counterparts, unconventional T cells recognize non-peptide antigens and have extraordinary adaptability, allowing them to respond quickly to emergent threats [131, 135]. In this section, we discuss the evolving understanding of how unconventional T lymphocytes contribute to the complex landscape of the cancer immune response, examining their roles in tumor surveillance, tumor microenvironment modulation, and potential as targets for novel immunotherapeutic strategies.

3.1 γδ T Cells in Cancer

γδ T cells in cancer can play both anti- and pro-tumor roles on the basis of their subsets, effector functions, and cytokine profiles [24]. Typically, IL-17A-producing γδ T cells are considered tumor-promoting subsets, also called regulatory γδ T cells [136–138], whereas subsets that secrete IFN-γ and TNF-α are involved in tumor cytotoxicity (Fig. 2). However, most of these studies were conducted in mice, and their γδ T-cell population is relatively prone to IL-17A production, which is known to promote tumor progression [137–139]. On the other hand, human γδ T cells are biased toward the type I cytotoxic phenotype, which is characterized by the release of the cytokines IFN-γ and TNF-α and the production of cytotoxic granules, which play an antitumor role [140, 141]. The correlation between tumor-infiltrating γδ T cells and clinical outcomes varies depending on the cancer type. However, in most cases, the presence of γδ T cells among tumor-infiltrating lymphocytes (TILs) has been associated with favorable prognosis in cancer patients. In fact, a pan-cancer study by Gentles et al. on transcriptomic signatures determining immune cell composition from 18,000 tumor samples revealed that γδ T cells, the most important immune cell subset, were correlated with a favorable prognosis [142]. Further studies revealed that the Vγ9Vδ2 T-cell gene signature within TILs was independent of both αβ T cells and the mutational load of tumors, confirming the functional and therapeutic relevance of Vγ9Vδ2 T cells in low-immunogenicity tumors [143]. Circulating γδ T-cell levels have been positively correlated with a better 5-year disease-free survival and overall survival in patients with acute leukemia who underwent bone marrow transplantation [144]. Tumor-infiltrating γδ T cells have also been linked to a positive prognosis in patients with melanoma and gastric cancer [145].

The type I effector characteristics of γδ T cells can be activated upon stimulation with IL-2 or IL-15 [146]. The antitumor functions of both human and mouse γδ T

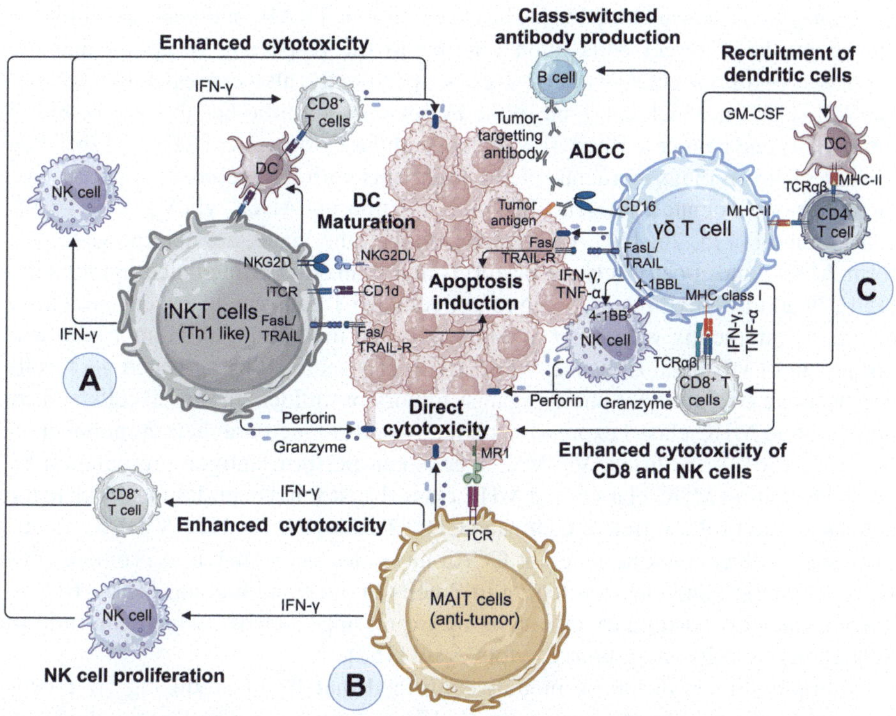

Fig. 2 Pro-Tumor Functions of Unconventional T Cells. (**a**) Regulatory iNKT cells promote tumor growth by enhancing CD4⁺ Treg and M2 tumor-associated macrophage (TAM) polarization and suppressing CD8⁺ T-cell functions. (**b**) Pro-tumor MAIT cells also promote tumor growth by releasing cytokines and chemokines such as IL-8, IL-17, and CXCL12. (**c**) Under certain circumstances, γδ T cells also perform pro-tumor functions. It can produce IL-17, which stimulates angiogenesis through the vascular endothelial growth factor (VEGF) axis. It can also suppress CD8⁺ T-cell activation through PD-L1–PD1 interactions. It also helps recruit pro-tumorigenic PMN-MDSCs and neutrophils in tumors that suppress CD8⁺ T-cell functions. (Image created with Biorender.com). CTLA4 Cytotoxic T lymphocyte attenuator 4, Gal Galectin, GM-CSF granulocyte-macrophage colony-stimulating factor, IFN-γ Interferon γ, IL Interleukin, MAIT mucosal-associated innate T cells, PD-L1 Programmed death ligand 1, PMN-MDSC Polymorphonuclear Myeloid-derived suppressor cells, PD1 Programmed cell death 1, ROS Reactive oxygen species, TAM tumor-associated macrophage, TGF-β Transforming growth factor-β, TNF-α Tumor necrosis factor-alpha, TIM3 T-cell immunoglobulin and mucin domain-containing protein 3, VEGF Vascular endothelial growth factor

cells include the secretion of type I (Th1) cytokines such as IFN-γ and TNF-α, the production of perforin and granzymes A and B, and the expression of death receptor ligands such as tumor necrosis factor-related apoptosis-inducing ligand (TRAIL) and first apoptosis signal ligand (FasL) [147]. Type I effector functions are elicited mainly upon TCR-based activation, and NKR signaling promotes cytotoxicity in γδ T cells. NKG2D signaling in γδ T cells has been reported to induce the release of cytotoxic granules and soluble TRAIL in vitro [85, 148–150]. The perforin and granzyme released by γδ T cells facilitate tumor killing by inducing apoptosis via the granzyme protease [149, 151]. The expression of TRAIL and FasL also induces the apoptosis of target cells upon binding to the respective receptors that are expressed on tumor cells. Vγ9Vδ2 T cells, specifically, also express high levels of FcγRIII (CD16), which can bind to the Fc regions of antibodies that are bound to tumor cells and trigger antibody-dependent cellular cytotoxicity (ADCC) [76, 152]. This mechanism plays a crucial role in treatments with monoclonal antibodies that target tumor-associated antigens (TAAs) by facilitating ADCC via Vγ9Vδ2 T cells against tumors [75, 76, 153]. Vδ1 T cells, on the other hand, are not known to perform ADCC and mostly rely on perforin- and granzyme B-mediated cytotoxicity [154]. In addition to directly lysing tumor cells, γδ T cells can also exert antitumor effects by interacting with other immune cells and modulating the tumor microenvironment (TME) against cancer cells. The IFN-γ secreted by activated γδ T cells can promote overall cytotoxicity against tumors by influencing αβ T cells and by upregulating MHC class I expression on tumor cells to increase their immunogenicity [155, 156]. In addition, Vγ9Vδ2 T cells can perform antigen presentation by cross-presenting MHC class I- and MHC class II-restricted peptides to stimulate the activation and proliferation of $CD8^+$ and $CD4^+$ T cells [69, 70, 157]. Vγ9Vδ2 T cells expressing the chemokine receptor CCR5 have also been shown to promote antibody production and class switching in B cells in mice and humans [158–161]. γδ T cells can also costimulate NK cells by expressing 4-1BBL, which can bind to 4-1BB on NK cells and promote cytotoxicity [162].

Multiple studies in mouse models confirmed that IL-17-producing γδ T cells promote tumor growth and metastasis. IL-17^+ γδ T cells can directly stimulate cancer cell proliferation and migration. They recruit and promote immunosuppressive myeloid cells such as polymorphonuclear myeloid-derived suppressor cells (PMN-MDSCs) [163] and neutrophils in the TME [137, 163–165]. These subsets of γδ T cells have also been linked to angiogenesis, metastasis, and poor survival in cancer patients. Although IL-17^+ γδ T cells are not found in human peripheral blood, their presence has been detected among TILs in patients with advanced stages of various cancers [163, 166, 167]. The pro-tumor roles of unconventional T lymphocytes are summarized in Fig. 3. The distinct antitumor and pro-tumor natures of γδ T cells have been temporally associated with different stages of cancer progression, including colorectal cancer, squamous cell carcinoma, and ovarian cancer, in both mouse models and humans [6, 166–169]. The regulatory phenotypes of γδ T cells are also characterized by the secretion of IL-10 and IL-4 [170, 171]. Even so, antitumor γδ T cells often exceed their regulatory subsets in the TME, given that human γδ T cells are more prone to a type I effector phenotype.

3.2 iNKT Cells in Cancer

NKT cells are CD1d restricted and respond quickly when antigens are presented through CD1d. The major group of NKT cells present in the body are iNKTs, which have potent antitumor functions. Since the discovery of iNKT cells 25 years ago, the

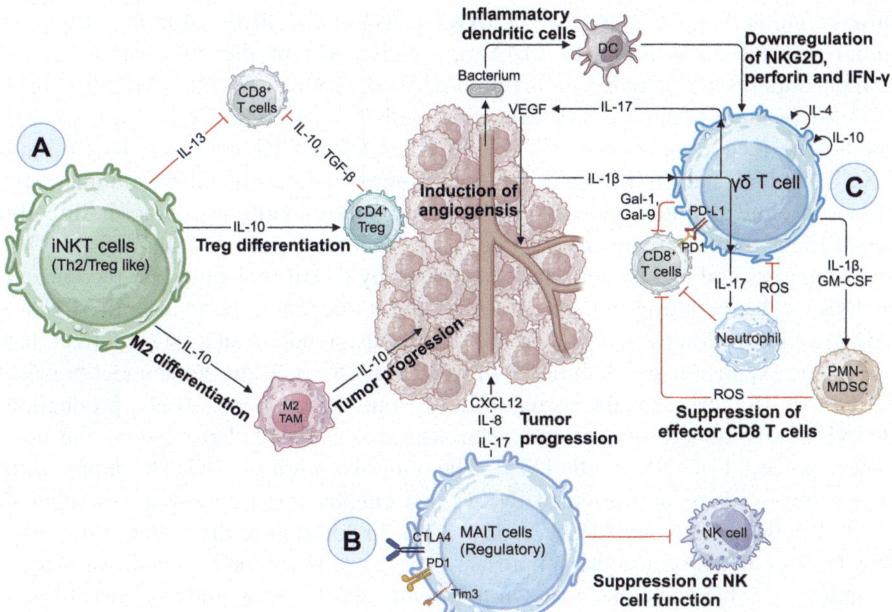

Fig. 3 Antitumor functions of unconventional T cells. (**a**) Most of the antitumor functions of iNKT cells are mediated by the direct recognition of lipid antigen-bound CD1d molecules expressed on cancer cells. The function of NKG2D is also further enhanced by the interaction of NKG2D on iNKT cells with its cognate ligand on cancer cells. After recognition, it kills cancer cells by releasing perforin and granzyme B or inducing their apoptosis through FasL or TRAIL-mediated pathways. iNKTs also promote the maturation of DCs and NK cells, which either help activate CD8+ T cells or directly kill cancer cells. (**b**) MAIT cells mostly perform antitumor functions by promoting the activities of CD8+ T cells and NK cells through the release of IFN-γ. It can also recognize cancer cells directly. (**c**) γδ T cells perform antitumor responses directly by engaging with tumor antigens via recognition through tumor antigen-bound antibodies by CD16, or they can promote the apoptosis of cancer cells through the interactions of the FasL/TRAIL ligand of γδ T cells with their cognate receptors on cancer cells. It also promotes the cytotoxicity of NK cells through signaling through 4-1BB and its ligand interaction. γδ T cells also present antigens through either MHC class I or II and activate CD8 and CD4 T cells, respectively. It also helps in the recruitment of DCs into the tumor microenvironment and helps in the further activation of CD4+ and CD8+ T cells in tumors. (Image created with BioRender.com). 4-1BB tumor necrosis factor receptor superfamily 9 or TNFRSF9, ADCC Antibody-dependent cellular toxicity, Fas First apoptosis signal, FasL First apoptosis signal ligand, GM-CSF Granulocyte-macrophage colony-stimulating factor, HLA-DR Human leukocyte antigen-DR, IL Interleukin, IFN-γ Interferon-gamma, iTCR invariant T-cell receptor, MAIT Mucosal-associated innate T lymphocyte, NKG2D natural killer group 2 member D, TAM tumor-associated macrophage, TNF-α tumor necrosis factor-alpha, TRAIL TNF-related apoptosis-inducing ligand, TRAIL-R TNF-related apoptosis-inducing ligand receptor, TCR T-cell receptor

sphingolipid α-GalCer has been used in more than 30 antitumor clinical trials with no optimum results, as α-GalCer is a classical ligand that activates iNKT cells and results in the release of cytokines, perforin and granzymes [172]. In addition, some cytokines produced by iNKT cells help in the maturation of DCs and the differentiation of αβ T cells (Fig. 2). iNKT cells are also involved in the killing of tumor-associated macrophages (TAMs), which leads to the elimination of the immunosuppressive activity of myeloid-derived suppressor cells (MDSCs) in a CD1d-dependent manner [173]. Compared with T cells, iNKT cells express many chemokine receptors, such as CCR1, CCR2, CCR3, CCR4, CCR5, CCR6, and CXCR3, which help iNKT cells infiltrate tumors and recruit other immune cells [174]. In clinical trials, increased infiltration of iNKT cells in tumors is directly proportional to the patient survival rate.

Exogenous and endogenous antigens bound by CD1d molecules are recognized by iNKT cells, resulting in their activation and immediate release of both Th1 and Th2 cytokines. Moreover, activated iNKT cells also mediate an adjuvant effect that causes the expansion and maturation of bystander DCs via cell-cell contact or soluble factors. Reciprocally, the secreted IL-12 from DCs promotes IFN-γ production by iNKT cells. As a result, systemic administration of IL-12 also enhances the anti-cancer potential of iNKT cells [175]. The administration of α-GalCer along with IL-12 improved the activation of iNKTs and enhanced the cytotoxic potential of $CD8^+$ T cells and NK cells [97, 99, 176, 177]. In addition to their supporting role, iNKT cells can also cause the apoptosis of cancer cells via the Fas/FasL pathway.

iNKT cells have also been shown to provide spontaneous immune surveillance (i.e., without exogenous Ag stimulation) against methylcholanthrene-induced sarcomas and in genetically engineered mouse tumor models of sarcoma, lymphoma, prostate adenocarcinoma, chronic lymphocytic leukemia (CLL), and pancreatic adenocarcinoma [178, 179]. This was established by the depletion of iNKT cells in invariant-chain TRAJ18 TCR chain ($J\alpha 18^{-/-}$) or CD1d ($CD1d^{-/-}$) knockout mice. In these mouse models of tumors, adoptive transfer of iNKT cells resulted in increased survival [178, 179]. Recently, conditional NKT cell depletion in mice was shown to regulate CTL cross-priming by DCs through a negative feedback loop [180].

In the case of humans, clinical studies in solid and hematological malignancies reported a reduced number of less functional iNKT cells [181]. A decrease in the number of iNKT cells also resulted in increased progression of prostate cancer, pancreatic cancer, oral cell squamous carcinoma, acute myeloid leukemia, and CLL. In contrast, an increased population of iNKT cells in circulation and in tumors resulted in improved survival in colorectal cancer patients [182]. In prostate and oral squamous cell carcinoma, iNKT cells exhibit defective IFN-γ production, which leads to a Th2-like cytokine profile [183, 184]. This cytokine defect can also be restored by the administration of IL-12 in vitro or α-GalCer or dendritic cells in vivo. Scientists have also reported that increased expression of CCL2 in iNKT cells promotes their infiltration in neuroblastoma [185]. In addition to tumors, iNKT cells respond to microbial infection in the lungs and support neutrophilic airway hyperactivity [186]. Despite being a smaller population, iNKT cells are potent enough to protect against many forms of cancer. Due to its anti-cancer potential, scientists are trying to weaponize it in different ways to combat cancer.

3.3 MAIT Cells in Cancer

The role of MAIT cells in cancer is a developing area of interest and investigation [187]. While our understanding of the precise activities of MAIT cells in cancer is still emerging, multiple studies indicate that they play diverse roles in tumor immunity [188, 189]. MAIT cells have been found in several tumor tissues, demonstrating that they can penetrate the tumor microenvironment [188, 190]. The number of MAIT cells within tumors varies between cancer types. MAIT cells have cytotoxic properties, suggesting that they can be used to kill cancer cells directly (Fig. 2) [191]. Their ability to recognize and respond to microbial antigens generated by some cancer cells may help eliminate contaminated or altered cells [191]. MAIT cells that have been activated can release cytokines such as interferon-gamma (IFN-γ) and tumor necrosis factor-alpha (TNF-α) [192, 193]. These cytokines can impact the tumor microenvironment, potentially increasing antitumor immune responses and limiting tumor growth. MAIT cells may interact with other immune cells, such as T cells, dendritic cells, and natural killer cells, impacting the overall immune response to malignancy [194]. This interaction can potentially alter the immunological landscape within the tumor microenvironment. Microbial products influence MAIT cell activation, and the gut microbiota composition has been connected to the activation status of MAIT cells [195]. Because the gut microbiota can influence systemic immune responses, including antitumor immunity, researchers are interested in learning how MAIT cells and the microbiome interact to influence cancer progression [191, 196]. This understanding will also help to identify the missing link between the role of the gut microbiome in cancer progression and, thereby, its effect on therapeutic responses to cancer.

4 Conclusions

Unconventional T cells develop from common lymphoid progenitor cells in the thymus, similar to conventional T cells, and perform a wide array of functions in the body. Owing to their non-MHC-restricted response pattern and ability to recognize varieties of antigens, which are sometimes elusive to conventional T cells, they can regulate the pathogenesis of different diseases, such as cancer. Despite being less common, these cells can affect cancer progression directly and regulate the functions of other immune cells in the tumor microenvironment. Unlike $\alpha\beta$ T cells, unconventional T cells also exhibit a dysfunctional response in the tumor microenvironment [197]. A deeper understanding of what causes the dysfunction of these cells in tissues will not only increase the understanding of the immune response to diseases, including cancer, but also help in the development of new therapeutic approaches to potentiate unconventional T cells to eradicate cancer.

Declaration of Interests The authors declare that they have no competing interests.

Funding Supports GL received the Swarna Jayanti Fellowship (DST/SJF/LSA-01/2017–18) from the Department of Science and Technology and a research grant from the Science and Engineering Research Board (CRG/2022/007108), Ministry of Science and Technology, Government of India. SK received a Senior Research Fellow (SRF) from the Council of Scientific and Industrial Research (CSIR), and SY received a Junior Research Fellowship (JRF) from the Department of Biotechnology (DBT), Ministry of Science and Technology, Government of India.

References

1. Robert C. A decade of immune-checkpoint inhibitors in cancer therapy. Nat Commun. 2020;11:3801.
2. Shiravand Y, Khodadadi F, Kashani SMA, Hosseini-Fard SR, Hosseini S, Sadeghirad H, Ladwa R, O'Byrne K, Kulasinghe A. Immune checkpoint inhibitors in cancer therapy. Curr Oncol. 2022;29:3044–60.
3. Jogalekar MP, Rajendran RL, Khan F, Dmello C, Gangadaran P, Ahn BC. CAR T-cell-based gene therapy for cancers: new perspectives, challenges, and clinical developments. Front Immunol. 2022;13:925985.
4. Zhang P, Zhang G, Wan X. Challenges and new technologies in adoptive cell therapy. J Hematol Oncol. 2023;16:97.
5. Zhang X, Zhu L, Zhang H, Chen S, Xiao Y. CAR-T cell therapy in hematological malignancies: current opportunities and challenges. Front Immunol. 2022;13:927153.
6. Mensurado S, Silva-Santos B. Battle of the γδ T cell subsets in the gut. Trends Cancer. 2022;8:881–3.
7. Courtney AN, Tian G, Metelitsa LS. Natural killer T cells and other innate-like T lymphocytes as emerging platforms for allogeneic cancer cell therapy. Blood. 2023;141:869–76.
8. Hadiloo K, Tahmasebi S, Esmaeilzadeh A. CAR-NKT cell therapy: a new promising paradigm of cancer immunotherapy. Cancer Cell Int. 2023;23:86.
9. Karmakar S, Mishra A, Pal P, Lal G. Effector and cytolytic function of natural killer cells in anticancer immunity. J Leukoc Biol. 2023;115:235–52.
10. Singh P, Szaraz-Szeles M, Mezei Z, Barath S, Hevessy Z. Age-dependent frequency of unconventional T cells in a healthy adult Caucasian population: a combinational study of invariant natural killer T cells, gamma-delta T cells, and mucosa-associated invariant T cells. Geroscience. 2022;44:2047–60.
11. Rosenberg SA, Restifo NP, Yang JC, Morgan RA, Dudley ME. Adoptive cell transfer: a clinical path to effective cancer immunotherapy. Nat Rev Cancer. 2008;8:299–308.
12. June CH. Adoptive T cell therapy for cancer in the clinic. J Clin Invest. 2007;117:1466–76.
13. Ingram Z, Madan S, Merchant J, Carter Z, Gordon Z, Carey G, Webb TJ. Targeting natural killer T cells in solid malignancies. Cells. 2021;10:1329.
14. de Araújo ND, Gama FM, de Souza BM, Ribeiro TLP, Alves FS, Xabregas LA, Tarragô AM, Malheiro A, Costa AG. Translating unconventional T cells and their roles in leukemia antitumor immunity. J Immunol Res. 2021;2021:6633824.
15. Kent A, Crump LS, Davila E. Beyond alphabeta T cells: NK, iNKT, and gammadeltaT cell biology in leukemic patients and potential for off-the-shelf adoptive cell therapies for AML. Front Immunol. 2023;14:1202950.
16. Cheng OJ, Lebish EJ, Jensen O, Jacenik D, Trivedi S, Cacioppo J, Aube J, Beswick EJ, Leung DT. 2024. MAIT cells modulate innate immune cells and inhibit colon cancer growth. bioRxiv.
17. Chan KF, Duarte JDG, Ostrouska S, Behren A. γδ T cells in the tumor microenvironment-interactions with other immune cells. Front Immunol. 2022;13:894315.

18. Li Y, Li G, Zhang J, Wu X, Chen X. The dual roles of human gammadelta T cells: anti-tumor or tumor-promoting. Front Immunol. 2020;11:619954.
19. Paul S, Lal G. Regulatory and effector functions of gamma-delta (gammadelta) T cells and their therapeutic potential in adoptive cellular therapy for cancer. Int J Cancer. 2016;139:976–85.
20. Hinshaw DC, Shevde LA. The tumor microenvironment innately modulates cancer progression. Cancer Res. 2019;79:4557–66.
21. Balta E, Wabnitz GH, Samstag Y. Hijacked immune cells in the tumor microenvironment: molecular mechanisms of immunosuppression and cues to improve T cell-based immunotherapy of solid tumors. Int J Mol Sci. 2021;22:5736.
22. Waldman AD, Fritz JM, Lenardo MJ. A guide to cancer immunotherapy: from T cell basic science to clinical practice. Nat Rev Immunol. 2020;20:651–68.
23. Paul S, Singh AK, Shilpi LG. Phenotypic and functional plasticity of gamma-delta (gammadelta) T cells in inflammation and tolerance. Int Rev Immunol. 2014;33:537–58.
24. Mensurado S, Blanco-Domínguez R, Silva-Santos B. The emerging roles of γδ T cells in cancer immunotherapy. Nat Rev Clin Oncol. 2023;20:178–91.
25. Zlatareva I, Wu Y. Local γδ T cells: translating promise to practice in cancer immunotherapy. Br J Cancer. 2023;129:393–405.
26. Ribot JC, Lopes N, Silva-Santos B. γδ T cells in tissue physiology and surveillance. Nat Rev Immunol. 2021;21:221–32.
27. Giri S, Lal G. Differentiation and functional plasticity of gamma-delta (γδ) T cells under homeostatic and disease conditions. Mol Immunol. 2021;136:138–49.
28. Bigby M, Markowitz JS, Bleicher PA, Grusby MJ, Simha S, Siebrecht M, Wagner M, Nagler-Anderson C, Glimcher LH. Most gamma delta T cells develop normally in the absence of MHC class II molecules. J Immunol. 1993;151:4465–75.
29. Deseke M, Prinz I. Ligand recognition by the γδ TCR and discrimination between homeostasis and stress conditions. Cell Mol Immunol. 2020;17:914–24.
30. Sheridan BS, Romagnoli PA, Pham QM, Fu HH, Alonzo F 3rd, Schubert WD, Freitag NE, Lefrançois L. γδ T cells exhibit multifunctional and protective memory in intestinal tissues. Immunity. 2013;39:184–95.
31. Sebestyen Z, Prinz I, Déchanet-Merville J, Silva-Santos B, Kuball J. Translating gammadelta (γδ) T cells and their receptors into cancer cell therapies. Nat Rev Drug Discov. 2020;19:169–84.
32. Maeda Y, Reddy P, Lowler KP, Liu C, Bishop DK, Ferrara JL. Critical role of host gammadelta T cells in experimental acute graft-versus-host disease. Blood. 2005;106:749–55.
33. Paul S, Shilpi LG. Role of gamma-delta (gammadelta) T cells in autoimmunity. J Leukoc Biol. 2015;97:259–71.
34. You H, Wang Y, Wang X, Zhu H, Zhao Y, Qin P, Liu X, Zhang M, Fu X, Xu B, Zhang Y, Wang Z, Gao Q. CD69(+) Vdelta1gammadelta T cells are anti-tumor subpopulations in hepatocellular carcinoma. Mol Immunol. 2024;172:76–84.
35. Flemming A. Engineered gamma-delta T cells show promise in bone tumors. Nat Rev Immunol. 2024;24:458.
36. Krangel MS, Yssel H, Brocklehurst C, Spits H. A distinct wave of human T cell receptor gamma/delta lymphocytes in the early fetal thymus: evidence for controlled gene rearrangement and cytokine production. J Exp Med. 1990;172:847–59.
37. Willcox CR, Davey MS, Willcox BE. Development and selection of the human Vγ9Vδ2^{+} T cell repertoire. Front Immunol. 2018;9:1501.
38. Sherwood AM, Desmarais C, Livingston RJ, Andriesen J, Haussler M, Carlson CS, Robins H. Deep sequencing of the human TCRγ and TCRβ repertoires suggests that TCRβ rearranges after αβ and γδ T cell commitment. Sci Transl Med. 2011;3:90ra61.
39. McVay LD, Carding SR. Extrathymic origin of human gamma delta T cells during fetal development. J Immunol. 1996;157:2873–82.
40. Wang H, Fang Z, Morita CT. Vgamma2 Vdelta2 T cell receptor recognition of prenyl pyrophosphates is dependent on all CDRs. J Immunol. 2010;184:6209–22.

41. Tanaka Y, Morita CT, Tanaka Y, Nieves E, Brenner MB, Bloom BR. Natural and synthetic non-peptide antigens recognized by human gamma delta T cells. Nature. 1995;375:155–8.
42. Hintz M, Reichenberg A, Altincicek B, Bahr U, Gschwind RM, Kollas AK, Beck E, Wiesner J, Eberl M, Jomaa H. Identification of (E)-4-hydroxy-3-methyl-but-2-enyl pyrophosphate as a major activator for human gamma-delta T cells in *Escherichia coli*. FEBS Lett. 2001;509:317–22.
43. Sandstrom A, Peigné CM, Léger A, Crooks JE, Konczak F, Gesnel MC, Breathnach R, Bonneville M, Scotet E, Adams EJ. The intracellular B30.2 domain of butyrophilin 3A1 binds phosphoantigens to mediate activation of human Vγ9Vδ2 T cells. Immunity. 2014;40:490–500.
44. Gu S, Sachleben JR, Boughter CT, Nawrocka WI, Borowska MT, Tarrasch JT, Skiniotis G, Roux B, Adams EJ. Phosphoantigen-induced conformational change of butyrophilin 3A1 (BTN3A1) and its implication on Vγ9Vδ2 T cell activation. Proc Natl Acad Sci USA. 2017;114:E7311–e20.
45. Peigné CM, Léger A, Gesnel MC, Konczak F, Olive D, Bonneville M, Breathnach R, Scotet E. The Juxtamembrane domain of Butyrophilin BTN3A1 controls phosphoantigen-mediated activation of human Vγ9Vδ2 T cells. J Immunol. 2017;198:4228–34.
46. Yang Y, Li L, Yuan L, Zhou X, Duan J, Xiao H, Cai N, Han S, Ma X, Liu W, Chen CC, Wang L, Li X, Chen J, Kang N, Chen J, Shen Z, Malwal SR, Liu W, Shi Y, Oldfield E, Guo RT, Zhang Y. A structural change in butyrophilin upon phosphoantigen binding underlies phosphoantigen-mediated Vγ9Vδ2 T cell activation. Immunity. 2019;50:1043–53.e5.
47. Nguyen K, Li J, Puthenveetil R, Lin X, Poe MM, Hsiao CC, Vinogradova O, Wiemer AJ. The butyrophilin 3A1 intracellular domain undergoes a conformational change involving the juxtamembrane region. FASEB J. 2017;31:4697–706.
48. Rigau M, Ostrouska S, Fulford TS, Johnson DN, Woods K, Ruan Z, McWilliam HEG, Hudson C, Tutuka C, Wheatley AK, Kent SJ, Villadangos JA, Pal B, Kurts C, Simmonds J, Pelzing M, Nash AD, Hammet A, Verhagen AM, Vairo G, Maraskovsky E, Panousis C, Gherardin NA, Cebon J, Godfrey DI, Behren A, Uldrich AP. Butyrophilin 2A1 is essential for phosphoantigen reactivity by gammadelta T cells. Science. 2020;367:eaay5516.
49. Scheper W, van Dorp S, Kersting S, Pietersma F, Lindemans C, Hol S, Heijhuurs S, Sebestyen Z, Gründer C, Marcu-Malina V, Marchant A, Donner C, Plachter B, Vermijlen D, van Baarle D, Kuball J. γδT cells elicited by CMV reactivation after allo-SCT cross-recognize CMV and leukemia. Leukemia. 2013;27:1328–38.
50. Déchanet J, Merville P, Bergé F, Bone-Mane G, Taupin JL, Michel P, Joly P, Bonneville M, Potaux L, Moreau JF. Major expansion of gammadelta T lymphocytes following cytomegalovirus infection in kidney allograft recipients. J Infect Dis. 1999;179:1–8.
51. Déchanet J, Merville P, Lim A, Retière C, Pitard V, Lafarge X, Michelson S, Méric C, Hallet MM, Kourilsky P, Potaux L, Bonneville M, Moreau JF. Implication of gammadelta T cells in the human immune response to cytomegalovirus. J Clin Invest. 1999;103:1437–49.
52. Scheper W, Sebestyen Z, Kuball J. Cancer immunotherapy using γδT cells: dealing with diversity. Front Immunol. 2014;5:601.
53. Tanaka Y, Kobayashi H, Terasaki T, Toma H, Aruga A, Uchiyama T, Mizutani K, Mikami B, Morita CT, Minato N. Synthesis of pyrophosphate-containing compounds that stimulate Vgamma2Vdelta2 T cells: application to cancer immunotherapy. Med Chem. 2007;3:85–99.
54. Wang RN, Wen Q, He WT, Yang JH, Zhou CY, Xiong WJ, Ma L. Optimized protocols for γδ T cell expansion and lentiviral transduction. Mol Med Rep. 2019;19:1471–80.
55. Burjanadzé M, Condomines M, Reme T, Quittet P, Latry P, Lugagne C, Romagne F, Morel Y, Rossi JF, Klein B, Lu ZY. *In vitro* expansion of gamma delta T cells with anti-myeloma cell activity by Phosphostim and IL-2 in patients with multiple myeloma. Br J Haematol. 2007;139:206–16.
56. Salot S, Laplace C, Saïagh S, Bercegeay S, Tenaud I, Cassidanius A, Romagne F, Dreno B, Tiollier J. Large scale expansion of gamma 9 delta 2 T lymphocytes: innacell gamma delta cell therapy product. J Immunol Methods. 2007;326:63–75.

57. Deniger DC, Moyes JS, Cooper LJ. Clinical applications of gamma delta T cells with multivalent immunity. Front Immunol. 2014;5:636.
58. Kondo M, Sakuta K, Noguchi A, Ariyoshi N, Sato K, Sato S, Sato K, Hosoi A, Nakajima J, Yoshida Y, Shiraishi K, Nakagawa K, Kakimi K. Zoledronate facilitates large-scale ex vivo expansion of functional gammadelta T cells from cancer patients for use in adoptive immunotherapy. Cytotherapy. 2008;10:842–56.
59. Kondo M, Izumi T, Fujieda N, Kondo A, Morishita T, Matsushita H, Kakimi K. Expansion of human peripheral blood gammadelta T cells using zoledronate. J Vis Exp. 2011;55:3182.
60. Lin JH. Bisphosphonates: a review of their pharmacokinetic properties. Bone. 1996;18:75–85.
61. Sicard H, Ingoure S, Luciani B, Serraz C, Fournié JJ, Bonneville M, Tiollier J, Romagné F. *In vivo* immunomanipulation of Vgamma9 Vdelta 2 T cells with a synthetic phosphoantigen in a preclinical nonhuman primate model. J Immunol. 2005;175:5471–80.
62. Bennouna J, Levy V, Sicard H, Senellart H, Audrain M, Hiret S, Rolland F, Bruzzoni-Giovanelli H, Rimbert M, Galea C, Tiollier J, Calvo F. Phase I study of bromohydrin pyrophosphate (BrHPP, IPH 1101), a Vgamma9Vdelta2 T lymphocyte agonist in patients with solid tumors. Cancer Immunol Immunother. 2010;59:1521–30.
63. Fournié JJ, Sicard H, Poupot M, Bezombes C, Blanc A, Romagné F, Ysebaert L, Laurent G. What lessons can be learned from γδ T cell-based cancer immunotherapy trials? Cell Mol Immunol. 2013;10:35–41.
64. Palakodeti A, Sandstrom A, Sundaresan L, Harly C, Nedellec S, Olive D, Scotet E, Bonneville M, Adams EJ. The molecular basis for modulation of human Vγ9Vδ2 T cell responses by CD277/butyrophilin-3 (BTN3A)-specific antibodies. J Biol Chem. 2012;287:32780–90.
65. Harly C, Guillaume Y, Nedellec S, Peigné CM, Mönkkönen H, Mönkkönen J, Li J, Kuball J, Adams EJ, Netzer S, Déchanet-Merville J, Léger A, Herrmann T, Breathnach R, Olive D, Bonneville M, Scotet E. Key implication of CD277/butyrophilin-3 (BTN3A) in cellular stress sensing by a major human γδ T-cell subset. Blood. 2012;120:2269–79.
66. De Gassart A, Le KS, Brune P, Agaugué S, Sims J, Goubard A, Castellano R, Joalland N, Scotet E, Collette Y, Valentin E, Ghigo C, Pasero C, Colazet M, Guillén J, Cano CE, Marabelle A, De Bonno J, Hoet R, Truneh A, Olive D, Frohna P. Development of ICT01, a first-in-class, anti-BTN3A antibody for activating Vγ9Vδ2 T cell-mediated antitumor immune response. Sci Transl Med. 2021;13:eabj0835.
67. Melandri D, Zlatareva I, Chaleil RAG, Dart RJ, Chancellor A, Nussbaumer O, Polyakova O, Roberts NA, Wesch D, Kabelitz D, Irving PM, John S, Mansour S, Bates PA, Vantourout P, Hayday AC. The γδTCR combines innate immunity with adaptive immunity by utilizing spatially distinct regions for agonist selection and antigen responsiveness. Nat Immunol. 2018;19:1352–65.
68. Altvater B, Pscherer S, Landmeier S, Kailayangiri S, Savoldo B, Juergens H, Rossig C. Activated human γδ T cells induce peptide-specific CD8+ T-cell responses to tumor-associated self-antigens. Cancer Immunol Immunother. 2012;61:385–96.
69. Brandes M, Willimann K, Bioley G, Levy N, Eberl M, Luo M, Tampe R, Levy F, Romero P, Moser B. Cross-presenting human gamma-delta T cells induce robust $CD8^+$ alpha-beta T cell responses. Proc Natl Acad Sci USA. 2009;106:2307–12.
70. Brandes M, Willimann K, Moser B. Professional antigen-presentation function by human gamma-delta T cells. Science. 2005;309:264–8.
71. Rincon-Orozco B, Kunzmann V, Wrobel P, Kabelitz D, Steinle A, Herrmann T. Activation of V gamma9 Vdelta2 T cells by NKG2D. J Immunol. 2005;175:2144–51.
72. Wrobel P, Shojaei H, Schittek B, Gieseler F, Wollenberg B, Kalthoff H, Kabelitz D, Wesch D. Lysis of a broad range of epithelial tumor cells by human gamma delta T cells: involvement of NKG2D ligands and T cell receptor- versus NKG2D-dependent recognition. Scand J Immunol. 2007;66:320–8.
73. Toutirais O, Cabillic F, Le Friec G, Salot S, Loyer P, Le Gallo M, Desille M, de La Pintiere CT, Daniel P, Bouet F, Catros V. DNAX accessory molecule-1 (CD226) promotes human hepatocellular carcinoma cell lysis by Vgamma9Vdelta2 T cells. Eur J Immunol. 2009;39:1361–8.

74. Angelini DF, Borsellino G, Poupot M, Diamantini A, Poupot R, Bernardi G, Poccia F, Fournié JJ, Battistini L. FcgammaRIII discriminates between 2 subsets of Vgamma9Vdelta2 effector cells with different responses and activation pathways. Blood. 2004;104:1801–7.
75. Gertner-Dardenne J, Bonnafous C, Bezombes C, Capietto AH, Scaglione V, Ingoure S, Cendron D, Gross E, Lepage JF, Quillet-Mary A, Ysebaert L, Laurent G, Sicard H, Fournié JJ. Bromohydrin pyrophosphate enhances antibody-dependent cell-mediated cytotoxicity induced by therapeutic antibodies. Blood. 2009;113:4875–84.
76. Tokuyama H, Hagi T, Mattarollo SR, Morley J, Wang Q, So HF, Moriyasu F, Nieda M, Nicol AJ. V gamma 9 V delta 2 T cell cytotoxicity against tumor cells is enhanced by monoclonal antibody drugs--rituximab and trastuzumab. Int J Cancer. 2008;122:2526–34.
77. Fulford TS, Soliman C, Castle RG, Rigau M, Ruan Z, Dolezal O, Seneviratna R, Brown HG, Hanssen E, Hammet A, Li S, Redmond SJ, Chung A, Gorman MA, Parker MW, Patel O, Peat TS, Newman J, Behren A, Gherardin NA, Godfrey DI, Uldrich AP. Vgamma9Vdelta2 T cells recognize butyrophilin 2A1 and 3A1 heteromers. Nat Immunol. 2024;25:1355.
78. Carding SR, Egan PJ. Gammadelta T cells: functional plasticity and heterogeneity. Nat Rev Immunol. 2002;2:336–45.
79. Ravens S, Schultze-Florey C, Raha S, Sandrock I, Drenker M, Oberdörfer L, Reinhardt A, Ravens I, Beck M, Geffers R, von Kaisenberg C, Heuser M, Thol F, Ganser A, Förster R, Koenecke C, Prinz I. Human γδ T cells are quickly reconstituted after stem-cell transplantation and show adaptive clonal expansion in response to viral infection. Nat Immunol. 2017;18:393–401.
80. Di Lorenzo B, Ravens S, Silva-Santos B. High-throughput analysis of the human thymic Vδ1$^+$ T cell receptor repertoire. Sci Data. 2019;6:115.
81. Spada FM, Grant EP, Peters PJ, Sugita M, Melián A, Leslie DS, Lee HK, van Donselaar E, Hanson DA, Krensky AM, Majdic O, Porcelli SA, Morita CT, Brenner MB. Self-recognition of CD1 by gamma/delta T cells: implications for innate immunity. J Exp Med. 2000;191:937–48.
82. Leslie DS, Vincent MS, Spada FM, Das H, Sugita M, Morita CT, Brenner MB. CD1-mediated gamma/delta T cell maturation of dendritic cells. J Exp Med. 2002;196:1575–84.
83. Uldrich AP, Le Nours J, Pellicci DG, Gherardin NA, McPherson KG, Lim RT, Patel O, Beddoe T, Gras S, Rossjohn J, Godfrey DI. CD1d-lipid antigen recognition by the γδ TCR. Nat Immunol. 2013;14:1137–45.
84. Luoma AM, Castro CD, Mayassi T, Bembinster LA, Bai L, Picard D, Anderson B, Scharf L, Kung JE, Sibener LV, Savage PB, Jabri B, Bendelac A, Adams EJ. Crystal structure of Vδ1 T cell receptor in complex with CD1d-sulfatide shows MHC-like recognition of a self-lipid by human γδ T cells. Immunity. 2013;39:1032–42.
85. Bauer S, Groh V, Wu J, Steinle A, Phillips JH, Lanier LL, Spies T. Activation of NK cells and T cells by NKG2D, a receptor for stress-inducible MICA. Science. 1999;285:727–9.
86. Poggi A, Venturino C, Catellani S, Clavio M, Miglino M, Gobbi M, Steinle A, Ghia P, Stella S, Caligaris-Cappio F, Zocchi MR. Vdelta1 T lymphocytes from B-CLL patients recognize ULBP3 expressed on leukemic B cells and upregulated by trans-retinoic acid. Cancer Res. 2004;64:9172–9.
87. Knight A, Mackinnon S, Lowdell MW. Human Vdelta1 gamma-delta T cells exert potent specific cytotoxicity against primary multiple myeloma cells. Cytotherapy. 2012;14:1110–8.
88. Mikulak J, Oriolo F, Bruni E, Roberto A, Colombo FS, Villa A, Bosticardo M, Bortolomai I, Lo Presti E, Meraviglia S, Dieli F, Vetrano S, Danese S, Della Bella S, Carvello MM, Sacchi M, Cugini G, Colombo G, Klinger M, Spaggiari P, Roncalli M, Prinz I, Ravens S, di Lorenzo B, Marcenaro E, Silva-Santos B, Spinelli A, Mavilio D. NKp46-expressing human gut-resident intraepithelial Vdelta1 T cell subpopulation exhibits high antitumor activity against colorectal cancer. JCI Insight. 2019;4:e125884.
89. Correia DV, Fogli M, Hudspeth K, da Silva MG, Mavilio D, Silva-Santos B. Differentiation of human peripheral blood Vδ1+ T cells expressing the natural cytotoxicity receptor NKp30 for recognition of lymphoid leukemia cells. Blood. 2011;118:992–1001.

90. Almeida AR, Correia DV, Fernandes-Platzgummer A, da Silva CL, da Silva MG, Anjos DR, Silva-Santos B. Delta one T cells for immunotherapy of chronic lymphocytic leukemia: clinical-grade expansion/differentiation and preclinical proof of concept. Clin Cancer Res. 2016;22:5795–804.
91. Sanz M, Mann BT, Ryan PL, Bosque A, Pennington DJ, Hackstein H, Soriano-Sarabia N. Deep characterization of human γδ T cell subsets defines shared and lineage-specific traits. Front Immunol. 2023;14:1148988.
92. Ebert LM, Meuter S, Moser B. Homing and function of human skin gamma-delta T cells and NK cells: relevance for tumor surveillance. J Immunol. 2006;176:4331–6.
93. Zakeri N, Hall A, Swadling L, Pallett LJ, Schmidt NM, Diniz MO, Kucykowicz S, Amin OE, Gander A, Pinzani M, Davidson BR, Quaglia A, Maini MK. Characterization and induction of tissue-resident gamma delta T cells to target hepatocellular carcinoma. Nat Commun. 2022;13:1372.
94. Brennan PJ, Tatituri RV, Brigl M, Kim EY, Tuli A, Sanderson JP, Gadola SD, Hsu FF, Besra GS, Brenner MB. Invariant natural killer T cells recognize lipid self antigen induced by microbial danger signals. Nat Immunol. 2011;12:1202–11.
95. Diana J, Lehuen A. NKT cells: friend or foe during viral infections? Eur J Immunol. 2009;39:3283–91.
96. Subleski JJ, Jiang Q, Weiss JM, Wiltrout RH. The split personality of NKT cells in malignancy, autoimmune and allergic disorders. Immunotherapy. 2011;3:1167–84.
97. Paul S, Chhatar S, Mishra A, Lal G. Natural killer T cell activation increases iNOS(+) CD206(−) M1 macrophage and controls the growth of solid tumor. J Immunother Cancer. 2019;7:208.
98. Baiu DC, Sharma A, Schehr JL, Basu J, Smith KA, Ohashi M, Johannsen EC, Kenney SC, Gumperz JE. Human CD4(+) iNKT cell adoptive immunotherapy induces anti-tumor responses against CD1d-negative EBV-driven B lymphoma. Immunology. 2024;172:627–40.
99. Hammond TC, Purbhoo MA, Kadel S, Ritz J, Nikiforow S, Daley H, Shaw K, van Besien K, Gomez-Arteaga A, Stevens D, Ortuzar W, Michelet X, Smith R, Moskowitz D, Masakayan R, Yigit B, Boi S, Soh KT, Chamberland J, Song X, Qin Y, Mishchenko I, Kirby M, Nasonenko V, Buffa A, Buell JS, Chand D, van Dijk M, Stebbing J, Exley MA. A phase 1/2 clinical trial of invariant natural killer T cell therapy in moderate-severe acute respiratory distress syndrome. Nat Commun. 2024;15:974.
100. Townsend MJ, Weinmann AS, Matsuda JL, Salomon R, Farnham PJ, Biron CA, Gapin L, Glimcher LH. T-bet regulates the terminal maturation and homeostasis of NK and Valpha14i NKT cells. Immunity. 2004;20:477–94.
101. Kovalovsky D, Uche OU, Eladad S, Hobbs RM, Yi W, Alonzo E, Chua K, Eidson M, Kim HJ, Im JS, Pandolfi PP, Sant'Angelo DB. The BTB-zinc finger transcriptional regulator PLZF controls the development of invariant natural killer T cell effector functions. Nat Immunol. 2008;9:1055–64.
102. Savage AK, Constantinides MG, Han J, Picard D, Martin E, Li B, Lantz O, Bendelac A. The transcription factor PLZF directs the effector program of the NKT cell lineage. Immunity. 2008;29:391–403.
103. Terashima A, Watarai H, Inoue S, Sekine E, Nakagawa R, Hase K, Iwamura C, Nakajima H, Nakayama T, Taniguchi M. A novel subset of mouse NKT cells bearing the IL-17 receptor B responds to IL-25 and contributes to airway hyperreactivity. J Exp Med. 2008;205:2727–33.
104. Watarai H, Sekine-Kondo E, Shigeura T, Motomura Y, Yasuda T, Satoh R, Yoshida H, Kubo M, Kawamoto H, Koseki H, Taniguchi M. Development and function of invariant natural killer T cells producing T(h)2- and T(h)17-cytokines. PLoS Biol. 2012;10:e1001255.
105. Stock P, Lombardi V, Kohlrautz V, Akbari O. Induction of airway hyperreactivity by IL-25 is dependent on a subset of invariant NKT cells expressing IL-17RB. J Immunol. 2009;182:5116–22.
106. Kim PJ, Pai SY, Brigl M, Besra GS, Gumperz J, Ho IC. GATA-3 regulates the development and function of invariant NKT cells. J Immunol. 2006;177:6650–9.

107. Rachitskaya AV, Hansen AM, Horai R, Li Z, Villasmil R, Luger D, Nussenblatt RB, Caspi RR. Cutting edge: NKT cells constitutively express IL-23 receptor and RORgammat and rapidly produce IL-17 upon receptor ligation in an IL-6-independent fashion. J Immunol. 2008;180:5167–71.
108. Michel ML, Keller AC, Paget C, Fujio M, Trottein F, Savage PB, Wong CH, Schneider E, Dy M, Leite-de-Moraes MC. Identification of an IL-17-producing NK1.1(neg) iNKT cell population involved in airway neutrophilia. J Exp Med. 2007;204:995–1001.
109. Coquet JM, Chakravarti S, Kyparissoudis K, McNab FW, Pitt LA, McKenzie BS, Berzins SP, Smyth MJ, Godfrey DI. Diverse cytokine production by NKT cell subsets and identification of an IL-17-producing CD4-NK1.1- NKT cell population. Proc Natl Acad Sci USA. 2008;105:11287–92.
110. Chang PP, Barral P, Fitch J, Pratama A, Ma CS, Kallies A, Hogan JJ, Cerundolo V, Tangye SG, Bittman R, Nutt SL, Brink R, Godfrey DI, Batista FD, Vinuesa CG. Identification of Bcl-6-dependent follicular helper NKT cells that provide cognate help for B cell responses. Nat Immunol. 2011;13:35–43.
111. King IL, Fortier A, Tighe M, Dibble J, Watts GF, Veerapen N, Haberman AM, Besra GS, Mohrs M, Brenner MB, Leadbetter EA. Invariant natural killer T cells direct B cell responses to cognate lipid antigen in an IL-21-dependent manner. Nat Immunol. 2011;13:44–50.
112. Monteiro M, Almeida CF, Caridade M, Ribot JC, Duarte J, Agua-Doce A, Wollenberg I, Silva-Santos B, Graca L. Identification of regulatory Foxp3+ invariant NKT cells induced by TGF-beta. J Immunol. 2010;185:2157–63.
113. Schipper HS, Rakhshandehroo M, van de Graaf SF, Venken K, Koppen A, Stienstra R, Prop S, Meerding J, Hamers N, Besra G, Boon L, Nieuwenhuis EE, Elewaut D, Prakken B, Kersten S, Boes M, Kalkhoven E. Natural killer T cells in adipose tissue prevent insulin resistance. J Clin Invest. 2012;122:3343–54.
114. Lynch L, Nowak M, Varghese B, Clark J, Hogan AE, Toxavidis V, Balk SP, O'Shea D, O'Farrelly C, Exley MA. Adipose tissue invariant NKT cells protect against diet-induced obesity and metabolic disorder through regulatory cytokine production. Immunity. 2012;37:574–87.
115. Bosma A, Abdel-Gadir A, Isenberg DA, Jury EC, Mauri C. Lipid-antigen presentation by CD1d(+) B cells is essential for the maintenance of invariant natural killer T cells. Immunity. 2012;36:477–90.
116. Meermeier EW, Harriff MJ, Karamooz E, Lewinsohn DM. MAIT cells and microbial immunity. Immunol Cell Biol. 2018;96:607–17.
117. Hackstein C-P, Klenerman P. Emerging features of MAIT cells and other unconventional T cell populations in human viral disease and vaccination. Semin Immunol. 2022;61–64:101661.
118. Corbett AJ, Awad W, Wang H, Chen Z. Antigen recognition by MR1-reactive T cells; MAIT cells, metabolites, and remaining mysteries. Front Immunol. 2020;11:1961.
119. Legoux F, Salou M, Lantz O. MAIT cell development and functions: the microbial connection. Immunity. 2020;53:710–23.
120. El Morr Y, Furstenheim M, Mestdagh M, Franciszkiewicz K, Salou M, Morvan C, Dupre T, Vorobev A, Jneid B, Premel V, Darbois A, Perrin L, Mondot S, Colombeau L, Bugaut H, du Halgouet A, Richon S, Procopio E, Maurin M, Philippe C, Rodriguez R, Lantz O, Legoux F. MAIT cells monitor intestinal dysbiosis and contribute to host protection during colitis. Sci Immunol. 2024;9:eadi8954.
121. Hinks TSC, Zhang XW. MAIT cell activation and functions. Front Immunol. 2020;11:1014.
122. Lamichhane R, Schneider M, de la Harpe SM, Harrop TWR, Hannaway RF, Dearden PK, Kirman JR, Tyndall JDA, Vernall AJ, Ussher JE. TCR- or cytokine-activated CD8+ mucosal-associated invariant T cells are rapid polyfunctional effectors that can coordinate immune responses. Cell Rep. 2019;28:3061–76.e5.
123. Kurioka A, Walker LJ, Klenerman P, Willberg CB. MAIT cells: new guardians of the liver. Clin Transl Immunol. 2016;5:e98.

124. Gebru YA, Choi MR, Raja G, Gupta H, Sharma SP, Choi YR, Kim HS, Yoon SJ, Kim DJ, Suk KT. Pathophysiological roles of mucosal-associated invariant T cells in the context of gut microbiota-liver axis. Microorganisms. 2021;9:296.
125. Eckle SB, Corbett AJ, Keller AN, Chen Z, Godfrey DI, Liu L, Mak JY, Fairlie DP, Rossjohn J, McCluskey J. Recognition of vitamin B precursors and byproducts by mucosal associated invariant T cells. J Biol Chem. 2015;290:30204–11.
126. Krawic JR, Ladd NA, Cansler M, McMurtrey C, Devereaux J, Worley A, Ahmed T, Froyd C, Kulicke CA, Swarbrick G, Nilsen A, Lewinsohn DM, Adams EJ, Hildebrand W. Multiple isomers of Photolumazine V bind MR1 and differentially activate MAIT cells. J Immunol. 2024;212(6):933–40.
127. Ito E, Inuki S, Izumi Y, Takahashi M, Dambayashi Y, Ciacchi L, Awad W, Takeyama A, Shibata K, Mori S, Mak JYW, Fairlie DP, Bamba T, Ishikawa E, Nagae M, Rossjohn J, Yamasaki S. Sulfated bile acid is a host-derived ligand for MAIT cells. Sci Immunol. 2024;9:eade6924.
128. Wong EB, Ndung'u T, Kasprowicz VO. The role of mucosal-associated invariant T cells in infectious diseases. Immunology. 2017;150:45–54.
129. Gold MC, Napier RJ, Lewinsohn DM. MR1-restricted mucosal-associated invariant T (MAIT) cells in the immune response to *Mycobacterium tuberculosis*. Immunol Rev. 2015;264:154–66.
130. Amini A, Pang D, Hackstein CP, Klenerman P. MAIT cells in barrier tissues: lessons from immediate neighbors. Front Immunol. 2020;11:584521.
131. Godfrey DI, Le Nours J, Andrews DM, Uldrich AP, Rossjohn J. Unconventional T cell targets for cancer immunotherapy. Immunity. 2018;48:453–73.
132. Dagher OK, Posey AD Jr. Forks in the road for CAR T and CAR NK cell cancer therapies. Nat Immunol. 2023;24:1994–2007.
133. Yi M, Li T, Niu M, Mei Q, Zhao B, Chu Q, Dai Z, Wu K. Exploiting innate immunity for cancer immunotherapy. Mol Cancer. 2023;22:187.
134. Trottein F, Paget C. Natural killer T cells and mucosal-associated invariant T cells in lung infections. Front Immunol. 2018;9:1750.
135. Singhal A, Mori L, De Libero G. T cell recognition of non-peptidic antigens in infectious diseases. Indian J Med Res. 2013;138:620–31.
136. Park JH, Lee HK. Function of γδ T cells in tumor immunology and their application to cancer therapy. Exp Mol Med. 2021;53:318–27.
137. Coffelt SB, Kersten K, Doornebal CW, Weiden J, Vrijland K, Hau CS, Verstegen NJM, Ciampricotti M, Hawinkels L, Jonkers J, de Visser KE. IL-17-producing γδ T cells and neutrophils conspire to promote breast cancer metastasis. Nature. 2015;522:345–8.
138. Rei M, Pennington DJ, Silva-Santos B. The emerging Protumor role of γδ T lymphocytes: implications for cancer immunotherapy. Cancer Res. 2015;75:798–802.
139. Papotto PH, Ribot JC, Silva-Santos B. IL-17(+) γδ T cells as kick-starters of inflammation. Nat Immunol. 2017;18:604–11.
140. Dimova T, Brouwer M, Gosselin F, Tassignon J, Leo O, Donner C, Marchant A, Vermijlen D. Effector Vγ9Vδ2 T cells dominate the human fetal γδ T-cell repertoire. Proc Natl Acad Sci USA. 2015;112:E556–65.
141. Gibbons DL, Haque SF, Silberzahn T, Hamilton K, Langford C, Ellis P, Carr R, Hayday AC. Neonates harbour highly active gamma-delta T cells with selective impairments in preterm infants. Eur J Immunol. 2009;39:1794–806.
142. Gentles AJ, Newman AM, Liu CL, Bratman SV, Feng W, Kim D, Nair VS, Xu Y, Khuong A, Hoang CD, Diehn M, West RB, Plevritis SK, Alizadeh AA. The prognostic landscape of genes and infiltrating immune cells across human cancers. Nat Med. 2015;21:938–45.
143. Tosolini M, Pont F, Poupot M, Vergez F, Nicolau-Travers ML, Vermijlen D, Sarry JE, Dieli F, Fournié JJ. Assessment of tumor-infiltrating TCRVγ9Vδ2 γδ lymphocyte abundance by deconvolution of human cancers microarrays. Onco Targets Ther. 2017;6:e1284723.
144. Godder KT, Henslee-Downey PJ, Mehta J, Park BS, Chiang KY, Abhyankar S, Lamb LS. Long term disease-free survival in acute leukemia patients recovering with increased

gammadelta T cells after partially mismatched related donor bone marrow transplantation. Bone Marrow Transplant. 2007;39:751–7.
145. Wang J, Lin C, Li H, Li R, Wu Y, Liu H, Zhang H, He H, Zhang W, Xu J. Tumor-infiltrating γδT cells predict prognosis and adjuvant chemotherapeutic benefit in patients with gastric cancer. Onco Targets Ther. 2017;6:e1353858.
146. Ribot JC, Ribeiro ST, Correia DV, Sousa AE, Silva-Santos B. Human γδ thymocytes are functionally immature and differentiate into cytotoxic type 1 effector T cells upon IL-2/IL-15 signaling. J Immunol. 2014;192:2237–43.
147. Lee D, Rosenthal CJ, Penn NE, Dunn ZS, Zhou Y, Yang L. Human γδ T cell subsets and their clinical applications for cancer immunotherapy. Cancers (Basel). 2022;14:3005.
148. Alexander AA, Maniar A, Cummings JS, Hebbeler AM, Schulze DH, Gastman BR, Pauza CD, Strome SE, Chapoval AI. Isopentenyl pyrophosphate-activated CD56+ {gamma}{delta} T lymphocytes display potent antitumor activity toward human squamous cell carcinoma. Clin Cancer Res. 2008;14:4232–40.
149. Todaro M, D'Asaro M, Caccamo N, Iovino F, Francipane MG, Meraviglia S, Orlando V, La Mendola C, Gulotta G, Salerno A, Dieli F, Stassi G. Efficient killing of human colon cancer stem cells by gammadelta T lymphocytes. J Immunol. 2009;182:7287–96.
150. Dokouhaki P, Schuh NW, Joe B, Allen CA, Der SD, Tsao MS, Zhang L. NKG2D regulates production of soluble TRAIL by ex vivo expanded human γδ T cells. Eur J Immunol. 2013;43:3175–82.
151. Zhao Y, Niu C, Cui J. Gamma-delta (γδ) T cells: friend or foe in cancer development? J Transl Med. 2018;16:3.
152. Couzi L, Pitard V, Sicard X, Garrigue I, Hawchar O, Merville P, Moreau JF, Déchanet-Merville J. Antibody-dependent anti-cytomegalovirus activity of human γδ T cells expressing CD16 (FcγRIIIa). Blood. 2012;119:1418–27.
153. Capietto AH, Martinet L, Fournié JJ. Stimulated γδ T cells increase the *in vivo* efficacy of trastuzumab in HER-2^{+} breast cancer. J Immunol. 2011;187:1031–8.
154. Fisher JP, Yan M, Heuijerjans J, Carter L, Abolhassani A, Frosch J, Wallace R, Flutter B, Capsomidis A, Hubank M, Klein N, Callard R, Gustafsson K, Anderson J. Neuroblastoma killing properties of Vδ2 and Vδ2-negative γδT cells following expansion by artificial antigen-presenting cells. Clin Cancer Res. 2014;20:5720–32.
155. Gao Y, Yang W, Pan M, Scully E, Girardi M, Augenlicht LH, Craft J, Yin Z. Gamma delta T cells provide an early source of interferon-gamma in tumor immunity. J Exp Med. 2003;198:433–42.
156. Riond J, Rodriguez S, Nicolau ML, al Saati T, Gairin JE. *In vivo* major histocompatibility complex class I (MHCI) expression on MHCIlow tumor cells is regulated by gamma-delta T and NK cells during the early steps of tumor growth. Cancer Immun. 2009;9:10.
157. Holmen Olofsson G, Idorn M, Carnaz Simões AM, Aehnlich P, Skadborg SK, Noessner E, Debets R, Moser B, Met Ö, Thor SP. Vγ9Vδ2 T cells concurrently kill cancer cells and cross-present tumor antigens. Front Immunol. 2021;12:645131.
158. Caccamo N, Battistini L, Bonneville M, Poccia F, Fournié JJ, Meraviglia S, Borsellino G, Kroczek RA, La Mendola C, Scotet E, Dieli F, Salerno A. CXCR5 identifies a subset of Vgamma9Vdelta2 T cells which secrete IL-4 and IL-10 and help B cells for antibody production. J Immunol. 2006;177:5290–5.
159. Rampoldi F, Ullrich L, Prinz I. Revisiting the interaction of gamma-delta T cells and B cells. Cells. 2020;9:743.
160. Bansal RR, Mackay CR, Moser B, Eberl M. IL-21 enhances the potential of human γδ T cells to provide B-cell help. Eur J Immunol. 2012;42:110–9.
161. Rezende RM, Lanser AJ, Rubino S, Kuhn C, Skillin N, Moreira TG, Liu S, Gabriely G, David BA, Menezes GB, Weiner HL. γδ T cells control humoral immune response by inducing T follicular helper cell differentiation. Nat Commun. 2018;9:3151.
162. Maniar A, Zhang X, Lin W, Gastman BR, Pauza CD, Strome SE, Chapoval AI. Human gamma-delta T lymphocytes induce robust NK cell-mediated antitumor cytotoxicity through CD137 engagement. Blood. 2010;116:1726–33.

163. Wu P, Wu D, Ni C, Ye J, Chen W, Hu G, Wang Z, Wang C, Zhang Z, Xia W, Chen Z, Wang K, Zhang T, Xu J, Han Y, Zhang T, Wu X, Wang J, Gong W, Zheng S, Qiu F, Yan J, Huang J. γδT17 cells promote the accumulation and expansion of myeloid-derived suppressor cells in human colorectal cancer. Immunity. 2014;40:785–800.
164. Hoshino H, Lötvall J, Skoogh BE, Lindén A. Neutrophil recruitment by interleukin-17 into rat airways *in vivo*. Role of tachykinins. Am J Respir Crit Care Med. 1999;159:1423–8.
165. Li J, Zeng Q, Zhang Y, Li X, Hu H, Miao X, Yang W, Zhang W, Song X, Mou L, Wang R. Neurokinin-1 receptor mediated breast cancer cell migration by increased expression of MMP-2 and MMP-14. Eur J Cell Biol. 2016;95:368–77.
166. Meraviglia S, Lo Presti E, Tosolini M, La Mendola C, Orlando V, Todaro M, Catalano V, Stassi G, Cicero G, Vieni S, Fourniè JJ, Dieli F. Distinctive features of tumor-infiltrating γδ T lymphocytes in human colorectal cancer. Onco Targets Ther. 2017;6:e1347742.
167. Chen X, Shang W, Xu R, Wu M, Zhang X, Huang P, Wang F, Pan S. Distribution and functions of γδ T cells infiltrated in the ovarian cancer microenvironment. J Transl Med. 2019;17:144.
168. Lo Presti E, Toia F, Oieni S, Buccheri S, Turdo A, Mangiapane LR, Campisi G, Caputo V, Todaro M, Stassi G, Cordova A, Moschella F, Rinaldi G, Meraviglia S, Dieli F. Squamous cell tumors recruit γδ T cells producing either IL17 or IFNγ depending on the tumor stage. Cancer Immunol Res. 2017;5:397–407.
169. Reis BS, Darcy PW, Khan IZ, Moon CS, Kornberg AE, Schneider VS, Alvarez Y, Eleso O, Zhu C, Schernthanner M, Lockhart A, Reed A, Bortolatto J, Castro TBR, Bilate AM, Grivennikov S, Han AS, Mucida D. TCR-Vγδ usage distinguishes protumor from antitumor intestinal γδ T cell subsets. Science. 2022;377:276–84.
170. Mao Y, Yin S, Zhang J, Hu Y, Huang B, Cui L, Kang N, He W. A new effect of IL-4 on human γδ T cells: promoting regulatory Vδ1 T cells via IL-10 production and inhibiting function of Vδ2 T cells. Cell Mol Immunol. 2016;13:217–28.
171. Hao J, Dong S, Xia S, He W, Jia H, Zhang S, Wei J, O'Brien RL, Born WK, Wu Z, Wang P, Han J, Hong Z, Zhao L, Yin Z. Regulatory role of Vγ1 γδ T cells in tumor immunity through IL-4 production. J Immunol. 2011;187:4979–86.
172. Fujii S, Shimizu K, Hemmi H, Steinman RM. Innate Valpha14(+) natural killer T cells mature dendritic cells, leading to strong adaptive immunity. Immunol Rev. 2007;220:183–98.
173. Ko HJ, Lee JM, Kim YJ, Kim YS, Lee KA, Kang CY. Immunosuppressive myeloid-derived suppressor cells can be converted into immunogenic APCs with the help of activated NKT cells: an alternative cell-based antitumor vaccine. J Immunol. 2009;182:1818–28.
174. Kim CH, Butcher EC, Johnston B. Distinct subsets of human Valpha24-invariant NKT cells: cytokine responses and chemokine receptor expression. Trends Immunol. 2002;23:516–9.
175. Cui J, Shin T, Kawano T, Sato H, Kondo E, Toura I, Kaneko Y, Koseki H, Kanno M, Taniguchi M. Requirement for Valpha14 NKT cells in IL-12-mediated rejection of tumors. Science. 1997;278:1623–6.
176. Morita M, Motoki K, Akimoto K, Natori T, Sakai T, Sawa E, Yamaji K, Koezuka Y, Kobayashi E, Fukushima H. Structure-activity relationship of alpha-galactosylceramides against B16-bearing mice. J Med Chem. 1995;38:2176–87.
177. Gebremeskel S, Clattenburg DR, Slauenwhite D, Lobert L, Johnston B. Natural killer T cell activation overcomes immunosuppression to enhance clearance of postsurgical breast cancer metastasis in mice. Onco Targets Ther. 2015;4:e995562.
178. Crowe NY, Coquet JM, Berzins SP, Kyparissoudis K, Keating R, Pellicci DG, Hayakawa Y, Godfrey DI, Smyth MJ. Differential antitumor immunity mediated by NKT cell subsets *in vivo*. J Exp Med. 2005;202:1279–88.
179. Cortesi F, Delfanti G, Grilli A, Calcinotto A, Gorini F, Pucci F, Luciano R, Grioni M, Recchia A, Benigni F, Briganti A, Salonia A, De Palma M, Bicciato S, Doglioni C, Bellone M, Casorati G, Dellabona P. Bimodal CD40/Fas-dependent crosstalk between iNKT cells and tumor-associated macrophages impairs prostate cancer progression. Cell Rep. 2018;22:3006–20.

180. Heuser-Loy C, Baumgart AK, Hackstein CP, Courreges CJF, Philipp MS, Thaiss CA, Holland T, Evaristo C, Garbi N, Kurts C. Conditional NKT cell depletion in mice reveals a negative feedback loop that regulates CTL cross-priming. J Immunol. 2024;212:35–42.
181. Xin G, Schauder DM, Jing W, Jiang A, Joshi NS, Johnson B, Cui W. Pathogen boosted adoptive cell transfer immunotherapy to treat solid tumors. Proc Natl Acad Sci. 2017;114:740–5.
182. Tachibana T, Onodera H, Tsuruyama T, Mori A, Nagayama S, Hiai H, Imamura M. Increased intratumor Valpha24-positive natural killer T cells: a prognostic factor for primary colorectal carcinomas. Clin Cancer Res. 2005;11:7322–7.
183. Tahir SM, Cheng O, Shaulov A, Koezuka Y, Bubley GJ, Wilson SB, Balk SP, Exley MA. Loss of IFN-gamma production by invariant NK T cells in advanced cancer. J Immunol. 2001;167:4046–50.
184. Singh AK, Shukla NK, Das SN. Altered invariant natural killer T cell subsets and its functions in patients with oral squamous cell carcinoma. Scand J Immunol. 2013;78:468–77.
185. Song L, Ara T, Wu HW, Woo CW, Reynolds CP, Seeger RC, DeClerck YA, Thiele CJ, Sposto R, Metelitsa LS. Oncogene MYCN regulates localization of NKT cells to the site of disease in neuroblastoma. J Clin Invest. 2007;117:2702–12.
186. Matangkasombut P, Pichavant M, DeKruyff RH, Umetsu DT. Natural killer T cells and the regulation of asthma. Mucosal Immunol. 2009;2:383–92.
187. Treiner E. Mucosal-associated invariant T cells in hematological malignancies: current knowledge, pending questions. Front Immunol. 2023;14:1160943.
188. O'Neill C, Cassidy FC, O'Shea D, Hogan AE. Mucosal associated invariant T cells in cancer-friend or foe? Cancers (Basel). 2021;13(7):1582.
189. Bacova B, Cierny J, Nemcekova L, Smetanova/Brozova J, Novak J. Systematic analysis of mucosal-associated invariant T cells in hematological malignancies. Scand J Immunol. 2024;99:e13364.
190. Sundström P, Szeponik L, Ahlmanner F, Sundquist M, Wong JSB, Lindskog EB, Gustafsson B, Quiding-Järbrink M. Tumor-infiltrating mucosal-associated invariant T (MAIT) cells retain expression of cytotoxic effector molecules. Oncotarget. 2019;10:2810–23.
191. Cogswell DT, Gapin L, Tobin HM, McCarter MD, Tobin RP. MAIT cells: partners or enemies in cancer immunotherapy? Cancers (Basel). 2021;13(7):1502.
192. Jo YG, Jin HM, Cho YN, Kim JC, Kee SJ, Park YW. Activation and impaired tumor necrosis factor-α production of circulating mucosal-associated invariant T cells in patients with trauma. J Innate Immun. 2019;11:506–15.
193. Huang W, Ye D, He W, He X, Shi X, Gao Y. Activated but impaired IFN-gamma production of mucosal-associated invariant T cells in patients with hepatocellular carcinoma. J Immunother Cancer. 2021;9:e003685.
194. Petley EV, Koay H-F, Henderson MA, Sek K, Todd KL, Keam SP, Lai J, House IG, Li J, Zethoven M, Chen AXY, Oliver AJ, Michie J, Freeman AJ, Giuffrida L, Chan JD, Pizzolla A, Mak JYW, McCulloch TR, Souza-Fonseca-Guimaraes F, Kearney CJ, Millen R, Ramsay RG, Huntington ND, McCluskey J, Oliaro J, Fairlie DP, Neeson PJ, Godfrey DI, Beavis PA, Darcy PK. MAIT cells regulate NK cell-mediated tumor immunity. Nat Commun. 2021;12:4746.
195. Krause JL, Schape SS, Schattenberg F, Muller S, Ackermann G, Rolle-Kampczyk UE, Jehmlich N, Pierzchalski A, von Bergen M, Herberth G. The activation of mucosal-associated invariant T (MAIT) cells is affected by microbial diversity and riboflavin utilization *in vitro*. Front Microbiol. 2020;11:755.
196. Jabeen MF, Hinks TSC. MAIT cells and the microbiome. Front Immunol. 2023;14:1127588.
197. Katsnelson EN, Spengler A, Domenico J, Couts KL, Loh L, Gapin L, McCarter MD, Tobin RP. Dysfunctional states of unconventional T-cell subsets in cancer. J Leukoc Biol. 2024;115:36–46.

Adoptive Cell Therapies Using Unconventional T Lymphocytes

Surojit Karmakar, Sourabh Yadav, Reshmi Suresh, and Girdhari Lal

Abstract Adoptive cell therapy (ACT) has emerged as a promising approach for the treatment of various cancers. The therapeutic regime of ACT involves the isolation and expansion of tumor-specific αβ T cells from cancer patients and the reinjection of these T cells into patients or the genetic modification of T-cell receptors (TCRs) to generate chimeric antigen receptor (CAR) T cells, which are subsequently used for ACT. In comparison, conventional ACT relies primarily on αβ T cells, recent research has expanded the scope of adoptive cellular therapies to include unconventional T-cell subsets, such as γδ T cells and invariant natural killer T (iNKT) cells or mucosal-associated invariant T (MAIT) cells. These unique T-cell populations offer distinct immunological properties and hold great promise for innovative approaches in cancer immunotherapy. γδ T cells, iNKT, and MAIT cells can recognize some unique antigen signatures on cancer cells and show greater flexibility and specificity in preferentially targeting malignant cells in the body. In this chapter, we discuss several approaches to harness the antitumor properties of unconventional T lymphocytes, such as their isolation, expansion, and genetic engineering techniques, to improve their efficacy in cancer therapeutics. We further provide insights into the opportunities and challenges of using these unconventional ACTs in recent advancements in cellular therapies to target cancer.

Keywords Gamma-delta T cells · Mucosal-associated invariant T (MAIT) cells · Natural killer (NK) cells · CAR-T cells · iNKT cells · Cancer

---

S. Karmakar · S. Yadav · R. Suresh · G. Lal (✉)
BRIC-National Centre for Cell Science (NCCS), Pune, Maharashtra, India
e-mail: glal@nccs.res.in

A. K. Mishra, S. K. Malonia (eds.), *Cell-based Immunotherapies for Cancer*,
https://doi.org/10.1007/978-3-031-88695-9_11

1 Introduction

The notion of adoptive cell therapies (ACTs) dates back to the 1980s, when Dr. Rosenberg and colleagues first described the use of IL-2-treated and expanded tumor-infiltrating lymphocytes (TILs) for the treatment of tumors in mice [1]. Later, in 1988, they initiated a clinical trial of IL-2-treated tumor-infiltrating lymphocytes (TILs) to treat patients with melanoma [2]. Since then, significant advances in ACTs have been made throughout the years, with a key breakthrough in 2002 involving the use of immune-depleting regimens before therapy to reestablish a healthy population of antitumor T cells in patients [3]. ACT involves the introduction of autologous, or in some instances, allogeneic immune cells into a patient's body that are selectively targeted against tumors or other pathological conditions [4, 5]. Before injecting immune cells into cancer patients, the ACT technique frequently involves in vitro sorting of specific immune cells, their activation, and their expansion (Fig. 1). Initially, to reduce the tumor burden, patients are given chemotherapy and/or radiation therapy, which also reduces local immune cell populations in the context of cancer treatment [6]. ACT is also critical in replenishing depleted immune cells in those settings. However, special measures must be taken with allogeneic cell transfers to avoid rejection by the recipient's immune system [7]. Although genetically modified T cells show promise in cancer immunotherapy, their efficacy is limited, particularly in solid tumors [8]. Researchers are working hard to overcome these obstacles to increase the effectiveness of ACT against solid tumors. One strategy involves the adoptive transfer of genetically modified cells, which can be especially successful against heterogeneous malignant cells expressing various tumor-associated antigens [9]. Solid tumors present additional obstacles because of the limited infiltration of adoptively transferred cells within the tumor and the immunosuppressive nature of the tumor microenvironment [10, 11]. Adoptive cellular therapies, such as chimeric antigen receptor (CAR)-T and CAR-NK cells, also have many inherent shortcomings that limit their efficiency. Some CAR-T cells are HLA-restricted and require autologous donors, sometimes limiting their bulk production [12, 13], and have been reported to have off-target effects. In contrast, although CAR-NK cells are off-the-shelf cellular therapies, they still have limitations, such as a lower division rate and survival rate in vivo, which certainly limits their efficacy [13]. As cancer immunotherapy has advanced, the focus has shifted to leveraging the unique properties of unconventional T cells for adoptive T-cell therapies [14]. γδ T cells, mucosal-associated invariant T (MAIT) cells, and natural killer T (NKT) cells are examples of unconventional T cells that have been explored in cancer therapeutics [15–19]. Owing to their non-MHC restrictions, they can be used as off-the-shelf CARs and overcome the limitations associated with the use of autologous donor cells for CAR development [20]. In addition, their longer persistence in vivo makes them attractive candidates for developing therapeutic regimens for ACTs that can potentially increase the breadth and effectiveness of cancer immunotherapies [21]. In the following section, we discuss various reports or mechanisms related to how these unconventional cells are utilized in ACTs in cancer, their advantages and disadvantages, and future aspects of their further utilization.

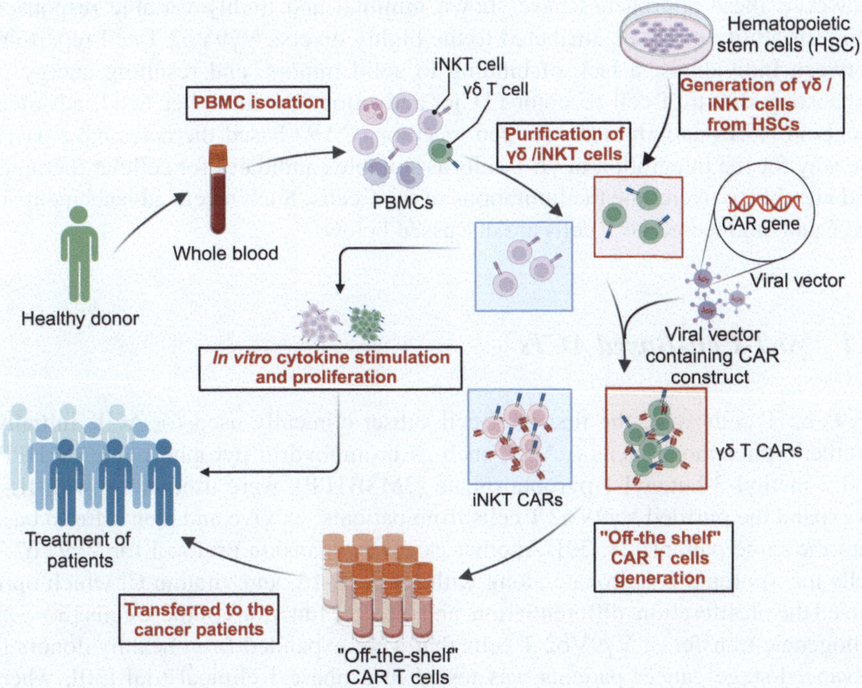

Fig. 1 Adoptive cellular therapy using unconventional T lymphocytes. iNKT cells or γδ T cells isolated from either the blood of healthy donors (syngeneic or allogeneic) or prepared from differentiating hematopoietic stem cells (HSCs). They are either adoptively transferred directly into cancer patients after activation and proliferation via different cytokines in vitro or transfected with viral vectors containing CAR constructs to generate CAR-T cells. These cells were subsequently expanded in vitro and adoptively transferred to cancer patients. These CARs are also called "off-the-shelf" CARs because they can be produced from the cells of allogeneic donors in bulk and stored and used whenever needed. (Image created with Biorender.com). CAR, chimeric antigen receptor; HSC, hematopoietic stem cell; iNKT, invariant natural killer T lymphocyte; PBMC, peripheral blood mononuclear cell

2 γδ T-Cell-Based Therapies

The robust and broad response of human γδ T cells and their phenotypic profile make them suitable therapeutic candidates for cancer immunotherapies [22–26]. They have become practical, especially in targeting hard-to-treat cancers that have reduced MHC class I expression and a low mutational burden and mostly where classical αβ T-cell-based therapies fail to eradicate tumor cells. Most clinical trials have focused on expanding the number of circulating Vγ9Vδ2 T cells in patients receiving amino bisphosphonates such as pamidronate and zoledronate. In vitro stimulation of isolated Vγ9Vδ2 T cells with bisphosphonate drugs followed by adoptive transfer into patients with malignancies has also been performed [27].

However, these approaches have shown minimal and highly variable responses. This variability could be attributed to the highly diverse Vγ9Vδ2 T-cell repertoire between individuals, a lack of binding to solid tumors, and resulting anergy or exhaustion due to T-cell receptor (TCR) activation. On the other hand, advancements in ACTs and chimeric antigen receptor (CAR)-based therapies have paved the way for the integration of γδ T cells as suitable candidates for cellular therapies and are able to overcome the limitations of αβ T cells. Such recent advancements in ACT and engineered γδ T cells are discussed below.

2.1 γδ T-Cell-Based ACTs

Vγ9Vδ2 T cells were the first γδ T-cell subset clinically used for ACT. Initially, synthetic phosphoantigens (pAgs), such as bromohydrin pyrophosphate (BrHPP) and 2-methyl-3-butenyl-1-pyrophosphate (2M3B1PP), were used along with IL-2 to expand the purified Vγ9Vδ2 T cells from patients' ex vivo and then infused back into the same patient [28, 29]. Another ex vivo expansion protocol for Vγ9Vδ2 T cells incorporated zoledronate along with IL-2, IL-15, and vitamin C, which optimized the proliferation, differentiation, and effector functions of these cells [30–32]. Allogeneic transfer of Vγ9Vδ2 T cells that were expanded from healthy donors to advanced-stage cancer patients was tested in a phase I clinical trial [30], where some of the patients showed increased overall survival without any adverse effects, such as GVHD, cytokine storm, or rejection, due to allogeneic transfer. Similar autologous ACTs have also been used in different cancers, and they all result in partial to exceptionally rare complete remission and low levels of adverse events [33, 34]. The general mechanism used to prepare unconventional T cells for adoptive transfer is summarized in Fig. 1.

The increasing therapeutic value of Vδ1 T cells has led to the development of Vδ1-enriched γδ T cells, called delta one T (DOT) cells, for adoptive transfer [35, 36]. The protocol involves a two-step process in which $V\delta1^+$ cells isolated from peripheral blood undergo selective expansion and differentiation into cytotoxic delta one T (DOT) cells. The expansion step is based on stimulation with TCR agonists and IL-4; the differentiation step is based on incubation with a combination of cytokines, primarily IL-15. This results in DOT cells expressing upregulated cytotoxicity receptors such as NKG2D, DNAM1, NKp30 (B7-H6), and NKp44, which significantly influences their cytotoxicity against tumor cells. Preclinical models of acute myeloid leukemia (AML) and chronic lymphocytic leukemia (CLL) transferred with DOT cells have shown significantly controlled tumor growth, improved survival, and prolonged presence of type I cytokines within the TME even two months post-infusion. DOT cells, such as GDX012 (NCT05001451), have improved in clinical trials for the treatment of AML [35–37]. Treating refractory or relapsed AML is a major therapeutic challenge. Recently, it has been shown that DOT cells use natural killer receptor (NKR) ligands and that pulmonary vascular resistance (PVR/CD155) on AML cells interacts with NKR and DNAM-1 on DOT cells, forms synapses between DOT cells and AML cells and drives cytolytic function [37]. These findings suggest that γδ T cells may be used across HLA barriers as ACTs and can be exploited as off-the-shelf products for their therapeutic use in patients.

2.2 *CAR γδ T Cells*

The generation of CAR γδ T cells has long been considered a suitable alternative for generating CAR-(αβ)T cells because of their safe profile in transfer settings and broad cytotoxic activity [26]. In one of the first attempts to generate such CAR γδ T cells, Vγ9Vδ2 T cells isolated from healthy donors were transduced with CD19- or GD2-specific CARs [38, 39]. Although these cells showed tumor-killing capacity in vitro, they could not survive in vivo without subsequent restimulation. Additionally, when combined with IL-2, CD19-specific polyclonal CAR γδ T cells control tumor growth in mouse xenograft leukemia models [40]. MUC1-Tn-specific CAR-γδ (Vγ9Vδ2) T cells were developed by isolating and expanding Vγ9Vδ2 T cells from healthy donors and expanding them with zoledronic acid and IL-2, followed by lentiviral transfection of the MUC1-Tn CAR. They exhibited greater cytotoxicity in vitro than CAR-(αβ) T cells did and suppressed gastric cancer growth in murine models. However, these cells have short persistence and lose their cytotoxicity upon prolonged exposure to target cells, and they require the subsequent addition of IL-2 to maintain their function [41]. To overcome the on-target off-tumor toxicity observed in CAR-T cell therapy, Schaft et al. developed an mRNA-based CAR transduction strategy in which CARs are transiently expressed via mRNA electroporation in γδ T cells. These cells showed potent cytotoxicity against melanoma cells with reduced cytokine production and related adverse effects [42]. A different strategy to construct NKG2DL-specific CARs uses the extracellular domain of NKG2D to construct CARs instead of conventional scFvs from monoclonal antibodies. The CAR construct was transfected into Vγ9Vδ2 T cells via mRNA electroporation, and these CAR γδ T cells showed enhanced cytotoxicity against several cultured solid tumor cell lines. Compared with no treatment, repeated infusions resulted in tumor regression in mouse xenograft models and prolonged survival [43]. On-target off-tumor toxicity leading to adverse effects in patients is a major concern in CAR-T-cell therapies. Fisher et al. developed a chimeric costimulatory receptor (CCR), an anti-GD2 CAR whose cytoplasmic domain is derived from the NKG2D adaptor DAP10, which is expressed on Vγ9Vδ2 T cells. The primary activation signal in these cells is derived from the endogenous γδ TCR, whereas the costimulatory signal is derived from the CAR when it binds to GD2 expressed on tumor cells. Therefore, the cytotoxic function of γδ T cells is deployed only when both phosphor-antigens and GD2 are present. These GD2-DAP10 CAR γδ T cells selectively lyse GD2-expressing neuroblastoma (NB) cells in vitro [44]. Capsomidis *et al.* demonstrated that GD2-specific CAR γδ T cells of both the Vδ1 and Vδ2 subsets showed enhanced cytotoxicity against GD2-positive NB cell lines and that (CAR^+) Vδ2 T cells retained their ability to cross-present tumor antigens to αβ T cells [45].

Vδ1 T cells are gaining traction as ideal CAR carriers because of their unique characteristics, such as enhanced homing, tissue tropism, and increased persistence in vivo. In addition, methods for large-scale expansion with simultaneous CAR transduction have been extensively studied over the past few years [35, 46]. DOT cells transduced with CARs specific for CD123 showed superior cytotoxicity against AML cells *in vitro* and *in vivo*. When combined with IL-15, a single dose of CD123 CAR-DOT suppressed tumor growth in AML xenograft models [24]. The first clinical trial with allogeneic Vδ1 CAR-T cells was based on a retrovirally transfected CD20 CAR against relapsed B-cell lymphoma (ADI-001), with four out of six patients experiencing complete remission (NCT04735471). In preclinical models, CD20 CAR-Vδ1 T (ADI-001) cells were found to elicit both CAR-dependent and CAR-independent functions against tumors without inducing GVHD. Moreover, ADI-001 evoked both innate and adaptive immune responses in a systemic manner [47]. In addition, therapeutics have developed CAR-Vδ1 T cells that target oncofetal proteins while simultaneously producing IL-15, which has demonstrated efficient antitumor immunity. These cells, which target glypican-3, which is exclusively expressed in hepatocellular carcinoma, effectively controlled tumor growth in a xenograft model with only a single infusion [48, 49]. With promising results in preclinical and clinical studies, CAR γδ T-cell therapy has been developed as an exciting avenue for cancer treatment. However, multiple biological, manufacturing, and translational barriers exist to overcome. For example, the impact of CAR expression on the biology of γδ T cells is still unexplored. The intracellular signaling events in γδ T cells following CAR interaction should be compared with those in αβ T cells. The molecular mechanisms underlying CAR γδ T-cell functions within the TME of different cancers need to be investigated to optimize their therapeutic use. The shorter persistence of adoptively transferred γδ T cells in vivo should be addressed to replace the multiple infusions of CAR γδ T cells, which could lead to immunogenicity of the CAR in the long run [50]. This also calls for the humanization of the receptor. Clinical studies also need to extensively evaluate the off-tumor toxicity and long-term side effects of CAR γδ T cells.

2.3 *γδ TCR-Engineered T (TEG) Cells*

In the most recent approach, researchers have combined the targeting ability of γδ T cells with the proliferation and memory formation of αβ T cells to effectively target tumors, independent of their mutational load. T cells engineered with defined γδ TCRs or TEGs are αβ T cells that are retrovirally transfected with tumor-reactive γδ TCRs ex vivo [51–54]. TEGs with Vγ9Vδ2 TCRs have been reported to show broad cytotoxicity against hematological and solid malignancies and induce dendritic cell (DC) maturation [51, 55, 56]. Moreover, activating signals through a Vγ9Vδ2 TCR expressed on αβ T cells are less susceptible to killer cell immunoglobulin-like receptor (KIR)-mediated inhibition. When $CD8^+$ and $CD4^+$ T cells express the Vγ9Vδ2 TCR, they are cytotoxic to tumors, and $CD4^+$ TEGs mediate the maturation of professional antigen-presenting cells [57]. $CD4^+$ TEGs can mediate dendritic cell maturation by sensing increased pAgs in immature DCs, and by detecting cognate antigens through the αβ TCR, they can promote epitope spreading [51, 58]. TEG002 is an autologous TEG being tested in a phase I clinical trial of relapsed or refractory multiple myeloma and AML (NCT04688853). Overall, γδ T cells are good candidates for generating ACT regimens in cancer. They are safe, have few off-target effects, and are efficient at killing cancer cells that are otherwise not recognized by conventional T cells. Although these cells have many limitations, scientists are working on enhancing their bulk production, cytotoxicity, and memory generation capacity to make them superior killers to eliminate cancer.

3 iNKT Cell-Based Cellular Therapies

Cell-based immunotherapy is being lauded as the future of cancer treatment, with a growing desire for the simple production of "off-the-shelf" cell products. iNKT cells are emerging as promising candidates for allogeneic cell therapy, particularly when genetically modified with CARs or gene-edited to minimize immunogenicity [19, 59]. The accumulation of iNKT cells in the tumor microenvironment drives a strong antitumor response in several cancers, including pancreatic cancer, suggesting that a therapeutic cell can be exploited as an ACT [18, 60, 61]. Owing to their low toxicity in allogeneic settings, invariant natural killer T (iNKT) cells are emerging as a preferred choice for producing "off-the-shelf" effector cells in cancer adoptive immunotherapy. iNKT cells intentionally downregulate MHC expression to avoid allogeneic rejection by the host immune system. In vitro production of iNKT cells from human hematopoietic stem cells with minimal HLA-I and HLA-II expression, followed by modification with chimeric antigen receptors (CARs), holds great potential for enhancing their antitumor actions. Ongoing clinical trials actively investigate options for commercially available allogeneic iNKT cell treatment. These techniques include CD19 CAR expression, IL-15 expression, and shRNAs targeting β2 microglobulin and CD74. Despite including lymphodepletion as part of this method, the results are positive. Notably, total MHC deletion in iNKT cells may result in a missing-self reaction by natural killer (NK) cells, resulting in the destruction of gene-edited iNKT cells. To counteract this impact, efforts include developing allogeneic tumor-redirected HLA-edited effectors that express NK inhibitory receptor ligands or "do not eat me" signals, such as CD47. These methods are intended to improve the long-term efficacy of adoptive cell treatment.

3.1 Cytokines That Regulate CAR-iNKT Function

In enhancing CAR-iNKT cell function, cytokines play a pivotal role, encompassing crucial aspects such as activation, proliferation, differentiation, and migration for optimal effector function [62]. The challenge of tumor relapse often stems from the inadequate persistence and proliferation of CAR-transduced effector cells. Several studies have underscored the importance of interleukin-15 (IL-15) in bolstering the longevity and effectiveness of CAR-T effector cells when co-expressed [63]. Additionally, some research has highlighted the relevance of STAT-5-related cytokines and immunotherapy cytokines such as IL-2 and IL-15 in this context. Recent preclinical investigations have revealed that antigen-dependent activation of STAT-3 and STAT-5 in CAR-T effector cells promotes their proliferation and prevents premature in vitro differentiation. These findings have improved in vivo persistence and antitumor efficacy, particularly in patients with lymphoblastic leukemia [64].

3.1.1 IL-21

Another cytokine of significance is IL-21, a member of the common-gamma chain (γc) cytokine family, which is known for its role in developing immune responses spanning innate and adaptive immunity against pathogens and tumors [65, 66]. IL-21, produced by both $CD4^+$ T cells and iNKT cells, enhances the production of perforin, optimizing the cytotoxicity of NK cells and assisting DCs in inducing IFN-γ production mediated by iNKT cells [65, 67]. Numerous studies have confirmed the ability of IL-21 to sustain the functionality of activated T cells, facilitate $CD8^+$ T-cell infiltration into tumors, and promote iNKT cell proliferation [68]. Additionally, it supports antitumor activity by selectively increasing the proliferation of $TIM3^-$ T-cell subsets while inhibiting $FOXP3^+$ Treg cells. Recent studies have shown promise in combining IL-21 with anti-PD-1 or anti-CTLA-4 therapies for enhanced antitumor efficacy in breast and colon cancer mouse models. Notably, iNKT cells, with high expression of CD62L, which is indicative of a memory phenotype, exhibit increased IL-21R mRNA expression in $CD62L^+$ iNKT subsets [69]. IL-21 also exerts anti-apoptotic effects by promoting Bcl-2 expression and driving iNKT cells toward Th1 transformation. Notably, compared with IL-2-amplified cells, IL-2/IL-21-amplified CAR-iNKT cells markedly improved survival, significantly extending the disease-free survival of treated mice [70].

3.1.2 IL-15

IL-15 plays a critical role in maintaining and developing memory $CD8^+$ T cells, NK cells, and iNKT cells. In iNKT cell development, IL-15 is important and is delivered via transport by parenchymal cells [71]. Studies have revealed that $CD4^-$ iNKT cells exhibit a more robust response to IL-15 due to the presence of IL-7. IL-15 promotes the production of IFN-γ, granzymes, and other effector molecules in iNKT cells [72, 73]. Researchers have identified the TBK-binding protein (Tbkbp1) as a facilitator of the autophagy-initiating kinase Ulk-1, which counteracts the inhibitory actions of mTORC1. Tbkbp1 deficiency hampers IL-15-mediated iNKT cell autophagy, resulting in increased ROS production, impaired Bcl-2 expression, and reduced iNKT cell survival [74]. Finally, IL-15, known for enhancing the long-term survival of $CD8^+$ T cells [75], assumes an adjuvant role by increasing poly-functional $CD4^+$ T-cell populations, potentially increasing treatment efficacy [76]. Further research is needed to ascertain its effects on iNKT or CAR-iNKT cells.

3.1.3 IL-7

IL-7 empowers immature human iNKT cells to influence cytokine expression and substantially impacts the memory phenotype [77]. Compared with IL-2 alone, IL-7 enhances Th2-type cytokines and selectively promotes the proliferation of $CD4^+$ iNKT cells [78]. The activation of IL-7R initiates a potent signaling cascade, increasing STAT-5 phosphorylation, which may represent a signaling pathway responsible for maintaining CAR-iNKT cell proliferation.

3.2 Allogenic Hematopoietic Stem Cell (HSC)-Engineered iNKT Cells

Human blood naturally contains a minute fraction of iNKT cells, ranging from 0.001% to 1%, posing a significant challenge in cultivating substantial numbers of allogeneic iNKT cells for effective cell-based therapies [79]. To overcome this hurdle, a Lenti/iNKT-sr39TK vector was employed to transduce human $CD34^+$ cells sourced from either cord blood (CB) or granulocyte-colony-stimulating factor (G-CSF)-mobilized human peripheral blood stem cells (PBSCs). These transduced cells were subsequently nurtured in a two-stage artificial thymic organoid (ATO)/α--galactosylceramide (α-GalCer) culture system in vitro.

α-GalCer, a synthetic agonist glycolipid ligand, was used to selectively activate iNKT cells [80]. Notably, lentivector transduction rates in HSCs frequently exceed 50%. Once transduced, the HSCs entered stage 1 ATO culture, where they underwent remarkable expansion, multiplying more than 100-fold over the course of 8 weeks, ultimately differentiating into fully functional human iNKT cells. At the end of stage 1 culture, ATOs were dissociated into individual cells, which were subsequently transferred to stage 2 α-GC expansion culture, extending the expansion by an additional 100- to 1000-fold and yielding a high-yield, high-purity Allo-HSC-iNKT cell product. Throughout the ATO/α-GC culture process, Allo-HSC-iNKT cells followed a typical developmental trajectory characterized by the expression of CD4/CD8 coreceptors [81]. They transitioned from the $CD4^-CD8^-$ (DN) to the $CD4^+CD8^+$ (DP) subset and finally to the $CD4CD8^{+/-}$ subset. Notably, $CD8^+CD4^-$ single-positive (SP) and DN human iNKT cells have emerged as proinflammatory and highly cytotoxic entities, rendering them particularly well suited for cancer immunotherapy—a highly promising avenue in cancer treatment [82–84].

3.3 Adoptive Cell Transfer of CAR-iNKT Cells

iNKT cells are engineered to express CARs specific for a particular protein *ex vivo* to treat cancer. The antigen-binding domain of CAR is a single variable fragment from the monoclonal antibody. Different CAR-iNKT cells are engineered to express specific antigen receptors against different cancers and are used in cancer therapeutics. Human iNKT cells engineered to express GD2-CARs against neuroblastoma. In this context, CAR-GD2 NKT cells target GD2-expressing cells, which are abundant in neuroblastoma. Its antitumor activity and persistence can be increased by expressing IL-15 with CAR [85]. To treat multiple myeloma cells, iNKT cells are engineered to co-express the CD38 antigen or B-cell maturation antigen (BCMA) CAR-iNKT cells, effectively eliminating multiple myeloma cells. BCMA-CARs or affinity-optimized CD28 CARs expressed by iNKT cells display a Th1 cytokine profile, indicating therapeutic utility. The costimulatory domain of CD38 CARs does not influence the cytotoxicity of iNKT cells. Nevertheless, when stimulated with CD1d^{+} DCs loaded with α-GalCer, stable expansion of CD38 CAR-iNKT cells and BCMA CAR-iNKT cells occurred [86]. In the case of B-cell lymphoma, CD1d expression is low, so CAR-iNKT is less effective; hence, it is co-expressed with CD19. CAR-iNKT cells can be developed for "off-the-self" use. Previous studies demonstrated CAR engineering of iNKT cells and their preclinical activity against neuroblastoma and CD1d^{-}CD19^{+} B-cell lymphoma [87, 88]. CAR-iNKT cells have shown excellent effects in phase I clinical studies (NCT03294954) in children with relapsed neuroblastoma. An interim study revealed no hazards associated with dose restrictions. CAR-iNKT cells also showed in vivo proliferation, active localization to tumor sites, and the ability to induce recovery in bone metastatic lesions.

Other ongoing phase I clinical trials (NCT03774654 and NCT04814004) are investigating the ability of donor-derived or allogenic iNKT cells expressing CD19 (CAR19) to help patients with CD19^{+} lymphoma and leukemia recover. These trials aim to further investigate the therapeutic potential of CAR-iNKT cells in various patient populations, signaling a step forward in the field of adoptive immunotherapy. While the adoptive transfer of iNKT cells shows great potential, it is not without restrictions. In the tumor microenvironment, lactic acid inhibits lipid synthesis in iNKT cells, lowering IFN-γ production. Furthermore, lactic acid inhibits mTORC1 activation, reducing peroxisome proliferator-activated receptor (PPAR) expression in iNKT cells. PPAR usually activates sterol regulatory element-binding protein 1 (SREBP1) transcription to regulate lipid synthesis in combination with promyelocytic leukemia zinc-finger protein (PLZF; the master transcription factor of iNKT cells). Notably, cholesterol contributes to increased T-cell receptor (TCR) signaling and IFN-γ production.

Lactic acid buildup in the tumor microenvironment inhibits IFN-γ production and cholesterol synthesis in tumor-infiltrating iNKT cells in cases such as hepatocellular carcinoma and mouse models. In the context of cancer, iNKT cells and dendritic cells (DCs) interact, resulting in reduced antitumor immune responses. A potential side effect of this therapy is an increased immune response and inflammation, often manifesting as cytokine release syndrome (cytokine storm). Recognizing these complexities underscores the importance of further research to refine iNKT cell adoptive transfer approaches, address limitations, and optimize therapeutic outcomes [89]. The list of ongoing or completed clinical trials using iNKT cells or γδ T cells is listed in Table 1.

Table 1 Clinical trials using iNKT and γδ T-cell-based adoptive cellular therapy in cancer

S. No.	NCT	Phase of trial	Type of cancer	Number of patients	Treatment	References
iNKT cell-based adoptive cellular therapy						
1.	02439788	–	Relapsed neuroblastoma	–	IC9-anti-GD2-CD28/OX40/CD3ζ CAR NKT (GINAKIT) cells.	[108, 109]
2.	03294954	Phase I	Recruiting: new CAR construct	24	Autologous Anti-GD2-CD28/CD3ζ-IL-15 CAR NKT (GINAKIT2) cells.	[72, 108, 109]
3.	03774654	Phase I	B-cell malignancy	48	Allogenic Anti-CD19-CD28/CD3ζ-IL-15 CAR NKT (ANCHOR) cells.	http://www.clinicaltrials.gov/
4.	04814004	Phase I	ALL B-cell lymphoma	20	Autologous CD19-CAR NKT expressing IL-15.	http://www.clinicaltrials.gov/
5	05487651	Phase I	Refractory/relapsed B-cell NHL or leukemia (ALL or CLL)	36	Allogenic MKT-cell expressing CD19-specific CAR and IL-15 in relapsed or refractory B-cell.	http://www.clinicaltrials.gov/
6.	03175679	Phase I	Relapsed/advanced hepatocellular carcinoma	–	iNKT cells, IL-2, Tegafur	http://www.clinicaltrials.gov/
7.	05400122	Phase I	Colorectal cancer and blood cancer	12	Combination with Interleukin-2 (IL2) and Transforming Growth Factor-Beta (TGF-β) Receptor 1 inhibitor Vactosertib	http://www.clinicaltrials.gov/
γδ T cells-based adoptive cellular therapy						
8.	03790072	Phase I	Refractory/relapsed Acute myeloid leukemia	10	Allogeneic γδ T cells	http://www.clinicaltrials.gov/ [110]

(continued)

Table 1 (continued)

S. No.	NCT	Phase of trial	Type of cancer	Number of patients	Treatment	References
9.	05400603	Phase I	Neuroblastoma	24	Allogenic γδ T cells in combination with Dinutuximab, Temozolomide, Irinotecan, and Zoledronate	http://www.clinicaltrials.gov/ [111, 112]
10.	UMIN C000000336	Phase I and II	Non-small cell lung cancer	30	Autologous-γδ T lymphocytes	UMIN-CTR Clinical Trial [113, 114]
11.	01728805	Phase III		372	KW-0761(other names-Mogamulizumab, POTELIGEO) and Vorinostat (other names ZOLINZA)	http://www.clinicaltrials.gov/ [115–117]
12.	03533816	Phase I	Leukemia and myelodysplastic	38	EAGD T-cell infusion	http://www.clinicaltrials.gov/ [118]
13.	05001451	Phase I	Acute myeloid leukemia	3	GDX012 suspension for IV infusion	http://www.clinicaltrials.gov/ [118]
14.	04243499	Phase I and II	Solid tumor, Adult hematopoietic/ lymphoid cancer	150	IV doses of ICT01 as monotherapy and in combination with a checkpoint inhibitor.	http://www.clinicaltrials.gov/ [118]

4 MAIT Cell-Based Adoptive Cellular Therapy in Cancer

MAIT cells play important roles in maintaining gut homeostasis. They are very efficient in recognizing some antigens expressed by microbes in the gut. Even though healthy mammalian cells do not express those antigens, they appear to be expressed in certain cancer cells. Recently, it has been shown that sulfated bile acid and cholic acid 7-sulfate (CA7S) produced by host cells bind to MHC class I-related protein 1 (MR1) on MAIT cells and promote their development and proliferation [90]. Therefore, MAIT cells can be repurposed to target cancer cells specifically with fewer off-target effects [15, 91]. Scientists are working to increase the efficacy of MAIT cells either by activating and propagating them in vitro or by genetically modifying them to specifically target and eliminate cancer cells with high efficiency. In the following sections, we discuss the different approaches used to weaponize MAIT cells and use them in ACTs to treat cancer.

4.1 *Peripheral Blood Nononuclear Cell (PBMC)-Derived MAIT Cells*

Current procedures for ex vivo multiplication of MAIT cells, which use MR1-tetramer-based artificial APCs or 5-OP-RU-loaded irradiated PBMCs as feeder cells, have exhibited modest fold changes in proliferation ranging from 60–200 [92, 93]. The apparent hyperproliferative nature of PBMC-MAIT cells is a substantial obstacle to their downstream use in cancer immunotherapy. Ex vivo cultured PBMC-MAIT cells exhibit low proliferation when stimulated with typical T-cell expansion cytokine mixtures or mitogens despite robust in vivo expansion upon activation [92–95]. Understanding the reasons underlying this discrepancy is critical for optimizing PBMC-MAIT cell culture methods, necessitating additional cellular and molecular research [96, 97]. Parrot et al. published an ex vivo expansion technique for human MAIT cells that uses immunomagnetic bead-based cell sorting with 5-OP-RU-loaded irradiated PBMC feeder cells. MACS-based enrichment of MAIT cells with 5-OP-RU-loaded MR1 tetramers was used in this approach, followed by coculture with irradiated PBMCs. IL-2 or IL-15 alone generated more growth than did IL-7 alone. Anti-CD3/CD28/CD2 stimulation did not significantly increase the yield but did increase the level of conventional T-cell impurities. This finding shows that TCR-dependent activation of conventional T cells reduces the percentage of MAIT cells in the final product [93].

Parrot et al. reported a remarkable 200-fold ex vivo growth of PBMC-MAIT cells, one of the largest reported. On the other hand, anti-CD3/CD28/CD2 stimulation increased the level of conventional T-cell impurities. MAIT and conventional $CD8^+$ T cells have unique TCR-dependent activation models, according to Slichter et al., with MAIT cells requiring a synergistic mixing of TCR signaling and inflammatory cytokines, including IL-12 and IL-18, for complete activation [93]. Priya and Brutkiewicz proposed an alternative methodology for creating fake APCs on the basis of MR1 tetramers by coating latex beads with 5-OP-RU-loaded MR1 tetramers and anti-CD28 antibodies. MAIT cell growth was approximately 74-fold greater with IL-2. The latex bead-based method offers clinical benefits such as less contamination risk and ease of quality control [92]. The density of MR1 molecules on the surface of natural PBMCs or manufactured APCs may affect the final expansion fold change of MAIT cells. Understanding how MR1 expression changes in response to MAIT cell antigens will help improve ex vivo MAIT cell culture and lead to the development of MAIT cell-based immunotherapy [97].

4.2 *Hematopoietic Stem Cell (HSC)-Derived MAIT Cells for ACTs*

TCRs such as the NY-ESO-1 TCR, MART1 TCR, and iNKT TCR have been used in HSC engineering for cancer immunotherapy [83, 98–100]. Previous studies used lentivector transduction of the iNKT TCR and a suicide gene to generate autologous and allogeneic HSC-derived iNKT cells [100, 101]. The HSC-engineered iNKT cells closely resembled their endogenous counterparts after a 6- to 10-week culture, successfully targeting tumor cells with reasonable safety and low immunogenicity. This HSC-derived iNKT cell treatment has demonstrated potential in treating various cancers, including hematologic and solid tumors [100].

The same technique can be used to create HSC-derived MAIT (HSC-MAIT) cells by transducing MAIT TCRs into HSCs and culturing them in an existing off-the-shelf HSC differentiation platform. The HSC-MAIT cell platform is strong and adaptable, allowing for the incorporation of new engineering concepts. Using retroviral transduction and CRISPR-Cas9 gene editing, previous research has shown the excellent efficacy of CAR engineering and HLA deletion in HSC-derived immune cells. Other genetic engineering methods, including IL-15 and IL-2 overexpression and checkpoint ablation (PD-1 and CTLA-4), can be added to HSC-MAIT cell products, suggesting translational potential in universal off-the-shelf cancer immunotherapy [15, 100, 102].

4.3 CAR-MAIT Cells

The innovative incorporation of innate MAIT cells into CAR-engineered cells shows potential in MAIT cell-based immunotherapy. In a simulated immunosuppressive tumor microenvironment, mesothelin-targeting CAR-MAIT cells produced from PBMCs demonstrated strong cytotoxicity against cancer cells, particularly tumor-associated macrophages (TAMs). In preclinical studies, CD19 CAR-engineered MAIT cells were engrafted successfully without generating graft-versus-host disease. However, the small number of MAIT cells (1–10%) in the peripheral circulation makes the large-scale synthesis of PBMC-derived MAIT cells difficult. To address this, genetically altered hematopoietic stem cells (HSCs) or induced pluripotent stem cells (iPSCs) have the potential to generate CAR-MAIT cells at a low cost. A stem cell culture platform previously used for BCMA CAR-HSC-derived iNKT cell production, which could be modified to generate HSC-derived CAR-MAIT cells. CRISPR-Cas9 genetic engineering approaches, such as knocking out HLA-I and HLA-II to inhibit host T-cell-mediated allo-rejection or targeting PD-1 and CTLA-4 to minimize immune checkpoint-mediated immunosuppression, could improve CAR-MAIT cell products [103–106].

5 Conclusions

Unconventional T cells such as iNKTs and γδ T cells participate in the antitumor immune response through direct tumor cytotoxicity and indirectly via interactions with TAMs and DCs, and the stimulation of NK cells and αβ T cells. However, the endogenous antitumor effects of these cells are limited because of their extremely low frequency and different regulatory mechanisms within the tumor microenvironment [107]. To overcome these limitations, scientists are exploring various means, such as in vivo and in vitro stimulation and propagation of these cells, along with the genetic modification of the cells to redirect their tumor-killing capacity. Redirected therapeutic strategies employ direct tumor cell killing by engineered CARs along with simultaneous harnessing of the intrinsic antitumor properties and tumor-homing abilities of these cells. CAR-T cells containing iNKTs or γδ T cells have shown durable responses in different cancer settings emerging from ongoing clinical trials, encouraging results. These cells also showed fewer off-target effects and caused significantly fewer toxicities. iNKTs and γδ T cells isolated from autologous sources or allogeneic sources, such as the PBMCs of healthy donors, cord blood, hematopoietic stem cells (HSCs), or iPSCs, are great candidates for the generation of off-the-shelf adoptive T-cell therapies against cancer because of their limited alloreactivity and much lower potential to generate GVHD. Although further research is needed to increase the persistence and memory of these cells within the tumor, the exploitation of these innate cells for the generation of an advanced immunotherapeutic regime paves the way for harnessing the unique broad-spectrum properties of other innate immune cells in future cell-based therapies for many forms of malignancies and autoimmune diseases.

Declaration of Interests The authors declare that they have no competing interests.

Funding Supports GL received the Swarna Jayanti Fellowship (DST/SJF/LSA-01/2017-18) from the Department of Science and Technology; the TATA Innovation Fellowship from the Department of Biotechnology (HRD-16012/6/2020-AFS-DBT); and a research grant from the Science and Engineering Research Board (CRG/2022/007108), Ministry of Science and Technology, Government of India. SK received a Senior Research Fellow (SRF) from the Council of Scientific and Industrial Research (CSIR), and SY received a Junior Research Fellowship (JRF) from the Department of Biotechnology (DBT), Ministry of Science and Technology, Government of India.

References

1. Rosenberg SA, Spiess P, Lafreniere R. A new approach to the adoptive immunotherapy of cancer with tumor-infiltrating lymphocytes. Science. 1986;233:1318–21.
2. Rosenberg SA, Packard BS, Aebersold PM, Solomon D, Topalian SL, Toy ST, Simon P, Lotze MT, Yang JC, Seipp CA, et al. Use of tumor-infiltrating lymphocytes and interleukin-2 in the immunotherapy of patients with metastatic melanoma. A preliminary report. N Engl J Med. 1988;319:1676–80.
3. Williams KM, Hakim FT, Gress RE. T cell immune reconstitution following lymphodepletion. Semin Immunol. 2007;19:318–30.
4. Rosenberg SA, Restifo NP, Yang JC, Morgan RA, Dudley ME. Adoptive cell transfer: a clinical path to effective cancer immunotherapy. Nat Rev Cancer. 2008;8:299–308.
5. Vivier E, Rebuffet L, Narni-Mancinelli E, Cornen S, Igarashi RY, Fantin VR. Natural killer cell therapies. Nature. 2024;626:727–36.
6. van Meir H, Nout RA, Welters MJ, Loof NM, de Kam ML, van Ham JJ, Samuels S, Kenter GG, Cohen AF, Melief CJ, Burggraaf J, van Poelgeest MI, van der Burg SH. Impact of

(chemo)radiotherapy on immune cell composition and function in cervical cancer patients. Onco Targets Ther. 2017;6:e1267095.
7. Ogonek J, Kralj Juric M, Ghimire S, Varanasi PR, Holler E, Greinix H, Weissinger E. Immune reconstitution after allogeneic hematopoietic stem cell transplantation. Front Immunol. 2016;7:507.
8. Li D, Li X, Zhou W-L, Huang Y, Liang X, Jiang L, Yang X, Sun J, Li Z, Han W-D, Wang W. Genetically engineered T cells for cancer immunotherapy. Signal Transduct Target Ther. 2019;4:35.
9. Restifo NP, Dudley ME, Rosenberg SA. Adoptive immunotherapy for cancer: harnessing the T cell response. Nat Rev Immunol. 2012;12:269–81.
10. Anderson KG, Stromnes IM, Greenberg PD. Obstacles posed by the tumor microenvironment to T cell activity: a case for synergistic therapies. Cancer Cell. 2017;31:311–25.
11. Chan JD, Lai J, Slaney CY, Kallies A, Beavis PA, Darcy PK. Cellular networks controlling T cell persistence in adoptive cell therapy. Nat Rev Immunol. 2021;21:769–84.
12. Janelle V, Delisle JS. T-cell dysfunction as a limitation of adoptive immunotherapy: current concepts and mitigation strategies. Cancers (Basel). 2021;13:598.
13. Rezvani K. Adoptive cell therapy using engineered natural killer cells. Bone Marrow Transplant. 2019;54:785–8.
14. La Manna MP, Orlando V, Tamburini B, Badami GD, Dieli F, Caccamo N. Harnessing unconventional T cells for immunotherapy of tuberculosis. Front Immunol. 2020;11:2107.
15. Li YR, Zhou K, Wilson M, Kramer A, Zhu Y, Dawson N, Yang L. Mucosal-associated invariant T cells for cancer immunotherapy. Mol Ther. 2023;31:631–46.
16. de Araújo ND, Gama FM, de Souza Barros M, Ribeiro TLP, Alves FS, Xabregas LA, Tarragô AM, Malheiro A, Costa AG. Translating unconventional T cells and their roles in leukemia antitumor immunity. J Immunol Res. 2021;2021:6633824.
17. Wang Y, Li YR. Harnessing chimeric antigen receptor-engineered invariant natural killer T cells: therapeutic strategies for cancer and the tumor microenvironment. Curr Pharm Biotechnol. 2024;25:2001. https://doi.org/10.2174/0113892010265228231116073012.
18. Tian C, Wang Y, Su M, Huang Y, Zhang Y, Dou J, Zhao C, Cai Y, Pan J, Bai S, Wu Q, Chen S, Li S, Xie D, Lv R, Chen Y, Wang Y, Fu S, Zhang H, Bai L. Motility and tumor infiltration are key aspects of invariant natural killer T cell anti-tumor function. Nat Commun. 2024;15:1213.
19. Baiu DC, Sharma A, Schehr JL, Basu J, Smith KA, Ohashi M, Johannsen EC, Kenney SC, Gumperz JE. Human CD4(+) iNKT cell adoptive immunotherapy induces anti-tumor responses against CD1d-negative EBV-driven B lymphoma. Immunology. 2024;172:627–40.
20. Martínez Bedoya D, Dutoit V, Migliorini D. Allogeneic CAR T cells: an alternative to overcome challenges of CAR T cell therapy in glioblastoma. Front Immunol. 2021;12:640082.
21. Zhang P, Zhang G, Wan X. Challenges and new technologies in adoptive cell therapy. J Hematol Oncol. 2023;16:97.
22. Choi H, Kim TG, Jeun SS, Ahn S. Human gamma-delta (gamma-delta) T cell therapy for glioblastoma: a novel alternative to overcome challenges of adoptive immune cell therapy. Cancer Lett. 2023;571:216335.
23. Li C, Zhou F, Wang J, Chang Q, Du M, Luo W, Zhang Y, Xu J, Tang L, Jiang H, Liu L, Kou H, Lu C, Liao D, Wu J, Wei Q, Ke S, Deng J, Liu C, Mei H, Hu Y. Novel CD19-specific gamma/delta TCR-T cells in relapsed or refractory diffuse large B-cell lymphoma. J Hematol Oncol. 2023;16:5.
24. Sanchez Martinez D, Tirado N, Mensurado S, Martinez-Moreno A, Romecin P, Gutierrez Aguera F, Correia DV, Silva-Santos B, Menendez P. Generation and proof-of-concept for allogeneic CD123 CAR-Delta One T (DOT) cells in acute myeloid leukemia. J Immunother Cancer. 2022;10:e005400.
25. Ma L, Feng Y, Zhou Z. A close look at current gamma-delta T-cell immunotherapy. Front Immunol. 2023;14:1140623.
26. Flemming A. Engineered gamma-delta T cells show promise in bone tumors. Nat Rev Immunol. 2024;24:458.

27. Lee D, Rosenthal CJ, Penn NE, Dunn ZS, Zhou Y, Yang L. Human γδ T cell subsets and their clinical applications for cancer immunotherapy. Cancers (Basel). 2022;14:3005.
28. Burjanadzé M, Condomines M, Reme T, Quittet P, Latry P, Lugagne C, Romagne F, Morel Y, Rossi JF, Klein B, Lu ZY. *In vitro* expansion of gamma delta T cells with anti-myeloma cell activity by Phosphostim and IL-2 in patients with multiple myeloma. Br J Haematol. 2007;139:206–16.
29. Kobayashi H, Tanaka Y, Yagi J, Minato N, Tanabe K. Phase I/II study of adoptive transfer of γδ T cells in combination with zoledronic acid and IL-2 to patients with advanced renal cell carcinoma. Cancer Immunol Immunother. 2011;60:1075–84.
30. Xu Y, Xiang Z, Alnaggar M, Kouakanou L, Li J, He J, Yang J, Hu Y, Chen Y, Lin L, Hao J, Li J, Chen J, Li M, Wu Q, Peters C, Zhou Q, Li J, Liang Y, Wang X, Han B, Ma M, Kabelitz D, Xu K, Tu W, Wu Y, Yin Z. Allogeneic Vγ9Vδ2 T-cell immunotherapy exhibits promising clinical safety and prolongs the survival of patients with late-stage lung or liver cancer. Cell Mol Immunol. 2021;18:427–39.
31. Kabelitz D, Cierna L, Juraske C, Zarobkiewicz M, Schamel WW, Peters C. Empowering gamma-delta T-cell functionality with vitamin C. Eur J Immunol. 2024;54:e2451028.
32. Inoue Y, Oda A, Maeda Y, Sumitani R, Oura M, Sogabe K, Maruhashi T, Takahashi M, Fujii S, Nakamura S, Miki H, Hiasa M, Teramachi J, Harada T, Abe M. Ex vivo expansion and activation of Vgamma9Vdelta2 T cells by CELMoDs in combination with zoledronic acid. Int J Hematol. 2024;119:626–30.
33. Hoeres T, Smetak M, Pretscher D, Wilhelm M. Improving the efficiency of Vγ9Vδ2 T-cell immunotherapy in cancer. Front Immunol. 2018;9:800.
34. Doki Y, Ajani JA, Kato K, Xu J, Wyrwicz L, Motoyama S, Ogata T, Kawakami H, Hsu CH, Adenis A, El Hajbi F, Di Bartolomeo M, Braghiroli MI, Holtved E, Ostoich SA, Kim HR, Ueno M, Mansoor W, Yang WC, Liu T, Bridgewater J, Makino T, Xynos I, Liu X, Lei M, Kondo K, Patel A, Gricar J, Chau I, Kitagawa Y. Nivolumab combination therapy in advanced esophageal squamous-cell carcinoma. N Engl J Med. 2022;386:449–62.
35. Almeida AR, Correia DV, Fernandes-Platzgummer A, da Silva CL, da Silva MG, Anjos DR, Silva-Santos B. Delta one T cells for immunotherapy of chronic lymphocytic leukemia: clinical-grade expansion/differentiation and preclinical proof of concept. Clin Cancer Res. 2016;22:5795–804.
36. Kabelitz D. Delta One T cells recognize AML via DNAM-1. Blood. 2024;143:1434–6.
37. Mensurado S, Condeco C, Sanchez-Martinez D, Shirley S, Coelho RML, Tirado N, Vinyoles M, Blanco-Dominguez R, Barros L, Galvao B, Custodio N, Gomes da Silva M, Menendez P, Silva-Santos B. CD155/PVR determines acute myeloid leukemia targeting by Delta One T cells. Blood. 2024;143:1488–95.
38. Rischer M, Pscherer S, Duwe S, Vormoor J, Jurgens H, Rossig C. Human gamma-delta T cells as mediators of chimeric-receptor redirected anti-tumor immunity. Br J Haematol. 2004;126:583–92.
39. Fowler D, Barisa M, Southern A, Nattress C, Hawkins E, Vassalou E, Kanouta A, Counsell J, Rota E, Vlckova P, Draper B, De Mooij T, Farkas A, Brezovjakova H, Baker AT, Scotlandi K, Manara MC, Tape C, Chester K, Anderson J, Fisher J. Payload-delivering engineered gamma-delta T cells display enhanced cytotoxicity, persistence, and efficacy in preclinical models of osteosarcoma. Sci Transl Med. 2024;16:eadg9814.
40. Deniger DC, Switzer K, Mi T, Maiti S, Hurton L, Singh H, Huls H, Olivares S, Lee DA, Champlin RE, Cooper LJ. Bispecific T-cells expressing polyclonal repertoire of endogenous γδ T-cell receptors and introduced CD19-specific chimeric antigen receptor. Mol Ther. 2013;21:638–47.
41. Zhai X, You F, Xiang S, Jiang L, Chen D, Li Y, Fan S, Han Z, Zhang T, An G, Zhang B, Chen Y, Meng H, Yang L. MUC1-Tn-targeting chimeric antigen receptor-modified Vγ9Vδ2 T cells with enhanced antigen-specific anti-tumor activity. Am J Cancer Res. 2021;11:79–91.
42. Harrer DC, Simon B, Fujii SI, Shimizu K, Uslu U, Schuler G, Gerer KF, Hoyer S, Dörrie J, Schaft N. RNA-transfection of γ/δ T cells with a chimeric antigen receptor or an α/β T-cell

receptor: a safer alternative to genetically engineered α/β T cells for the immunotherapy of melanoma. BMC Cancer. 2017;17:551.
43. Ang WX, Ng YY, Xiao L, Chen C, Li Z, Chi Z, Tay JC, Tan WK, Zeng J, Toh HC, Wang S. Electroporation of NKG2D RNA CAR improves Vγ9Vδ2 T cell responses against human solid tumor xenografts. Mol Ther Oncolytics. 2020;17:421–30.
44. Fisher J, Abramowski P, Wisidagamage Don ND, Flutter B, Capsomidis A, Cheung GW, Gustafsson K, Anderson J. Avoidance of on-target off-tumor activation using a co-stimulation-only chimeric antigen receptor. Mol Ther. 2017;25:1234–47.
45. Capsomidis A, Benthall G, Van Acker HH, Fisher J, Kramer AM, Abeln Z, Majani Y, Gileadi T, Wallace R, Gustafsson K, Flutter B, Anderson J. Chimeric antigen receptor-engineered human gamma delta T cells: enhanced cytotoxicity with retention of cross presentation. Mol Ther. 2018;26:354–65.
46. Ferry GM, Agbuduwe C, Forrester M, Dunlop S, Chester K, Fisher J, Anderson J, Barisa M. A simple and robust single-step method for CAR-Vδ1 γδT cell expansion and transduction for cancer immunotherapy. Front Immunol. 2022;13:863155.
47. Nishimoto KP, Barca T, Azameera A, Makkouk A, Romero JM, Bai L, Brodey MM, Kennedy-Wilde J, Shao H, Papaioannou S, Doan A, Masri C, Hoang NT, Tessman H, Ramanathan VD, Giner-Rubio A, Delfino F, Sharma K, Bray K, Hoopes M, Satpayev D, Sengupta R, Herrman M, Abbot SE, Aftab BT, An Z, Panuganti S, Hayes SM. Allogeneic CD20-targeted γδ T cells exhibit innate and adaptive antitumor activities in preclinical B-cell lymphoma models. Clin Transl Immunol. 2022;11:e1373.
48. Makkouk A, Yang XC, Barca T, Lucas A, Turkoz M, Wong JTS, Nishimoto KP, Brodey MM, Tabrizizad M, Gundurao SRY, Bai L, Bhat A, An Z, Abbot S, Satpayev D, Aftab BT, Herrman M. Off-the-shelf Vdelta1 gamma delta T cells engineered with glypican-3 (GPC-3)-specific chimeric antigen receptor (CAR) and soluble IL-15 display robust antitumor efficacy against hepatocellular carcinoma. J Immunother Cancer. 2021;9:e003441.
49. Nishida T, Kataoka H. Glypican 3-targeted therapy in hepatocellular carcinoma. Cancers (Basel). 2019;11:1339.
50. Ganapathy T, Radhakrishnan R, Sakshi S, Martin S. CAR γδ T cells for cancer immunotherapy. Is the field more yellow than green? Cancer Immunol Immunother. 2023;72:277–86.
51. Marcu-Malina V, Heijhuurs S, van Buuren M, Hartkamp L, Strand S, Sebestyen Z, Scholten K, Martens A, Kuball J. Redirecting αβ T cells against cancer cells by transfer of a broadly tumor-reactive γδT-cell receptor. Blood. 2011;118:50–9.
52. Paul S, Lal G. Regulatory and effector functions of gamma-delta (gamma-delta) T cells and their therapeutic potential in adoptive cellular therapy for cancer. Int J Cancer. 2016;139:976–85.
53. Davies DM, Pugliese G, Parente Pereira AC, Whilding LM, Larcombe-Young D, Maher J. Engineering a dual specificity gamma-delta T-cell receptor for cancer immunotherapy. Biology (Basel). 2024;13(3):196.
54. Yuan M, Wang W, Hawes I, Han J, Yao Z, Bertaina A. Advancements in gamma-delta T cell engineering: paving the way for enhanced cancer immunotherapy. Front Immunol. 2024;15:1360237.
55. Straetemans T, Kierkels GJJ, Doorn R, Jansen K, Heijhuurs S, Dos Santos JM, van Muyden ADD, Vie H, Clemenceau B, Raymakers R, de Witte M, Sebestyén Z, Kuball J. GMP-grade manufacturing of T cells engineered to express a defined γδTCR. Front Immunol. 2018;9:1062.
56. Johanna I, Straetemans T, Heijhuurs S, Aarts-Riemens T, Norell H, Bongiovanni L, de Bruin A, Sebestyen Z, Kuball J. Evaluating *in vivo* efficacy – toxicity profile of TEG001 in humanized mice xenografts against primary human AML disease and healthy hematopoietic cells. J Immunother Cancer. 2019;7:69.
57. Gründer C, van Dorp S, Hol S, Drent E, Straetemans T, Heijhuurs S, Scholten K, Scheper W, Sebestyen Z, Martens A, Strong R, Kuball J. γ9 and δ2CDR3 domains regulate functional avidity of T cells harboring γ9δ2TCRs. Blood. 2012;120:5153–62.

58. Sebestyen Z, Prinz I, Déchanet-Merville J, Silva-Santos B, Kuball J. Translating gamma-delta (γδ) T cells and their receptors into cancer cell therapies. Nat Rev Drug Discov. 2020;19:169–84.
59. Zhang H, Chen S, Zhang Y, Tian C, Pan J, Wang Y, Bai S, Wu Q, Su M, Xie D, Fu S, Li S, Zhang J, Chen Y, Zhu S, Qian Y, Bai L. Antigen priming induces functional reprogramming in iNKT cells via metabolic and epigenetic regulation: an insight into iNKT cell-based anti-tumor immunotherapy. Cancer Immunol Res. 2023;11:1598–610.
60. Li J, Moresco P, Fearon DT. Intratumoral NKT cell accumulation promotes antitumor immunity in pancreatic cancer. Proc Natl Acad Sci USA. 2024;121:e2403917121.
61. Boonchalermvichian C, Yan H, Gupta B, Rubin A, Baker J, Negrin RS. invariant Natural Killer T cell therapy as a novel therapeutic approach in hematological malignancies. Front Transplant. 2024;3:1353803.
62. Hombach AA, Geumann U, Gunther C, Hermann FG, Abken H. IL7-IL12 engineered Mesenchymal Stem Cells (MSCs) Improve A CAR T cell attack against colorectal cancer cells. Cells. 2020;9:873.
63. Sockolosky JT, Trotta E, Parisi G, Picton L, Su LL, Le AC, Chhabra A, Silveria SL, George BM, King IC, Tiffany MR, Jude K, Sibener LV, Baker D, Shizuru JA, Ribas A, Bluestone JA, Garcia KC. Selective targeting of engineered T cells using orthogonal IL-2 cytokine-receptor complexes. Science. 2018;359:1037–42.
64. Kagoya Y, Tanaka S, Guo T, Anczurowski M, Wang CH, Saso K, Butler MO, Minden MD, Hirano N. A novel chimeric antigen receptor containing a JAK-STAT signaling domain mediates superior antitumor effects. Nat Med. 2018;24:352–9.
65. Spolski R, Leonard WJ. IL-21 is an immune activator that also mediates suppression via IL-10. Crit Rev Immunol. 2010;30:559–70.
66. Liu Y, Dang Y, Zhang C, Liu L, Cai W, Li L, Fang L, Wang M, Xu S, Wang G, Zheng J, Li H. IL-21-armored B7H3 CAR-iNKT cells exert potent antitumor effects. iScience. 2024;27:108597.
67. Coquet JM, Kyparissoudis K, Pellicci DG, Besra G, Berzins SP, Smyth MJ, Godfrey DI. IL-21 is produced by NKT cells and modulates NKT cell activation and cytokine production. J Immunol. 2007;178:2827–34.
68. Maeda M, Yanagawa Y, Iwabuchi K, Minami K, Nakamaru Y, Takagi D, Fukuda S, Onoé K. IL-21 enhances dendritic cell ability to induce interferon-gamma production by natural killer T cells. Immunobiology. 2007;212:537–47.
69. Yi JS, Cox MA, Zajac AJ. Interleukin-21: a multifunctional regulator of immunity to infections. Microbes Infect. 2010;12:1111–9.
70. Deng S, Sun Z, Qiao J, Liang Y, Liu L, Dong C, Shen A, Wang Y, Tang H, Fu YX, Peng H. Targeting tumors with IL-21 reshapes the tumor microenvironment by proliferating PD-1intTim-3-CD8+ T cells. JCI Insight. 2020;5:e132000.
71. Castillo EF, Acero LF, Stonier SW, Zhou D, Schluns KS. Thymic and peripheral microenvironments differentially mediate development and maturation of iNKT cells by IL-15 transpresentation. Blood. 2010;116:2494–503.
72. Crowe NY, Coquet JM, Berzins SP, Kyparissoudis K, Keating R, Pellicci DG, Hayakawa Y, Godfrey DI, Smyth MJ. Differential antitumor immunity mediated by NKT cell subsets *in vivo*. J Exp Med. 2005;202:1279–88.
73. Gordy LE, Bezbradica JS, Flyak AI, Spencer CT, Dunkle A, Sun J, Stanic AK, Boothby MR, He YW, Zhao Z, Van Kaer L, Joyce S. IL-15 regulates homeostasis and terminal maturation of NKT cells. J Immunol. 2011;187:6335–45.
74. Zhu L, Xie X, Zhang L, Wang H, Jie Z, Zhou X, Shi J, Zhao S, Zhang B, Cheng X, Sun S-C. TBK-binding protein 1 regulates IL-15-induced autophagy and NKT cell survival. Nat Commun. 2018;9:2812.
75. Cheng L, Du X, Wang Z, Ju J, Jia M, Huang Q, Xing Q, Xu M, Tan Y, Liu M, Du P, Su L, Wang S. Hyper-IL-15 suppresses metastatic and autochthonous liver cancer by promoting tumor-specific $CD8^+$ T cell responses. J Hepatol. 2014;61:1297–303.

76. Ding ZC, Habtetsion T, Cao Y, Li T, Liu C, Kuczma M, Chen T, Hao Z, Bryan L, Munn DH, Zhou G. Adjuvant IL-7 potentiates adoptive T cell therapy by amplifying and sustaining polyfunctional antitumor CD4+ T cells. Sci Rep. 2017;7:12168.
77. Liu Y, Wang G, Chai D, Dang Y, Zheng J, Li H. iNKT: a new avenue for CAR-based cancer immunotherapy. Transl Oncol. 2022;17:101342.
78. Trujillo-Ocampo A, Cho HW, Clowers M, Pareek S, Ruiz-Vazquez W, Lee SE, Im JS. IL-7 during antigenic stimulation using allogeneic dendritic cells promotes expansion of CD45RA(−)CD62L(+)CD4(+) invariant NKT cells with Th-2 biased cytokine production profile. Front Immunol. 2020;11:567406.
79. Bosma A, Abdel-Gadir A, Isenberg DA, Jury EC, Mauri C. Lipid-antigen presentation by CD1d(+) B cells is essential for the maintenance of invariant natural killer T cells. Immunity. 2012;36:477–90.
80. Paul S, Chhatar S, Mishra A, Lal G. Natural killer T cell activation increases iNOS(+) CD206(−) M1 macrophage and controls the growth of solid tumor. J Immunother Cancer. 2019;7:208.
81. Cui J, Shin T, Kawano T, Sato H, Kondo E, Toura I, Kaneko Y, Koseki H, Kanno M, Taniguchi M. Requirement for Valpha14 NKT cells in IL-12-mediated rejection of tumors. Science. 1997;278:1623–6.
82. Morita M, Motoki K, Akimoto K, Natori T, Sakai T, Sawa E, Yamaji K, Koezuka Y, Kobayashi E, Fukushima H. Structure-activity relationship of alpha-galactosylceramides against B16-bearing mice. J Med Chem. 1995;38:2176–87.
83. Li YR, Zeng S, Dunn ZS, Zhou Y, Li Z, Yu J, Wang YC, Ku J, Cook N, Kramer A, Yang L. Off-the-shelf third-party HSC-engineered iNKT cells for ameliorating GvHD while preserving GvL effect in the treatment of blood cancers. iScience. 2022;25:104859.
84. Li YR, Zhou Y, Kim YJ, Zhu Y, Ma F, Yu J, Wang YC, Chen X, Li Z, Zeng S, Wang X, Lee D, Ku J, Tsao T, Hardoy C, Huang J, Cheng D, Montel-Hagen A, Seet CS, Crooks GM, Larson SM, Sasine JP, Wang X, Pellegrini M, Ribas A, Kohn DB, Witte O, Wang P, Yang L. Development of allogeneic HSC-engineered iNKT cells for off-the-shelf cancer immunotherapy. Cell Rep Med. 2021;2:100449.
85. Rotolo A, Caputo VS, Holubova M, Baxan N, Dubois O, Chaudhry MS, Xiao X, Goudevenou K, Pitcher DS, Petevi K, Kachramanoglou C, Iles S, Naresh K, Maher J, Karadimitris A. Enhanced anti-lymphoma activity of CAR19-iNKT cells underpinned by dual CD19 and CD1d targeting. Cancer Cell. 2018;34:596–610.e11.
86. Ihara F, Sakurai D, Takami M, Kamata T, Kunii N, Yamasaki K, Iinuma T, Nakayama T, Motohashi S, Okamoto Y. Regulatory T cells induce CD4(−) NKT cell anergy and suppress NKT cell cytotoxic function. Cancer Immunol Immunother. 2019;68:1935–47.
87. Poels R, Drent E, Lameris R, Katsarou A, Themeli M, van der Vliet HJ, de Gruijl TD, van de Donk N, Mutis T. Preclinical evaluation of invariant Natural Killer T cells modified with CD38 or BCMA chimeric antigen receptors for multiple myeloma. Int J Mol Sci. 2021;22(3):1096.
88. O'Neal J, Cooper ML, Ritchey JK, Gladney S, Niswonger J, Gonzalez LS, Street E, Haas GJ, Carter A, Amayta PN, Gao F, Lee BH, Choi D, Berrien-Elliott M, Zhou A, Fehniger TA, Rettig MP, DiPersio JF. Anti-myeloma efficacy of CAR-iNKT is enhanced with a long-acting IL-7, rhIL-7-hyFc. Blood Adv. 2023;7:6009–22.
89. Shou D, Wen L, Song Z, Yin J, Sun Q, Gong W. Suppressive role of myeloid-derived suppressor cells (MDSCs) in the microenvironment of breast cancer and targeted immunotherapies. Oncotarget. 2016;7:64505–11.
90. Ito E, Inuki S, Izumi Y, Takahashi M, Dambayashi Y, Ciacchi L, Awad W, Takeyama A, Shibata K, Mori S, Mak JYW, Fairlie DP, Bamba T, Ishikawa E, Nagae M, Rossjohn J, Yamasaki S. Sulfated bile acid is a host-derived ligand for MAIT cells. Sci Immunol. 2024;9:eade6924.
91. Zhang B, Chen P, Zhu J, Lu Y. The quantity, function and anti-tumor effect of Mucosal associated invariant T cells in patients with bladder cancer. Int Immunopharmacol. 2024;133:111892.
92. Priya R, Brutkiewicz RR. MR1 tetramer-based artificial APCs expand MAIT cells from human peripheral blood that effectively kill glioblastoma cells. Immunohorizons. 2021;5:500–11.

93. Parrot T, Healy K, Boulouis C, Sobkowiak MJ, Leeansyah E, Aleman S, Bertoletti A, Sallberg Chen M, Sandberg JK. Expansion of donor-unrestricted MAIT cells with enhanced cytolytic function suitable for TCR redirection. JCI Insight. 2021;6:e140074.
94. Dusseaux M, Martin E, Serriari N, Péguillet I, Premel V, Louis D, Milder M, Le Bourhis L, Soudais C, Treiner E, Lantz O. Human MAIT cells are xenobiotic-resistant, tissue-targeted, CD161hi IL-17-secreting T cells. Blood. 2011;117:1250–9.
95. Godfrey DI, Koay HF, McCluskey J, Gherardin NA. The biology and functional importance of MAIT cells. Nat Immunol. 2019;20:1110–28.
96. Yan J, Allen S, McDonald E, Das I, Mak JYW, Liu L, Fairlie DP, Meehan BS, Chen Z, Corbett AJ, Varelias A, Smyth MJ, Teng MWL. MAIT cells promote tumor initiation, growth, and metastases via tumor MR1. Cancer Discov. 2020;10:124–41.
97. Ruf B, Catania VV, Wabitsch S, Ma C, Diggs LP, Zhang Q, Heinrich B, Subramanyam V, Cui LL, Pouzolles M, Evans CN, Chari R, Sakai S, Oh S, Barry CE 3rd, Barber DL, Greten TF. Activating mucosal-associated invariant T cells induces a broad antitumor response. Cancer Immunol Res. 2021;9:1024–34.
98. Themeli M, Kloss CC, Ciriello G, Fedorov VD, Perna F, Gonen M, Sadelain M. Generation of tumor-targeted human T lymphocytes from induced pluripotent stem cells for cancer therapy. Nat Biotechnol. 2013;31:928–33.
99. Snauwaert S, Verstichel G, Bonte S, Goetgeluk G, Vanhee S, Van Caeneghem Y, De Mulder K, Heirman C, Stauss H, Heemskerk MH, Taghon T, Leclercq G, Plum J, Langerak AW, Thielemans K, Kerre T, Vandekerckhove B. *In vitro* generation of mature, naive antigen-specific CD8(+) T cells with a single T-cell receptor by agonist selection. Leukemia. 2014;28:830–41.
100. Li YR, Dunn ZS, Garcia G Jr, Carmona C, Zhou Y, Lee D, Yu J, Huang J, Kim JT, Arumugaswami V, Wang P, Yang L. Development of off-the-shelf hematopoietic stem cell-engineered invariant natural killer T cells for COVID-19 therapeutic intervention. Stem Cell Res Ther. 2022;13:112.
101. Zhu Y, Smith DJ, Zhou Y, Li YR, Yu J, Lee D, Wang YC, Di Biase S, Wang X, Hardoy C, Ku J, Tsao T, Lin LJ, Pham AT, Moon H, McLaughlin J, Cheng D, Hollis RP, Campo-Fernandez B, Urbinati F, Wei L, Pang L, Rezek V, Berent-Maoz B, Macabali MH, Gjertson D, Wang X, Galic Z, Kitchen SG, An DS, Hu-Lieskovan S, Kaplan-Lefko PJ, De Oliveira SN, Seet CS, Larson SM, Forman SJ, Heath JR, Zack JA, Crooks GM, Radu CG, Ribas A, Kohn DB, Witte ON, Yang L. Development of hematopoietic stem cell-engineered invariant Natural Killer T cell therapy for cancer. Cell Stem Cell. 2019;25:542–57.e9.
102. Ruf B, Bruhns M, Babaei S, Kedei N, Ma L, Revsine M, Benmebarek MR, Ma C, Heinrich B, Subramanyam V, Qi J, Wabitsch S, Green BL, Bauer KC, Myojin Y, Greten LT, McCallen JD, Huang P, Trehan R, Wang X, Nur A, Murphy Soika DQ, Pouzolles M, Evans CN, Chari R, Kleiner DE, Telford W, Dadkhah K, Ruchinskas A, Stovroff MK, Kang J, Oza K, Ruchirawat M, Kroemer A, Wang XW, Claassen M, Korangy F, Greten TF. Tumor-associated macrophages trigger MAIT cell dysfunction at the HCC invasive margin. Cell. 2023;186:3686–705.e32.
103. Li YR, Brown J, Yu Y, Lee D, Zhou K, Dunn ZS, Hon R, Wilson M, Kramer A, Zhu Y, Fang Y, Yang L. Targeting immunosuppressive tumor-associated macrophages using innate T cells for enhanced antitumor reactivity. Cancers (Basel). 2022;14:2749.
104. Bohineust A, Tourret M, Derivry L, Caillat-Zucman S. Mucosal-associated invariant T (MAIT) cells, a new source of universal immune cells for chimeric antigen receptor (CAR)-cell therapy. Bull Cancer. 2021;108:S92–s5.
105. Dogan M, Karhan E, Kozhaya L, Placek L, Chen X, Yigit M, Unutmaz D. Engineering human MAIT cells with chimeric antigen receptors for cancer immunotherapy. J Immunol. 2022;209:1523–31.
106. Peng Q, Huang R, Wang H, Xiao H, Wang Y, Zhai Z, Wang Z. Immune characteristics and prognostic implications of mucosal-associated invariant T cells in acute myeloid leukemia. Cancer Immunol Immunother. 2023;72:4399–414.
107. Butterfield LH, Najjar YG. Immunotherapy combination approaches: mechanisms, biomarkers and clinical observations. Nat Rev Immunol. 2023;24:399.

108. Shelburne N, Bevans M. Non-myeloablative allogeneic hematopoietic stem cell transplantation. Semin Oncol Nurs. 2009;25:120–8.
109. Ingram Z, Madan S, Merchant J, Carter Z, Gordon Z, Carey G, Webb TJ. Targeting Natural Killer T cells in solid malignancies. Cells. 2021;10:1329.
110. Vydra J, Cosimo E, Lesny P, Wanless RS, Anderson J, Clark AG, Scott A, Nicholson EK, Leek M. A Phase I Trial of Allogeneic gamma-delta T lymphocytes from haploidentical donors in patients with refractory or relapsed acute myeloid leukemia. Clin Lymphoma Myeloma Leuk. 2023;23:e232–e9.
111. Becker SA, Petrich BG, Yu B, Knight KA, Brown HC, Raikar SS, Doering CB, Spencer HT. Enhancing the effectiveness of γδ T cells by mRNA transfection of chimeric antigen receptors or bispecific T cell engagers. Mol Ther Oncolytics. 2023;29:145–57.
112. Branella GM, Lee JY, Okalova J, Parwani KK, Alexander JS, Arthuzo RF, Fedanov A, Yu B, McCarty D, Brown HC, Chandrakasan S, Petrich BG, Doering CB, Spencer HT. Ligand-based targeting of c-kit using engineered gamma-delta T cells as a strategy for treating acute myeloid leukemia. Front Immunol. 2023;14:1294555.
113. Kakimi K, Matsushita H, Murakawa T, Nakajima J. γδ T cell therapy for the treatment of non-small cell lung cancer. Transl Lung Cancer Res. 2013;3:23–33.
114. Nakajima J, Murakawa T, Fukami T, Goto S, Kaneko T, Yoshida Y, Takamoto S, Kakimi K. A phase I study of adoptive immunotherapy for recurrent non-small-cell lung cancer patients with autologous γδ T cells. Eur J Cardiothorac Surg. 2010;37:1191–7.
115. Kim YH, Bagot M, Pinter-Brown L, Rook AH, Porcu P, Horwitz SM, Whittaker S, Tokura Y, Vermeer M, Zinzani PL, Sokol L, Morris S, Kim EJ, Ortiz-Romero PL, Eradat H, Scarisbrick J, Tsianakas A, Elmets C, Dalle S, Fisher DC, Halwani A, Poligone B, Greer J, Fierro MT, Khot A, Moskowitz AJ, Musiek A, Shustov A, Pro B, Geskin LJ, Dwyer K, Moriya J, Leoni M, Humphrey JS, Hudgens S, Grebennik DO, Tobinai K, Duvic M, Abhyankar S, Akilov O, Alpdogan O, Beylot-Barry M, Boh E, Caballero D, Cowan R, Dreno B, Dummer R, Fenske T, Foss F, Fukuhara N, Giri P, Habe K, Hamada T, Hatake K, Iida S, Ishikawa O, Iversen L, Kiyohara E, Koga H, Korman N, Kuss BJ, Lamar Z, Lansigan F, Lechowicz MJ, Lerner A, Magnolo N, Mark L, Miyagaki T, Munoz J, Nicolay JP, Nishiwaki K, Okamoto H, Ohtsuka M, Pacheco T, Querfeld C, Rapini RP, Sano S, Tanaka M, Tharp MD, Uehara J, Wada H, Wells J, Wilcox RA, William B, Yonekura K. Mogamulizumab versus vorinostat in previously treated cutaneous T-cell lymphoma (MAVORIC): an international, open-label, randomized, controlled phase 3 trial. Lancet Oncol. 2018;19:1192–204.
116. Cowan RA, Scarisbrick JJ, Zinzani PL, Nicolay JP, Sokol L, Pinter-Brown L, Quaglino P, Iversen L, Dummer R, Musiek A, Foss F, Ito T, Rosen J-P, Medley MC. Efficacy and safety of mogamulizumab by patient baseline blood tumor burden: a post hoc analysis of the MAVORIC trial. J Eur Acad Dermatol Venereol. 2021;35:2225–38.
117. Sokol L, Duvic M, Musiek A, Kim E, Eradat H, Elmets C, Shinohara M, Fisher D, Poligone B, Pro B, Reddy N, Halwani A, Herr F, Ito T, Kim Y. TCL-139: safety of mogamulizumab in mycosis fungoides and Sézary syndrome: final results from the Phase 3 MAVORIC Study. Clin Lymphoma Myeloma Leuk. 2020;20:S253.
118. Saura-Esteller J, de Jong M, King LA, Ensing E, Winograd B, de Gruijl TD, Parren P, van der Vliet HJ. Gamma delta T-cell based cancer immunotherapy: past-present-future. Front Immunol. 2022;13:915837.

The Emerging Role of CAR-NK Cells: Next-Generation Cancer Immunotherapy

Apoorva Narain and Rikesh Kumar Dubey

Abstract The use of the body's own defenses to combat cancer represents a cornerstone of cancer immunotherapy. Enhancing the innate functionality of the immune system enhances its ability to identify and eradicate cancerous cells within the body. Among the array of cancer immunotherapy approaches, chimeric antigen receptor (CAR)-natural killer (NK) cell therapy has emerged as a promising contender. CARs, engineered synthetic receptors, redirect NK cells to identify and eliminate cells expressing specific antigens. NK cells, which are integral to the innate immune response, exhibit rapid and potent cytotoxicity against cancer cells without requiring prior sensitization or peptide antigen recognition. Activated NK cells execute targeted cell death through perforin and granzyme release, alongside the secretion of proinflammatory cytokines such as interferon-gamma (IFN-gamma) and tumor necrosis factor (TNF-alpha). Unlike T cells, NK cells possess inherent cytotoxicity, offering advantages in cancer immunotherapy. When equipped with CARs, NK cells synergistically target and destroy transformed cells. The robust antitumor activity and favorable safety profile of these compounds have spurred significant interest in CAR-NK-cell immunotherapy. To overcome limitations associated with autologous NK cells, including reduced efficacy and patient-specific production requirements, allogeneic NK cells have emerged as promising platforms for CAR engineering. NK cells have demonstrated efficacy against certain hematological cancers and show promise in preventing metastasis, as evidenced by animal studies and clinical observations. Preclinical investigations highlight the ability of NK cells to target and eliminate host lymphohematopoietic cells involved in donor cell rejection. Overall, advancements in CAR-NK cell technology offer valuable

A. Narain (✉)
Versiti Blood Research Institute, Milwaukee, WI, USA
e-mail: apoorvanarain.micro@gmail.com

R. K. Dubey (✉)
Division of Medical Research, SRM Institute of Science and Technology, Kattankulathur, Tamil Nadu, India
e-mail: rikeshkd@srmist.edu.in

A. K. Mishra, S. K. Malonia (eds.), *Cell-based Immunotherapies for Cancer*,
https://doi.org/10.1007/978-3-031-88695-9_12

insights into the intricacies of cancer immunotherapy and hold promise for further therapeutic development.

Keywords Immunotherapy · CAR-NK cells · CAR-T cells · Tumor microenvironment

1 Introduction

Natural killer (NK) cells, innate lymphocytes with cytotoxic capabilities, can eliminate cancerous or virally infected cells. Their function is finely tuned by a range of germline-encoded receptors, which can either trigger activating signals or inhibit responses, thereby ensuring self-tolerance while eliciting NK-cell activity. NK cells employ two main pathways to exert their cytotoxic effects. They can release cytotoxic granules containing perforin and granzymes, or they can induce apoptosis through death receptor-mediated mechanisms by expressing TRAIL and/or Fas ligand (FasL) to engage TRAIL-R1/-R2 or CD95/FasL on the surface of target cells [1].

The cytotoxic activity mediated by NK cells or T cells is fundamental to various immunotherapeutic approaches against cancer. For example, antibody-based therapies such as rituximab, which target CD20 in non-Hodgkin's lymphoma, rely on NK-cell-mediated antibody-dependent cellular cytotoxicity (ADCC) [2]. Chimeric antigen receptor T (CAR-T) and NK-cell therapies aim to harness the direct cytotoxic potential of these cells against tumors, whereas checkpoint inhibitor therapy enhances the activity of cytotoxic T cells [3].

As we understand the molecular mechanisms governing NK-cell activity, their potential application in cancer immunotherapy has become increasingly apparent. NK cells play pivotal roles in both cellular and humoral immune responses through distinct activation mechanisms, contributing to early host defense against tumors and viruses. They execute these dual functions primarily through contact-dependent cytotoxicity and cytokine production for immune modulation. The perforin (Prf1)- and granzyme B (GzmB)-mediated pathways predominantly mediate target cell apoptosis, whereas cytokines such as interferon-γ (IFN-γ) and tumor necrosis factor-α (TNF-α) regulate immune responses.

Compared with those of T and B cells, our understanding of the checkpoints and developmental stages guiding NK-cell lineage commitment remains incomplete. Nevertheless, ongoing research is elucidating the developmental processes that steer hematopoietic stem cells (HSCs) toward NK-cell differentiation [4]. Recent experimental evidence highlights the dependence of NK-cell differentiation on specific cytokines, temporally induced transcription factors, and microRNA (miRNA)-based gene expression [5]. Furthermore, the development of protocols for in vitro NK-cell differentiation on the basis of cytokine reconstitution offers a promising strategy for expanding NK cells for therapeutic purposes [6].

Advancements in manipulating NK-cell activation and development have instilled hope that these cells could serve as potent immunotherapeutic agents against cancers and other diseases.

2 Mechanisms of NK-Cell Cytotoxicity Against Tumors

Natural killer (NK) cells, a subset of innate immune cells, exhibit spontaneous cytolytic activity against stressed cells such as tumors and virus-infected cells. Upon activation, NK cells release chemokines (CCL1, CCL2, CCL3, CCL4, CCL5, and CXCL8) and various cytokines, including granulocyte–macrophage colony–stimulating factor (GM-CSF), tumor necrosis factor-α (TNF-α), and interferon-γ (IFN-γ), which modulate the function of other immune cells. In mice, strains such as C57BL/6, FVB/N, and NZB NK cells are identified as CD3 − NK1.1+ cells, whereas strains such as BALB/c, CBA/J, AKR, and others are identified as CD3 − CD49b + cells. In humans, NK cells are identified as CD3 − CD56+ cells, constituting a small percentage of lymphocytes in peripheral blood and peripheral blood mononuclear cells (PBMCs) [7]. Under physiological conditions, NK cells are distributed in various tissues, including the gut, skin, lungs, liver, joints, breasts, uterus, and kidney. They constitute a significant portion of hepatic and pulmonary lymphocytes. During early pregnancy, a specific subset of NK cells controls fetal–maternal interface development, constituting a substantial portion of uterine lymphoid cells [8].

NK-cell development primarily occurs in the bone marrow but can also occur in the liver and thymus, where it progresses through stages of development and the acquisition of specific receptors. All NK receptors are germline encoded and independent of RAG-mediated recombination. Various factors, including cell-intrinsic signals (transcription factors) and external signals (growth factors and cytokines), regulate NK-cell development [7]. NK cells play crucial roles in innate immunity, shaping the initial immune response to viral infections, tumors, and organ transplantation.

2.1 Inhibitory Receptors on NK Cells

Upon ligand engagement, inhibitory receptors on NK cells trigger phosphorylation in their cytoplasmic tails through immunoreceptor tyrosine-based inhibitory motifs (ITIMs), recruiting phosphatases such as Src homology-containing tyrosine phosphatase 1 & 2 (SHP-1/2) and SH2 domain-containing inositol-5-phosphatase (SHIP) to neutralize activating signals. SHP-1 and SHP-2 dephosphorylate immunoreceptor tyrosine-based activation motif (ITAM)-bearing Vav-1 molecules by inhibiting downstream signaling [9].

2.2 Activating Receptors on NK Cells

NK-cell activation occurs via the recognition of stress-induced molecules by activating receptors. These receptors signal through immunoreceptor tyrosine-based activation motifs (ITAMs), triggering phosphorylation by Src family tyrosine kinases [10]. Subsequent signaling involves the activation of proteins such as phosphatidylinositol-3-OH kinase [PI(3)K], Zap70, Syk, SLP-76, and Shc, leading to actin cytoskeleton reorganization, calcium elevation, and the release of cytolytic granules containing perforin and granzymes, together with chemokine and cytokine gene transcription [7].

2.3 *Cytotoxic and Effector Immune Response of NK Cells*

NK-cell cytotoxicity involves several steps: immunological synapse formation, microtubule organizing center (MTOC) polarization, secretory lysosome docking with the plasma membrane, and fusion with target cells, culminating in degranulation. This process releases cytotoxic molecules such as perforin and granzymes, which, along with granulysin, mediate target cell killing [7]. The family of human granzymes includes granzymes A, B, H, K, and M, with A and B being extensively studied. NK cells release multiple granzymes during degranulation, collectively contributing to target cell death.

2.4 *NK-Cell Degranulation and Granzyme Entry into Target Cells*

NK-cell killing involves integrin-mediated adhesion, immunologic synapse formation, and lytic granule fusion with the target cell membrane [11]. Perforin creates pores in the target cell membrane, thereby facilitating granzyme entry. Granzymes enter target cells via endocytosis facilitated by perforin-mediated membrane damage and calcium influx [12–15].

2.5 *FasL-Mediated and TRAIL-Mediated Cytotoxicity*

TRAIL is a type II transmembrane protein that is homologous to FasL and TNF. FasL binds to CD95 receptors on target cells, activating apoptotic signaling cascades [16]. TRAIL expression can be induced on NK-cell surfaces by cytokine stimulation. TRAIL binds to various receptors, inducing apoptosis or NF-κB signaling [17]. Both FasL and TRAIL lead to targeted cell apoptosis, as they are cleaved for surface expression and stored in secretory granules [1, 18, 19]. While freshly isolated NK cells do not express detectable levels of TRAIL on their surface, stimulation with IL-12, IL-15 or IL-2 can induce their surface expression [20]. TRAIL-mediated cytotoxicity is important for the NK-cell-mediated control of viral cancer and infection, especially in the liver [21]. Apoptosis is caused primarily by the TRAIL-R1 and TRAIL-R2 receptors, whereas osteoprotegerin, TRAIL-R3, and TRAIL-R4 induce mostly NF-κB signaling. They are likely decoy receptors that limit the binding of TRAIL to proapoptotic receptors [22, 23].

3 Tumor Microenvironment and NK Cells

NK cells have shown promise in treating various cancers, but their effectiveness against solid tumors remains inadequate. This shortfall largely stems from the immunomodulatory nature of the tumor microenvironment (TME) [24]. The expression of ligands for NK-activating receptors (NK-ARs) and the downregulation of inhibitory ligands on cancer cells can trigger NK cells to release IFN-γ and TNF-α. Consequently, disrupting the interaction between NK-ARs and their ligands or enhancing NK-AR–ligand binding could offer a viable strategy for eliciting an antitumor response. Notably, depleting NK-cell populations before tumor transplantation could lead to a more aggressive phenotype with metastatic tumors.

Tbx21, also known as T-bet, is a transcription factor crucial for NK-cell differentiation. Studies involving Tbx21−/− mice intravenously injected with colorectal carcinoma or melanoma cells have demonstrated increased susceptibility to metastasis. Additionally, the invasion of primary tumors by NK cells and their migration to metastatic sites rely on heparanase. When exposed to methylcholanthrene, a carcinogen, mice that specifically lack heparanase in NK cells exhibit increased susceptibility to mammary carcinoma, prostate carcinoma, lymphoma, and metastatic melanoma [25]. Moreover, NK cells play a pivotal role in controlling tumor development and dissemination by eliminating cancer stem cells (CSCs), a subset of cells crucial for tumor generation and evolution. CSCs typically exhibit low levels of PD-L1, CD54, and PD-L1 and high expression of CD44 [26, 27].

Accumulating evidence suggests that the TME produces soluble modulators that negatively influence NK-cell maturation, proliferation, and effector function. These immunosuppressive factors may act directly on NK cells or indirectly by stimulating regulatory T cells (Tregs), antigen-presenting cells (APCs), and myeloid-derived suppressor cells (MDSCs) to produce additional immunosuppressive molecules. Notably, TGF-β, a cytokine prevalent in the TME, inhibits the expansion and function of effector cells while promoting Treg proliferation. It impairs NK-cell function directly or indirectly through cell–cell contact between NK cells and other cytokine-producing cells [28]. TGF-β limits NK-cell cytotoxicity and IFN-γ production by inhibiting the T-bet transcription factor (SMAD3) and downregulating the expression of NKG2D and NKp30 [29].

3.1 Tumor-Derived Metabolites

The TME also accumulates tumor-derived metabolites such as adenosine and lactate, which hinder antitumor responses. The adenosine concentration increases during tumor hypoxia, modulating immune responses [24]. Hypoxia induces AMP and ATP release, with ectonucleotidases CD39 and CD73 converting extracellular ATP to AMP and then to adenosine [30]. Stimulation of NK cells via the adenosine A2A receptor (A2AR) suppresses their effector functions, as evidenced by the inhibition of cytotoxic activity in mouse lymphokine-activated killer (LAK) cells by adenosine [31].

The exposure of NK cells to lactic acid suppresses IFNγ production following PMA/ionomycin stimulation. Lactic acid inhibits the upregulation of nuclear factor of activated T cells (NFAT), which is crucial for IFNγ transcription. Additionally, lactic acid uptake leads to impaired energy metabolism and intracellular acidification, as observed in liver-resident NK cells, promoting apoptosis [32]. The accumulation of lactate in the TME results from the metabolic reprogramming of tumors, favoring glycolytic metabolism over oxidative phosphorylation (OXPHOS) [24].

3.2 Glucose Restriction

NK cells utilize glucose for ATP and NADPH generation and as a carbon source for biomolecules. They express GLUT1, GLUT3, and GLUT4, with some studies reporting GLUT8 and HMIT expression. Upon cytokine stimulation, NK cells increase GLUT1 expression, which is correlated with increased glucose uptake and glycolysis during activation [24].

In the TME, metabolic competition promotes cancer progression by impairing antigen-specific responses of tumor-infiltrating T cells, and limited glucose dampens T-cell antitumor activity. Thus, tumor-driven glucose restriction in the TME may diminish NK-cell glycolysis and impair their antitumor functions.

4 Chimeric Antigen Receptors (CARs)

CARs, or chimeric antigen receptors, are engineered receptors designed to recognize specific antigens, enabling immune cells to target these antigens more effectively. In cancer immunotherapy, CARs aim to quickly create tumor-targeting cells, bypassing the slower processes associated with traditional immunization [33]. When CARs are incorporated into immune cells, these modified cells act as "living drugs," offering both immediate and long-lasting therapeutic effects. CARs can target any cell surface molecule, thus overcoming tolerance to self-antigens and addressing gaps in the natural T-cell antigen recognition repertoire. They can be used with various T-cell subsets, T-cell progenitors, and other immune cells, such as natural killer (NK) cells. CARs provide a broader range of functional effects than do transduced T-cell receptors (TCRs), where the affinity of a TCR for an antigen primarily determines T-cell fate. Both CARs and TCRs have unique strengths and weaknesses. CARs are limited to recognizing cell surface antigens, whereas TCRs can recognize both cell surface and intracellular proteins. However, CARs do not require antigen processing and presentation by HLAs, making them more widely applicable to patients with diverse HLA types [33].

CARs interact with molecules that do not require peptide processing or HLA expression for recognition, thus targeting antigens across any HLA background, unlike TCRs, which need to match the patient's haplotype. Moreover, CARs can target tumor cells with downregulated HLA expression or impaired proteasomal antigen processing, mechanisms contributing to tumor escape from TCR-mediated immunity. Another advantage of the broad applicability of CARs is their ability to bind not only to proteins but also to carbohydrate and glycolipid structures, broadening the range of potential targets. The antigen-binding component of the CAR is largely defined by the signaling components integrated into the CAR's cytoplasmic domain. Optimal epitopes for CAR-targeting selection rely on several factors, including the epitope's location and distance from the cell surface, the structure of the spacer region between the single-chain variable fragment (scFv) and the transmembrane region, and the length and protrusion of the CAR from the T-cell membrane. While CARs typically target highly expressed antigens, minimal thresholds for CAR sensitivity remain uncertain compared with those of TCRs [33].

CARs were initially engineered for expression in T cells, and several generations have since been developed, each enhancing CAR-T-cell efficacy in tumor recognition. The first-generation CAR-T cells exhibited antitumor reactivity but had limited proliferation and short in vivo persistence, requiring exogenous IL-2 for an effective response [33, 34]. Second-generation CARs improved T-cell proliferation, cytotoxicity, and persistence by incorporating costimulatory domains such as CD28, OX-40, or 4-1BB, although no single domain has proven universally superior [33, 35, 36]. Third-generation 'triple-decker' CARs integrated two costimulatory domains to enhance potency, yet their efficacy remains comparable to that of second-generation CARs [33]. Fourth-generation CARs, or TRUCKs (T cells redirected for universal cytokine-mediated killing), introduce cytokine transgenes to enhance efficacy while reducing systemic toxicity. Fifth-generation CARs incorporate membrane receptors to activate CAR-T cells and promote memory T-cell formation, offering improved safety and therapeutic potential, with strategies such as lenalidomide-gated CARs further refining control and therapeutic windows [36–39].

5 CAR-NK Cells vs CAR-T Cells: Advantages and Limitations

The CAR configuration is crucial for activating CAR-transduced cells, as the CARs used in CAR-NK cells resemble those used in CAR-T cells [40]. A standard CAR has four components. The extracellular binding domain grants specificity to CAR-modified effector cells, enabling them to target tumor-associated antigens (TAAs). The hinge region links the extracellular binding domain with the transmembrane domain. Intracellular signaling domains, which vary across CAR generations, dictate activation signal strength and influence killing activity. Compared with those engineered with CD3ζ-based CARs, NK cells engineered with DAP12-based CARs demonstrate superior performance [41]. Virus-based transduction methods enable stable integration into CAR-NK cell genomes, with retroviral vector transduction exhibiting high efficacy (43–93% in primary NK cells) [40]. However, insertional mutagenesis and its deleterious impacts limit its clinical application. RNA transfection methods are more cost-effective with greater gene transfer efficiency, although they typically result in transient expression of CAR constructs lasting approximately 3–5 days. While this method's narrow therapeutic window is a limitation, its transient nature may reduce CAR-related side effects such as on-target off-tumor effects [42–44] (Table 1).

In terms of safety, CAR-NK cells have advantages over CAR-T cells. CAR-T-cell therapy often leads to cytokine release syndrome (CRS) and neurotoxicity, which are mediated by proinflammatory cytokines such as TNFa, IL-1, and IL-6 [45]. Conversely, CAR-NK cells secrete a distinct spectrum of cytokines (e.g., IFN-γ and GM-CSF). Unlike CAR-T cells, which may induce life-threatening graft-versus-host disease (GVHD), NK cells can prevent GVHD by eliminating recipient antigen-presenting cells and cytotoxic T lymphocytes, thus increasing safety in clinical applications.

Moreover, CAR-NK cells may exhibit superior efficacy in tumor cell eradication compared with CAR-T cells [40]. By retaining natural cytotoxicity even if targeted tumor antigens are downregulated, NK cells employ various mechanisms, including natural cytotoxicity, antibody-dependent cell-mediated cytotoxicity (ADCC), TRAIL, and FAS/FasL pathways. Additionally, CAR-NK cells feature specialized costimulatory molecules (e.g., DAP10, DAP12, and 2B4) alongside shared domains such as CD28 and 4-1BB, enhancing NK-cell signaling specificity. Manufacturing CAR-NK cells is more convenient than generating CAR-T cells because of the absence of GVHD risk, allowing for donor selection from matched or HLA-mismatched donors and thus improving final product quality [46].

However, challenges persist in optimizing CAR-NK cell function and efficacy. Sensitivity to freezing and thawing reduces NK-cell antitumor capacity and survival rates [47, 48]. Immunosuppressive factors in the tumor microenvironment (TME), including cytokines (e.g., TGF-beta), inhibit CAR-NK cell function. Inhibitory receptors (e.g., TIGIT, PD-1, and NKG2A) also contribute to CAR-NK cell dysfunction [49, 50]. Strategies involving cytokine gene transduction and inhibitory gene knockout show promise. Identifying tumor-specific TAAs is crucial to avoid on-target off-tumor effects.

Novel CAR designs facilitate CAR-NK cell infiltration into tumor sites, bolstered by cytokines, immune checkpoint inhibitors, and monoclonal antibodies [51, 52]. Lymphoid-depleting chemotherapy and radiation before CAR-NK-cell infusion reduce the tumor burden and increase the effector-to-target ratios [53]. Preclinical studies and clinical trials have validated the safety and efficacy of CAR-NK-cell therapy in patients with hematological malignancies [54].

Table 1 Clinical trials for CAR-NK cell immunotherapy

NO. NCT	Other name/ID numbers	Start date	Phase	Disease	Target	Sponsor locations
NCT00995137	NKCD19 R01CA113482 NCI-2011-01226	October 2009	I	B-lineage acute lymphoblastic leukemia	CD19	St. Jude Children's Research Hospital
NCT01974479	NKCARCD19	September 2013	I	B-lineage acute lymphoblastic leukemia	CD20	National University Health System, Singapore
NCT03056339	2016-0641 NCI-2018-01221	June 21, 2017	I and II	B-lymphoid malignancies	CD19	M.D. Anderson Cancer Center
NCT03383978	EudraCT 2016–000225-39	December 1, 2017	I	Glioblastoma	HER2	Johann Wolfgang Goethe University Hospital
NCT03415100	NRC-NK-01	January 2, 2018	I	Metastatic solid tumors	NKG2D	The Third Affiliated Hospital of Guangzhou Medical University
NCT03656705	CNK-101	September 29, 2018	I	Non-small cell lung carcinoma	PD-1	Xinxiang medical university
NCT03692663	TABP EIC-01	December, 2018	Early I	Castration-resistant prostate cancer	PSMA	Allife Medical Science and Technology Co., Ltd.
NCT03824964	CD19/CD22 CAR-NK-BJZL-01	February 1, 2019	Early I	Relapsed or refractory B-cell lymphoma	CD19/ CD22	Allife Medical Science and Technology Co., Ltd.
NCT03692767	CD22 CAR-NK-BJZL-01	March 2019	Early I	Relapsed and refractory B-cell lymphoma	CD22	Allife Medical Science and Technology Co., Ltd.
NCT03690310	CD19 CAR-NK-BJZL-01	March 2019	Early I	Relapsed and refractory B-cell lymphoma	CD19	Allife Medical Science and Technology Co., Ltd.
NCT03692637	Mesothelin Car-NK-HNRM-01	March 2019	Early I	Epithelial ovarian cancer	Mesothelin	Allife Medical Science and Technology Co., Ltd.
NCT04245722	FT596–101	March 19, 2020	I	B-cell lymphoma, chronic lymphocytic leukemia	CD19	Fate Therapeutics

Table 1 (continued)

NO. NCT	Other name/ID numbers	Start date	Phase	Disease	Target	Sponsor locations
NCT04623944	NKX101–101	September 21, 2020	I	Adults With AML or MDS	NKG2D	Nkarta Inc.
NCT05215015	IBR733-T01 WX-IBR-7	November 30, 2020	Early I	Acute myeloid leukemia	CD33/ CLL1	Wuxi People's Hospital
NCT04639739	CAR-NK for NHL	December 17, 2020	Early I	Relapsed or refractory B-cell non-Hodgkin lymphoma	CD19	Xinqiao Hospital of Chongqing
NCT04747093	ITNK-2021	January 29, 2021	I and II	B-Cell malignancies	CD19	Nanfang Hospital of Southern Medical University
NCT04796675	CAR-NK-CD19 cells	April 10, 2021	I	B-lymphoid malignancies	CD19	Wuhan Union Hospital, China
NCT04887012	IR2021002168	May 1, 2021	I	Refractory or relapsed B-cell non-Hodgkin lymphoma	CD19	Second Affiliated Hospital, School of Medicine, Zhejiang University
NCT05020678	NKX019–101	August 20, 2021	I	Adults with B-cell cancers	CD19	Nkarta Inc.
NCT05137275	IBR854–03	November 24, 2021	Early I	Locally advanced or metastatic solid tumors	5 T4	Shanghai East Hospital
NCT05008536	BCMA NK for MM	October 1, 2021	Early I	Relapsed or refractory multiple myeloma	BCMA	Xinqiao Hospital of Chongqing
NCT05247957	CAR-NK-001	October 13, 2021	I	Relapsed or refractory acute myeloid leukemia	NKG2D	Hangzhou Cheetah Cell Therapeutics Co., Ltd
NCT05213195	CAR-NK-002	December 10, 2021	I	Refractory metastatic colorectal cancer	NKG2D	Zhejiang University
NCT04847466	10000096, 000096-C	December 14, 2021	II	Recurrent or metastatic gastric or head and neck cancer	PD-L1	National Cancer Institute (NCI)
NCT05008575	CD33 CAR-NK-AML	December 23, 2021	I	Relapsed or refractory acute myeloid leukemia	CD33	Xinqiao Hospital of Chongqing

NCT05194709	IBR854-T01, WX-IBR-8	December 30, 2021	Early I	Advanced solid tumors	5 T4	Wuxi People's Hospital
NCT05379647	NK-002 (QN-019a)	November 4, 2021	I	B-cell malignancies	CD19	Zhejiang University
NCT05182073	FT576-101	November 24, 2021	I	Multiple myeloma	BCMA	Fate Therapeutics
NCT05110742	2021-0526	June 30, 2022	I and II	Relapse or refractory hematological malignancies	CD5	M.D. Anderson Cancer Center
NCT05092451	2021-0386	August 1, 2022	I and II	Relapse or refractory hematological malignancies	CD70	M.D. Anderson Cancer Center
NCT05336409	CNTY-101-111-01	December 2022	I	Relapsed or refractory CD19-positive B-cell malignancies	CD19	Century Therapeutics, Inc.

Adapted from Li et al. [54]

6 Manufacture of CAR-NK Cells

The production of chimeric antigen receptor-natural killer (CAR-NK) cells encompasses several pivotal stages, integrating genetic engineering and cell culture methodologies. Representing a promising avenue in cancer immunotherapy, CAR-NK cells harness the innate cytotoxicity of NK cells, augmented with the specificity conferred by CARs [52]. The manufacturing process of CAR-NK cells is intricate and comprises multiple steps that require meticulous optimization to ensure the development of a safe, scalable, and efficacious therapeutic product. Ongoing advancements in genetic engineering, cell culture, and immunotherapy have continually increased the potential of CAR-NK cells in combating various forms of cancer.

6.1 Isolation of NK Cells

NK cells are derived from diverse sources, including peripheral blood, cord blood, and stem cell sources, such as induced pluripotent stem cells (iPSCs) or hematopoietic stem cells (HSCs), and employ differentiation protocols.

6.2 Activation and Expansion of NK Cells

Following isolation, NK cells undergo ex vivo activation and expansion to increase their numbers and functional capabilities. This process, which is conducted under good manufacturing practice (GMP) conditions, aims to produce a sufficient quantity of cells for therapeutic use [55–57]. Cytokines such as interleukin-2 (IL-2), IL-15, and IL-21, along with irradiated feeder cells, often genetically modified K562 cells expressing membrane-bound IL-21 or 4-1BBL, are employed to stimulate NK-cell proliferation and activation [58–60].

6.3 Genetic Modification to Express CAR

Genetic engineering techniques are utilized to introduce the CAR construct into NK cells. The enhancement of transduction efficiency involves the use of lentiviral and retroviral vectors for stable integration of the CAR gene into the NK-cell genome [61]. Nonviral methods such as transposon systems (e.g., Sleeping Beauty or PiggyBac) and CRISPR/Cas9 may also be employed [62]. The CAR construct typically comprises an extracellular domain (commonly a single-chain variable fragment derived from an antibody), Hinge and transmembrane domains, and intracellular signaling domains (e.g., CD3ζ and costimulatory domains such as CD28 or 4-1BB) [63–67].

6.4 Expansion and Quality Control

After genetic modification, CAR-NK cells undergo further expansion to attain the required cell dose, with rigorous monitoring of cell viability and purity, CAR expression, and functional assays to ensure efficacy and safety.

6.5 Cryopreservation

CAR-NK cells are often cryopreserved to maintain viability and function until they are used for clinical administration, and controlled-rate freezing and storage in liquid nitrogen are employed.

6.6 Clinical Administration

The final CAR-NK cell product is thawed and administered to patients, typically via intravenous infusion, with dosing and treatment combinations tailored to the specific cancer type and patient condition.

7 Role of CAR-NK Cells in Tumors

CAR-NK-cell-mediated immunotherapy has made significant strides recently, experiencing exponential growth and emerging as a viable alternative for treating patients with metastatic malignancies [64]. Despite extensive exploration of the role of CAR-NK cells in cancer immunotherapy, many of their applications remain relatively limited and are largely confined to the preclinical stage in tumor models [65, 66].

7.1 Hematological Malignancies

Initially, CAR-NK cells were generated to overcome hematological malignancies such as myeloma, lymphoma, and leukemia, similar to CAR-T-cell-based immunotherapy. Hematological malignancies, accounting for approximately 7% of all newly diagnosed cancers, are a group of malignant clonal diseases that originate in the hematopoietic system [40]. Allogeneic hematopoietic stem cell transplantation (allo-HSCT) is often regarded as the primary curative approach for these malignancies and represents the moniker "the ancestor of immunotherapy" in this field. While advancements in allo-HSCT have improved overall survival rates, relapse of the underlying malignancy remains a major challenge post-HSCT. Recent studies have highlighted the better efficacy of CD19 CAR-NK cells in treating lymphoid malignancies than CAR-T-cell-based cellular immunotherapy [64]. Both CAR-T cells and, more recently, CAR-NK cells have demonstrated efficacy against liquid tumors, particularly CD19+ lymphoid-derived cancers. Ongoing clinical studies are investigating the translation of these promising results to solid tumors.

A groundbreaking preclinical study presented a novel approach to immunotherapy utilizing engineered cord blood-derived NK cells, which are known for their ease of production and remarkable efficacy while incorporating safety measures to mitigate toxicity [67–69]. This study involved transducing cord blood-derived NK cells via a retroviral vector carrying genes encoding an inducible caspase-9-based suicide gene (iC9), IL-15 and CAR-CD19. It efficiently killed CD19-expressing cell lines and primary leukemia cells in vitro and significantly extended survival in a xenograft Raji lymphoma murine model.

In parallel, the first large-scale trial (phases 1 and 2) of HLA-mismatched CAR-UC-NK-cell-based immunotherapy in combination with lymphodepletion chemotherapy was conducted in 11 patients with CD19+ chronic lymphocytic lymphoma (CLL) and B-cell lymphoid tumors and demonstrated promising safety and clinical outcomes. The response rate was 73%, with 8 out of 11 patients showing a response to treatment and a complete remission rate of 64%, which was observed in 7 patients [70]. These responses were rapid, occurring within 30 days after infusion, and the infused CAR-NK cells persisted and expanded at low levels for at least one year. This large-scale trial demonstrated that most patients respond to treatment with CAR-NK cells without experiencing toxic side effects.

7.2 *Lymphoma*

B-cell-originated lymphomas, which are classified among hematological B-cell malignancies, are distinguished by the presence of one or more common B-cell antigens. CAR-T-cell therapy has emerged as a viable approach for managing B-cell malignancies [71]. These antigens, such as CD19, CD20, and CD22, which are targeted in CAR-T-cell therapy, are also utilized as target antigens in CAR-NK-cell therapy [40]. Numerous preclinical and clinical trials have been undertaken to evaluate the effectiveness of CAR-NK cells for treating B-cell malignancies. Published clinical trial results and preclinical studies have indicated that CAR-NK cells could be a suitable option for treating lymphoma. Compared with CAR-T cells, CAR-NK cells have demonstrated promising outcomes, particularly in B-cell lymphoma, with significantly lower risks of GVHD, cytokine release, and neurotoxicity [52].

7.3 Myeloma

Multiple myeloma (MM) is a type of blood cancer characterized by the excessive growth of malignant plasma cells [72]. Although myeloma cells are vulnerable to being killed by natural killer (NK) cells, they develop mechanisms to evade NK-cell surveillance by avoiding detection and suppressing NK-cell function. MM constitutes approximately 10% of hematological malignancies and often presents with hypercalcemia, renal failure, anemia, and bone lesions.

Preclinical trials have demonstrated that FT576 NK cells, a type of CAR-NK cell derived from induced pluripotent stem cell (iNK) cells, enhance cytotoxicity and persistence, prevent self-fratricide, and mitigate antigen loss when combined with other monoclonal antibodies, such as anti-CD38 mAbs, in MM treatment. Additionally, NK cells engineered to express a CD138 CAR exhibit enhanced cytotoxicity against primary MM cells and CD138+ human MM cell lines [73, 74]. These engineered NK cells also show increased secretion of IFN-γ and granzyme B, along with elevated expression of CD107a, both in vitro and in xenograft models [75].

7.4 Leukemia

Chimeric antigen receptor (CAR) T cells directed against the B-cell differentiation antigen CD19 have shown significant clinical effectiveness in treating patients with B-cell lymphomas and leukemias. However, the application of CAR-mediated redirection of NK cells has been less investigated, and there are currently no available clinical data for this approach [70, 76].

Studies have shown that exposing CD19-positive targets to CAR-NK-92 cells leads to the formation of conjugates between NK cells and cancer cells, resulting in NK-cell degranulation and selective cytotoxicity against established B-cell leukemia and lymphoma cells. CAR-NK cells have also displayed targeted killing of primary pre-B-ALL blasts resistant to parental NK-92 cells. Additionally, the integration of additional costimulatory molecules, such as 2B4 or CD28, can enhance all aspects of the CD19 CAR-NK cell activation response to B-cell leukemia cells [77, 78].

Research by Oelsner et al. demonstrated the successful transduction of cytokine-induced killer (CIK) cells with a lentiviral vector encoding a CAR carrying a CD28-CD3ζ-CD19 structure, suggesting the potential of CAR-modified CIK cells as a promising immunotherapeutic strategy for treating relapsed/refractory B-ALL [79]. Other studies have confirmed that CAR-modified NK92 cells can exert antileukemic effects, with CAR-NK cells proving to be not only as effective as CAR-T cells but also less toxic [80].

For example, a preclinical study using UC-NK cells and fourth-generation iCas9. Compared with nontransduced UC-NK cells, the CAR-19-CD28-ζ-IL15 plasmid, which includes a suicide gene, demonstrated superior antitumor efficacy and prolonged persistence in a Raji lymphoma xenograft model. Additionally, the initial large-scale trial of HLA-mismatched CAR-UC-NK-cell-based immunotherapy combined with lymphodepleting chemotherapy in 11 patients with CD19+ CLL and B-cell lymphoid tumors showed promising safety and clinical outcomes within 30 days. The trial reported a response rate of 73% and a complete remission rate of 64% [69, 70].

However, as highlighted by Karadimitris, CAR-UC-NK cells may lead to a high rate of nondurable clinical responses. This suggests initial favorable toxicity and efficacy, but uncertainty remains regarding the long-term durability of CAR-NK-based immunotherapy for leukemia. While CAR-T-cell therapies have substantially advanced the treatment of hematological malignancies, their effectiveness against solid tumors has been limited, possibly due to insufficient homing capabilities and an immunosuppressive tumor microenvironment. [81]. Solid tumors such as breast cancer (BC), ovarian cancer, and glioblastoma have been extensively researched to assess the potential of CAR-NK-cell therapy.

7.5 Glioblastoma

Glioblastoma (GBM) is widely recognized as one of the most challenging cancers to treat because of its high resistance to therapy [82]. In recent years, there has been a shift toward immunotherapies for GBM as alternatives to conventional treatments. Among these immunotherapies, natural killer (NK) cell-based approaches hold promise for GBM treatment owing to their unique properties, such as unrestricted targeting by antigen–antibody reactions and their ability to penetrate deeply into the tumor microenvironment. In particular, chimeric antigen receptor (CAR)-NK cells have emerged as significant contributors to GBM immunotherapy [82–84].

The utilization of NK cells in therapy is supported by human studies on brain tumors, where activated CD16-positive NK cells have been found to be significantly linked to improved survival in glioblastoma patients [85]. Notably, a decrease in the number of activated NK cells has been correlated with the transition from low-grade to high-grade brain tumors, indicating a potential role for NK cells in tumor surveillance.

Compared with single-target NK cells, CAR-NK cells designed to recognize both EGFR and EGFRvIII have demonstrated stronger antitumor effects [86]. These findings suggest that the incorporation of CAR technology can increase the targeting specificity and efficacy of NK cells against GBM, potentially offering a promising avenue for improving treatment outcomes in this challenging disease [87].

7.6 Breast Cancer

Breast cancer (BC) is the leading cause of cancer-related mortality in women. Despite advancements, current treatments often fail to cure metastatic BC, with patients facing an average survival of approximately 2 years postdiagnosis [88–90].

The emergence of chimeric antigen receptors (CARs) has sparked optimism for the use of NK/CAR-NK cell therapy, whether autologous or allogeneic, in cancer treatment.

Like those in glioblastoma, breast cancer targets such as HER2, EGFR, and EGFRvIII have garnered attention. Modified NK-92 cells equipped with CARs capable of recognizing both EGFR and EGFRvIII have shown increased cytotoxicity and increased production of IFN-γ in breast cancer cells [88]. Additionally, NK-92 cells engineered to target HER2 have demonstrated comparable antitumor effects, suggesting potential avenues for improving outcomes in breast cancer treatment [91].

7.7 *Ovarian Cancer*

Ovarian cancer represents the leading cause of gynecological cancer-related mortality in the developed world, with chemoresistance and disease recurrence posing major challenges to patient survival. A recent study highlighted the potential of third-generation CAR therapy targeting CD44, a marker widely expressed on ovarian cancer cells that is associated with cancer stem cell-like properties and intraperitoneal tumor spread. An engineered CAR (CD44NK) has demonstrated potent and specific cytotoxic activity against CD44-positive ovarian cancer cell lines (SKOV3 and OVCAR3) as well as primary ovarian cancer cells obtained from ascites [92].

Furthermore, human leukocyte antigen G (HLA-G) acts as a tumor-associated antigen (TAA) expressed on many solid tumors, as well as on specific immune-privileged and adult tissues 56. Researchers such as Jan et al. have investigated CAR-NK cell therapy targeting HLA-G. Their study evaluated the use of CAR-NK cells in combination with low-dose chemotherapy and demonstrated that pretreatment with low-dose chemotherapy can induce the upregulation of HLA-G expression. This enhancement potentially enhances the antitumor cytotoxicity of HLA-G CAR-NK cells [93, 94].

Furthermore, CAR-NK-92 cells targeting CD24, CD44, or CD133 have demonstrated significant antitumor effects in preclinical studies, suggesting the potential of CAR-NK cell therapy in combating ovarian cancer [92, 95, 96].

8 Timeline of the Development of CAR-NK Cells

NK cells have advantages over T cells in the development of cellular therapy because of their ability to kill tumors without prior sensitization [97, 98]. When coupled with exogenous recognition molecules such as CAR, NK cells form a synergistic system capable of immediately recognizing and eliminating transformed cells. The pioneering work of Tran et al. (1995) demonstrated the development of the first CAR-NK cells, leveraging prior CAR technology designed for T cells, which was first reported by Miller et al. in 1987 [99]. Studies, such as the one conducted by Chu et al. in 2014, demonstrated the efficacy of modified NK cells with CAR-targeting antigens present on tumors, such as CS1 in multiple myeloma, in mouse models [99].

Further advancements culminated in the first clinical trial utilizing the CAR-NK cell line NK 92, as reported by Tang et al. in 2018. Encouraging results from this phase 1 trial indicated the safety and efficacy of CD33 CAR-T cells in patients with relapsed and refractory acute myeloid leukemia (AML) [100]. A significant breakthrough occurred in 2020 with the study by Liu et al., who reported successful phase I and II trials demonstrating the efficacy and tolerability of anti-CD19 CAR-NK cells derived from cord blood in treating B-cell malignancies. This trial revealed a rapid response within 30 days of infusion, with the infused CAR-NK cells exhibiting expansion and persistence for at least one year [70].

9 Challenges Associated with CAR-NK Cell Therapy

While CAR-NK-cell therapy offers distinct advantages over T-cell therapy, it also presents several limitations. One major constraint is the short lifespan of NK cells, which typically lasts only 7–14 days. Moreover, their inability to persist without cytokine support restricts their efficacy [101]. The incorporation of a transgene encoding an exogenous cytokine such as IL-15 can increase CAR-NK cell stability, but this can also lead to unwanted adverse effects and the suppression of NK-cell effector function by activating other immune subsets, such as regulatory T cells.

Additionally, NK cells constitute a small portion of peripheral blood leukocytes, posing a challenge in generating enough NK cells for adoptive immunotherapy. Ex vivo expansion and activation are often necessary, but the tumor microenvironment presents a formidable barrier to the efficacy of CAR-NK-cell therapy, with its immunosuppressive factors hindering optimal immune cell function. Researchers are exploring strategies to overcome these hurdles, such as combining CAR engineering with checkpoint deletion to increase efficacy. Blocking checkpoints to bolster NK-cell function has also been extensively investigated [102].

NK-92, a well-established NK-cell line, is commonly used as a source for CAR-NK cell therapies, offering easier manufacturing of off-the-shelf products. However, its derivation from a tumor cell line poses potential tumorigenicity risks. Moreover, approaches to gene transfer in NK cells are limited, as gene transduction can randomly integrate DNA into the target cell genome, leading to off-target effects such as the silencing of essential or tumor suppressor genes [52]. Lentivirus-based transduction is challenging in NK cells because of their innate resistance, resulting in low transgene expression levels [103, 104]. Nonviral vectors are being explored for their efficacy and safety, although their overall benefits remain unclear [105].

10 Conclusion and Future Directions

CAR-NK-cell therapy presents tremendous promise for the treatment of numerous cancers, particularly those where conventional NK-cell therapy is limited. Encouraging results from the initial large-scale trial of CAR-NK cells demonstrated both safety and significant clinical activity in patients with CD19+ CLL and B-cell lymphoma [70]. Despite the advantages of CAR-NK cells over CAR-T cells in cancer immunotherapy, they encounter similar challenges. Tumor heterogeneity, a hostile tumor microenvironment, and antigen loss pose major hurdles, necessitating strategies to address these obstacles to maximize CAR-NK therapy efficacy [52].

Recent advancements in genetic engineering, such as CRISPR-based modifications and the incorporation of novel genes to modulate the tumor microenvironment within CAR constructs, have yielded promising outcomes [106–108]. These innovations have led to the development of novel NK-cell products with potent antitumor activity and minimal toxicity upon infusion into patients. By employing advanced approaches such as combinatorial and multitarget strategies, the remaining challenges in CAR-NK immunotherapy can be overcome, potentially leading to significant improvements in cancer patient survival. Overall, CAR-NK cells represent a promising type of cancer immunotherapy. Ongoing preclinical and clinical studies are poised to unveil the full potential of CAR-NK therapy soon.

References

1. Prager I, Watzl C. Mechanisms of natural killer cell-mediated cellular cytotoxicity. J Leukoc Biol. 2019;105:1319.
2. Rezvani AR, Maloney DG. Rituximab resistance. Best Pract Res Clin Haematol. 2011;24(2):203–16.
3. Yang K, Zhao Y, Sun G, Zhang X, Cao J, Shao M, et al. Clinical application and prospect of immune checkpoint inhibitors for CAR-NK cell in tumor immunotherapy. Front Immunol. 2023;13:1081546.
4. Narni-Mancinelli E, Ugolini S, Vivier E. Tuning the threshold of natural killer cell responses. Curr Opin Immunol. 2013;25:53.
5. Leong JW, Sullivan RP, Fehniger TA. microRNA management of NK-cell developmental and functional programs. Eur J Immunol. 2014;44:2862.
6. Yoon SR, Kim TD, Choi I. Understanding of molecular mechanisms in natural killer cell therapy. Exp Mol Med. 2015;47:e141.
7. Paul S, Lal G. The molecular mechanism of natural killer cells function and its importance in cancer immunotherapy. Front Immunol. 2017;8:1124.
8. King A, Balendran N, Wooding P, Loke YW. Cd3– Leukocytes present in the human uterus during early placentation: phenotypic and morphologic characterization of the Cd56++ population. Dev Immunol. 1991;1(3):169.
9. Lanier LL. Up on the tightrope: Natural killer cell activation and inhibition. Nat Immunol. 2008;9:495.
10. Page A, Chuvin N, Valladeau-Guilemond J, Depil S. Development of NK cell-based cancer immunotherapies through receptor engineering. Cell Mol Immunol. 2024;21:315.

11. Ogbomo H, Timm-McCann M, Barnes T, Xiang RF, Jamil K, Ganguly A, et al. Granule-dependent NK cell killing of cryptococcus requires kinesin to reposition the cytolytic machinery for directed cytotoxicity. Cell Rep. 2018;24(11):3017.
12. De Saint Basile G, Ménasché G, Fischer A. Molecular mechanisms of biogenesis and exocytosis of cytotoxic granules. Nat Rev Immunol. 2010;10:568.
13. Simone CB, Henkart P. Permeability changes induced in erthrocyte ghost targets by antibody-dependent cytotoxic effector cells: evidence for membrane pores. J Immunol. 1980;124(2):954.
14. Bird CH, Sun J, Ung K, Karambalis D, Whisstock JC, Trapani JA, et al. Cationic sites on granzyme B contribute to cytotoxicity by promoting its uptake into target cells. Mol Cell Biol. 2005;25(17):7854.
15. Kurowska M, Goudin N, Nehme NT, Court M, Garin J, Fischer A, et al. Terminal transport of lytic granules to the immune synapse is mediated by the kinesin-1/Slp3/Rab27a complex. Blood. 2012;119(17):3879.
16. Kirkin V, Cahuzac N, Guardiola-Serrano F, Huault S, Lückerath K, Friedmann E, et al. The Fas ligand intracellular domain is released by ADAM10 and SPPL2a cleavage in T-cells. Cell Death Differ. 2007;14(9):1678.
17. Zuccato E, Blott EJ, Holt O, Sigismund S, Shaw M, Bossi G, et al. Sorting of Fas ligand to secretory lysosomes is regulated by mono-ubiquitylation and phosphorylation. J Cell Sci. 2007;120(1):191.
18. Lettau M, Qian J, Kabelitz D, Janssen O. Activation-dependent FasL expression in T lymphocytes and natural killer cells. Signal Transduct. 2004;4(5–6):206–11.
19. He JS, Gong DE, Ostergaard HL. Stored Fas ligand, a mediator of rapid CTL-mediated killing, has a lower threshold for response than degranulation or newly synthesized Fas ligand. J Immunol. 2010;184(2):555.
20. Wang ZH, Li W, Dong H, Han F. Current state of NK cell-mediated immunotherapy in chronic lymphocytic leukemia. Front Oncol. 2023;12:1077436.
21. Björkström NK, Strunz B, Ljunggren HG. Natural killer cells in antiviral immunity. Nat Rev Immunol. 2022;22:112.
22. McGrath EE. The tumor necrosis factor-related apoptosis-inducing ligand and lung cancer: still following the right TRAIL? J Thorac Oncol. 2011;6(6):983.
23. De Miguel D, Lemke J, Anel A, Walczak H, Martinez-Lostao L. Onto better TRAILs for cancer treatment. Cell Death Differ. 2016;23:733.
24. Terrén I, Orrantia A, Vitallé J, Zenarruzabeitia O, Borrego F. NK cell metabolism and tumor microenvironment. Front Immunol. 2019;10:2278.
25. Melaiu O, Lucarini V, Cifaldi L, Fruci D. Influence of the tumor microenvironment on NK cell function in solid tumors. Front Immunol. 2020;10:3038.
26. Kaur K, Nanut MP, Ko MW, Safaie T, Kos J, Jewett A. Natural killer cells target and differentiate cancer stem-like cells/undifferentiated tumors: strategies to optimize their growth and expansion for effective cancer immunotherapy. Curr Opin Immunol. 2018;51:170.
27. Luna JI, Grossenbacher SK, Murphy WJ, Canter RJ. Targeting cancer stem cells with natural killer cell immunotherapy. Expert Opin Biol Ther. 2017;17:313.
28. Wilson EB, El-Jawhari JJ, Neilson AL, Hall GD, Melcher AA, Meade JL, et al. Human tumour immune evasion via TGF-β blocks NK cell activation but not survival allowing therapeutic restoration of anti-tumour activity. PLoS One. 2011;6(9):e22842.
29. Liu S, Ren J, ten Dijke P. Targeting TGFβ signal transduction for cancer therapy. Signal Transduct Target Ther. 2021;6(1):8.
30. Schäkel L, Mirza S, Winzer R, Lopez V, Idris R, Al-Hroub H, et al. Protein kinase inhibitor ceritinib blocks ectonucleotidase CD39 – a promising target for cancer immunotherapy. J Immunother Cancer. 2022;10(8):e004660.
31. Raskovalova T, Huang X, Sitkovsky M, Zacharia LC, Jackson EK, Gorelik E. Gs protein-coupled adenosine receptor signaling and lytic function of activated NK cells. J Immunol. 2005;175(7):4383.

32. Brand A, Singer K, Koehl GE, Kolitzus M, Schoenhammer G, Thiel A, et al. LDHA-associated lactic acid production blunts tumor immunosurveillance by T and NK cells. Cell Metab. 2016;24(5):657.
33. Sadelain M, Brentjens R, Rivière I. The basic principles of chimeric antigen receptor design, vol. 3. Cancer Discov; 2013. p. 388.
34. Lattime EC, Gerson SL. Gene therapy of cancer: Translational approaches from preclinical studies to clinical implementation. 3rd ed. San Diego: Academic Press; 2013.
35. Ajina A, Maher J. Strategies to address chimeric antigen receptor tonic signaling. Mol Cancer Ther. 2018;17:1795.
36. Tomasik J, Jasiński M, Basak GW. Next generations of CAR-T cells – new therapeutic opportunities in hematology? Front Immunol. 2022;13
37. Khanali J, Azangou-Khyavy M, Boroomand-Saboor M, Ghasemi M, Niknejad H. JAK/STAT-dependent Chimeric Antigen Receptor (CAR) expression: a design benefiting from a dual AND/OR gate aiming to increase specificity, reduce tumor escape and affect tumor microenvironment. Front Immunol. 2021;12:638639.
38. Mishra AK, Gupta A, Dagar G, Das D, Chakraborty A, Haque S, et al. CAR-T-cell therapy in multiple myeloma: B-Cell Maturation Antigen (BCMA) and beyond. Vaccines. 2023;11:1721.
39. Jan M, Scarfò I, Larson RC, Walker A, Schmidts A, Guirguis AA, et al. Reversible ON-And OFF-switch chimeric antigen receptors controlled by lenalidomide. Sci Transl Med. 2021;13(575)
40. Lu H, Zhao X, Li Z, Hu Y, Wang H. From CAR-T cells to CAR-NK cells: a developing immunotherapy method for hematological malignancies. Front Oncol. 2021;11:720501.
41. Töpfer K, Cartellieri M, Michen S, Wiedemuth R, Müller N, Lindemann D, et al. DAP12-based activating chimeric antigen receptor for NK cell tumor immunotherapy. J Immunol. 2015;194(7):3201.
42. Schmidt P, Raftery MJ, Pecher G. Engineering NK cells for CAR therapy—recent advances in gene transfer methodology. Front Immunol. 2021;11:611163.
43. Oei VYS, Siernicka M, Graczyk-Jarzynka A, Hoel HJ, Yang W, Palacios D, et al. Intrinsic functional potential of NK-Cell subsets constrains retargeting driven by chimeric antigen receptors. Cancer Immunol Res. 2018;6(4):467.
44. Xiao L, Cen D, Gan H, Sun Y, Huang N, Xiong H, et al. Adoptive transfer of NKG2D CAR mRNA-engineered natural killer cells in colorectal cancer patients. Mol Ther. 2019;27(6):1114–25.
45. Adkins SCAR. T-cell therapy: adverse events and management. J Adv Pract Oncol. 2019;10(4):21–8.
46. Zhang L, Chu J, Yu J, Wei W. Cellular and molecular mechanisms in graft-versus-host disease. J Leukoc Biol. 2016;99(2):279.
47. Van Ostaijen-Ten Dam MM, Prins HJ, Boerman GH, Vervat C, Pende D, Putter H, et al. Preparation of cytokine-activated NK cells for use in adoptive cell therapy in cancer patients: protocol optimization and therapeutic potential. J Immunother. 2016;39(2):90.
48. Klingemann H. Challenges of cancer therapy with natural killer cells. Cytotherapy. 2015;17:245.
49. Lian GY, Mak TSK, Yu XQ, Lan HY. Challenges and recent advances in NK cell-targeted immunotherapies in solid tumors. Int J Mol Sci. 2022;23:164.
50. Laskowski TJ, Biederstädt A, Rezvani K. Natural killer cells in antitumour adoptive cell immunotherapy. Nat Rev Cancer. 2022;22:557.
51. Daher M, Rezvani K. Outlook for new CAR-based therapies with a focus on car NK cells: what lies beyond CAR-engineered T cells in the race against cancer. Cancer Discov. 2021;11:45.
52. Xie G, Dong H, Liang Y, Ham JD, Rizwan R, Chen J. CAR-NK cells: a promising cellular immunotherapy for cancer. EBioMedicine. 2020;59:102975.

53. Murakami T, Nakazawa T, Natsume A, Nishimura F, Nakamura M, Matsuda R, et al. Novel human NK cell line carrying CAR targeting EGFRvIII induces antitumor effects in glioblastoma cells. Anticancer Res. 2018;38(9):5049.
54. Li H, Song W, Li Z, Zhang M. Preclinical and clinical studies of CAR-NK-cell therapies for malignancies. Front Immunol. 2022;13:992232.
55. Becker PSA, Suck G, Nowakowska P, Ullrich E, Seifried E, Bader P, et al. Selection and expansion of natural killer cells for NK cell-based immunotherapy. Cancer Immunol Immunother. 2016;65(4):477.
56. Heipertz EL, Zynda ER, Stav-Noraas TE, Hungler AD, Boucher SE, Kaur N, et al. Current perspectives on "off-the-shelf" allogeneic NK and CAR-NK cell therapies. Front Immunol. 2021;12:732135.
57. Childs RW, Berg M. Bringing natural killer cells to the clinic: ex vivo manipulation. Hematology Am Soc Hematol Educ Program. 2013;2013:234.
58. Phan MTT, Lee SH, Kim SK, Cho D. Expansion of NK cells using genetically engineered K562 feeder cells. Methods Mol Biol. 2016;1441:167–74.
59. Ojo EO, Sharma AA, Liu R, Moreton S, Checkley-Luttge MA, Gupta K, et al. Membrane bound IL-21 based NK cell feeder cells drive robust expansion and metabolic activation of NK cells. Sci Rep. 2019;9(1):14916.
60. Gurney M, Kundu S, Pandey S, O'Dwyer M. Feeder cells at the Interface of natural killer cell activation, expansion and gene editing. Front Immunol. 2022;13:802906.
61. Soldierer M, Bister A, Haist C, Thivakaran A, Cengiz SC, Sendker S, et al. Genetic engineering and enrichment of human NK cells for CAR-enhanced immunotherapy of hematological malignancies. Front Immunol. 2022;13:13.
62. Labbé RP, Vessillier S, Rafiq QA. Lentiviral vectors for T cell engineering: clinical applications, bioprocessing and future perspectives. Viruses. 2021;13:1528.
63. Monjezi R, Miskey C, Gogishvili T, Schleef M, Schmeer M, Einsele H, et al. Enhanced CAR T-cell engineering using non-viral Sleeping Beauty transposition from minicircle vectors. Leukemia. 2017;31(1):186.
64. Zhang L, Meng Y, Feng X, Han Z. CAR-NK cells for cancer immunotherapy: from bench to bedside. Biomark Res. 2022;10:12.
65. Xu J, Niu T. Natural killer cell-based immunotherapy for acute myeloid leukemia. J Hematol Oncol. 2020;13:167.
66. Zhang L, Liu M, Yang S, Wang J, Feng X, Han Z. Natural killer cells: of-the-shelf cytotherapy for cancer immunosurveillance. Am J Cancer Res. 2021;11(4):1770.
67. Li YR, Dunn ZS, Yu Y, Li M, Wang P, Yang L. Advancing cell-based cancer immunotherapy through stem cell engineering. Cell Stem Cell. 2023;30:592.
68. Liu E, Tong Y, Dotti G, Savoldo B, Muftuoglu M, Kondo K, et al. Cord blood derived natural killer cells engineered with a chimeric antigen receptor targeting CD19 and expressing IL-15 have long term persistence and exert potent anti-leukemia activity. Blood. 2015;126(23):3091.
69. Liu E, Tong Y, Dotti G, Shaim H, Savoldo B, Mukherjee M, et al. Cord blood NK cells engineered to express IL-15 and a CD19-targeted CAR show long-term persistence and potent antitumor activity. Leukemia. 2018;32(2):520.
70. Liu E, Marin D, Banerjee P, Macapinlac HA, Thompson P, Basar R, et al. Use of CAR-transduced natural killer cells in CD19-positive lymphoid tumors. N Engl J Med. 2020;382(6):545.
71. Huang J, Huang X, Huang J. CAR-T cell therapy for hematological malignancies: limitations and optimization strategies. Front Immunol. 2022;13:1019115.
72. Abduh MS. An overview of multiple myeloma: a monoclonal plasma cell malignancy's diagnosis, management, and treatment modalities. Saudi J Biol Sci. 2024;31:103920.
73. Bjordahl R, Gaidarova S, Goodridge JP, Mahmood S, Bonello G, Robinson M, et al. FT576: a novel multiplexed engineered off-the-shelf natural killer cell immunotherapy for the dual-targeting of CD38 and BCMA for the treatment of multiple myeloma. Blood. 2019;134(Supplement_1):3214.

74. Goodridge JP, Bjordahl R, Mahmood S, Reiser J, Gaidarova S, Blum R, et al. FT576: multi-specific off-the-shelf CAR-NK cell therapy engineered for enhanced persistence, avoidance of self-fratricide and optimized mab combination therapy to prevent antigenic escape and elicit a deep and durable response in multiple myeloma. Blood. 2020;136(Supplement 1):4.
75. Jiang H, Zhang W, Shang P, Zhang H, Fu W, Ye F, et al. Transfection of chimeric anti-CD138 gene enhances natural killer cell activation and killing of multiple myeloma cells. Mol Oncol. 2014;8(2):297.
76. Fousek K, Watanabe J, Joseph SK, George A, An X, Byrd TT, et al. CAR T-cells that target acute B-lineage leukemia irrespective of CD19 expression. Leukemia. 2021;35(1):75.
77. Altvater B, Landmeier S, Pscherer S, Temme J, Schweer K, Kailayangiri S, et al. 2B4 (CD244) signaling by recombinant antigen-specific chimeric receptors costimulates natural killer cell activation to leukemia and neuroblastoma cells. Clin Cancer Res. 2009;15(15):4857.
78. Oelsner S, Friede ME, Zhang C, Wagner J, Badura S, Bader P, et al. Continuously expanding CAR NK-92 cells display selective cytotoxicity against B-cell leukemia and lymphoma. Cytotherapy. 2017;19(2):235.
79. Oelsner S, Wagner J, Friede ME, Pfirrmann V, Genßler S, Rettinger E, et al. Chimeric antigen receptor-engineered cytokine-induced killer cells overcome treatment resistance of pre-B-cell acute lymphoblastic leukemia and enhance survival. Int J Cancer. 2016;139(8):1799.
80. Romanski A, Uherek C, Bug G, Seifried E, Klingemann H, Wels WS, et al. CD19-CAR engineered NK-92 cells are sufficient to overcome NK cell resistance in B-cell malignancies. J Cell Mol Med. 2016;20(7):1287.
81. Karadimitris A. Cord blood CAR-NK cells: favorable initial efficacy and toxicity but durability of clinical responses not yet clear. Cancer Cell. 2020;37:426.
82. Hosseinalizadeh H, Habibi Roudkenar M, Mohammadi Roushandeh A, Kuwahara Y, Tomita K, Sato T. Natural killer cell immunotherapy in glioblastoma. Discover Oncol. 2022;13:113.
83. Burger MC, Zhang C, Harter PN, Romanski A, Strassheimer F, Senft C, et al. CAR-engineered NK cells for the treatment of glioblastoma: turning innate effectors into precision tools for cancer immunotherapy. Front Immunol. 2019;10:2683.
84. Morimoto T, Nakazawa T, Maeoka R, Nakagawa I, Tsujimura T, Matsuda R. Natural killer cell-based immunotherapy against glioblastoma. Int J Mol Sci. 2023;24:2111.
85. Mostafa H, Pala A, Högel J, Hlavac M, Dietrich E, Westhoff MA, et al. Immune phenotypes predict survival in patients with glioblastoma multiforme. J Hematol Oncol. 2016;9(1):77.
86. Lu J, Li H, Chen Z, Fan L, Feng S, Cai X, et al. Identification of 3 subpopulations of tumor-infiltrating immune cells for malignant transformation of low-grade glioma. Cancer Cell Int. 2019;19(1):265.
87. Genßler S, Burger MC, Zhang C, Oelsner S, Mildenberger I, Wagner M, et al. Dual targeting of glioblastoma with chimeric antigen receptor-engineered natural killer cells overcomes heterogeneity of target antigen expression and enhances antitumor activity and survival. Onco Targets Ther. 2016;5(4):e1119354.
88. Kiaei SZF, Nouralishahi A, Ghasemirad M, Barkhordar M, Ghaffari S, Kheradjoo H, et al. Advances in natural killer cell therapies for breast cancer. Immunol Cell Biol. 2023;101:705.
89. Salemme V, Centonze G, Cavallo F, Defilippi P, Conti L. The crosstalk between tumor cells and the immune microenvironment in breast cancer: implications for immunotherapy. Front Oncol. 2021;11:610303.
90. Hu Y, Tian Z, Zhang C. Natural killer cell-based immunotherapy for cancer: advances and prospects. Engineering. 2019;5(1):106.
91. Zhang C, Burger MC, Jennewein L, Genßler S, Schönfeld K, Zeiner P, et al. ErbB2/HER2-specific NK cells for targeted therapy of glioblastoma. J Natl Cancer Inst. 2016;108(5) https://doi.org/10.1093/jnci/djv375.
92. Klapdor R, Wang S, Morgan MA, Zimmermann K, Hachenberg J, Büning H, et al. NK cell-mediated eradication of ovarian cancer cells with a novel chimeric antigen receptor directed against CD44. Biomedicines. 2021;9(10):1339.

93. Menier C, Rabreau M, Challier JC, Le Discorde M, Carosella ED, Rouas-Freiss N. Erythroblasts secrete the nonclassical HLA-G molecule from primitive to definitive hematopoiesis. Blood. 2004;104(10):3153.
94. Anna F, Bole-Richard E, Lemaoult J, Escande M, Lecomte M, Certoux JM, et al. First immunotherapeutic CAR-T cells against the immune checkpoint protein HLA-G. J Immunother Cancer. 2021;9(3):e001998.
95. Klapdor R, Wang S, Hacker U, Büning H, Morgan M, Dörk T, et al. Improved killing of ovarian cancer stem cells by combining a novel chimeric antigen receptor-based immunotherapy and chemotherapy. Hum Gene Ther. 2017;28(10):886.
96. Klapdor R, Wang S, Morgan M, Dörk T, Hacker U, Hillemanns P, et al. Characterization of a novel third-generation anti-CD24-CAR against ovarian cancer. Int J Mol Sci. 2019;20(3):660.
97. Fabian KP, Hodge JW. The emerging role of off-the-shelf engineered natural killer cells in targeted cancer immunotherapy, vol. 23. Mol Ther Oncolytics; 2021. p. 266.
98. Moscarelli J, Zahavi D, Maynard R, Weiner LM. The next generation of cellular immunotherapy: chimeric antigen receptor-natural killer cells. Transpl Cell Ther. 2022;28:650.
99. Raftery MJ, Franzen AS, Pecher G. CAR NK. Ann Rev Cancer Biol. 2023;7:229.
100. Tang X, Yang L, Li Z, Nalin AP, Dai H, Xu T, et al. Erratum: first-in-man clinical trial of CAR NK-92 cells: safety test of CD33-CAR NK-92 cells in patients with relapsed and refractory acute myeloid leukemia. Am J Cancer Res. 2018;8(9):1899.
101. Rennert PD. Novel immunotherapeutic approaches to the treatment of cancer: drug development and clinical application. Springer; 2016.
102. Khawar MB, Sun H. CAR-NK cells: from natural basis to design for kill. Front Immunol. 2021;12:707542.
103. Carlsten M, Childs RW. Genetic manipulation of NK cells for cancer immunotherapy: techniques and clinical implications. Front Immunol. 2015;6:266.
104. Savan R, Chan T, Young HA. Lentiviral gene transduction in human and mouse NK cell lines. Methods Mol Biol. 2010;612:209–21.
105. Ramamoorth M, Narvekar A. Non viral vectors in gene therapy – an overview. J Clin Diagn Res. 2015;9:GE01–6.
106. Chen XZ, Guo R, Zhao C, Xu J, Song H, Yu H, et al. A novel anti-cancer therapy: CRISPR/Cas9 gene editing. Front Pharmacol. 2022;13:939090.
107. Xiao-Jie L, Hui-Ying X, Zun-Ping K, Jin-Lian C, Li-Juan J. CRISPR-Cas9: a new and promising player in gene therapy. J Med Genet. 2015;52(5):289.
108. Ding S, Liu J, Han X, Tang M. CRISPR/Cas9-mediated genome editing in cancer therapy. Int J Mol Sci. 2023;24:16325.

Dendritic Cell-Based Immunotherapies for Cancer

Shubham Dutta

Abstract Dendritic cells (DCs) stand out as a highly promising tool for triggering immune responses in cancer immunotherapy. DCs are recognized as potent antigen-presenting cells (APCs) which excel in generating robust immune reactions and play a pivotal role in connecting the innate and adaptive arms of the immune system. Consequently, DC-based cancer immunotherapy seeks to leverage these distinctive qualities to enhance the battle against cancer. Over the past decades, extensive research has focused on developing immunotherapeutic strategies against cancer through vaccination. To achieve this goal, a comprehensive understanding of DC immunobiology, the regulation of innate and adaptive immune systems, the tumor microenvironment (TME), and the integration of cutting-edge scientific advances is imperative to unlock their significant antitumor immunotherapeutic potential. This chapter concentrates on delving into various aspects of DC immunobiology, including their origin, localization, unique properties, distinct subsets, and their connection to innate and adaptive immunity. Additionally, it explores the contemporary concept of cancer immunotherapy and sheds light on insights derived from clinical trials involving DC vaccines. Ultimately, the chapter outlines future perspectives for this burgeoning field.

Keywords Cancer · Immunotherapy · Cell-based therapy · Immunity · CAR-T · Synthetic biology · Genetic engineering

Abbreviations

ACT	Adoptive cell therapy
APC	Antigen-presenting cells

S. Dutta (✉)
Department of Medicine, Division of Innate Immunity, UMass Chan Medical School, Worcester, MA, USA
e-mail: Shubham.Dutta@umassmed.edu

A. K. Mishra, S. K. Malonia (eds.), *Cell-based Immunotherapies for Cancer*,
https://doi.org/10.1007/978-3-031-88695-9_13

CAR-T Cells	Chimeric antigen receptor T-cells
CD	Cluster of differentiation
CISH	Cytokine-inducible SH2-containing protein
CPI	Checkpoint inhibitors
CRS	Cytokine release syndrome
CT	Computed tomography
CTL	Cytotoxic T lymphocyte
CXCR	Chemokine receptor
EC	European Commission
FDA	United States Food and Drug Administration
G-CSF	Granulocyte colony-stimulating factor
HNSCC	Head and neck squamous cell carcinoma
IL	Interleukin
mbIL15	Membrane-bound interleukin-15
MHC	Major histocompatibility complex
MRI	Magnetic resonance imaging
NFAT	Nuclear factor of activated T-cells
NSCLC	Non-small cell lung cancer
PD-1	Programmed cell death protein 1
TAAs	Tumor-associated antigens
TCR	T-cell receptor
TIL	Tumor-infiltrating lymphocytes
TIME	Tumor immune microenvironment
TME	Tumor microenvironment
TNF	Tumor necrosis factor
TRAIL	TNF-related apoptosis-inducing ligand
WHO	World Health Organization

1 Introduction

Dendritic cells (DCs) originate from pluripotent hematopoietic stem cells (HSCs) in the bone marrow, belonging to a category of antigen-presenting cells (APCs) alongside B-cells and macrophages. First identified in 1973 by Canadian scientists Ralph Steinman and Zanvil Cohn, DCs were initially an undefined cell type in mouse spleens [1]. They were later termed 'dendritic cells' due to their characteristic features of multiple pseudopodia-like cytoplasmic protrusions during maturation (Fig. 1). Serving as sentinel cells, DCs are ubiquitously distributed throughout the body, found in the mucosal surfaces, skin, interstitial tissues, peripheral blood, lymphoid and non-lymphoid tissues [2]. Studying DCs is crucial because they are potent in presenting antigens to T-cells. They play a central role in the immune system by initiating inflammatory responses to pathogens. This leads to efficient T-cell activation and subsequent B-cell activation. Additionally, DCs continuously present tissue-derived self-antigens to $CD4^{+}$ and $CD8^{+}$ T-cells, leading to tolerance against

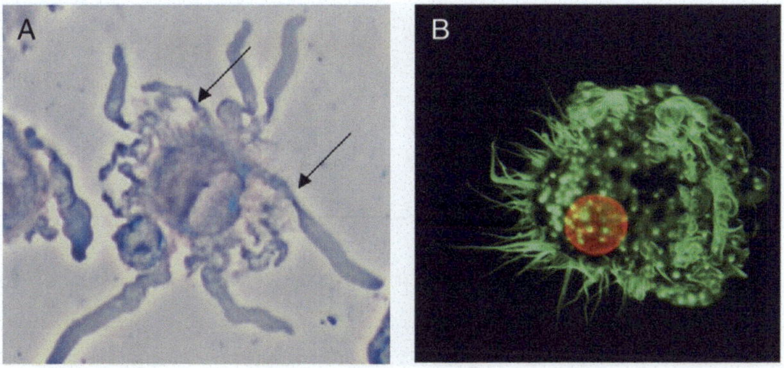

Fig. 1 (**a**) An image of a dendritic cell. The arrows show pseudopodia-like cytoplasmic protrusions during maturation [3]; (**b**) A confocal image of a human dendritic cell. The nucleus is shown in red. (Image by: Karla Daniels, Central Microscopy Research Facility, University of Iowa)

self-antigens. They are integral to the development of an effective adaptive immune response, serving as a critical link between the innate and adaptive immune systems.

2 Role of Dendritic Cells in the Immune System

Dendritic cells (DCs) differentiate from pluripotent hematopoietic stem cells (HSCs) in the bone marrow, belonging to a category of antigen-presenting cells (APCs) alongside B-cells and macrophages (Fig. 2). The role of DCs in the immune system involves phagocytosing, processing, and presenting antigens to both major histocompatibility complex (MHC) Class I and II molecules on T-cells. Initially generated in the bone marrow through hematopoiesis, immature DCs migrate from the bone marrow to non-lymphoid tissues, where they actively monitor the tissues for foreign antigens [4]. Recognition of foreign antigens by DCs is facilitated through specialized surface receptors such as retinoic acid-inducible gene-I (RIG-I) and 'pattern recognition receptors' (PRRs), such as toll-like receptors (TLR) and C-type lectin receptor (CLR). These receptors bind to distinct components on the foreign antigen (e.g., microbes), recognizing a conserved region called a 'pathogen-associated molecular pattern' (PAMP), such as lipopeptides, lipopolysaccharides (LPS), or nucleic acids (viral or bacterial RNA & DNA). This recognition leads to the internalization of the foreign antigen after which they migrate to the lymph nodes where they present antigens to T-cells. Mature DCs lose their antigen-uptake capacity and become antigen-presenting cells (APCs). This leads to the expression of MHC-I and MHC-II molecules on their surface, upregulation of several costimulatory surface molecules like CD40, CD80, and CD86, and the production of immunostimulatory cytokines. Upon interaction with naïve $CD4^{+}$ T-cells, DCs play a crucial role in differentiating these T-cells into antigen-specific helper T-cells (T_H) (such as T_H1, T_H2, T_H17) and T-follicular helper cells (T_{FH}). The specific type of

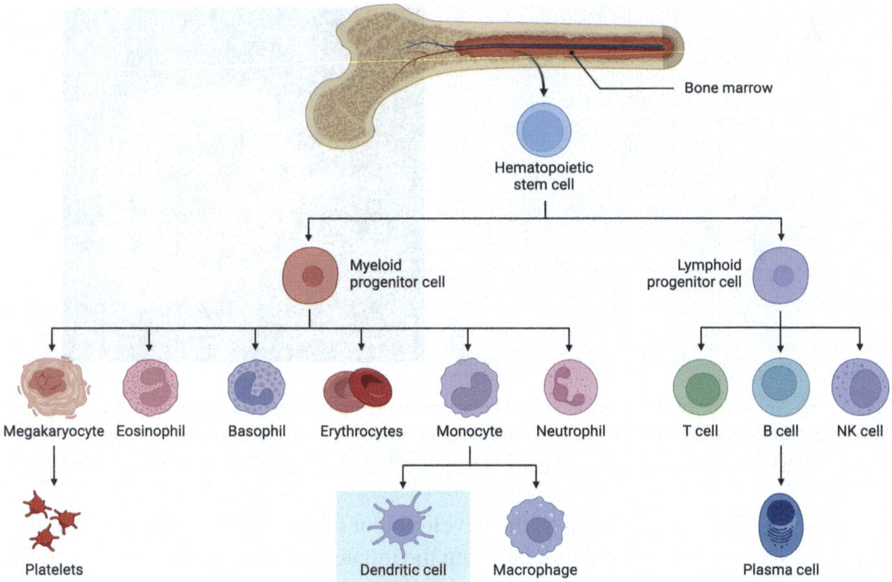

Fig. 2 Hematopoietic family tree showing how the mature cells in the various lineages are generated by self-generating hematopoietic stem cells.

$T_H$cell generated depends on factors such as the type of captured antigen (bacterial, viral, etc.) and the types of costimulatory molecules and interleukins (ILs) expressed by the DCs. This process ultimately results in the proliferation and clonal expansion of T-cells, playing a vital role in B-cell development and antibody production [5].

3 Dendritic Cell Types

DCs play a crucial role as immune sentinels, essential for initiating and regulating immune responses. This diverse population exhibits phenotypic and functional variations depending on their location within the body and their specific immunological functions. In a nonactivated or 'steady state', immature DCs continually survey their local environment, actively seeking foreign antigens for presentation to T-cells. These immature DCs have elevated expression of CD11c, intermediate expression of MHC-II, and limited expression of surface costimulatory molecules such as CD25, CD40, CD69, CD80, CD83, and CD86. When 'steady state' DCs capture antigens, they migrate to lymph nodes to present the antigens to T-cells. These 'mature' DCs exhibit increased expression of MHC-II and costimulatory markers although these DCs may not be fully activated. Complete activation of DCs relies on their recognition of 'danger signals' that are accomplished through pattern recognition receptors (PRRs). DCs can be generally classified into three distinct subsets based on their maturation/activation status:

(a) Immature, nonactivated DCs, 'steady state' DCs found in the spleen, exhibit high levels of CD11c, and low-to-intermediate expression of MHC-II and costimulatory markers. In the absence of prior activation, these DCs do not generate inflammatory cytokines but can stimulate naïve T-cells. Plasmacytoid (p) DCs express lower levels of CD11c and MHC-II, are weak stimulators of naïve T-cells, and do not produce pro-inflammatory cytokines in the 'steady state'.
(b) Mature but nonactivated, and potentially tolerogenic, migratory DCs exhibit intermediate CD11c expression and higher levels of MHC-II and costimulatory markers on their surface and does not produce inflammatory cytokines.
(c) Mature, activated DCs have encountered 'danger signals' in response to an invading pathogen or damaged self. Depending on the DC subset these DCs produce substantial amounts of inflammatory cytokines very high levels of MHC-II and costimulatory molecules on their surface.

DCs recognize pathogen-associated signals through pattern recognition receptors (PRRs) located on their surface. Since the identification of the first mammalian toll-like receptors (TLRs) that activated the innate immune system [6], the innate immune system has demonstrated plasticity in responding to invading pathogens. These activating 'danger signals' recognized by DCs are the following:

(a) Pathogen-associated molecular patterns (PAMPs), which are evolutionarily conserved molecules associated with pathogens (e.g., LPS, bacterial and viral nucleic acids), not typically found within eukaryotes.
(b) Damage-associated molecular patterns (DAMPs), such as intracellular proteins released by body's own cells undergoing necrosis.
(c) Inflammatory cytokines.

The response to danger signals triggers alterations in the phenotype and morphology of DCs. These modifications encompass an augmented expression of MHC molecules on the cell surface, upregulation of costimulatory markers, cytokine and chemokine release, and the release of cellular proteases. Microscopy, flow cytometry analysis, or assays measuring soluble protein excretion, such as ELISA or Multiplexed Bead-based Immunoassays, can effectively detect all these changes induced by maturation or activation.

4 Antigen Presentation by Dendritic Cells

Immature dendritic cells recognize pathogen-associated molecular patterns (PAMPs)—structures that are evolutionarily conserved, such as microbial LPS, carbohydrates, nucleic acids, and intermediates of viral replication. This recognition is facilitated through pattern recognition receptors (PRRs). Various PRRs participate in the innate recognition of pathogens, including nucleotide-binding oligomerization-domain (NOD-like) receptors, C-type lectin receptors (CLRs), toll-like receptors

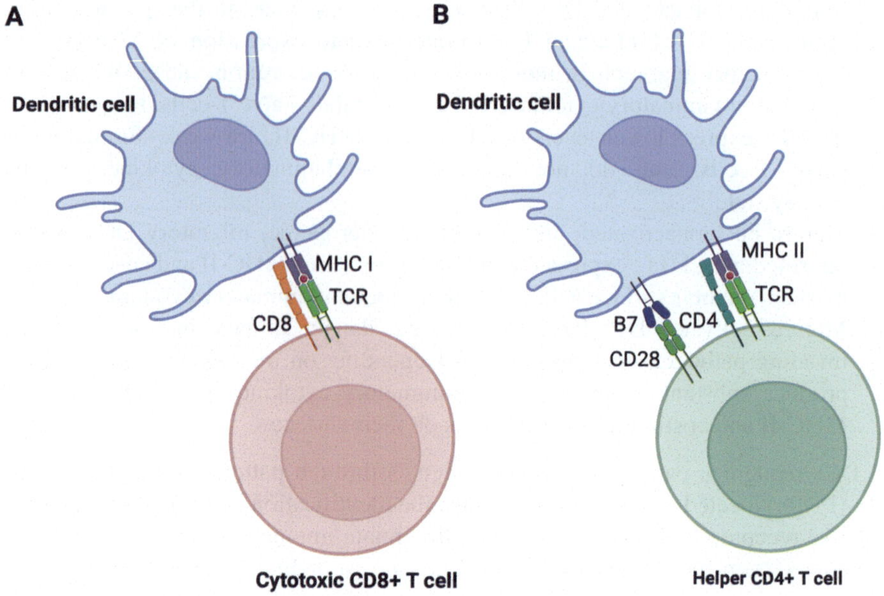

Fig. 3 An illustration showing dendritic cells presenting antigens to both MHC Class I (**a**) and II (**b**) molecules on T-cells

(TLRs), RIG-I-like helicases, & active protein kinase (PKR) [7]. Various mechanisms such as macropinocytosis, endocytosis, and receptor-mediated phagocytosis are employed to capture foreign antigens after antigen recognition [8–10]. Specifically, in the case of receptor-mediated phagocytosis, it involves the engulfment of pathogens such as bacterial cells. This process requires actin re-modeling to create a cup-shaped structure around the foreign particle, which subsequently closes to form a phagosome. The various processes of antigen capture by DCs are facilitated by numerous receptors that transport the antigen to processing compartments [8–10]. DCs convert proteins into peptides, presenting them on major histocompatibility complex (MHC) molecules, specifically MHC class I and II (Fig. 3) [8, 9]. In the case of lipid antigens, their processing differs as they are loaded onto non-classical MHC molecules belonging to the CD1 family [8]. Following antigen uptake and processing, DCs present antigens in the following ways:

(a) Via MHC-II to $CD4^{+}$ T lymphocytes (exogenous route): This route typically occurs when exogenous peptides are presented by DCs through MHC-II molecules. These peptides are derived from proteins that have undergone endocytosis and degradation by acid-dependent proteases in endosomes [9, 11].
(b) Via MHC-I to $CD8^{+}$ T lymphocytes (endogenous route): In this route DCs present intracellular antigens associated through MHC class I molecules. For instance, during a viral infection, DCs can present viral peptides, allowing the immune system to recognize and activate $CD8^{+}$ T lymphocytes, leading to the elimination of infected cells [9, 11].

(c) Via cross-presentation: This involves presenting exogenous antigens on MHC-I molecules, ultimately stimulating $CD8^+$ T lymphocytes or cytotoxic T-cells ($T_C$cells). Phagocytosis is a critical process for cross-presentation, and it is noteworthy that this capability is a distinctive feature of DCs, particularly specific subsets such as $CD8^+$ DCs and migratory $CD103^+$ DCs [9, 11, 12].

5 Dendritic Cell Isolation and Culture

5.1 Methods for Isolating Dendritic Cells

DCs make up only about 0.2% of human blood mononuclear cells and can be isolated and enriched using several methods. One method is to use density gradient centrifugation over metrizamide to isolate dendritic cells from human mononuclear cells. Mononuclear cells are isolated from leukapheresis pack or buffy coat preparation by Ficoll-Paque density gradient and the T-cells are depleted using magnetic beads conjugated with anti-TCRα/β monoclonal antibodies (mAbs). The T-cell depleted cell suspension is incubated overnight in RPMI1640 media at 37 °C on tissue-culture plates. This allows the monocytes to adhere to the plastic and helps the DCs to further differentiate into mature DCs in suspension. The non-adherent T-cells are gently removed, and the process is repeated a second time to further enrich DCs. The resulting cell suspension is subjected to gradient centrifugation with a sterile 14.5% metrizamide solution. Cells at the interface of the top layer (RPMI1640) and bottom metrizamide layer are carefully removed which contains 20–80% dendritic cells and is largely free of lymphocytes [13, 14]. Another method is to isolate adherent monocytes (described above) and incubate them with tumor necrosis factor (TNF)-α, GM-CSF, and IL-4 for 5–7 days. The differentiation process occurs without T-cell proliferation, making the quantity of monocytes a critical factor for dendritic cell recovery. As monocytes are more abundant than dendritic cells, this approach can lead to higher yields compared to the previous protocol [15]. Highly purified dendritic cell preparations can be obtained from these populations by a process known as magnetic-activated cell sorting (MACS). In this protocol mononuclear cells are incubated with a mix of anti-CD3, -CD14, -CD19, and -CD56 monoclonal antibodies conjugated with magnetic beads. Using a magnetic separation apparatus lymphocytes, monocytes, and natural-killing (NK) cells are depleted and the resulting cell suspension is incubated overnight in RPMI1640 media at 37 °C on tissue-culture plates. The non-adherent T-cells are gently removed and incubated with anti-CD83 antibody conjugated to magnetic beads. The resulting CD83-positive population contains a highly enriched population of DCs for various downstream applications [13, 14].

5.2 Maturation of Dendritic Cells

DC maturation is a crucial step that should ideally precede vaccination. However, there is currently no consensus on the most suitable methods to generate robust immunostimulatory DCs. Various combinations of maturation stimuli have been explored, including pro-inflammatory cytokines, CD40L/CD154), and toll-like receptor agonists [16–18]. The utilization of toll-like receptor agonists leads to elevated IL-12 levels, resulting in potent activation of DCs and subsequent effective immune activation. The maturation process is pivotal for efficient vaccine production because mature DCs generally exhibit enhanced expression of costimulatory molecules and increased production of cytokines and chemokines. In contrast, immature DCs lack the ability to induce antigen-specific responses and may potentially foster the differentiation of regulatory T (Treg)-cells.

5.3 Optimization of Dendritic Cell Yield

In order to increase DC yield, it is crucial to start with a large volume of human blood mononuclear cells. Small blood volumes are inadequate for dendritic cell isolation. When dealing with limited blood volumes it is recommended to generate dendritic cells from monocytes. Freshly isolated mononuclear cells are ideal, but satisfactory results can be achieved using 24-hour-old blood preparations stored on ice. Leukocyte-enriched ‘leukopaks’ obtained within the last 24 hours can provide a significant number of mononuclear cells, typically ranging from 2 to 12×10^8 cells per leukopak. Alternatively, ‘buffy coats’ from donated blood units can be used. Care must be taken with isolation procedures, as it may affect neutrophils within leukopaks, making it difficult to remove neutrophils which eventually will impact DC purity. A critical step of DC isolation is the removal of non-adherent T-cells. The tissue-culture plate must be washed thoroughly with warm media using a pipette with sufficient force to remove non-adherent T-cells. The cell morphology can be roughly assessed through phase-contrast microscopy. DCs are large and irregularly shaped with long membrane processes and can be easily identified (Fig. 1). For higher dendritic cell purity, two metrizamide density gradient centrifugation steps can be performed. Contaminating B lymphocytes and monocytes can be further depleted through adherence to Ig-coated plates. Lineage-associated monoclonal antibodies (mAbs) can also be employed for depletion using magnetic beads or panning. Depleting contaminating cells through F_C receptor-mediated procedures is effective but may limit dendritic cell heterogeneity, considering dendritic cells express CD32 and CD64 F_C receptors [19].

5.4 *Antigen Loading onto Dendritic Cells*

Techniques for Loading Antigens onto Dendritic Cells

Loading DCs with peptides, tumor cells, or intact proteins represents the most common method, typically conducted prior to maturation (Fig. 4). In this approach, peptides are directly loaded on either MHC-I or MHC-II molecules on the surface of the DCs. Conversely, tumor cells or intact proteins need to be processed and presented by the DCs to activate $CD4^{+}$ and $CD8^{+}$ T-cells. The primary drawback associated with peptide usage is the requirement to identify the patient's haplotype and the specific peptides that would bind to these particular haplotypes. Contrastingly, utilizing tumor cells or intact proteins is advantageous as it is not restricted to particular haplotypes [16, 20, 21]. The use of viral vectors presents an appealing alternative for loading antigenic material onto dendritic cells (DCs), as it enables gene insertion encoding tumor antigens or intact proteins, while allowing for the removal of virulence or replication factor genes. In some instances, the vector itself may promote DC maturation, eliminating the need for an additional maturation process. Another advantage lies in the ability to incorporate genes encoding costimulatory cytokines, thereby enhancing DC immunogenicity. However, the existence of preexisting immunity against the viral vector might diminish the patients' capacity to induce in vivo responses, which makes safety a major concern. Lentivirus-based vectors has been shown to be significantly less immunogenic due to the removal of the viral protein-encoding genes. Additionally, lentivirus-based vectors offer specific advantages, including the potential to activate the innate immune system

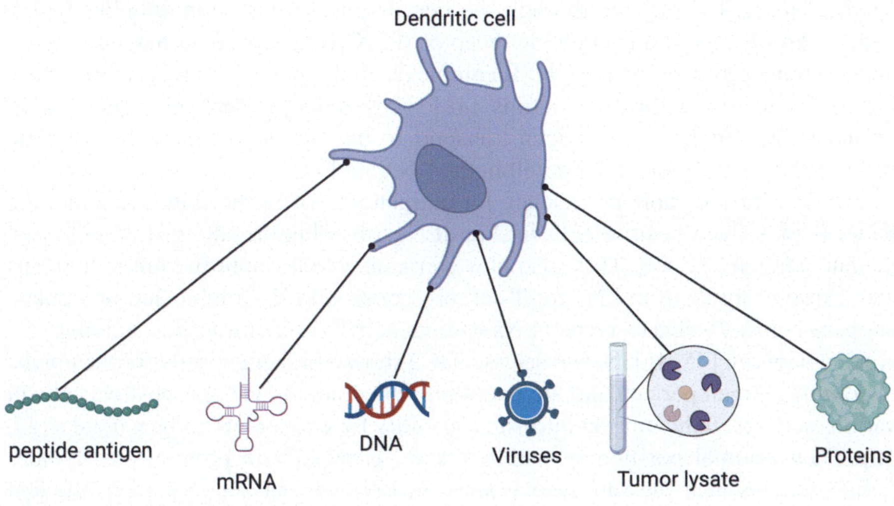

Fig. 4 Distinct DC vaccine loading approaches have been tested in clinical trials: loading of DCs with peptides, proteins, and tumor lysates; mRNA transfection; and delivery of DNA and the use of viral vectors

through cytoplasmic or endosomal molecules such as TLRs, RIG-I, and PKR [16, 22–25].

An alternative and appealing approach involves loading DCs with messenger RNA (mRNA) which encodes tumor-associated antigens (TAAs), a method proven to elicit $CD4^+$ and $CD8^+$ T-cell responses. These mRNAs, characterized by a short half-life and not integrated into the host genome, can be directly loaded onto DCs without the need for viral vectors or knowledge of the patient's haplotypes [7, 16]. Furthermore, mRNA transfection allows for the presentation of several antigenic epitopes, along with loading options involving maturation stimuli (such as CD40L) or cytokines. Electroporation has been identified as the most efficient method for introducing mRNA into DCs without the requirement for additional reagents [16, 26–28].

Enhancing Antigen Presentation Efficiency

Dendritic cells (DCs), acting as vigilant sentinels in tissues, continuously survey their local environment, capturing antigens for presentation to $CD4^+$ $T_H$cells or $CD8^+$ $T_C$cells on MHC-I or MHC-II molecules, respectively [29, 30]. The uptake of exogenous antigen is accomplished through various mechanisms, including phagocytosis [31, 32], receptor-mediated endocytosis [33–35], or micropinocytosis [36, 37], depending on the DC subtype and activation state [33]. For instance, Langerhans cells (LC) and $CX3CR1^+$ macrophages at barrier sites like the intestinal epithelium [38, 39] utilize dendritic projections to sample antigens. Dermal cDC2 can access epicutaneously applied antigen through hair follicles [40]. Following antigen uptake, human LC migrate through the dermis and then to skin-draining lymph nodes (skin-dLNs) in a chemokine receptor (CXCR) 4-dependent manner, subsequently transferring antigen-MHC-II complexes to dermal cDC through direct contact or indirectly within the dermis [41]. This transfer mechanism potentially enhances the efficiency of antigen transport to the LN, as dermal cDCs migrate faster and disperse more widely within the LN than LCs.

cDC2, with a notable proficiency for antigen uptake in the skin, constitute the majority of antigen-positive cells in skin-dLNs following the administration of particulate antigen [42, 43]. They also play a crucial role in capturing tumor antigens and transporting them to LNs for T-cell presentation [44]. Soluble and particulate antigens below 70 kDa can reach LNs without active cellular transport, relying on a conduit network [45, 46]. Nevertheless, cDC2 show a higher capacity for the uptake of soluble antigen per cell and are overrepresented among antigen-positive cells in various settings, indicating an intrinsic capability for exogenous antigen uptake [47, 48]. Their optimal positioning within tissues, especially in proximity to lymph-borne antigens near the subcapsular sinus in LNs, further supports their efficient antigen capture [43, 47].

In contrast, cDC1, situated deep within the LN paracortex [47], excel at capturing cell-associated antigens and dead cells through specific receptors like Clec9A, DEC205, Axl, and TIM3 [49]. This subtype predominantly processes cell-associated antigens via cross-presentation on MHC-I, crucial for antiviral and antitumor immunity [49]. Monoclonal antibodies targeting receptors such as CLEC9a and DEC205 have successfully enhanced antigen uptake and cross-presentation in vaccination contexts [50–52]. Additionally, both monocyte-derived DCs (moDC) and cDC2 can engage in cross-presentation, indicating a degree of redundancy in this pathway, particularly notable in humans [53–55].

DCs constitute the primary population of antigen-presenting cells in vivo, playing a crucial role in initiating antigen-specific activation and expansion of naive $CD4^+$ T-cells through interactions involving peptide–MHC-II binding with the T-cell receptor (TCR) and costimulatory signaling [50, 56]. Among DC subtypes, cDC2 stands out for its remarkable efficiency in processing exogenous antigens for presentation on MHC-II, resulting in superior $CD4^+$ T-cell proliferation compared to cDC1. This specialization is believed to be influenced by the expression of Irf4 in cDC2, as IRF4 has been linked to enhanced peptide–MHC-II complex formation in these cells [56].

Beyond facilitating $CD4^+$ T-cell activation and proliferation, the communication between DCs and $CD4^+$ T-cells during antigen presentation plays a crucial role in determining the differentiation fate of T helper (T_H) cells. Several factors, including the strength and duration of costimulatory signals and interactions between peptide-MHC-II and TCR, are implicated in regulating the T_H cell differentiation program [57].

5.5 Strategies for Dendritic Cell Activation

Use of Adjuvants in Dendritic Cell Activation

An effective adjuvant should be designed to specifically target DCs to enhance antigen presentation and activate immune responses. DCs, a type of white blood cell, are highly proficient in capturing, processing, and presenting antigens to T-cells. When appropriately activated, DCs engage with $CD4^+$ T lymphocytes through surface receptors like MHC-II, CD80, and/or CD86, and they release cytokines such as IL-12, initiating T-cell activation.

In mice, DCs are identified by the expression of CD11c and can be categorized into broad populations based on CD8a and CD4 expression, resulting in $CD8a^+$, $CD4^+$, and double-negative (DN) DCs ($CD8a^-CD4^-$) subsets [58, 59]. Conversely, in humans, DC subpopulations are not differentiated by CD4 and CD8 expression. DCs also express pattern recognition molecules, specifically toll-like receptors (TLRs), that act as activation signals. The TLR family comprises 11 members (TLR1–11) that recognize several PAMPs [60], inducing the maturation and migration of DCs to lymph nodes which in turn promote immune responses. While TLR2 and TLR4 mRNAs are present in all murine DC subsets [61], the expression of TLR9 protein has also been demonstrated [62, 63]. TLR2 is notable for recognizing a broad range of lipid ligands derived from different microbe types, including bacteria. It forms heterodimers with either TLR6 or TLR1 for recognizing di-acylated or tri-acylated lipids, respectively. However, it is suggested that TLR2 may also function as homodimers for signaling, as seen in the recognition of synthetic lipopeptides.

Cytokine-Based Activation Methods

A novel approach to antigen loading involves direct antigen targeting to dendritic cells (DCs) in vivo, aiming to promote anti-tumorigenic immune responses [65]. This strategy presents a promising option for DC immunotherapy, bypassing the costly and labor-intensive ex vivo DC generation process. It allows to produce vaccines on a larger scale, eliminating the need for customized vaccines for each patient.

Early methods for in vivo DC targeting included engineering irradiated tumor cells to secrete granulocyte-macrophage colony-stimulating factor (GM-CSF) [66]. GM-CSF was employed to recruit and promote the function of antigen-presenting cells (APCs). The strategies employed included the use of autologous tumor cells modified to produce GM-CSF via gene transfer mediated by retroviruses or adenoviruses, allogeneic tumor cell lines that had been genetically altered to continuously produce GM-CSF, or a combination of autologous tumor cells with cell lines that secrete GM-CSF. Clinical studies utilizing these techniques have shown they can attract dendritic cells, granulocytes, macrophages, and T-cells to the site of vaccination. Typically, patients exhibited delayed-type hypersensitivity responses that involved $CD4^+$ and $CD8^+$ T-cells, eosinophils, and macrophages following the vaccinations [67–69]. Examination of tumor samples from individuals who received the vaccine showed significant tumor cell death and infiltration by cytotoxic $CD4^+$ and $CD8^+$ T-cells and plasma cells, which produced strong cytokine responses upon reactivation, demonstrating the effective generation of immune responses specific to the tumor.

Extended GM-CSF production within the tumor environment is linked to disease progression in some experimental conditions, including a Phase III study that tested the immunization with irradiated GM-CSF-producing allogeneic prostate cancer cells in patients with metastatic disease resistant to hormone therapy [66]. This advancement could be due to the immune system becoming tolerant because of long-term administration of GM-CSF, which could lead to the recruitment of myeloid suppressor cells or cause myeloid precursors to develop into immature, tolerogenic dendritic cells [70, 71].

Newer strategies are being developed that aim at dendritic cell-specific molecules, such as Fc receptors, CD40, and C-type lectin receptors (CLRs). CLRs are of particular interest because they are expressed differently across DC subsets and include DEC205, DC-SIGN, mannose receptor (MR), and Dectin-1. These receptors play a crucial role in the identification and capture of glycosylated self-antigens and pathogens for antigen presentation, as well as in the movement of DCs, interactions between DCs and T-cells, and the activation of subsequent immune responses

[71]. Initial experiments that targeted antigens to DCIR2 and DEC205 found that without activating signals for DCs, immune tolerance was induced, while the simultaneous administration of DC-activating signals was required to trigger immune responses [35]. Later experiments that targeted antigens to various CLRs along with a DC-activating signal were successful in eliciting strong $CD4^+$ and $CD8^+$ T-cell responses [52, 72]. Additionally, directing antigens toward CLRs has been shown to improve antibody responses [73]. While most of this research has been conducted in mice, there are emerging studies in humans focusing on MR [74] and DC-SIGN [75], showing promising results in the activation of naive and memory tumor-specific T-cell responses. Nevertheless, further research is necessary to adapt this promising approach for clinical use in humans.

Genetic Modification for Enhanced Functionality

Although there is knowledge about signals for dendritic cell (DC) maturation and activation, the optimal signal to activate T-cells in vivo remains unclear. Current clinical trials of cancer immunotherapy use DCs from various sources, including those isolated directly from unmobilized or Flt3 ligand-mobilized peripheral blood, or generated in vitro from $CD34^+$ progenitors, $CD14^+$ monocytes, or adherent peripheral blood mononuclear cells. Immature DCs are loaded with antigens in various forms, such as peptides, proteins, tumor lysates, apoptotic bodies, or tumor cell fusions (Fig. 4). Gene transfer through DNA or messenger RNA encoding the antigen has also been investigated for antigen loading and processing. These loaded DCs may undergo further processing by maturation with second signals like tumor necrosis factor-α, CD40 ligand, or monocyte-conditioned media before administration to patients via different routes, such as intravenous, intradermal, subcutaneous, or direct intralymphatic or intranodal injection.

Mature DCs within secondary lymphoid organs express high levels of costimulatory molecules, facilitating antigen presentation. However, several issues persist regarding gene modification of DCs. These include questions about uniform maturation, the persistence of molecule expression, the potential for enhanced function with higher levels of expression, and the broader impact of genetic modification on DC function. Studies suggest that using exogenous expression systems like viral vectors engineered to express crucial molecules can optimize the delivery and processing of antigens. One study demonstrates the effectiveness of their strategy for modifying DCs and improving their in vitro performance [76]. However, uncertainties remain, including the optimal strategy for gene expression or repression, concerns about the viral vector system's potential impact on DC functions, and questions about the efficiency of migration for matured DCs and whether fully matured DCs should be administered directly or matured in vivo. Despite these uncertainties, Hodge et al.'s experiments show that DC function can be manipulated through genetic modification, paving the way for further exploration of their biologic activity and role in eliciting authentic antitumor responses in patients.

5.6 Dendritic Cell Vaccine Design

In recent years, there has been clinical promise in reprogramming the immune system against cancer. Dendritic cells (DCs) emerge as a compelling target for immunotherapy due to their capacity to capture and present tumor-associated antigens (TAAs) through various mechanisms, thereby initiating robust effector responses against the tumor [77–79]. Beyond direct antigen presentation, other intrinsic properties of DCs play a crucial role in immunotherapy. These include their ability to migrate between lymphoid and non-lymphoid tissues, regulate cytokine and chemokine gradients, and control inflammation which are essential for achieving systemic and enduring antitumor effects. Personalized vaccines involving patient-derived DCs manipulated ex vivo have been extensively investigated to harness these features. Typically, these therapies are developed by isolating hematopoietic stem and progenitor cells or monocytes from peripheral blood. These cells undergo treatment with recombinant cytokines to induce differentiation, are stimulated for maturation, and are loaded with TAAs in various forms. This comprehensive process has been employed in numerous preclinical and clinical studies [80].

Design, Manufacturing, and Quality Control of Dendritic Cell Vaccines

The treatment landscape for various malignant diseases is evolving, with tumor vaccines gaining prominence in therapeutic strategies. The effectiveness of these innovative products is being rigorously evaluated through numerous global clinical trials [81]. Cell-based vaccines in the European Union must align with a specified definition. According to this definition, somatic cell therapy involves the use of autologous (patient-derived), allogeneic (from a different human subject), or xenogeneic (originating in animals) cells whose characteristics have been significantly modified [82]. Somatic cell therapy medicinal products encompass cells that are genetically modified, unless the genetic modification is unrelated to therapeutic, diagnostic, or preventive objectives, as is the case with primary tumor cells immortalized through gene transfer [82].

Currently, tumor vaccines are not covered by the European Pharmacopoeia, and specific guidelines are lacking for both tumor and therapeutic vaccines. Good manufacturing practice (GMP) principles should be adhered to during the manufacturing of cell-based tumor vaccines in the EU. The EU is also preparing guidelines for GMP inspections of gene therapy and vaccine products. The European Medicines Agency published a "points to consider" document on the manufacture and quality control of somatic cell therapy products, which may be subject to modification in the future to incorporate emerging technologies like tissue engineering. This document also addresses the genetic manipulation of cells and should be read alongside the guidance on the quality, preclinical, and clinical aspects of gene-transfer medicinal products, which deals with specific issues related to gene-transfer methodologies involving plasmids or viruses and addresses safety concerns associated with certain gene-transfer approaches.

Challenges and Limitations

Dendritic cells (DCs) are typically found in an immature state in circulation and peripheral tissues. Upon receiving maturation signals, they undergo changes such as upregulation of chemokine receptors, increased surface expression of MHC molecules, and upregulation of costimulatory models. These changes facilitate their migration to lymph nodes, enhance antigen presentation, and amplify T-cell responses. The type of maturation signals determines the phenotypes of DCs, influencing their interactions with T-cells and the cytokines they secrete [83]. While DCs play a crucial role in activating the immune system, they can also induce immune tolerance, potentially hindering effective vaccine strategies. Studies have linked immature DCs to the promotion of regulatory T-cell development, contributing to peripheral self-tolerance. In some cases, immature DCs induced nonproliferative responses and IL-10 secretion, characteristic of a Treg population. The use of antigen-pulsed immature DCs has even led to antigen-specific immune suppression, inhibiting pre-existing antigen-specific T-cell function [84]. To address these challenges, vaccines need to incorporate signals that ensure full maturation and activation of DCs before administration. Attempts to exploit immature DCs to promote anergy have been made in transplantation and autoimmunity settings. However, in cancer vaccines, ensuring full maturation and appropriate activation of DCs is crucial to overcome immune tolerance barriers. Researchers have also demonstrated that inappropriately activated DCs, even with mature features, can induce T-cell tolerance. Understanding the signals that lead to tolerogenic DC states is essential for directing the choice of maturation signals and targeting factors present in the tumor microenvironment (TME) that mediate tolerance. Moreover, maturation alone may not be sufficient for immune activation, as demonstrated in studies where mature DCs led to the expansion of Treg populations in myeloma patients. Confirming DC maturity and phenotype through cell-surface markers and cytokine secretion before vaccine administration is crucial for cancer vaccines to overcome the challenge of immune tolerance [85].

5.7 Clinical Trials and Outcomes

Effective cancer vaccines can function as preventive agents against cancers linked to infectious diseases, such as hepatitis B virus or human papillomavirus, and serve as onco-therapeutic agents. The latter approach relies on the recognition of specific tumor-associated antigens (TAAs) by $CD3^+$ T-cells within the host body. Vaccination can enhance the existing immune response against TAAs or induce a de novo response. Dendritic cells (DCs), known for their effectiveness as antigen-presenting cells (APCs), play a crucial role in both major histocompatibility complex class I (MHC-I) and II (MHC-II) presentation to $CD8^+$ and $CD4^+$ T-cells, respectively. Dendritic cells exhibit migratory capabilities between lymphoid and non-lymphoid tissues, modulating cytokine and chemokine gradients to regulate inflammation and lymphocyte homing.

In ongoing clinical trials (January 2021–October 2024), the most common cancer type being targeted is breast cancer, followed by gliomas (GBM), and trials enrolling patients with various solid tumors (Table 1 and Fig. 5). Sipuleucel-T, the first United States Food and Drug Administration (FDA)-approved DC vaccine therapy for metastatic prostate cancer, demonstrated increased average survival but was shown to be sub-optimal with respect to tumor size reduction or halted tumor progression. Standard response evaluation criteria in solid tumors (RECIST) criteria, based on cytotoxic effects, may not be suitable for assessing outcomes in cancer immunotherapy. A novel version of RECIST has been proposed to evaluate vaccination as a valid alternative to standard cancer therapies.

Current dendritic cell vaccination methods involve standard vaccination and in vivo targeting of dendritic cells. Standard vaccination uses antigens with an adjuvant, lacking precise targeting in vivo DC targeting injects anti-DC antibodies with antigens, triggering strong immunity with an appropriate maturation stimulus. DC vaccination entails the transfer of dendritic cells that are generated ex vivo, loaded with TAAs and activated with pro-inflammatory cytokines before re-injection into the host.

Despite promising ex vivo effects in many DC vaccines, clinical efficacy, especially in late-stage cancer, remains modest. Different routes and methodologies in preparing DCs for clinical trials contribute to variations in efficiency. Preclinical studies are underway to develop next-generation DC vaccines, aiming to enhance effectiveness by boosting immunogenicity through different maturation cocktails and improving effector T lymphocyte function.

5.8 Combinatorial Therapies

Synergy with Other Immunotherapeutic Approaches

Cytotoxic treatments can exert various positive effects on the immune system, ranging from the simple release of tumor antigens due to cancer cell lysis to more complex immunological impacts. The release of tumor antigens enhances the uptake and presentation of a diverse array of antigens, promoting T-cell activation. The immunological effects induced by cytotoxic actions involve the increased expression of immuno-stimulatory molecules (e.g., DAMPs), heightened tumor antigen expression, decreased proliferation of suppressor cells, and promote proliferation and activation of cytotoxic T lymphocytes (CTLs) [36]. In a clinical example, 26 patients with different types of advanced and treatment-refractory cancers underwent combined therapy involving radiation, immature dendritic cells, keyhole limpet hemocyanin, and T-cells. The initial treatment successfully eliminated metastatic and recurrent tumors in 21 out of 26 patients, with half of them exhibiting a complete response and no evidence of disease recurrence. These encouraging outcomes

Table 1 Overview of clinical trials registered on clinicaltrials.gov between January 2022 and October 2024 testing dendritic cell-based immunotherapy in cancer patients

NCT number	Study status	Conditions	Combinatorial treatment
NCT05809752	Recruiting	Triple-negative breast cancer	Single agent
NCT06435910	Recruiting	Multiple myeloma	Single agent
NCT05504707	Recruiting	Triple-negative breast cancer	Single agent
NCT04912765	Recruiting	Hepatocellular carcinoma	Nivolumab
NCT06253234	Recruiting	WHO Grade III/IV gliomas	Single agent
NCT05773859	Recruiting	Ovarian carcinoma	Single agent
NCT05767684	Recruiting	Solid tumor	Nivolumab
NCT04879888	Completed	Breast cancer	Single agent
NCT05964361	Recruiting	Pancreas cancer, ovarian cancer, liver cancer	Single agent
NCT05378464	Recruiting	Breast cancer	Pepinemab, CAR-T
NCT06152367	Completed	Stage III/IV malignant melanoma	Proleukin
NCT04911621	Active (not recruiting)	High-grade glioma	Temozolomide
NCT05127824	Recruiting	Renal cell carcinoma	Cabozantinib
NCT04999943	Recruiting	Myelodysplastic syndromes	Single agent
NCT05195619	Recruiting	Non-small cell lung cancer	Cyclophosphamide
NCT04968366	Active (not recruiting)	Glioblastoma	Temozolomide
NCT05799612	Recruiting	Angiosarcoma	Pegylated-IFN a-2A, filgrastim
NCT04739527	Active (not recruiting)	Ovarian cancer	Single agent
NCT04963413	Terminated	Glioblastoma	GM-CSF
NCT05765084	Recruiting	Malignant pleural mesothelioma	Atezolizumab, platinum/pemetrexed
NCT05007496	Completed	COVID-19	Single agent
NCT05325632	Recruiting	Breast cancer	Trastuzumab, paclitaxel
NCT05920798	Recruiting	Fallopian tube carcinosarcoma	Pembrolizumab
NCT06253494	Recruiting	Endometrial cancer	Pembrolizumab, N-803, lenvatinib
NCT05344209	Recruiting	Non-small cell lung cancer	Anti-PD-1/PD-L1
NCT05631886	Recruiting	Solid tumor	Abraxane, cyclophosphamide, anti-PD-1 monoclonal antibody, anti-CTLA4 monoclonal antibody
NCT05631899	Recruiting	Solid tumor	Abraxane, cyclophosphamide, anti-PD-1 monoclonal antibody, anti-CTLA4 monoclonal antibody
NCT06100705	Recruiting	Prostate cancer	Testosterone cypionate, Sipuleucel-T

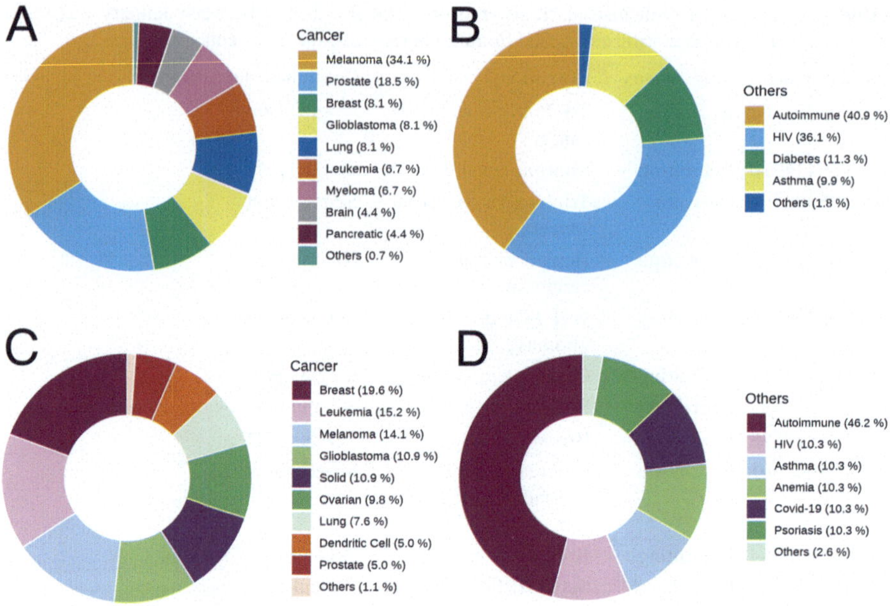

Fig. 5 Overview of completed (**a**, **b**) and ongoing (**c**, **d**) clinical trials of dendritic cell vaccination for therapy in cancer and other diseases

provide an impetus for further exploration of research into combining conventional therapies with DC-based antitumor immunotherapy [62]. Additionally, low doses of cyclophosphamide and paclitaxel have been demonstrated to stimulate DC maturation. As a result, these agents are employed in combination with DC vaccines to enhance their efficacy [64].

Tumors can create an immunosuppressive environment by expressing negative costimulatory molecules and secreting factors that inhibit both innate and adaptive immunity. This immunosuppressive milieu employs various mechanisms to evade immune surveillance, including the loss of tumor antigen expression, alterations in MHC molecules, induction of Treg cells, expression of inhibitory ligands, lack of co-stimulation, presence of indoleamine 2,3-dioxygenase promoting Treg cell generation, and production of immunosuppressive cytokines such as IL-10, IL-6, and TGF-β. Overcoming this tolerance and suppression within the tumor microenvironment is crucial for enhancing the immunogenicity and effectiveness of DC vaccines in vivo [15].

Combining DC-based vaccines with the blockade of inhibitory signals holds significant promise for eliciting a more robust immune response against cancer. Disrupting the interaction between the PD-1 receptor on activated T-cells and its ligand PD-L1 which is overexpressed on tumors and dendritic cells showed improved in vitro immune responses. Strategies to interfere with this mechanism include the use of anti-PD-1 antibodies or small interfering RNA (siRNA) to silence PD-L1 and PD-L2 ligands on dendritic cells [65].

Another valuable target is the cytotoxic T lymphocyte-associated antigen 4 (CTLA-4), which inhibits T-cell activation. Blocking CTLA-4 with antibodies, such as ipilimumab, has demonstrated therapeutic efficacy and received FDA approval for treating metastatic melanoma in 2011 [43].

The JAK/STAT signaling pathway, which is crucial for proper dendritic cell (DC) function, can be negatively regulated by the suppressors of the cytokine signaling (SOCS) family and glucocorticoid-induced leucine zipper (GILZ). Targeting these inhibitory components, such as by using siRNA to silence SOCS expression, represents a promising strategy to enhance DC functionality and improve immune responses. By relieving the inhibitory effects of SOCS and GILZ on JAK/STAT signaling in DCs, their antigen presentation, cytokine production, and T-cell stimulatory capacities may be augmented, providing additional avenues for immunomodulation [86]. Inhibiting GILZ has potential therapeutic benefits by altering DC maturation in response to TLR agonists and CD40L, as demonstrated in studies. Furthermore, GILZ expression in macrophages within Burkitt lymphomas has been linked to immune system failure to reject the tumor [67].

Immunotherapeutic strategies directed at inhibiting immuno-suppressive cytokines like TGF-β and IL-10 are also under exploration to enhance vaccine effectiveness. IL-10 receptor antibodies have shown promise in enhancing specific immune responses and IL-12 production [68]. Additionally, inhibiting TGF-β has been associated with suppressing Treg cell proliferation and increasing the number of TAA-specific T-cells [69].

Prospects for Personalized Combinatorial Therapies

Alternative cancer immunotherapies exhibiting promise encompass cellular transfer of tumor-infiltrating lymphocytes (TILs), genetically modified T-cells, and antibodies targeting immune checkpoint inhibitors (CPI) [87]. Adoptive cellular therapy (ACT) involves isolation and expansion of antigen-specific T-cells ex vivo for later transfer back to patients [88]. Despite its success in hematologic malignancies and melanoma, adoptive cell therapy (ACT)'s efficacy against most solid tumors is limited, primarily due to T-cells' inability to function and persist in vivo. Obstacles include the necessity for tumor resection, challenges in isolating or expanding TILs adequately, and the standard lymphodepletion procedure before transfer, causing side effects in roughly 50% of patients [88].

An alternative approach is the genetic engineering of T-cells to express chimeric antigen receptors (CARs), linking an antigen-binding domain to an intracellular T-cell signaling domain [89]. Prominent results have been achieved with CAR T-cells targeting CD19 in B-cell malignancies, albeit with significant toxicity, including cytokine release syndrome (CRS) and neurotoxicity. In contrast, DC-based immunotherapies boast a favorable safety profile, observed over two decades in numerous trials, with minimal toxicity and rare severe immunotoxicity reactions [90].

DC vaccines have demonstrated tolerability, inducing only localized reactions at injection sites and occasional systemic effects like fever and malaise. In contrast, immune checkpoint blockade can lead to autoimmune complications affecting various organs. CTLA4 blockade, for instance, may disrupt the effective inhibition of autoreactive T-cells, resulting in immune-related adverse events affecting the liver, skin, endocrine glands, and bowel [90]. In summary, immunotherapy holds promise for improving tumor outcomes. Monoclonal antibodies (mAbs), dendritic cell vaccines, immune checkpoint blockade, and adoptive and genetically engineered T-cell therapy have shown encouraging results. Future directions involve integrating several approaches with standard chemotherapy and radiotherapy, determining optimal timing for immunotherapy initiation, and devising strategies to maximize effectiveness while limiting toxicity.

6 Conclusion and Future Perspectives

Recent T-cell therapy successes have sparked interest in enhancing antitumor T-cell immunity for cancer therapy. Dendritic cells (DCs) stand out as the most effective antigen-presenting cells (APCs) capable of activating naive T-cells and eliciting immune memory responses against cancer. Despite often being dysfunctional or tolerogenic in the tumor microenvironment (TME), a deeper understanding of the regulation of DCs in this context holds therapeutic potential in various clinical settings. Exploring how different dendritic cell subsets may elicit distinct functional immune responses in cancer is a notable area of interest. For instance, the cDC1 subset is associated with the induction of cancer-controlling immunity and increased survival in specific cancer types [91–94]. On the other hand, monocyte-derived DCs (MoDCs) play a fundamental role during treatment with immunogenic cell death-inducing chemotherapy agents and radiation therapy [95–97], while cDC2s are crucial for inducing antitumor $CD4^+$ T-cell immunity [44, 98]. Although DCs can enhance the effectiveness of established cancer therapies, the development of efficient vaccination strategies requires a deeper understanding of dendritic cell biology and functions. Progress in preclinical studies encourages the exploration of DCs in more effective therapeutic treatments through clinical trials. Strategies to achieve this involve administration in conjunction with neo-antigens, mobilization of endogenous DCs, and the use of stimulating adjuvants. A more precise and refined targeting of DCs could further enhance the efficacy of these approaches. DC vaccination approaches show promise, especially in delaying or preventing relapse and metastasis following debulking surgical procedures. Overall, there is a need to gain more insights into how specific DC subsets with specialized functions can be optimally exploited to coordinate effective immune responses against cancer.

References

1. Chen P, Liu X, Sun Y, Zhou P, Wang Y, Zhang Y. Dendritic cell targeted vaccines: recent progresses and challenges. Hum Vaccin Immunother. 2016;12(3):612–22.
2. Rehman Z, Umar S, Meng C, Ullah Z, Riaz F, Rehman S, et al. Dendritic cell harmonised immunity to poultry pathogens; a review. Worlds Poult Sci J. 2017;73(3):581–90.
3. Tan YF, Leong CF, Cheong SK. Observation of dendritic cell morphology under light, phase-contrast or confocal laser scanning microscopy. Malays J Pathol. 2010;32(2):97–102.
4. Steinman RM. Decisions about dendritic cells: past, present, and future. Annu Rev Immunol. 2012;30:1–22.
5. Wykes M, MacPherson G. Dendritic cell–b-cell interaction: dendritic cells provide b cells with CD40-independent proliferation signals and CD40-dependent survival signals. Immunology. 2000;100(1):1–3.
6. Medzhitov R, Preston-Hurlburt P, Janeway CA Jr. A human homologue of the drosophila toll protein signals activation of adaptive immunity. Nature. 1997;388(6640):394–7.
7. Van Brussel I, Berneman ZN, Cools N, et al. Optimizing dendritic cell-based immunotherapy: tackling the complexity of different arms of the immune system. Mediat Inflamm. 2012;2012:1.
8. Palucka K, Banchereau J. Cancer immunotherapy via dendritic cells. Nat Rev Cancer. 2012;12(4):265–77.
9. Steinman RM, Banchereau J. Taking dendritic cells into medicine. Nature. 2007;449(7161):419–26.
10. Borghaei H, Smith MR, Campbell KS. Immunotherapy of cancer. Eur J Pharmacol. 2009;625(1–3):41–54.
11. Boudreau JE, Bonehill A, Thielemans K, Wan Y. Engineering dendritic cells to enhance cancer immunotherapy. Mol Ther. 2011;19(5):841–53.
12. Joffre OP, Segura E, Savina A, Amigorena S. Cross-presentation by dendritic cells. Nat Rev Immunol. 2012;12(8):557–69.
13. Nair S, Archer GE, Tedder TF. Isolation and generation of human dendritic cells. Curr Protoc Immunol. 2012;99(1):7–32.
14. Tedder TF, Jansen PJ. Isolation and generation of human dendritic cells. Curr Protoc Immunol. 1997;23(1):7–32.
15. Zhou LJ, Tedder TF. CD14+ blood monocytes can differentiate into functionally mature CD83+ dendritic cells. Proc Natl Acad Sci. 1996;93(6):2588–92.
16. Sabado RL, Bhardwaj N. Directing dendritic cell immunotherapy towards successful cancer treatment. Immunotherapy. 2010;2(1):37–56.
17. Amedei A, Benagiano M, della Bella C, Niccolai E, D'Elios MM, et al. Novel immunotherapeutic strategies of gastric cancer treatment. Biomed Res Int. 2011;2011:437348.
18. Nicolette C, Healey D, Tcherepanova I, Whelton P, Monesmith T, Coombs L, et al. Dendritic cells for active immunotherapy: optimizing design and manufacture in order to develop commercially and clinically viable products. Vaccine. 2007;25:B47–60.
19. Fanger NA, Wardwell K, Shen L, Tedder TF, Guyre PM. Type I (CD64) and type II (CD32) Fc gamma receptor-mediated phagocytosis by human blood dendritic cells. J Immunol (Baltimore, MD: 1950). 1996;157(2):541–8.
20. O'Neill D, Bhardwaj N. Generation of autologous peptide-and protein-pulsed dendritic cells for patient-specific immunotherapy. In: Adoptive immunotherapy: methods and protocols. Springer Science & Business Media; 2005. p. 97–112.
21. Schnurr M, Chen Q, Shin A, Chen W, Toy T, Jenderek C, et al. Tumor antigen processing and presentation depend critically on dendritic cell type and the mode of antigen delivery. Blood. 2005;105(6):2465–72.
22. Schroers R, Sinha I, Segall H, Schmidt-Wolf IG, Rooney CM, Brenner MK, et al. Transduction of human PBMC-derived dendritic cells and macrophages by an HIV-1-based lentiviral vector system. Mol Ther. 2000;1(2):171–9.

23. Dyall J, Latouche JB, Schnell S, Sadelain M. Lentivirus-transduced human monocyte-derived dendritic cells efficiently stimulate antigen-specific cytotoxic T lymphocytes. Blood J Am Soc Hematol. 2001;97(1):114–21.
24. Lizée G, Gonzales MI, Topalian SL. Lentivirus vector-mediated expression of tumor-associated epitopes by human antigen presenting cells. Hum Gene Ther. 2004;15(4):393–404.
25. Yang L, Yang H, Rideout K, Cho T, Joo KI, Ziegler L, et al. Engineered lentivector targeting of dendritic cells for in vivo immunization. Nat Biotechnol. 2008;26(3):326–34.
26. Nair SK, Morse M, Boczkowski D, Cumming RI, Vasovic L, Gilboa E, et al. Induction of tumor-specific cytotoxic T lymphocytes in cancer patients by autologous tumor RNA-transfected dendritic cells. Ann Surg. 2002;235(4):540.
27. Gilboa E, Vieweg J. Cancer immunotherapy with mRNA-transfected dendritic cells. Immunol Rev. 2004;199(1):251–63.
28. Heiser A, Maurice MA, Yancey DR, Coleman DM, Dahm P, Vieweg J. Human dendritic cells transfected with renal tumor RNA stimulate polyclonal T-cell responses against antigens expressed by primaryand metastatic tumors. Cancer Res. 2001;61(8):3388–93.
29. Doyle C, Strominger JL. Interaction between CD4 and class II MHC molecules mediates cell adhesion. Nature. 1987;330(6145):256–9.
30. Engleman EG, Benike C, Grumet FC, Evans R. Activation of human T lymphocyte subsets: helper and suppressor/cytotoxic T cells recognize and respond to distinct histocompatibility antigens. J Immunol (Baltimore, MD: 1950). 1981;127(5):2124–9.
31. Reis e Sousa C, Austyn JM. Phagocytosis of antigens by langerhans cells. In: Dendritic cells in fundamental and clinical immunology. Springer Science & Business Media; 1993. p. 199–204.
32. Hoffmann E, Kotsias F, Visentin G, Bruhns P, Savina A, Amigorena S. Autonomous phagosomal degradation and antigen presentation in dendritic cells. Proc Natl Acad Sci. 2012;109(36):14556–61.
33. Garrett WS, Chen LM, Kroschewski R, Ebersold M, Turley S, Trombetta S, et al. Developmental control of endocytosis in dendritic cells by Cdc42. Cell. 2000;102(3):325–34.
34. Platt CD, Ma JK, Chalouni C, Ebersold M, Bou-Reslan H, Carano RA, et al. Mature dendritic cells use endocytic receptors to capture and present antigens. Proc Natl Acad Sci. 2010;107(9):4287–92.
35. Bonifaz LC, Bonnyay DP, Charalambous A, Darguste DI, Fujii SI, Soares H, et al. In vivo targeting of antigens to maturing dendritic cells via the DEC-205 receptor improves T cell vaccination. J Exp Med. 2004;199(6):815–24.
36. Sallusto F, Cella M, Danieli C, Lanzavecchia A. Dendritic cells use macropinocytosis and the mannose receptor to concentrate macromolecules in the major histocompatibility complex class II compartment: downregulation by cytokines and bacterial products. J Exp Med. 1995;182(2):389–400.
37. Norbury CC, Chambers BJ, Prescott AR, Ljunggren HG, Watts C. Constitutive macropinocytosis allows TAP-dependent major histocompatibility compex class I presentation of exogenous soluble antigen by bone marrow-derived dendritic cells. Eur J Immunol. 1997;27(1):280–8.
38. Niess JH, Brand S, Gu X, Landsman L, Jung S, McCormick BA, et al. CX3CR1-mediated dendritic cell access to the intestinal lumen and bacterial clearance. Science. 2005;307(5707):254–8.
39. Rescigno M, Urbano M, Valzasina B, Francolini M, Rotta G, Bonasio R, et al. Dendritic cells express tight junction proteins and penetrate gut epithelial monolayers to sample bacteria. Nat Immunol. 2001;2(4):361–7.
40. Tordesillas L, Lozano-Ojalvo D, Dunkin D, Mondoulet L, Agudo J, Merad M, et al. PDL2+ CD11b+ dermal dendritic cells capture topical antigen through hair follicles to prime LAP+ tregs. Nat Commun. 2018;9(1):5238.
41. Yao C, Kaplan DH. Langerhans cells transfer targeted antigen to dermal DC and acquire MHC-II in vivo. J Invest Dermatol. 2018;138(7):1665.
42. Deckers J, Sichien D, Plantinga M, Van Moorleghem J, Vanheerswynghels M, Hoste E, et al. Epicutaneous sensitization to house dust mite allergen requires interferon regulatory factor 4–dependent dermal dendritic cells. J Allergy Clin Immunol. 2017;140(5):1364–77.

43. Gerner MY, Torabi-Parizi P, Germain RN. Strategically localized dendritic cells promote rapid T cell responses to lymph-borne particulate antigens. Immunity. 2015;42(1):172–85.
44. Binnewies M, Mujal AM, Pollack JL, Combes AJ, Hardison EA, Barry KC, et al. Unleashing type-2 dendritic cells to drive protective antitumor CD4+ T cell immunity. Cell. 2019;177(3):556–71.
45. Sixt M, Kanazawa N, Selg M, Samson T, Roos G, Reinhardt DP, et al. The conduit system transports soluble antigens from the afferent lymph to resident dendritic cells in the T cell area of the lymph node. Immunity. 2005;22(1):19–29.
46. Gretz JE, Norbury CC, Anderson AO, Proudfoot AE, Shaw S. Lymph-borne chemokines and other low molecular weight molecules reach high endothelial venules via specialized conduits while a functional barrier limits access to the lymphocyte microenvironments in lymph node cortex. J Exp Med. 2000;192(10):1425–40.
47. Gerner MY, Casey KA, Kastenmuller W, Germain RN. Dendritic cell and antigen dispersal landscapes regulate T cell immunity. J Exp Med. 2017;214(10):3105–22.
48. Gilfillan CB, Kuhn S, Baey C, Hyde EJ, Yang J, Ruedl C, et al. Clec9A+ dendritic cells are not essential for antitumor CD8+ T cell responses induced by poly I: C immunotherapy. J Immunol. 2018;200(8):2978–86.
49. Gutiérrez-Martínez E, Planès R, Anselmi G, Reynolds M, Menezes S, Adiko AC, et al. Cross-presentation of cell-associated antigens by MHC class I in dendritic cell subsets. Front Immunol. 2015;6:363.
50. Dudziak D, Kamphorst AO, Heidkamp GF, Buchholz VR, Trumpfheller C, Yamazaki S, et al. Differential antigen processing by dendritic cell subsets in vivo. Science. 2007;315(5808):107–11.
51. Lehmann CH, Baranska A, Heidkamp GF, Heger L, Neubert K, Lühr JJ, et al. DC subset–specific induction of T cell responses upon antigen uptake via Fcγ receptors in vivo. J Exp Med. 2017;214(5):1509–28.
52. Caminschi I, Proietto AI, Ahmet F, Kitsoulis S, Shin Teh J, Lo JC, et al. The dendritic cell subtype-restricted c-type lectin Clec9A is a target for vaccine enhancement. Blood J Am Soc Hematol. 2008;112(8):3264–73.
53. den Haan JM, Bevan MJ. Constitutive versus activation-dependent cross-presentation of immune complexes by CD8+ and CD8- dendritic cells in vivo. J Exp Med. 2002;196(6):817–27.
54. Ballesteros-Tato A, León B, Lund FE, Randall TD. Temporal changes in dendritic cell subsets, cross-priming and costimulation via CD70 control CD8+ T cell responses to influenza. Nat Immunol. 2010;11(3):216–24.
55. Baker K, Qiao SW, Kuo TT, Aveson VG, Platzer B, Andersen JT, et al. Neonatal Fc receptor for IgG (FcRn) regulates cross-presentation of IgG immune complexes by CD8- CD11b+ dendritic cells. Proc Natl Acad Sci. 2011;108(24):9927–32.
56. Vander Lugt B, Khan AA, Hackney JA, Agrawal S, Lesch J, Zhou M, et al. Transcriptional programming of dendritic cells for enhanced MHC class II antigen presentation. Nat Immunol. 2014;15(2):161–7.
57. Van Panhuys N. TCR signal strength alters T–DC activation and interaction times and directs the outcome of differentiation. Front Immunol. 2016;7:6.
58. McLellan AD, Kapp M, Eggert A, Linden C, Bommhardt U, Bröcker EB, et al. Anatomic location and T-cell stimulatory functions of mouse dendritic cell subsets defined by CD4 and CD8 expression. Blood J Am Soc Hematol. 2002;99(6):2084–93.
59. Vremec D, Pooley J, Hochrein H, Wu L, Shortman K. CD4 and CD8 expression by dendritic cell subtypes in mouse thymus and spleen. J Immunol. 2000;164(6):2978–86.
60. Takeda K, Kaisho T, Akira S. Toll-like receptors. Annu Rev Immunol. 2003;21(1):335–76.
61. Edwards AD, Diebold SS, Slack EM, Tomizawa H, Hemmi H, Kaisho T, et al. Toll-like receptor expression in murine DC subsets: lack of TLR7 expression by CD8α+ DC correlates with unresponsiveness to imidazoquinolines. Eur J Immunol. 2003;33(4):827–33.
62. Boonstra A, Asselin-Paturel C, Gilliet M, Crain C, Trinchieri G, Liu YJ, et al. Flexibility of mouse classical and plasmacytoid-derived dendritic cells in directing T helper type 1 and 2 cell

development: dependency on antigen dose and differential toll-like receptor ligation. J Exp Med. 2003;197(1):101–9.
63. Boonstra A, Rajsbaum R, Holman M, Marques R, Asselin-Paturel C, Pereira JP, et al. Macrophages and myeloid dendritic cells, but not plasmacytoid dendritic cells, produce IL-10 in response to MyD88-and TRIF-dependent TLR signals, and TLR-independent signals. J Immunol. 2006;177(11):7551–8.
64. Shortman K, Lahoud MH, Caminschi I. Improving vaccines by targeting antigens to dendritic cells. Exp Mol Med. 2009;41(2):61–6.
65. Tacken PJ, Torensma R, Figdor CG. Targeting antigens to dendritic cells in vivo. Immunobiology. 2006;211(6–8):599–608.
66. Jinushi M, Tahara H. Cytokine gene-mediated immunotherapy: current status and future perspectives. Cancer Sci. 2009;100(8):1389–96.
67. Luiten RM, Kueter EW, Mooi W, Gallee MP, Rankin EM, Gerritsen WR, et al. Immunogenicity, including vitiligo, and feasibility of vaccination with autologous GM-CSF–transduced tumor cells in metastatic melanoma patients. J Clin Oncol. 2005;23(35):8978–91.
68. Small EJ, Sacks N, Nemunaitis J, Urba WJ, Dula E, Centeno AS, et al. Granulocyte macrophage colony-stimulating factor–secreting allogeneic cellular immunotherapy for hormone-refractory prostate cancer. Clin Cancer Res. 2007;13(13):3883–91.
69. Higano CS, Corman JM, Smith DC, Centeno AS, Steidle CP, Gittleman M, et al. Phase 1/2 dose-escalation study of a GM-CSF-secreting, allogeneic, cellular immunotherapy for metastatic hormone-refractory prostate cancer. Cancer. 2008;113(5):975–84.
70. Filipazzi P, Valenti R, Huber V, Pilla L, Canese P, Iero M, et al. Identification of a new subset of myeloid suppressor cells in peripheral blood of melanoma patients with modulation by a granulocyte-macrophage colony-stimulation factor–based antitumor vaccine. J Clin Oncol. 2007;25(18):2546–53.
71. Sica A, Bronte V, et al. Altered macrophage differentiation and immune dysfunction in tumor development. J Clin Invest. 2007;117(5):1155–66.
72. Carter RW, Thompson C, Reid DM, Wong SY, Tough DF. Preferential induction of CD4+ T cell responses through in vivo targeting of antigen to dendritic cell-associated C-type lectin-1. J Immunol. 2006;177(4):2276–84.
73. Boscardin SB, Hafalla JC, Masilamani RF, Kamphorst AO, Zebroski HA, Rai U, et al. Antigen targeting to dendritic cells elicits long-lived T cell help for antibody responses. J Exp Med. 2006;203(3):599–606.
74. Ramakrishna V, Treml JF, Vitale L, Connolly JE, O'Neill T, Smith PA, et al. Mannose receptor targeting of tumor antigen pmel17 to human dendritic cells directs anti-melanoma T cell responses via multiple HLA molecules. J Immunol. 2004;172(5):2845–52.
75. Tacken PJ, de Vries IJM, Gijzen K, Joosten B, Wu D, Rother RP, et al. Effective induction of naive and recall T-cell responses by targeting antigen to human dendritic cells via a humanized anti–DC-SIGN antibody. Blood. 2005;106(4):1278–85.
76. Hodge JW, Rad AN, Grosenbach DW, Sabzevari H, Yafal AG, Gritz L, et al. Enhanced activation of T cells by dendritic cells engineered to hyperexpress a triad of costimulatory molecules. J Natl Cancer Inst. 2000;92(15):1228–39.
77. Steinman RM, Nussenzweig MC. Avoiding horror autotoxicus: the importance of dendritic cells in peripheral T cell tolerance. Proc Natl Acad Sci. 2002;99(1):351–8.
78. de Vries IJM, Lesterhuis WJ, Scharenborg NM, Engelen LP, Ruiter DJ, Gerritsen MJP, et al. Maturation of dendritic cells is a prerequisite for inducing immune responses in advanced melanoma patients. Clin Cancer Res. 2003;9(14):5091–100.
79. Jonuleit H, Giesecke-Tuettenberg A, Tüting T, Thurner-Schuler B, Stuge TB, Paragnik L, et al. A comparison of two types of dendritic cell as adjuvants for the induction of melanoma-specific T-cell responses in humans following intranodal injection. Int J Cancer. 2001;93(2):243–51.
80. Mastelic-Gavillet B, Balint K, Boudousquie C, Gannon PO, Kandalaft LE. Personalized dendritic cell vaccines—recent breakthroughs and encouraging clinical results. Front Immunol. 2019;10:766.

81. Laureano RSVI, Sprooten J. Trial watch: dendritic cell (DC)-based immunotherapy for cancer. Oncoimmunology. 2022;11(1):2096363.
82. European Commission. Commission directive 2003/63/EC of 25 june 2003 amending directive 2001/83/EC of the european parliament and of the council on the community code relating to medicinal products for human use (text with EEA relevance). European Commission.
83. Steinman RM, Pack M, Inaba K. Dendritic cell development and maturation. Adv Exp Med Biol. 1997;417:1–6.
84. Mellman I, Steinman RM. Dendritic cells: specialized and regulated antigen processing machines. Cell. 2001;106(3):255–8.
85. Sabado RL, Meseck M, Bhardwaj N. Dendritic cell vaccines. Methods Mol Biol. 2016;1403:763–77.
86. Jinushi M, Hodi FS, Dranoff G. Enhancing the clinical activity of granulocyte-macrophage colony-stimulating factor-secreting tumor cell vaccines. Immunol Rev. 2008;222(1):287–98.
87. Abbott M, Ustoyev Y. Cancer and the immune system: the history and background of immunotherapy. Semin Oncol Nurs. 2019;35(5):150923.
88. Chan JD, Lai J, Slaney CY, Kallies A, Beavis PA, Darcy PK. Cellular networks controlling T cell persistence in adoptive cell therapy. Nat Rev Immunol. 2021;21(12):769–84.
89. Guo J, Kent A, Davila E. Chimeric non-antigen receptors in T cell-based cancer therapy. J Immunother Cancer. 2021;9(8):e002628.
90. Sabado RL, Balan S, Bhardwaj N. Dendritic cell-based immunotherapy. Cell Res. 2017 Jan;27(1):74–95.
91. Salmon H, Idoyaga J, Rahman A, Leboeuf M, Remark R, Jordan S, et al. Expansion and activation of CD103+ dendritic cell progenitors at the tumor site enhances tumor responses to therapeutic PD-L1 and BRAF inhibition. Immunity. 2016;44(4):924–38.
92. Sánchez-Paulete AR, Cueto FJ, Martínez-López M, Labiano S, Morales-Kastresana A, Rodríguez-Ruiz ME, et al. Cancer immunotherapy with immunomodulatory anti-CD137 and anti–PD-1 monoclonal antibodies requires BATF3-dependent dendritic cells. Cancer Discov. 2016;6(1):71–9.
93. Wculek SK, Amores-Iniesta J, Conde-Garrosa R, Khouili SC, Melero I, Sancho D. Effective cancer immunotherapy by natural mouse conventional type-1 dendritic cells bearing dead tumor antigen. J Immunother Cancer. 2019;7:1–16.
94. Theisen DJ, Davidson JT IV, Briseño CG, Gargaro M, Lauron EJ, Wang Q, et al. WDFY4 is required for cross-presentation in response to viral and tumor antigens. Science. 2018;362(6415):694–9.
95. Ma Y, Adjemian S, Mattarollo SR, Yamazaki T, Aymeric L, Yang H, et al. Anticancer chemotherapy-induced intratumoral recruitment and differentiation of antigen-presenting cells. Immunity. 2013;38(4):729–41.
96. Casares N, Pequignot MO, Tesniere A, Ghiringhelli F, Roux S, Chaput N, et al. Caspase-dependent immunogenicity of doxorubicin-induced tumor cell death. J Exp Med. 2005;202(12):1691–701.
97. Apetoh L, Ghiringhelli F, Tesniere A, Obeid M, Ortiz C, Criollo A, et al. Toll-like receptor 4–dependent contribution of the immune system to anticancer chemotherapy and radiotherapy. Nat Med. 2007;13(9):1050–9.
98. Laoui D, Keirsse J, Morias Y, Van Overmeire E, Geeraerts X, Elkrim Y, et al. The tumour microenvironment harbours ontogenically distinct dendritic cell populations with opposing effects on tumour immunity. Nat Commun. 2016;7(1):13720.

Stem Cells: Allogeneic Cells for Cellular Immunotherapy

Krishan Gopal Jain, Anil Kumar, and Manu Dalela

Abstract Cellular immunotherapy, particularly with chimeric antigen receptor (CAR)-T cells, has revolutionized cancer treatment approaches. However, the use of patient-derived (autologous) immune cells for therapy presents significant challenges, including limited availability, high costs, and delayed treatment times. To address these limitations, therapies involving allogeneic sources of immune cells are emerging as a promising alternative, offering an off-the-shelf solution for scalable and efficient cancer immunotherapy. Stem cells, particularly hematopoietic stem cells (HSCs) and pluripotent stem cells (PSCs), provide an expandable and renewable source for generating therapeutic immune cells. These cells can be genetically engineered to enhance their tumor-targeting abilities and reduce the risk of immune rejection. Despite their potential, challenges such as graft-versus-host disease (GVHD), immune clearance, and manufacturing complexities remain key hurdles to widespread clinical application. This chapter explores the current landscape of allogeneic stem cells in immunotherapy, focusing on their advantages over autologous sources, current gene-editing strategies, and recent technological advancements that may overcome existing limitations. Through a comprehensive examination of both the opportunities and challenges in this field, we provide insights into the future direction of stem cell-based immunotherapies and their potential impact on cancer treatment paradigms.

Keywords Hematopoietic stem cells · Graft-versus-host disease · Allogeneic · Autologous · Induced pluripotent stem cells · Gene editing · Chimeric antigen receptor

---

K. G. Jain (✉)
Division of Nephrology, Indiana University School of Medicine, Indianapolis, IN, USA
e-mail: krjain@iu.edu

A. Kumar
School of Engineering and Sciences, GD Goenka University, Gurugram, Haryana, India

M. Dalela
DBT Center of Excellence for Stem Cell Research, All India Institute of Medical Sciences, New Delhi, India

A. K. Mishra, S. K. Malonia (eds.), *Cell-based Immunotherapies for Cancer*,
https://doi.org/10.1007/978-3-031-88695-9_14

1 Introduction

Cell therapy has been hindered by the lack of availability of a source of cells for therapy. Primary autologous cells are not a good fit for cell therapy, as they have limited ex vivo expansion potential and gradually become exhausted following cell culture. Preclinical studies and clinical studies have shown that terminally differentiated effector primary T cells exhibit poor persistence and fail at tumor regression [1, 2]. Rosenberg et al. performed numerous trials for testing autologous tumor-infiltrating lymphocyte (TIL) therapy in certain cancers [3–5]. Primary allogeneic cells derived from healthy donors hold promise as accessible, readily administrable products, but they also have certain limitations, including scarcity of donors and rejection by the host immune system in addition to limited in vitro expansion potential [6, 7]. The advancement and success of cell therapy relies on a self-renewing cell population that can be employed to generate universal donor cells to meet the demand. The extensive expansion potential of stem cells makes them ideal candidates because of their potential for an unlimited supply of cells for immunotherapy. Stem cells are unspecialized cells that have unique self-renewal properties and can differentiate into multiple cell types. These two unique properties make them potential sources for cell-based therapies. Stem cell-derived products such as differentiated cells (e.g., cardiomyocytes) and exosomes have been investigated for the treatment of diseases such as muscular dystrophy, heart disease, Parkinson's disease, Alzheimer's disease, spinal cord injuries, diabetes, and cancer [8]. Only a few stem cell-derived products have been approved for clinical use, such as Prochymal, which is used for the treatment of acute GVHD, and Holoclar, which is used to repair injured corneas [9]. Recent advances in stem cell methods such as 3D organoids and genetic engineering technologies such as clustered regularly interspaced short palindromic repeats (CRISPR) hold promise for the development of advanced stem cell-based therapies to shift current paradigms for several diseases. One of the most desirable areas is cancer, where therapies that rely on genetically engineered stem cells are beginning to enter the clinic and show encouraging signs of safety and efficacy [10]. Two major stem cell resources, HSCs and PSCs, have been used to develop therapeutic immune cells and could serve as renewable sources of allogeneic universal donor cells. Compared with HSCs, PSCs can be utilized as an 'unlimited' cell source to derive immune cells for the generation of therapeutic products.

Hematopoietic stem cell (HSC) transplantation (HSCT) has been used for decades for the treatment of blood cancer [7]. Research and an improved understanding of the role of the immune system in preventing cancer growth have inspired clinical scientists to utilize immune cells to combat cancer [11]. Immune cells such as lymphocytes have been used in several clinical trials to treat end-stage solid tumors, and positive responses have been reported [1, 12]. Later, advancements in gene therapy tools enabled scientists to genetically modify immune cells to utilize them to combat cancer [13]. In the last decade, chimeric antigen receptor (CAR)-T-cell therapy has achieved incomparable results in the treatment of blood tumors. Although the Food and Drug Administration (FDA) has approved six CAR-T-cell therapy products for the treatment of hematological malignancies, they are still not successful in treating solid tumors [14–18]. There are several limitations, and there is still a very long way to go before the development of immune cell therapy for treating all kinds of cancers. The major obstacles include cost; the use of autologous sources of cells, which delays the treatment process; the lack of availability of ready-to-use cells for the treatment of patients with urgent needs; reduced quality; and the reduced therapeutic potential of cells during ex vivo expansion [19–24]. Ex vivo expansion causes aging, introduces variations, and induces the differentiation of patient-derived T cells. Accumulating evidence has revealed that younger and less differentiated T cells are correlated with improved clinical outcomes. This impacts the therapeutic effects of T cells [25–27]. In many patients, CAR-T-cell therapy has been associated with relapse and adverse effects, leading to specialized treatment. These include cytokine release syndrome (CRS) and neurotoxicity [28–30]. Although allogeneic cell therapies have the potential to overcome the problems associated with autologous therapies, they also present their own set of challenges [31, 32]. Many questions related to safety and efficiency have not yet been answered and will continue to be of significant consideration as allogeneic cell immunotherapies progress. Cell banks can be generated for storing well-characterized healthy donor-derived allogeneic cells for easy administration. However, allogeneic cells are subject to life-threatening GVHD and may be rejected by the recipient's immune system [32]. To make allogeneic cell therapy safe and successful, it is important to overcome GVHD and graft rejection. This can be achieved by either selecting non-alloreactive T cells or engineering donor T cells to generate universal CAR-T cells [33]. Another approach includes the use of alternative effector cells such as macrophages, invariant natural killer T (iNKT) cells, and naturally occurring "Universal" donor T cells, such as γδ T cells, mucosal-associated invariant T (MAIT) cells, and virus-specific T cells (VSTs) [6, 32, 34]. These cells do not require HLA matching and recognize lipids, metabolites, or phosphoantigens presented in the context of non-polymorphic receptors on target cells [35, 36]. The approved chimeric antigen receptor (CAR-T) cells recognize and bind to only one tumor target on cancer cells. However, approximately 50% of patients experience relapse or refractory disease due to the evolution of antigen escape and antigen heterogeneity in cancer cells. This makes it reasonable to develop CAR-T cells capable of recognizing more than one tumor-related antigen among malignant cells. This can be achieved by using two pooled CAR-T cells with different antigen-binding specificities or by engineering single CAR-T cells capable of recognizing two tumor-associated antigens on single cancer cells [37]. However, the clinical efficacy of dual-targeting CAR-T-cell therapy has been validated in a few trials. It is quite challenging to manufacture dual-targeting CAR-T cells because of the limitations of the current gene-editing techniques. The relatively small cloning size of lentiviral and retroviral vectors limits the number of genes introduced in a single production round. In addition, increased inconsistency in the batch manufacturing of viral vectors, low transduction efficiency into mature immune cells, and high manufacturing failure rates affect cell yield and quality [38, 39]. Furthermore, elimination of allogeneic cells by the host immune system may limit the efficacy of cellular immunotherapy [40]. CD19 CAR-T-cell therapy for B-cell malignancies presents a unique situation. This treatment often results in a deficiency of B cells. Consequently, the risk of humoral-mediated immune rejection against allogeneic cell therapy is reduced [41, 42]. A study investigated the efficacy of allogeneic natural killer (NK) cells engineered to express CD19 CAR and IL-15 for treating CD19-positive lymphoid tumors [43]. The treatment showed a promising 73% response rate. The infused CAR-NK cells could expand and persist at low levels for at least 1 year. This extended persistence is thought to be due to two factors: first, the lymphodepletion regimen that creates a favorable environment, and second, the engineered expression of IL-15 by the CAR-NK cells themselves. Unfortunately, despite the presence of these therapeutic cells, tumor relapse has still been observed [43]. Allogeneic cell therapy holds promise for treating solid tumors, but overcoming host rejection remains a hurdle [44]. One strategy, HLA cloaking, disables HLA genes via gene editing, enabling immune evasion. However, this approach increases susceptibility to NK cell attack. To address this, additional modifications, such as HLA-E overexpression, are needed. A key challenge in cell therapy development is creating off-the-shelf, safe, and scalable cell products that can induce long-lasting responses against solid tumors. In this book, we discuss stem cells as potential allogeneic cell sources for cell therapies. In addition, we discuss stem cell sources, gene-editing methods, challenges in stem cell engineering and manufacturing, and the future directions of stem cell-based therapeutic platforms and clinical trials.

2 Allogeneic Cell Sources for Therapeutic Applications

Stem cells can play a crucial role in the development of therapeutic approaches for various medical conditions. Among stem cell resources, there are two main types of stem cell resources used for developing therapeutic cells: hematopoietic stem cells (HSCs) and pluripotent stem cells (PSCs).

1. *Hematopoietic Stem Cells (HSCs)*: HSCs are multipotent cells that can differentiate into various blood cell types. HSCs are essential for replenishing blood cells and maintaining the immune system [45, 46]. HSCs are appealing targets for gene therapies because of their accessibility for ex vivo genetic modification. This approach enables sustained transgene expression in circulating blood cells throughout an individual's lifetime. The recent FDA approval of Casgevy marked a significant milestone in the field of gene therapy. Casgevy is the first cell-based gene therapy for treating sickle cell disease (SCD) in patients aged 12 years and older. It utilizes CRISPR/Cas9 to modify patients' HSCs. These modified cells increase the production of fetal hemoglobin, preventing the sickling of red blood cells in SCD patients [47]. The sources of HSCs include the following:

    (i) *Bone Marrow*: In adults, HSCs primarily inhabit the bone marrow [48]. Aspirated bone marrow provides HSCs for transplantation. The bone marrow is the main site for hematopoiesis. The bone marrow offers a unique microenvironment that provides niches to support self-renewal and the differentiation of HSCs [48]. HSCs can be identified by monoclonal antibodies directed to surface markers [49]. For example, CD34 and CD38 are commonly used as markers to distinguish HSCs from other cell types [50]. HSCs exhibit unique properties related to drug efflux pumps. These pumps actively expel fluorescent dyes, allowing researchers to identify HSCs on the basis of their low dye retention. The Hoechst 33342 dye efflux assay is a well-known technique for isolating HSCs. By combining information from surface markers, dye efflux, and metabolic properties, HSCs can be isolated with high purity [51].

(ii) *Umbilical Cord Blood (UCB)*: Umbilical cord blood (UCB) is considered a valuable source of HSCs for allogeneic transplantation [52]. HSCs within UCB proliferate and differentiate more rapidly than those extracted from bone marrow or peripheral blood [53]. While HLA-matched siblings are preferred donors for hematopoietic stem cell transplantation (HSCT), only approximately 30% of patients have such donors available [54]. Unrelated donor HSCT success depends on HLA matching between the donor or cord blood unit and the recipient [55]. An optimal adult donor should match the recipient at the HLA-A, HLA-B, HLA-C, and HLA-DRB1 loci (an 8/8 high-resolution HLA match). A single mismatch at one of these loci (a 7/8 HLA match) reduces the 5-year overall survival rate by approximately 8 percentage points. However, in certain clinical situations where alternative therapies offer little chance of a cure, a 7/8 match is still considered suitable [56]. Cord blood (CB) requirements are less stringent, with acceptable matching standards for HLA-A and HLA-B antigen levels and high-resolution matching at HLA-DRB1, even with one or two mismatched loci [54]. Collecting UCB is easier, poses less risk to the donor (pregnant mother), has a lower infection risk, and importantly, the lack of criteria for human leucocyte antigen typing results in fewer GVHD cases [57]. The use of UCB has significantly increased access to allografts for racial and ethnic minorities [58, 59]. However, a key limitation in using UCB grafts for adults is the concern that they have an insufficient number of precursor cells compared with bone marrow grafts.

(iii) *Peripheral Blood*: Due to its cost-effectiveness, less invasive procedures, faster engraftment, higher HSC yields, and shorter hospital stays, mobilized peripheral blood has become the preferred source for HSC transplantation, replacing bone marrow [60]. G-CSF treatment significantly increases (~50–100 times) HSC mobilization in peripheral blood after 4–5 days. In some patients, G-CSF alone fails to mobilize enough HSCs and requires more than one apheresis procedure. In those cases, the administration of G-CSF combined with plerixafor (AMD3100) or GROβ promotes HSC mobilization from the BM to the peripheral blood for HSC transplantation [61, 62]. G-CSF triggers a complex process involving various cells and mechanisms, including neutrophils, macrophages, and mesenchymal stem cells [63]. The interaction between HSCs and niche cells plays a crucial role in retaining HSPCs within the bone marrow [64]. The administration of G-CSF activates neutrophils, leading to the release of serine proteases such as neutrophil elastase, cathepsin G (CG), dipeptidyl peptidase I (DPPI), and matrix metalloprotease-9. These proteases breakdown interaction molecules such as vascular cell adhesion molecule 1 (VCAM-1), stromal-derived factor-1 (SDF-1 or CXCL12), and c-Kit (CD117), disrupting their interaction with very late antigen-4 (VLA-4), CXC chemokine receptor-4 (CXCR4), and stem cell factor. This disruption of the interaction between HSCs and niche cells leads to HSC mobilization [65, 66].

2. *Pluripotent Stem Cells (PSCs)*: Conventional sources of HSCs, such as cord blood, bone marrow, and peripheral blood, have limitations due to the restricted number of transplantable cells and inefficient expansion strategies [67]. Additionally, there are other hurdles, including finding a good donor match (HLA matching), avoiding GVHD, viral contamination, and immune reactions. Because of these limitations, researchers are looking for alternative sources of HSCs to improve regenerative medicine treatments. PSCs have the remarkable ability to differentiate into any cell type in the body [67]. PSCs are widely used to generate HSCs/progenitor cells and mature immune cells. PSCs also face limitations, and strategies have been developed to improve the production of immune cells. When formed into embryoid bodies, PSCs show improved differentiation potential and immune cell production [68–70]. Two main types of PSCs are as follows:

   (i) *Embryonic Stem Cells (ESCs)*: ESCs are undifferentiated cells that can be maintained indefinitely in culture. ESCs are pluripotent and can give rise to all cell lineages [71]. The coculture of human ESCs with the murine bone marrow cell line S17 or the yolk sac endothelial cell line C166 differentiates them into hematopoietic precursor cells. When cultured on semisolid media in the presence of hematopoietic growth factors, these hematopoietic precursor cells form characteristic myeloid, erythroid, and megakaryocyte colonies [69]. However, the manipulation of human ES cells raises some ethical issues and immunoreactions.

   (ii) *Induced Pluripotent Stem Cells (iPSCs)*: iPSCs are generated by reprogramming adult cells, such as skin cells or fibroblasts, back into a pluripotent state, similar to embryonic stem cells [72]. PBMC-derived immune cells such as T cells, NK cells, iNKT cells, and MAIT cells can be reprogrammed into iPSCs. These PBMC-derived iPSCs can be genetically engineered with CARs before their differentiation into T cells, NK cells, iNKT cells, and MAIT cells for cellular therapy. iPSCs constitute an ideal source of stem cells for unlimited supply for future cell-based therapy and regenerative medicine [73].

3 Advantages of Allogeneic Cell Sources Over Autologous Cell Sources

Allogenic cell therapy has always been an extremely attractive source of readily available cells for transplantation at a reduced cost. Allogeneic NK cells can be prepared from numerous sources, including peripheral blood mononuclear cells (PBMCs), cord blood, and PSCs [74]. Cord blood (CB) is an attractive, allogeneic, off-the-shelf source of NK cells for immunotherapy [43]. Allogeneic NK cells from cord blood can be safely administered without the need for full HLA matching. Furthermore, allogeneic NK cells have a proven record of safety after infusion for adoptive immunotherapy in patients with cancer, which eliminates the need to produce a unique CAR product for each patient [75, 76]. Unlike autologous CAR-T cells, which have drawbacks such as long expansion times and complex steps, allogeneic NK cells eliminate the need for patient-specific CAR products. Additionally, autologous CAR-T cells face economic challenges related to production costs and transportation from the laboratory to the bedside [73, 77]. The most significant feature of CAR-T cells is their continuous expansion and persistence in the patient's body, which results in a long-term antitumor response.

HSC-derived iNKT cells reconstitute another potentially promising cell population for cancer immunotherapy. Initially, the first generation of iNKT cells was created by genetically modifying autologous HSCs to express the iNKT T-cell receptor (TCR), which were then reintroduced into the patient, potentially offering a lifelong supply of therapeutic iNKT cells. However, this method is costly and challenging for all patients in need. This led to the development of allogeneic HSC-iNKT cells, which are produced in high yield and purity. These allogeneic cells possess inherent cancer-killing abilities and a low risk of GVHD [78]. To further enhance their therapeutic potential, they can be engineered to express CARs and have HLA classes I and II ablated [79]. Additionally, HSC-iNKT cells can alter the tumor microenvironment by effectively targeting and eliminating tumor-associated macrophages and other immunosuppressive cells [73].

In the allogeneic production of therapeutic cells, the starting material can be from healthy donors, and we can expect a better-quality control product. Furthermore, allogenic therapeutic cell production technology provides a ready-to-use "off-the-shelf" immunotherapeutic. Figure 1 lists the differences between autologous and allogenic cell therapy [73]. This will ensure the multidose availability of immunotherapeutic cells for a single patient or different patients. Further production can be scaled up to establish immunotherapeutic cell banks, similar to how blood banks operate, and to significantly reduce the cost of this type of treatment, making it more affordable and accessible to individuals [80].

Although allogeneic CAR therapies hold great promise, they are associated with significant risks and challenges [73]. These differences arise from the immune differences between the donor and recipient, leading to potential risk to both the donor cells and the patient. If the donor cells attack the recipient's healthy tissues, GVHD

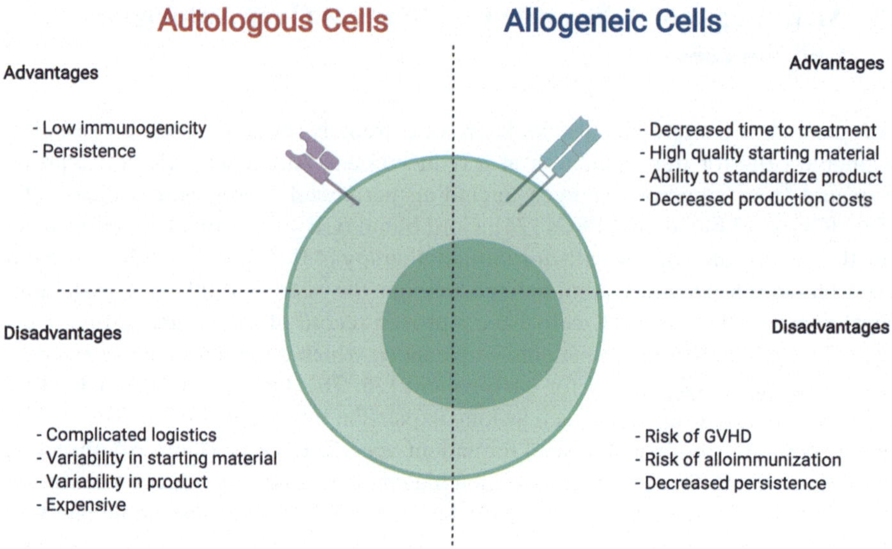

Fig. 1 Comparison between autologous and allogeneic cells. (Image Courtesy: Caldwell et al. [73])

can develop. On the other hand, if the recipient's immune system targets donor cells, the therapy might be rejected, reducing its effectiveness.

4 In Vitro and In Vivo Systems for Stem Cell Expansion and Differentiation

The first step in cell therapy is to expand cells in vitro without compromising their differentiation potential. In the last two decades, several stem cell expansion and differentiation systems have been developed [81, 82]. The in vivo system includes the bone marrow–liver–thymus (BLT) mouse model, and the in vitro system includes both feeder-dependent cultures and feeder-free cultures [83–86]. The feeder-dependent cultures include the OP9-DL system and artificial thymic organoid (ATO) system, and the feeder-free cultures include the HSC-iNKT culture system [82, 87, 88]. The ATO culture system was developed to support T-cell differentiation from HSCs or PSCs. The ATO system can also be used to create custom-made T cells. By starting with genetically engineered stem cells, scientists can design antigen-specific T cells that exhibit a vigorous response to antigen stimulation. The presence of human embryonic mesodermal progenitors (hEMPs) and Notch signaling within the ATO increases T-cell yield by capturing all the potential T-cell precursors within the organoid and specifically promoting T-cell differentiation over other blood cell types [87]. This could lead to more powerful and targeted treatments for diseases such as cancer [87, 88]. The modified ATO system has the potential to generate large numbers of antigen-specific T cells for cell-based therapies. The OP9-DL system relies on genetically engineered helper cells (OP9-DL1) that trigger Notch signaling to instruct stem cells to become T cells or NK cells [89, 90]. For T-cell differentiation, pluripotent stem cells are first coaxed into hematopoietic stem lineages with other helper cells (C3H10T1/2) [91]. Then, coculture with OP9-DL1 activated Notch signaling for T-cell development [91]. Finally, these committed T cells are exposed to immune system cells (PBMCs) for maturation and proliferation [92, 93]. To differentiate into NK cells, pluripotent stem cells are first grown with or without stromal cells in the presence of specific growth factors (SCF, VEGF, BMP4) to induce hematopoietic differentiation [94, 95]. These early precursor cells are then exposed to a different set of growth factors (IL-3, IL-15, IL-7, SCF, and FLT3L) to trigger the development of the precursors into mature NK cells. Finally, these developing NK cells are exposed to special cell lines that provide additional signals (through membrane-bound IL-15 or mbIL-21) to promote their growth and function [86, 96, 97].

The humanized BLT mouse model was created by transplanting human HSCs, liver tissue, and fetal thymus tissue into immunodeficient mice. This "humanized" mouse provides an ideal environment for the development of human immune cells [98]. Genetically modified HSCs can be transplanted into BLT mice to generate TCR-engineered antigen-specific T cells [99]. However, a major challenge with the original BLT model is that human T cells attack the mouse itself (GVHD) because they are not "educated" to tolerate the host [100]. To address this issue, sublethally irradiated neonatal mice were used as a new BLT model. This allows human T cells to mature within the mouse thymus, creating tolerance and preventing GVHD. This improved model allows for longer-term studies of human T cells in mice [101, 102]. However, the application of these systems is limited to research only because of the animal origin of the cellular product. To address safety concerns and consider off-the-shelf clinical application of the cellular products, human-derived feeder cells (e.g., foreskin fibroblasts and mesenchymal stem cells) were utilized as feeders for human embryonic stem cell (ESC) and iPSC cultures [103–106]. In addition, feeder-free systems have been developed to meet clinically applicable and scalable requirements. These feeder-free culture systems were used to produce allogeneic hematopoietic stem cell-engineered invariant natural killer T (HSC-iNKT) cells with high yield and purity [97]. In these systems, plate-bound DLL4 and VCAM-1 induce T-cell commitment from HSCs [107, 108]. The resulting allogeneic HSC-engineered iNKT cells demonstrated potent antitumor efficacy and antiviral capacity [97]. They could also be further modified with CARs for enhanced tumor targeting and gene editing via CRISPR-Cas9 to remove surface major histocompatibility complex (MHC) molecules [109]. Another feeder-free differentiation culture system enables the large-scale production of iPSC-derived T cells for cancer immunotherapy [110]. For T-cell therapy via iPSCs to be safe for patients, the cells must be completely free of viruses. This can be very expensive or even impossible to achieve with current methods that rely on feeder cells. Therefore, developing a system in which iPSCs can differentiate into T cells without feeder cells is crucial. While feeder-free systems already exist for growing the early stages of these cells, there is a critical gap in technology for the feeder-free differentiation and expansion of T cells themselves. For the clinical translation of iPSC-based T-cell therapy, scientists need a way to grow these immune cells in a clean and controlled environment that is completely free of feeder cells. However, the low differentiation efficiency and poor scalability of current methods may compromise their utility. Furthermore, improvements are needed to increase the stability and efficiency of immune cell production.

5 Gene-Editing Methods for Stem Cell Engineering

The main objective of engineering immune cells for cancer immunotherapy is to increase their ability to specifically target tumor cells. The two most widely used stem cell engineering approaches to increase immune cell specificity are TCRs and CARs [22]. Tumor antigen-specific TCRs are typically obtained from patient-derived, tumor-responsive T-cell clones [22, 111], humanized mouse models [112, 113], or phage display technology [114, 115]. In contrast, chimeric antigen receptors (CARs) use a single chain variable fragment (scFv) derived from monoclonal antibodies or immunoglobulin heavy and light chains. These CAR-T cells were able to kill tumor cells in vitro and were clinically safe but could not produce sufficient costimulatory signals for effective target cell killing in vivo. To improve costimulation efficiency, a CD28 ligand or 4-1BB was incorporated to develop second-generation CAR-T cells. These CAR-T cells effectively kill cancer cells in vivo and were approved for the treatment of many forms of hematological malignancies. TCRs are dependent on CD3 coexpression and are MHC restricted for their function [116–118]. In contrast, synthetic CARs function independently of major histocompatibility complex (MHC) restriction and can be applied to various immune cells, such as NK cells, macrophages, and myeloid cells, to increase their antitumor capabilities [119–121].

Knockout of TCR genes prevents GVHD, whereas knockout of MHC-related genes reduces allorejection. Knock-in of the CD19 CAR gene results in uniform CAR expression on T cells, a lack of endogenous TCR expression, and increased T-cell potency. Additionally, gene-editing efficiency is greater in stem cells than in mature immune cells and requires fewer materials. Other gene-editing strategies, such as incorporating IL-15 or introducing HLA-E/G, enhance immune cell antitumor capacity and grant resistance to host NK cell-mediated allorejection. Disruption of CD52 can make cells resistant to lymphodepleting drugs. By integrating somatic cell reprogramming, CAR engineering, and gene-editing technologies, we can increase the production of engineered cytotoxic T cells with specific antigen recognition ability. Further genetic modifications, including the elimination of HLA molecules and inhibitory receptors such as PD1, allow us to envision the use of hESCs or iPSC-derived T cells as readily available 'off-the-shelf' cellular therapies [122, 123].

To engineer immune cells, retroviral or lentiviral vectors are used to introduce TCRs and CARs into mature PBMCs or stem cells. Gene-editing technologies such as CRISPR-Cas9, zinc-finger nucleases (ZFNs), and transcription activator-like effector nucleases (TALENs) are also employed for efficient gene knockout, site-specific knockin, and genome-wide screening in mature immune cells and stem cells [124–127]. ZFNs are engineered proteins that combine a zinc-finger DNA-binding domain with a FokI endonuclease domain. The zinc-finger domain can be customized to bind specific DNA sequences, directing the FokI endonuclease to create a double-strand break (DSB) at a targeted location in the genome [127, 128]. This DSB can then be repaired by the cell's natural repair mechanisms, such as nonhomologous end joining (NHEJ) or homology-directed repair, leading to permanent genetic modifications. Despite their precision, ZFNs have several limitations, including a relatively narrow targeting range and the complex, time-consuming process required to design new zinc-finger domains for different DNA sequences [129]. TALENs, on the other hand, are composed of transcription activator-like effector (TALE) proteins fused to the FokI endonuclease [128]. TALE proteins are easier to engineer than zinc-finger domains because their DNA-binding specificity is determined by simple, modular repeats that can be easily customized to target virtually any DNA sequence [130]. TALENs also create DSBs at specific genomic locations, which can be repaired by NHEJ or homology-directed repair, resulting in genetic modifications. TALENs are known for their high cleavage activity and low cytotoxicity, making them powerful and versatile tools for genome editing [127]. CRISPR-Cas9 is particularly effective for cellular reprogramming because of its ability to target multiple genes and endogenous loci [131]. It uses an inactivated Cas9 protein guided by a 20-base pair single RNA sequence to specific genome regions, enabling precise gene activation or silencing [132]. The main advantage of the CRISPR/Cas9 system is its simplicity, as DNA-binding specificity is determined solely by the sgRNA, eliminating the need for complex protein engineering. This efficiency has led to its widespread adoption, replacing older gene-editing technologies such as TALENs and ZFNs, and has revolutionized the rapid development of targeted genetic therapies [133].

Researchers have developed nonviral and nonintegrating methods to increase the safety of virus-mediated reprogramming for iPSCs. Initial approaches, such as piggyback, DNA minicircles, and microRNAs, were inefficient. More effective nonintegrating methods have been developed, including Sendai vectors, mRNA reprogramming, self-replicating RNA, and episomal vectors, each with distinct advantages and challenges [134, 135]. Self-replicating RNA relies on the Venezuelan equine encephalitis virus's positive-sense, single-stranded RNA backbone, mimicking cellular mRNA without a DNA intermediate. However, coagents are needed to suppress the immune response and retain viral RNA components for several passages, complicating iPSC selection. mRNA reprogramming involves repeated daily transfections for up to 17 days, making it labor-intensive and expensive. This method also involves interferon production in transfected cells, which impacts reprogramming efficiency and presents potential immunological concerns. Additionally, somatic suspension cells are resistant to repeated transfection processes. Sendai virus reprogramming is popular and robust but requires numerous cell divisions to dilute the cell line free of contaminating viral proteins and the vector, necessitating extensive quality control measures to confirm safety.

Episomal reprogramming is considered ideal for creating clinical-grade, safe, nonviral, and nonintegrating iPSCs [134]. According to the Global Alliance for iPSC Therapies, episomal reprogramming is the most common approach for producing clinical-grade iPSCs owing to the rapid clearance of transgenes from targeted cells. Episomal vectors are typically active for 17–21 days before becoming undetectable owing to dilution and instability caused by cell division. However, the episomal reprogramming efficiency is relatively low compared with that of other nonintegrating methods. To compensate, researchers have utilized factors such as c-Myc or a combination of l-Myc and Lin28 [132]. For example, Yu et al. reported no colony formation using an episomal reprogramming strategy that delivered Oct4, Sox2, c-Myc, Nanog, Lin28, and Klf4 but achieved a reprogramming efficiency of approximately 0.0006% after adding an SV40 large T-antigen gene [136]. Similarly, Okita et al. reported comparable reprogramming efficiency with p53 suppression combined with l-Myc and Lin28 heterologous expression [137].

Recent advancements have demonstrated that reprogramming small molecules without oncogenes can produce fully reprogrammed iPSC colonies with high efficiency and safety. This method, which is applicable to both adherent and suspension cells, reduces neoplastic risk and eliminates viral immunogenicity. For example, iPSCs have been successfully derived from human foreskin fibroblasts, cord blood-derived CD34+ cells, cord blood-derived mononuclear cells, and peripheral blood mononuclear cells from patients with cystic fibrosis and alpha-1 antitrypsin deficiency. These iPSC colonies, formed without oncogenes, were fully reprogrammed on the basis of SSEA4 expression, exceeding the 70% threshold recommended by the Global Alliance in iPSC Therapy.

6 Challenges of Gene Editing in Stem Cells for Immune Cell Therapy

A significant limitation in gene editing is that human cells typically opt for the less accurate NHEJ pathway to fix double-strand breaks (DSBs) rather than the precise homologous DNA repair pathway. This choice results in unintentional insertions and deletions instead of exact modifications. Although NHEJ-mediated gene disruption can be beneficial for inducing loss-of-function events, clinical treatments usually require precise editing to produce gain-of-function effects on specific genes [138]. Another challenge of genome editing in human cells is short conversion tracts, meaning that modifications are limited to areas close to the double-strand break (DSB), even when homology-directed repair (HDR) is used. This limits the introduction of complex genetic changes. However, using Cpf1, a CRISPR effector similar to Cas9 but cleaving DNA further from the PAM sequence and creating a single-stranded overhang, may improve the rate of HDR over NHEJ events [139]. Gene-editing methods, such as CRISPR and TALENs, have improved transfection efficiency. CRISPR/Cas9 is particularly effective because of its ability to target specific genes and its cost-effectiveness. Despite their broad applications, designer nucleases such as CRISPR/Cas9 are not considered safe for gene therapy because of significant off-target effects and difficulties in gene knock-ins [133, 138]. Efforts to reduce these off-target effects by developing new Cas9 variants and enhancing gRNA design have been largely ineffective. Additionally, CRISPR/Cas9 requires a PAM sequence near the target site, and the commonly used Cas9 from the bacteria *Streptococcus pyogenes* (SpCas9) is challenging to package into Adeno associated virus (AAV) vectors owing to its size. CRISPR-induced double-strand breaks can also lead to apoptosis and cellular toxicity. These issues highlight the technical and functional limitations of current gene-editing systems, raising concerns about their real-world application. However, it has limitations, such as potential off-target effects [125]. TALENs, while more specific and less cytotoxic, are harder to design. The design process is complex and time-consuming, as each TALEN must be custom-built to recognize specific DNA sequences. Delivery into cells is also difficult due to the large size of TALEN proteins, which can hinder their efficiency, especially in vivo. Additionally, the cost of designing and producing TALENs is greater than that of CRISPR, limiting their accessibility. While TALENs offer high specificity and low cytotoxicity, there is still a risk of off-target effects. These factors have contributed to their limited adoption in research settings [130].

7 Challenges in the Commercial Manufacturing of Stem Cells for Clinical Use

Commercial manufacturing of iPSC-derived therapies involves complex processes and stringent quality controls at various stages, including patient recruitment, tissue procurement, cell isolation, expansion, scaling up to produce sufficient doses for clinical use, differentiation, phenotype characterization, cryopreservation, and cold chain management.

Scalability One of the critical challenges for future success in commercializing allogeneic cell products is establishing a scalable manufacturing technology that can reliably reproduce the yield and quality of PSC-derived products generated from small-scale R&D at larger scales sufficient for commercial manufacturing for clinical use [140]. In vitro bioassays and computational models are essential for evaluating cell potency during the manufacturing scale-up process. Automation, closed systems, and data-centric processes are crucial for making autologous iPSC-based cell therapy manufacturing scalable, cost-effective, and efficient. The Miltenyi CliniMACS Prodigy and Quantum Cell Expansion System (QES) are key instruments. The Prodigy now handles GMP-compliant iPSC differentiation, whereas the QES ensures large-scale production with consistent cell metabolism and yield through advanced bioreactor technology [141].

Safety and Standardization Ensuring that stem cell products are consistent, safe, and meet quality standards is critical for clinical use. In addition to the adverse events associated with CAR-T-cell therapy, such as CRS and neurotoxicity, stem cell-derived therapies have other safety concerns that must be addressed [118]. These include tumorigenicity, GVHD, and the presence of undifferentiated cells in the final cell product [123]. It is very important to ensure that stem cells differentiate into the desired cell type and mature properly. This process must be tightly controlled to produce functional cells that can integrate into the patient's body and perform their intended function. The development of reliable protocols for the differentiation of stem cells into specific cell types is complex and time-consuming. Even if stem cells differentiate into the desired cell type, they may not fully mature or function correctly. Ensuring proper maturation is critical for the success of therapy.

The self-renewal ability of stem cells, particularly PSCs, raises concerns about their potential to form tumors [142]. Even a small number of remaining undifferentiated PSCs in the final product can lead to teratoma formation when implanted in a patient [143]. Additionally, tumorigenic mutations can occur during cell expansion, leading to genetic changes such as chromosomal abnormalities, copy number variations, and single nucleotide mutations [144]. In products using iPSCs, reprogramming factors pose a risk of causing tumors. Yamanaka factors, especially c-Myc, are often overexpressed and are known to initiate cancers. Although current practices use Sendai virus transduction for episomal expression of reprogramming factors, ensuring that these factors are not integrated as transgenes before clinical development is crucial. Finally, suicide switches can be incorporated into stem cell-derived therapies to eliminate cells if tumors form [145].

GVHD is another significant safety concern associated with T-cell-based allogeneic cell therapies. This can lead to the destruction of the transplanted cells and failure of the treatment [146]. Scientists have developed several genetic modification methods to prevent GVHD. Techniques such as CRISPR/Cas9 can be used to modify stem cells to make them less recognizable to the immune system [147]. One such approach is to knock out the endogenous TCR, which is used to create universal allogeneic T-cell therapies such as UCART [123, 146]. Another strategy involves inserting genes into the T-cell TRAC locus to mitigate the risk of GVHD [147]. Interestingly, some immune cell populations, such as germline-encoded natural killer (NK) receptors and TCRs on innate T cells, do not require genetic manipulation to prevent GVHD [123]. Numerous clinical trials have confirmed the safety of these allogeneic cell types for cancer immunotherapy.

Quality Control and Consistency of the Manufactured Product

To ensure the quality of allogeneic stem cell-based products, various in vitro and in vivo tests have been conducted to assess tumorigenicity [148]. Karyotyping is a traditional method used to detect chromosomal abnormalities, such as deletions, duplications, or rearrangements, and any clones with these alterations are discarded [149, 150]. Next-generation sequencing can identify single nucleotide changes and other genomic alterations, although the impact of these smaller genetic changes on the final cell product is still debated and requires further research [141, 151]. Additionally, flow cytometry and quantitative reverse-transcriptase PCR (qRT–PCR) or droplet digital PCR are used to evaluate the quality and differentiation status of stem cells in the final product [152, 153]. With an increasing number of cell therapy products derived from hPSCs or other cell types harboring tumorigenic potential progressing toward the clinic, there is a need to improve current strategies for tumorigenicity assessment and ultimately minimize undifferentiated and transformed hPSCs in final cell therapy products.

Cost The high cost of large-scale stem cell production currently makes these therapies expensive and out of reach for many patients. The highest consumable cost in GLP-specific cell product development is the high cost of the growth factors required to support cell expansion in vitro. According to one study, the estimated cost of the growth factors required to produce 7 billion neuro progenitor cells (NPCs) from iPSCs is $21,165,500.00 [124]. Another significant cost factor is the number of cell doses that can be produced from a single batch of off-the-shelf cell products. Furthermore, expensive testing is required for post-manufacturing quality control tests before genetically modified cells are released for clinical use. To reduce the cost of stem cell therapies, several strategies can be employed: developing universal cells via an HLA cloaking approach that can be used for multiple patients, establishing large cell banks, enhancing manufacturing processes, and using engineered viruses to fix cells within the body without the need for in vitro expansion [154]. Both CRISPR and TALENs are expected to enhance iPSC reprogramming, improving differentiation efficiency and reducing the time and cost of producing iPSC-derived cell therapies [124]. Improving the yield of stem cell production and the efficiency of differentiation processes can reduce costs. This includes optimizing culture conditions and automation to scale up production. Additionally, expanding insurance coverage, streamlining regulatory pathways, and increasing public and private funding can also help make these therapies more affordable and accessible.

Regulatory Approval Regulatory bodies worldwide, including the Global Alliance in iPSC Therapy, the European Medicines Agency (EMA), the US FDA, the Japan Ministry of Health, Labor and Welfare (MHLW), etc., have introduced specific legislation and guidelines to support the development of cell therapy medicinal products intended for human use. These guidelines provide guidelines for genetic fidelity, stability, safety, cell potency testing, and efficacy aspects that are considered for clinical trial approval and marketing authorization [124, 148]. In the United States, cellular therapy products are regulated under 21 CFR Part 1271 (Human Cells, Tissues, and Cellular and Tissue-Based Products), 21 Part 600 (Biological Products: General) and 21 CFR Part 610 (General Biological Products Standards), and drug manufacturing requirements are defined in 21 CFR Part 211 (Current Good Manufacturing Practice) [155]. Navigating the regulatory landscape to gain approval for stem cell therapies is complex and time-consuming [156]. Simplifying regulatory pathways for stem cell therapies can reduce the time and cost associated with bringing these treatments to market.

8 Future Directions

While iPSC technology may eventually enable autologous cell therapies for any cell type, the complex and costly process of iPSC reprogramming and differentiation currently limits the widespread use of personalized iPSC precision medicine. Therefore, stem cell-derived therapies are likely to remain allogeneic. The success of these therapies has boosted the development of CAR-T-cell therapy to target diverse types of cancer. Currently, 27 cell therapies have been approved by the FDA for the treatment of various diseases, including malignancies such as pre-B-ALL, chronic lymphocytic leukemia (CLL), and non-Hodgkin's lymphoma, and more than 150 CAR-T-cell-related clinical trials are ongoing (per ClinicalTrials.gov). Nevertheless, we believe that continued investment in the optimization of these allogeneic strategies is warranted on the basis of the current data and that allogeneic cell products will usher in a new era of cell therapy. To further advance CAR-T-cell-based therapy, the next task involves improving the scale-up, availability, consistency, and safety of CAR-T cells. Recently, a French biotech company treefrog developed a method to manufacture 15 billion cells in a single batch in a bioreactor with exponential amplification of 276-fold per week. Stem cell engineering for immune cell therapy holds immense promise for treating a wide range of diseases. However, significant challenges must be overcome to fully realize its potential. Addressing these challenges requires a multidisciplinary approach involving scientists, clinicians, ethicists, and policymakers. By working together, we can overcome these obstacles and benefit patients in need of stem cell therapy.

Bibliography

1. Rosenberg SA, Yang JC, Sherry RM, Kammula US, Hughes MS, Phan GQ, et al. Durable complete responses in heavily pretreated patients with metastatic melanoma using T-cell transfer immunotherapy. Clin Cancer Res. 2011;17(13):4550–7.
2. Singh N, Perazzelli J, Grupp SA, Barrett DM. Early memory phenotypes drive T cell proliferation in patients with pediatric malignancies. Sci Transl Med. 2016;8(320):320ra3.
3. Rosenberg SA, Restifo NP. Adoptive cell transfer as personalized immunotherapy for human cancer. Science. 2015;348(6230):62–8.
4. Dudley ME, Rosenberg SA. Adoptive-cell-transfer therapy for the treatment of patients with cancer. Nat Rev Cancer. 2003;3(9):666–75.
5. Rosenberg SA, Yannelli JR, Yang JC, Topalian SL, Schwartzentruber DJ, Weber JS, et al. Treatment of patients with metastatic melanoma with autologous tumor-infiltrating lymphocytes and interleukin 2. JNCI J Natl Cancer Inst. 1994;86(15):1159–66.
6. Perez C, Gruber I, Arber C. Off-the-shelf allogeneic T cell therapies for cancer: opportunities and challenges using naturally occurring "universal" donor T cells. Front Immunol. 2020;11:583716.
7. Singh AK, McGuirk JP. Allogeneic stem cell transplantation: a historical and scientific overview. Cancer Res. 2016;76(22):6445–51.
8. Wilson JM. A history lesson for stem cells. Science. 2009;324(5928):727–8.

9. Zhang K, Cheng K. Stem cell-derived exosome versus stem cell therapy. Nat Rev Bioeng. 2023;1(9):608–9.
10. Li YR, Dunn ZS, Yu Y, Li M, Wang P, Yang L. Advancing cell-based cancer immunotherapy through stem cell engineering. Cell Stem Cell. 2023;30(5):592–610.
11. Yewdell JW, Del Val M. Immunodominance in TCD8+ responses to viruses: cell biology, cellular immunology, and mathematical models. Immunity. 2004;21(2):149–53.
12. Radvanyi LG. Tumor-infiltrating lymphocyte therapy: addressing prevailing questions. Cancer J (United States). 2015;21(6):450–64.
13. June CH. Adoptive T cell therapy for cancer in the clinic. J Clin Invest. 2007;117(6):1466–76.
14. Kansagra A, Farnia S, Majhail N. Expanding access to chimeric antigen receptor T-cell therapies: challenges and opportunities. Am Soc Clin Oncol Educ Book. 2020;40:e27–34.
15. Mirzaei HR, Rodriguez A, Shepphird J, Brown CE, Badie B. Chimeric antigen receptors T cell therapy in solid tumor: challenges and clinical applications. Front Immunol. 2017;8(DEC):1850.
16. Roddie C, O'Reilly M, Dias Alves Pinto J, Vispute K, Lowdell M. Manufacturing chimeric antigen receptor T cells: issues and challenges. Cytotherapy. 2019;21(3):327–40.
17. Srivastava S, Riddell SR. Chimeric antigen receptor T cell therapy: challenges to bench-to-bedside efficacy. J Immunol. 2018;200(2):459–68.
18. Xia A-L, Wang X-C, Lu Y-J, Lu X-J, Sun B. Chimeric-antigen receptor T (CAR-T) cell therapy for solid tumors: challenges and opportunities. Oncotarget. 2017;8(52):90521.
19. Lim WA, June CH. The principles of engineering immune cells to treat cancer. Cell. 2017;168(4):724–40.
20. Mikkilineni L, Kochenderfer JN. Chimeric antigen receptor T-cell therapies for multiple myeloma. Blood. 2017;130(24):2594–602.
21. Ramos CA, Heslop HE, Brenner MK. CAR-T cell therapy for lymphoma. Annu Rev Med. 2016;67(1):165–83.
22. Themeli M, Rivière I, Sadelain M. New cell sources for T cell engineering and adoptive immunotherapy. Cell Stem Cell. 2015;16(4):357–66.
23. Ruella M, Xu J, Barrett DM, Fraietta JA, Reich TJ, Ambrose DE, et al. Induction of resistance to chimeric antigen receptor T cell therapy by transduction of a single leukemic B cell. Nat Med. 2018;24(10):1499–503.
24. Good CR, Aznar MA, Kuramitsu S, Samareh P, Agarwal S, Donahue G, et al. An NK-like CAR T cell transition in CAR T cell dysfunction. Cell. 2021;184(25):6081–100.e26.
25. Deng Q, Han G, Puebla-Osorio N, Ma MCJ, Strati P, Chasen B, et al. Characteristics of anti-CD19 CAR T cell infusion products associated with efficacy and toxicity in patients with large B cell lymphomas. Nat Med. 2020;26(12):1878–87.
26. Fraietta JA, Lacey SF, Orlando EJ, Pruteanu-Malinici I, Gohil M, Lundh S, et al. Determinants of response and resistance to CD19 chimeric antigen receptor (CAR) T cell therapy of chronic lymphocytic leukemia. Nat Med. 2018;24(5):563–71.
27. Turtle CJ, Hanafi L-A, Berger C, Gooley TA, Cherian S, Hudecek M, et al. CD19 CAR–T cells of defined CD4+:CD8+ composition in adult B cell ALL patients. J Clin Invest. 2016;126(6):2123–38.
28. Shah NN, Zhu F, Taylor C, Schneider D, Krueger W, Worden A, et al. A phase 1 study with point-of-care manufacturing of dual targeted, tandem anti-CD19, anti-CD20 chimeric antigen receptor modified T (CAR-T) cells for relapsed, refractory, non-Hodgkin lymphoma. Blood. 2018;132(Supplement 1):4193.
29. Benjamin R, Graham C, Yallop D, Jozwik A, Ciocarlie O, Jain N, et al. Preliminary data on safety, cellular kinetics and anti-leukemic activity of UCART19, an allogeneic anti-CD19 CAR T-cell product, in a pool of adult and pediatric patients with high-risk CD19+ relapsed/refractory B-cell acute lymphoblastic leukemia. Blood. 2018;132(Supplement 1):896.
30. Murthy H, Iqbal M, Chavez JC, Kharfan-Dabaja MA. Cytokine release syndrome: current perspectives. Immunotargets Ther. 2019;8:43–52.

31. Depil S, Duchateau P, Grupp SA, Mufti G, Poirot L. 'Off-the-shelf' allogeneic CAR T cells: development and challenges. Nat Rev Drug Discov. 2020;19(3):185–99.
32. Ruella M, Kenderian SS. Next-generation chimeric antigen receptor T-cell therapy: going off the shelf. BioDrugs. 2017;31(6):473–81.
33. Ren J, Liu X, Fang C, Jiang S, June CH, Zhao Y. Multiplex genome editing to generate universal CAR T cells resistant to PD1 inhibition. Clin Cancer Res. 2017;23(9):2255–66.
34. Chaidos A, Patterson S, Szydlo R, Chaudhry MS, Dazzi F, Kanfer E, et al. Graft invariant natural killer T-cell dose predicts risk of acute graft-versus-host disease in allogeneic hematopoietic stem cell transplantation. Blood. 2012;119(21):5030–6.
35. Li YR, Zhou K, Wilson M, Kramer A, Zhu Y, Dawson N, et al. Mucosal-associated invariant T cells for cancer immunotherapy. Mol Ther. 2023;31(3):631–46.
36. Bae E-A, Seo H, Kim I-K, Jeon I, Kang C-Y. Roles of NKT cells in cancer immunotherapy. Arch Pharm Res. 2019;42(7):543–8.
37. Xie B, Li Z, Zhou J, Wang W. Current status and perspectives of dual-targeting chimeric antigen receptor T-cell therapy for the treatment of hematological malignancies. Cancers (Basel). 2022;14(13):3230.
38. Bulcha JT, Wang Y, Ma H, Tai PWL, Gao G. Viral vector platforms within the gene therapy landscape. Signal Transduct Target Ther. 2021;6(1):53.
39. Ellis GI, Sheppard NC, Riley JL. Genetic engineering of T cells for immunotherapy. Nat Rev Genet. 2021;22(7):427–47.
40. Weber EW, Maus MV, Mackall CL. The emerging landscape of immune cell therapies. Cell. 2020;181(1):46–62.
41. Cascalho M, Platt JL. Basic mechanisms of humoral rejection. Pediatr Transplant. 2005;9(1):9–16.
42. Montgomery RA, Cozzi E, West LJ, Warren DS. Humoral immunity and antibody-mediated rejection in solid organ transplantation. Semin Immunol. 2011;23(4):224–34.
43. Liu E, Marin D, Banerjee P, Macapinlac HA, Thompson P, Basar R, et al. Use of CAR-transduced natural killer cells in CD19-positive lymphoid tumors. N Engl J Med. 2020;382(6):545–53.
44. Figueiredo C, Blasczyk R. A future with less HLA: potential clinical applications of HLA-universal cells. Tissue Antigens. 2015;85(6):443–9.
45. Hawley RG, Ramezani A, Hawley TS. Hematopoietic stem cells. Methods Enzymol. 2006;419:149–79.
46. Notta F, Doulatov S, Laurenti E, Poeppl A, Jurisica I, Dick JE. Isolation of single human hematopoietic stem cells capable of long-term multilineage engraftment. Science. 2011;333(6039):218–21.
47. Laurent M, Geoffroy M, Pavani G, Guiraud S. CRISPR-based gene therapies: from preclinical to clinical treatments. Cells. 2024;13(10):800.
48. Dzierzak E, Bigas A. Blood development: hematopoietic stem cell dependence and independence. Cell Stem Cell. 2018;22(5):639–51.
49. Rix B, Maduro AH, Bridge KS, Grey W. Markers for human haematopoietic stem cells: The disconnect between an identification marker and its function. Front Physiol. 2022;13:1009160.
50. Weissman IL, Shizuru JA. The origins of the identification and isolation of hematopoietic stem cells, and their capability to induce donor-specific transplantation tolerance and treat autoimmune diseases. Blood. 2008;112(9):3543–53.
51. Mayani H. The regulation of hematopoietic stem cell populations. F1000Res. 2016;5:F1000-Faculty.
52. Nordin F, Idris MRM, Mahdy ZA, Wahid SFA. Preeclampsia in pregnancy affecting the stemness and differentiation potency of haematopoietic stem cell of the umbilical cord blood. BMC Pregnancy Childbirth. 2020;20(1):399.
53. Ballen KK, Gluckman E, Broxmeyer HE. Umbilical cord blood transplantation: the first 25 years and beyond. Blood. 2013;122(4):491–8.

54. Gragert L, Eapen M, Williams E, Freeman J, Spellman S, Baitty R, et al. HLA match likelihoods for hematopoietic stem-cell grafts in the U.S. registry. N Engl J Med. 2014;371(4):339–48.
55. Horan J, Wang T, Haagenson M, Spellman SR, Dehn J, Eapen M, et al. Evaluation of HLA matching in unrelated hematopoietic stem cell transplantation for nonmalignant disorders. Blood. 2012;120(14):2918–24.
56. Lee SJ, Klein J, Haagenson M, Baxter-Lowe LA, Confer DL, Eapen M, et al. High-resolution donor-recipient HLA matching contributes to the success of unrelated donor marrow transplantation. Blood. 2007;110(13):4576–83.
57. Rocha V, Wagner JE Jr, Sobocinski KA, Klein JP, Zhang MJ, Horowitz MM, et al. Graft-versus-host disease in children who have received a cord-blood or bone marrow transplant from an HLA-identical sibling. Eurocord and International Bone Marrow Transplant Registry Working Committee on Alternative Donor and Stem Cell Sources. N Engl J Med. 2000;342(25):1846–54.
58. Dehn J, Spellman S, Hurley CK, Shaw BE, Barker JN, Burns LJ, et al. Selection of unrelated donors and cord blood units for hematopoietic cell transplantation: guidelines from the NMDP/CIBMTR. Blood. 2019;134(12):924–34.
59. Barker JN, Byam CE, Kernan NA, Lee SS, Hawke RM, Doshi KA, et al. Availability of cord blood extends allogeneic hematopoietic stem cell transplant access to racial and ethnic minorities. Biol Blood Marrow Transplant. 2010;16(11):1541–8.
60. Chang H-H, Liou Y-S, Sun D-S. Hematopoietic stem cell mobilization. Tzu Chi Med J. 2022;34(3):270.
61. Liles WC, Broxmeyer HE, Rodger E, Wood B, Hübel K, Cooper S, et al. Mobilization of hematopoietic progenitor cells in healthy volunteers by AMD3100, a CXCR4 antagonist. Blood. 2003;102(8):2728–30.
62. Hoggatt J, Singh P, Tate TA, Chou B-K, Datari SR, Fukuda S, et al. Rapid mobilization reveals a highly engraftable hematopoietic stem cell. Cell. 2018;172(1):191–204.e10.
63. Suárez-Álvarez B, López-Vázquez A, López-Larrea C. Mobilization and homing of hematopoietic stem cells. Adv Exp Med Biol. 2012;741:152–70.
64. Crees ZD, Rettig MP, Jayasinghe RG, Stockerl-Goldstein K, Larson SM, Arpad I, et al. Motixafortide and G-CSF to mobilize hematopoietic stem cells for autologous transplantation in multiple myeloma: a randomized phase 3 trial. Nat Med. 2023;29(4):869–79.
65. Winkler IG, Pettit AR, Raggatt LJ, Jacobsen RN, Forristal CE, Barbier V, et al. Hematopoietic stem cell mobilizing agents G-CSF, cyclophosphamide or AMD3100 have distinct mechanisms of action on bone marrow HSC niches and bone formation. Leukemia. 2012;26(7):1594–601.
66. Bonig H, Papayannopoulou T. Mobilization of hematopoietic stem/progenitor cells: general principles and molecular mechanisms. Methods Mol Biol. 2012;904:1–14.
67. Lim WF, Inoue-Yokoo T, Tan KS, Lai MI, Sugiyama D. Hematopoietic cell differentiation from embryonic and induced pluripotent stem cells. Stem Cell Res Ther. 2013;4(3):71.
68. Vodyanik MA, Bork JA, Thomson JA, Slukvin II. Human embryonic stem cell-derived CD34+ cells: efficient production in the coculture with OP9 stromal cells and analysis of lymphohematopoietic potential. Blood. 2005;105(2):617–26.
69. Kaufman DS, Hanson ET, Lewis RL, Auerbach R, Thomson JA. Hematopoietic colony-forming cells derived from human embryonic stem cells. Proc Natl Acad Sci USA. 2001;98(19):10716–21.
70. Ng ES, Davis R, Stanley EG, Elefanty AG. A protocol describing the use of a recombinant protein-based, animal product-free medium (APEL) for human embryonic stem cell differentiation as spin embryoid bodies. Nat Protoc. 2008;3(5):768–76.
71. Thomson JA, Itskovitz-Eldor J, Shapiro SS, Waknitz MA, Swiergiel JJ, Marshall VS, et al. Embryonic stem cell lines derived from human blastocysts. Science. 1998;282(5391):1145–7.
72. Takahashi K, Yamanaka S. Induction of pluripotent stem cells from mouse embryonic and adult fibroblast cultures by defined factors. Cell. 2006;126(4):663–76.
73. Caldwell KJ, Gottschalk S, Talleur AC. Allogeneic CAR cell therapy-more than a pipe dream. Front Immunol. 2020;11:618427.

74. Veluchamy JP, Kok N, van der Vliet HJ, Verheul HMW, de Gruijl TD, Spanholtz J. The rise of allogeneic natural killer cells as a platform for cancer immunotherapy: recent innovations and future developments. Front Immunol. 2017;8:631.
75. Nguyen R, Wu H, Pounds S, Inaba H, Ribeiro RC, Cullins D, et al. A phase II clinical trial of adoptive transfer of haploidentical natural killer cells for consolidation therapy of pediatric acute myeloid leukemia. J Immunother Cancer. 2019;7(1):81.
76. Geller MA, Cooley S, Judson PL, Ghebre R, Carson LF, Argenta PA, et al. A phase II study of allogeneic natural killer cell therapy to treat patients with recurrent ovarian and breast cancer. Cytotherapy. 2011;13(1):98–107.
77. Thommen DS, Schumacher TN. T cell dysfunction in cancer. Cancer Cell. 2018;33(4):547–62.
78. Rotolo A, Caputo VS, Holubova M, Baxan N, Dubois O, Chaudhry MS, et al. Enhanced anti-lymphoma activity of CAR19-iNKT cells underpinned by dual CD19 and CD1d targeting. Cancer Cell. 2018;34(4):596–610.e11.
79. Heczey A, Courtney AN, Montalbano A, Robinson S, Liu K, Li M, et al. Anti-GD2 CAR-NKT cells in patients with relapsed or refractory neuroblastoma: an interim analysis. Nat Med. 2020;26(11):1686–90.
80. Caldwell KJ, Gottschalk S, Talleur AC. Allogeneic CAR cell therapy—more than a pipe dream. Front Immunol. 2021;11:618427.
81. Li Y-R, Dunn ZS, Zhou Y, Lee D, Yang L. Development of stem cell-derived immune cells for off-the-shelf cancer immunotherapies. Cells. 2021;10(12):3497.
82. Li YR, Zhou Y, Kramer A, Yang L. Engineering stem cells for cancer immunotherapy. Trends Cancer. 2021;7(12):1059–73.
83. Zhu Y, Smith DJ, Zhou Y, Li Y-R, Yu J, Lee D, et al. Development of hematopoietic stem cell-engineered invariant natural killer T cell therapy for cancer. Cell Stem Cell. 2019;25(4):542–57.e9.
84. Smith DJ, Lin LJ, Moon H, Pham AT, Wang X, Liu S, et al. Propagating humanized BLT mice for the study of human immunology and immunotherapy. Stem Cells Dev. 2016;25(24):1863–73.
85. Ciofani M, Zuniga-Pflücker JC. Notch promotes survival of pre–T cells at the β-selection checkpoint by regulating cellular metabolism. Nat Immunol. 2005;6(9):881–8.
86. Shimasaki N, Jain A, Campana D. NK cells for cancer immunotherapy. Nat Rev Drug Discov. 2020;19(3):200–18.
87. Montel-Hagen A, Seet CS, Li S, Chick B, Zhu Y, Chang P, et al. Organoid-induced differentiation of conventional T cells from human pluripotent stem cells. Cell Stem Cell. 2019;24(3):376–89.e8.
88. Seet CS, He C, Bethune MT, Li S, Chick B, Gschweng EH, et al. Generation of mature T cells from human hematopoietic stem and progenitor cells in artificial thymic organoids. Nat Methods. 2017;14(5):521–30.
89. Ciofani M, Knowles GC, Wiest DL, von Boehmer H, Zúñiga-Pflücker JC. Stage-specific and differential notch dependency at the αβ and γδ T lineage bifurcation. Immunity. 2006;25(1):105–16.
90. Besseyrias V, Fiorini E, Strobl LJ, Zimber-Strobl U, Dumortier A, Koch U, et al. Hierarchy of Notch-Delta interactions promoting T cell lineage commitment and maturation. J Exp Med. 2007;204(2):331–43.
91. Takayama N, Nishikii H, Usui J, Tsukui H, Sawaguchi A, Hiroyama T, et al. Generation of functional platelets from human embryonic stem cells in vitro via ES-sacs, VEGF-promoted structures that concentrate hematopoietic progenitors. Blood. 2008;111(11):5298–306.
92. Awong G, La Motte-Mohs RN, Zúñiga-Pflücker JC. Generation of pro-T cells in vitro: potential for immune reconstitution. Semin Immunol. 2007;19(5):341–9.
93. Nishimura T, Nakauchi H. Generation of antigen-specific T cells from human induced pluripotent stem cells. Methods Mol Biol. 2019;1899:25–40.
94. Li Y, Hermanson DL, Moriarity BS, Kaufman DS. Human iPSC-derived natural killer cells engineered with chimeric antigen receptors enhance anti-tumor activity. Cell Stem Cell. 2018;23(2):181–92.e5.

95. Knorr DA, Ni Z, Hermanson D, Hexum MK, Bendzick L, Cooper LJ, et al. Clinical-scale derivation of natural killer cells from human pluripotent stem cells for cancer therapy. Stem Cells Transl Med. 2013;2(4):274–83.
96. Woll PS, Martin CH, Miller JS, Kaufman DS. Human embryonic stem cell-derived NK cells acquire functional receptors and cytolytic activity. J Immunol. 2005;175(8):5095–103.
97. Li Y-R, Dunn ZS, Garcia G, Carmona C, Zhou Y, Lee D, et al. Development of off-the-shelf hematopoietic stem cell-engineered invariant natural killer T cells for COVID-19 therapeutic intervention. Stem Cell Res Ther. 2022;13(1):112.
98. Kitchen SG, Bennett M, Galić Z, Kim J, Xu Q, Young A, et al. Engineering antigen-specific T cells from genetically modified human hematopoietic stem cells in immunodeficient mice. PLoS One. 2009;4(12):e8208.
99. Smith DJ, Liu S, Ji S, Li B, McLaughlin J, Cheng D, et al. Genetic engineering of hematopoietic stem cells to generate invariant natural killer T cells. Proc Natl Acad Sci USA. 2015;112(5):1523–8.
100. Greenblatt MB, Vrbanac V, Tivey T, Tsang K, Tager AM, Aliprantis AO. Graft versus host disease in the bone marrow, liver and thymus humanized mouse model. PLoS One. 2012;7(9):e44664.
101. Lan P, Tonomura N, Shimizu A, Wang S, Yang Y-G. Reconstitution of a functional human immune system in immunodeficient mice through combined human fetal thymus/liver and CD34+ cell transplantation. Blood. 2006;108(2):487–92.
102. Melkus MW, Estes JD, Padgett-Thomas A, Gatlin J, Denton PW, Othieno FA, et al. Humanized mice mount specific adaptive and innate immune responses to EBV and TSST-1. Nat Med. 2006;12(11):1316–22.
103. Yu G, Kamano Y, Wang F, Okawa H, Yatani H, Egusa H, editors. Feeder cell sources and feeder-free methods for human iPS cell culture. Interface oral health science 2014; 2015; Tokyo: Springer Japan.
104. Lu Z, Zhu W, Yu Y, Jin D, Guan Y, Yao R, et al. Derivation and long-term culture of human parthenogenetic embryonic stem cells using human foreskin feeders. J Assist Reprod Genet. 2010;27(6):285–91.
105. Park Y, Kim JH, Lee SJ, Choi IY, Park SJ, Lee SR, et al. Human feeder cells can support the undifferentiated growth of human and mouse embryonic stem cells using their own basic fibroblast growth factors. Stem Cells Dev. 2011;20(11):1901–10.
106. Unger C, Gao S, Cohen M, Jaconi M, Bergstrom R, Holm F, et al. Immortalized human skin fibroblast feeder cells support growth and maintenance of both human embryonic and induced pluripotent stem cells. Hum Reprod. 2009;24(10):2567–81.
107. Huijskens MJAJ, Walczak M, Koller N, Briedé JJ, Senden-Gijsbers BLMG, Schnijderberg MC, et al. Technical advance: ascorbic acid induces development of double-positive T cells from human hematopoietic stem cells in the absence of stromal cells. J Leukoc Biol. 2014;96(6):1165–75.
108. Shukla S, Langley MA, Singh J, Edgar JM, Mohtashami M, Zúñiga-Pflücker JC, et al. Progenitor T-cell differentiation from hematopoietic stem cells using Delta-like-4 and VCAM-1. Nat Methods. 2017;14(5):531–8.
109. Li Y-R, Zhou Y, Kim YJ, Zhu Y, Ma F, Yu J, et al. Development of allogeneic HSC-engineered iNKT cells for off-the-shelf cancer immunotherapy. Cell Rep Med. 2021;2(11):100449.
110. Iriguchi S, Yasui Y, Kawai Y, Arima S, Kunitomo M, Sato T, et al. A clinically applicable and scalable method to regenerate T-cells from iPSCs for off-the-shelf T-cell immunotherapy. Nat Commun. 2021;12(1):430.
111. Johnson LA, Heemskerk B, Powell DJ Jr, Cohen CJ, Morgan RA, Dudley ME, et al. Gene transfer of tumor-reactive TCR confers both high avidity and tumor reactivity to nonreactive peripheral blood mononuclear cells and tumor-infiltrating lymphocytes. J Immunol. 2006;177(9):6548–59.
112. Cohen CJ, Zheng Z, Bray R, Zhao Y, Sherman LA, Rosenberg SA, et al. Recognition of fresh human tumor by human peripheral blood lymphocytes transduced with a bicistronic retroviral vector encoding a murine anti-p53 TCR. J Immunol. 2005;175(9):5799–808.

113. Parkhurst MR, Joo J, Riley JP, Yu Z, Li Y, Robbins PF, et al. Characterization of genetically modified T-cell receptors that recognize the CEA:691-699 peptide in the context of HLA-A2.1 on human colorectal cancer cells. Clin Cancer Res. 2008;15(1):169–80.
114. Li Y, Moysey R, Molloy PE, Vuidepot A-L, Mahon T, Baston E, et al. Directed evolution of human T-cell receptors with picomolar affinities by phage display. Nat Biotechnol. 2005;23(3):349–54.
115. Varela-Rohena A, Molloy PE, Dunn SM, Li Y, Suhoski MM, Carroll RG, et al. Control of HIV-1 immune escape by CD8 T cells expressing enhanced T-cell receptor. Nat Med. 2008;14(12):1390–5.
116. Lee DW, Barrett DM, Mackall C, Orentas R, Grupp SA. The future is now: chimeric antigen receptors as new targeted therapies for childhood cancer. Clin Cancer Res. 2012;18(10):2780–90.
117. Porter DL, Levine BL, Kalos M, Bagg A, June CH. Chimeric antigen receptor-modified T cells in chronic lymphoid leukemia. N Engl J Med. 2011;365(8):725–33.
118. Lee DW, Kochenderfer JN, Stetler-Stevenson M, Cui YK, Delbrook C, Feldman SA, et al. T cells expressing CD19 chimeric antigen receptors for acute lymphoblastic leukaemia in children and young adults: a phase 1 dose-escalation trial. Lancet. 2015;385(9967):517–28.
119. Zhang L, Tian L, Dai X, Yu H, Wang J, Lei A, et al. Pluripotent stem cell-derived CAR-macrophage cells with antigen-dependent anti-cancer cell functions. J Hematol Oncol. 2020;13(1):153.
120. Zhang C, Hu Y, Shi C. Targeting natural killer cells for tumor immunotherapy. Front Immunol. 2020;11:60.
121. Adair JE, Kubek SP, Kiem H-P. Hematopoietic stem cell approaches to cancer. Hematol Oncol Clin North Am. 2017;31(5):897–912.
122. Zhang J, Hu Y, Yang J, Li W, Zhang M, Wang Q, et al. Non-viral, specifically targeted CAR-T cells achieve high safety and efficacy in B-NHL. Nature. 2022;609(7926):369–74.
123. Morgan MA, Büning H, Sauer M, Schambach A. Use of cell and genome modification technologies to generate improved "off-the-shelf" CAR T and CAR NK cells. Front Immunol. 2020;11:1965.
124. Moy AB, Kamath A, Ternes S, Kamath J. The challenges to advancing induced pluripotent stem cell-dependent cell replacement therapy. Med Res Arch. 2023;11(11):4784.
125. Bhardwaj A, Nain V. TALENs-an indispensable tool in the era of CRISPR: a mini review. J Genet Eng Biotechnol. 2021;19(1):125.
126. Urnov FD, Miller JC, Lee YL, Beausejour CM, Rock JM, Augustus S, et al. Highly efficient endogenous human gene correction using designed zinc-finger nucleases. Nature. 2005;435(7042):646–51.
127. Gaj T, Gersbach CA, Barbas CF 3rd. ZFN, TALEN, and CRISPR/Cas-based methods for genome engineering. Trends Biotechnol. 2013;31(7):397–405.
128. Urnov FD, Rebar EJ, Holmes MC, Zhang HS, Gregory PD. Genome editing with engineered zinc finger nucleases. Nat Rev Genet. 2010;11(9):636–46.
129. Händel EM, Cathomen T. Zinc-finger nuclease based genome surgery: it's all about specificity. Curr Gene Ther. 2011;11(1):28–37.
130. Mussolino C, Alzubi J, Fine EJ, Morbitzer R, Cradick TJ, Lahaye T, et al. TALENs facilitate targeted genome editing in human cells with high specificity and low cytotoxicity. Nucleic Acids Res. 2014;42(10):6762–73.
131. Cong L, Ran FA, Cox D, Lin S, Barretto R, Habib N, et al. Multiplex genome engineering using CRISPR/Cas systems. Science. 2013;339(6121):819–23.
132. Weltner J, Balboa D, Katayama S, Bespalov M, Krjutškov K, Jouhilahti EM, et al. Human pluripotent reprogramming with CRISPR activators. Nat Commun. 2018;9(1):2643.
133. Ding Q, Regan SN, Xia Y, Oostrom LA, Cowan CA, Musunuru K. Enhanced efficiency of human pluripotent stem cell genome editing through replacing TALENs with CRISPRs. Cell Stem Cell. 2013;12(4):393–4.
134. Schlaeger TM, Daheron L, Brickler TR, Entwisle S, Chan K, Cianci A, et al. A comparison of non-integrating reprogramming methods. Nat Biotechnol. 2015;33(1):58–63.

135. Warren L, Manos PD, Ahfeldt T, Loh YH, Li H, Lau F, et al. Highly efficient reprogramming to pluripotency and directed differentiation of human cells with synthetic modified mRNA. Cell Stem Cell. 2010;7(5):618–30.
136. Yu J, Vodyanik MA, Smuga-Otto K, Antosiewicz-Bourget J, Frane JL, Tian S, et al. Induced pluripotent stem cell lines derived from human somatic cells. Science. 2007;318(5858):1917–20.
137. Okita K, Matsumura Y, Sato Y, Okada A, Morizane A, Okamoto S, et al. A more efficient method to generate integration-free human iPS cells. Nat Methods. 2011;8(5):409–12.
138. Hockemeyer D, Jaenisch R. Induced pluripotent stem cells meet genome editing. Cell Stem Cell. 2016;18(5):573–86.
139. Zetsche B, Gootenberg JS, Abudayyeh OO, Slaymaker IM, Makarova KS, Essletzbichler P, et al. Cpf1 is a single RNA-guided endonuclease of a class 2 CRISPR-Cas system. Cell. 2015;163(3):759–71.
140. Pigeau GM, Csaszar E, Dulgar-Tulloch A. Commercial scale manufacturing of allogeneic cell therapy. Front Med. 2018;5:233.
141. Madrid M, Sumen C, Aivio S, Saklayen N. Autologous induced pluripotent stem cell–based cell therapies: promise, progress, and challenges. Curr Protoc. 2021;1(3):e88.
142. Lee AS, Tang C, Rao MS, Weissman IL, Wu JC. Tumorigenicity as a clinical hurdle for pluripotent stem cell therapies. Nat Med. 2013;19(8):998–1004.
143. Sato Y, Bando H, Di Piazza M, Gowing G, Herberts C, Jackman S, et al. Tumorigenicity assessment of cell therapy products: the need for global consensus and points to consider. Cytotherapy. 2019;21(11):1095–111.
144. Andrews PW, Ben-David U, Benvenisty N, Coffey P, Eggan K, Knowles BB, et al. Assessing the safety of human pluripotent stem cells and their derivatives for clinical applications. Stem Cell Rep. 2017;9(1):1–4.
145. Chakravarti D, Caraballo LD, Weinberg BH, Wong WW. Inducible gene switches with memory in human T cells for cellular immunotherapy. ACS Synth Biol. 2019;8(8):1744–54.
146. Stenger D, Stief TA, Kaeuferle T, Willier S, Rataj F, Schober K, et al. Endogenous TCR promotes in vivo persistence of CD19-CAR-T cells compared to a CRISPR/Cas9-mediated TCR knockout CAR. Blood. 2020;136(12):1407–18.
147. Song P, Zhang Q, Xu Z, Shi Y, Jing R, Luo D. CRISPR/Cas-based CAR-T cells: production and application. Biomark Res. 2024;12(1):54.
148. Sullivan S, Stacey GN, Akazawa C, Aoyama N, Baptista R, Bedford P, et al. Quality control guidelines for clinical-grade human induced pluripotent stem cell lines. Regen Med. 2018;13(7):859–66.
149. Andrews PW, Baker D, Benvinisty N, Miranda B, Bruce K, et al. Points to consider in the development of seed stocks of pluripotent stem cells for clinical applications: International Stem Cell Banking Initiative (ISCBI). Regen Med. 2015;10(sup2):1–44.
150. Amps K, Andrews PW, Anyfantis G, Armstrong L, Avery S, Baharvand H, et al. Screening ethnically diverse human embryonic stem cells identifies a chromosome 20 minimal amplicon conferring growth advantage. Nat Biotechnol. 2011;29(12):1132–44.
151. Hu Y, Huang J. The chimeric antigen receptor detection toolkit. Front Immunol. 2020;11:1770.
152. Kuroda T, Yasuda S, Matsuyama S, Tano K, Kusakawa S, Sawa Y, et al. Highly sensitive droplet digital PCR method for detection of residual undifferentiated cells in cardiomyocytes derived from human pluripotent stem cells. Regen Ther. 2015;2:17–23.
153. Kuroda T, Yasuda S, Kusakawa S, Hirata N, Kanda Y, Suzuki K, et al. Highly sensitive in vitro methods for detection of residual undifferentiated cells in retinal pigment epithelial cells derived from human iPS cells. PLoS One. 2012;7(5):e37342.
154. Yamanaka S. Pluripotent stem cell-based cell therapy—promise and challenges. Cell Stem Cell. 2020;27(4):523–31.
155. Mendicino M, Fan Y, Griffin D, Gunter KC, Nichols K. Current state of U.S. Food and Drug Administration regulation for cellular and gene therapy products: potential cures on the horizon. Cytotherapy. 2019;21(7):699–724.
156. Sullivan S, Fairchild PJ, Marsh SGE, Müller CR, Turner ML, Song J, et al. Haplobanking induced pluripotent stem cells for clinical use. Stem Cell Res. 2020;49:102035.

CAR-Ms: Macrophage-Based Cellular Immunotherapies for Solid Tumors

Romana Rashid, Shahid Banday, Anshul Sharma, and Alok K. Mishra

Abstract Chimeric antigen receptor (CAR)-Ms represents an innovative frontier in cancer immunotherapy, offering solutions to the challenges associated with treating solid tumors. Unlike CAR-T cells, CAR-Ms exploit the natural tumor-infiltrating properties and adaptability of macrophages, enabling them to navigate the hostile TME and achieve targeted cytotoxic effects. This chapter explores the multifaceted role of macrophages in cancer, emphasizing their dual functionality as tumor-supporting or tumoricidal cells and how these properties are harnessed and reprogrammed for therapeutic purposes. The engineering of CAR-Ms involves the integration of tumor-specific antigen recognition domains, macrophage-optimized signaling domains, and prophagocytic enhancements, enabling these cells to directly kill tumor cells and remodel the TME. Early clinical studies with unmodified macrophages established the safety and feasibility of CAR-M-cell therapy, setting the stage for advanced CAR-M-cell technologies. Despite their potential, CAR-Ms face challenges such as efficient gene delivery, short macrophage lifespans, and maintaining functionality in the immunosuppressive TME. With promising preclinical data and ongoing clinical trials, CAR-Ms are poised to complement existing therapies, including checkpoint inhibitors and CAR-T cells, transforming the treatment

R. Rashid
Division of Innate Immunity, UMass Chan Medical School, Worcester, MA, USA

S. Banday
Department of Molecular, Cell and Cancer Biology, UMass Chan Medical School, Worcester, MA, USA

A. Sharma
Department of Medicine, Division of Hematology and Oncology, Medical College of Wisconsin, Milwaukee, WI, USA

A. K. Mishra (✉)
Department of Molecular, Cell and Cancer Biology, UMass Chan Medical School, Worcester, MA, USA

Department of Hematopoietic Biology & Malignancy, Division of Cancer Medicine, The University of Texas MD Anderson Cancer Center, Houston, TX, USA
e-mail: AKMishra1@mdanderson.org

A. K. Mishra, S. K. Malonia (eds.), *Cell-based Immunotherapies for Cancer*,
https://doi.org/10.1007/978-3-031-88695-9_15

landscape for solid tumors and offering new hope for patients with refractory cancers.

Keywords Chimeric antigen receptor macrophages (CAR-Ms) · Tumor microenvironment (TME) · Tumor-associated macrophages (TAMs) · Phagocytosis · Solid tumors · Immunotherapy · Macrophage engineering

Abbreviations

CAR	Chimeric antigen receptor
CAR-Ms	Chimeric antigen receptor macrophages
CCL2	C-C motif chemokine ligand 2
CRS	Cytokine release syndrome
CSF-1	Colony-stimulating factor 1
ECM	Extracellular matrix
EGFR	Epidermal growth factor receptor
HER2	Human epidermal growth factor receptor 2
IFN-γ	Interferon-gamma
IL-4	Interleukin-4
IL-10	Interleukin-10
IL-12	Interleukin-12
LPS	Lipopolysaccharides
MMPs	Matrix metalloproteinases
mRNA	Messenger ribonucleic acid
NK	Natural killer
ROS	Reactive oxygen species
scFv	Single-chain variable fragment
TAMs	Tumor-associated macrophages
TGF-β	Transforming growth factor-beta
TME	Tumor microenvironment
TNF-α	Tumor necrosis factor-alpha
VEGF	Vascular endothelial growth factor

1 Introduction

Cancer remains one of the leading causes of mortality worldwide, despite significant advances in therapeutic strategies. In recent years, immunotherapy has emerged as a groundbreaking approach that leverages the body's immune system to recognize and eliminate cancer cells. Among immunotherapy modalities, chimeric antigen receptor (CAR)-T-cell therapy has revolutionized the treatment of hematological malignancies, resulting in unprecedented clinical outcomes. However, the extension

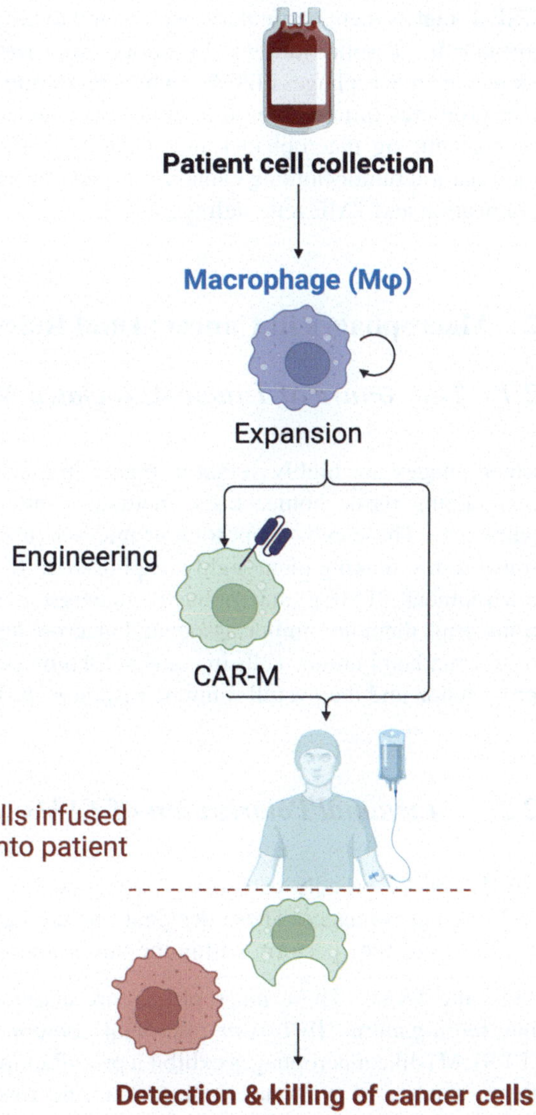

Fig. 1 Schematic representation of CAR-macrophage-based immunotherapy for cancer. The illustration outlines the critical steps: isolating monocytes from patients, expanding them, genetically engineering them to enhance therapeutic efficacy, reinfusing them into the patient, and detecting tumor cell elimination through phagocytosis

of CAR-T-cell technology to solid tumors has encountered formidable challenges, including the immunosuppressive tumor microenvironment (TME), physical barriers, and tumor antigen heterogeneity. These limitations have prompted researchers to explore innovative approaches beyond T cells, leading to the development of CAR-Ms [1–4] (Fig. 1).

As key components of the innate immune system, macrophages play critical roles in immune surveillance, inflammation, and tissue repair [5–7]. Their natural ability to infiltrate solid tumors, driven by chemotactic signals such as CCL2 and

CSF-1, makes them particularly well suited to target the dense and hostile microenvironment of solid tumors. In tumors, macrophages differentiate into tumor-associated macrophages (TAMs), which predominantly adopt an M2-like phenotype that promotes tumor growth, angiogenesis, and immune evasion. The concept of reprogramming macrophages into CAR-Ms offers a dual advantage: harnessing their natural tumor-homing capabilities and reorienting their function toward tumor eradication and TME remodeling [5–11].

2 Macrophages in Cancer: Dual Roles in Immunity

2.1 The Nature of Tumor-Associated Macrophages

Macrophages are highly versatile innate immune cells that play critical roles in maintaining tissue homeostasis, mediating inflammation, and defending against pathogens. These cells adapt to their microenvironment by assuming distinct functional states, making them highly responsive to local cues. Within the tumor microenvironment (TME), macrophages undergo a process of reprogramming that transforms them into tumor-associated macrophages (TAMs). TAMs are among the most abundant immune cells in many solid tumors and significantly influence tumor progression and the overall immune response [8, 12–14].

2.2 Functional Polarization of TAMs

TAMs can be broadly categorized into two functional phenotypes: M1-like and M2-like macrophages. These designations are based on their activation states and the functions they perform within the tumor microenvironment.

M1-Like TAMs These macrophages are classically activated by stimuli such as interferon-gamma (IFN-γ) or microbial components such as lipopolysaccharides (LPS). M1-like macrophages exhibit a proinflammatory phenotype characterized by the production of cytokines such as tumor necrosis factor-alpha (TNF-α), interleukin-12 (IL-12), and reactive oxygen species (ROS). These molecules enable M1-like TAMs to perform antitumor functions by promoting the recruitment and activation of cytotoxic immune cells, such as CD8+ T cells and natural killer (NK) cells. M1-like macrophages also directly kill tumor cells and suppress their proliferation. Despite their antitumor potential, M1-like TAMs are often outnumbered and outcompeted by M2-like macrophages in solid tumors [13, 15, 16].

M2-Like TAMs In contrast, M2-like macrophages are alternatively activated by signals such as interleukin-4 (IL-4), interleukin-10 (IL-10), and interleukin-13 (IL-13), which are abundant in the TME. M2-like TAMs exhibit an anti-inflammatory phenotype, producing cytokines and growth factors such as transforming growth factor-beta (TGF-β), vascular endothelial growth factor (VEGF), and matrix metalloproteinases

(MMPs). These molecules support tumor progression by promoting angiogenesis, remodeling the extracellular matrix, and suppressing immune responses. M2-like TAMs also inhibit the activation of cytotoxic T cells and NK cells, further weakening the immune system's ability to combat tumor growth [4, 6, 13, 16–19].

2.3 Recruitment and Polarization of TAMs in Solid Tumors

Solid tumors create an immunosuppressive microenvironment that favors the recruitment and polarization of M2-like TAMs. Tumors secrete chemotactic factors such as C-C motif chemokine ligand 2 (CCL2) and colony-stimulating factor 1 (CSF-1) to attract circulating monocytes, which differentiate into TAMs upon entering the TME. These monocyte-derived TAMs are then reprogrammed by tumor-derived signals, including hypoxia and cytokines, to adopt an M2-like phenotype.

This predominance of M2-like TAMs contributes significantly to the immunosuppressive nature of the TME. By suppressing proinflammatory immune responses and enhancing tissue remodeling, M2-like TAMs create a supportive niche for tumor growth and metastasis. Additionally, M2-like TAMs release immunosuppressive molecules, such as IL-10 and TGF-β, which further inhibit the activity of antitumor immune cells, including M1-like macrophages, dendritic cells, and effector T cells [6, 14, 20–22].

2.4 Implications for Therapy

The plasticity of TAMs presents both challenges and opportunities for cancer therapy. While the predominance of M2-like TAMs poses a significant barrier to effective immune responses, the potential to reprogram these cells toward an M1-like phenotype offers a promising therapeutic avenue. Strategies targeting TAMs include blocking their recruitment, neutralizing immunosuppressive signals, or directly reprogramming M2-like TAMs into M1-like macrophages. These approaches aim to restore immune activation and improve the efficacy of existing therapies, including checkpoint inhibitors and adoptive cell therapies. Understanding the dual nature of TAMs and their dynamic role in the TME is critical to developing novel therapeutic strategies aimed at modulating their behavior to favor antitumor immunity [13, 17, 23, 24].

3 Macrophages as Trojan Horses

One of the key advantages of macrophages in cancer immunotherapy is their natural ability to infiltrate solid tumors, a feature that distinguishes them from other immune cells, such as T cells. Macrophages are inherently equipped to respond to a wide range of chemotactic signals secreted by tumors, allowing them to navigate the complex and often hostile environment of solid tumors. This intrinsic ability positions macrophages as ideal candidates for chimeric antigen receptor (CAR)-based therapies, particularly in the context of solid malignancies where other immune cells struggle to penetrate the tumor microenvironment (TME).

3.1 Chemotactic Navigation of Macrophages in Tumors

Tumors actively recruit macrophages to their microenvironment by secreting a variety of chemotactic factors, such as C-C motif chemokine ligand 2 (CCL2), colony-stimulating factor 1 (CSF-1), and vascular endothelial growth factor (VEGF). These factors enable macrophages to migrate toward the tumor core, even in the presence of physical barriers such as a dense extracellular matrix (ECM) or high interstitial fluid pressure. This capability provides macrophages with a unique advantage over T cells, which often struggle to infiltrate solid tumors and effectively reach cancer cells in such challenging environments. The innate tumor-homing ability of macrophages makes them a compelling "Trojan Horse" for delivering targeted therapies to solid tumors [25]. By reprogramming macrophages into CAR-Ms, researchers aim to transform these immune cells from passive tumor-supporting cells into active tumor-killing agents. This reprogramming involves the integration of CAR constructs that enable macrophages to recognize and attack tumor-specific antigens. The dual strategy of leveraging the natural infiltration ability of macrophages while enhancing their cytotoxic functionality through CAR engineering is a novel and powerful approach for treating solid tumors [26, 27].

3.2 Unique Properties of Macrophages for Their Therapeutic Utility in Solid Tumors

1. *Tumor Penetration*: Macrophages efficiently migrate into the tumor core, even in regions with low oxygen levels (hypoxia) and high immunosuppression, conditions that are typically prohibitive for T cells and other immune cells. This allows CAR-Ms to access and target cancer cells that are otherwise shielded from conventional therapies [13].
2. *Adaptability*: Macrophages are highly plastic and can adapt to diverse microenvironments. By engineering CAR-Ms, researchers can redirect this adaptability to counteract tumor-derived signals that would otherwise suppress immune responses [4, 8, 13, 26].
3. *Immunomodulation*: In addition to direct tumor killing, CAR-Ms have the ability to reshape the TME by secreting cytokines and chemokines that recruit and activate other immune cells. This secondary effect amplifies the immune response against tumors and can overcome some of the limitations of current immunotherapies [7, 26, 28, 29].

4 Early Clinical Studies with Unmodified Nacrophages

In the early 1980s, researchers began investigating the use of ex vivo activated macrophages as potential cancer treatments, with the goal of leveraging their natural ability to migrate to tumors and modulate the tumor microenvironment to stimulate the body's immune response. A key breakthrough came from Andreesen's group in Germany, who pioneered the use of monocyte-derived macrophage therapy in patients with advanced cancers unresponsive to standard treatments. The macrophages were activated with IFNγ to induce an antitumor M1 phenotype before being administered to 15 patients. While therapy did not lead to regression of the primary tumors, several patients experienced stable disease for up to 6 months. The side effects were minimal, with only low-grade fever and localized discomfort at the injection site.

Despite promising early results in preclinical studies showing antitumor activity, the clinical effectiveness of these nonengineered macrophages remains limited. A head-to-head comparison with the Bacillus Calmette-Guerin (BCG) vaccine in bladder cancer revealed no significant improvement in tumor control with macrophage therapy. This lack of efficacy was likely due to the inability of macrophages to consistently recognize tumor-associated antigens or their potential shift from the IFNγ-induced M1 phenotype to the tumor-promoting M2 phenotype within the immunosuppressive TME. Nevertheless, a phase I trial in which cytotoxic macrophages were delivered via hepatic artery infusion in patients with metastatic liver disease demonstrated the safety and tolerability of this approach, with only mild flu-like symptoms reported. Moreover, radiolabeled macrophages persist at metastatic sites for more than 1 week, confirming the feasibility of intravenous macrophage delivery without significant adverse effects. While these early efforts did not yield widespread clinical success, they were instrumental in shaping the future of macrophage-based cancer therapies, paving the way for more advanced approaches such as CAR macrophages [23, 29–32].

4.1 Evolution of CAR Macrophages

The conceptualization of CAR-T macrophages builds upon the foundational success of CAR-T cells. CARs are synthetic receptors engineered to target tumor-specific antigens and activate immune cells. For macrophages, CAR engineering involves not only antigen-specific targeting but also the activation of macrophage-specific effector functions such as phagocytosis, cytokine secretion, and immune modulation. Unlike CAR-T cells, which rely predominantly on cytotoxic killing, CAR-Ms exhibit multifaceted mechanisms of action, including direct tumor killing by phagocytosis, TME remodeling, and immune activation (Fig. 2).

The development of chimeric antigen receptor (CAR) macrophages has evolved significantly from the first to the second generation, enhancing their therapeutic

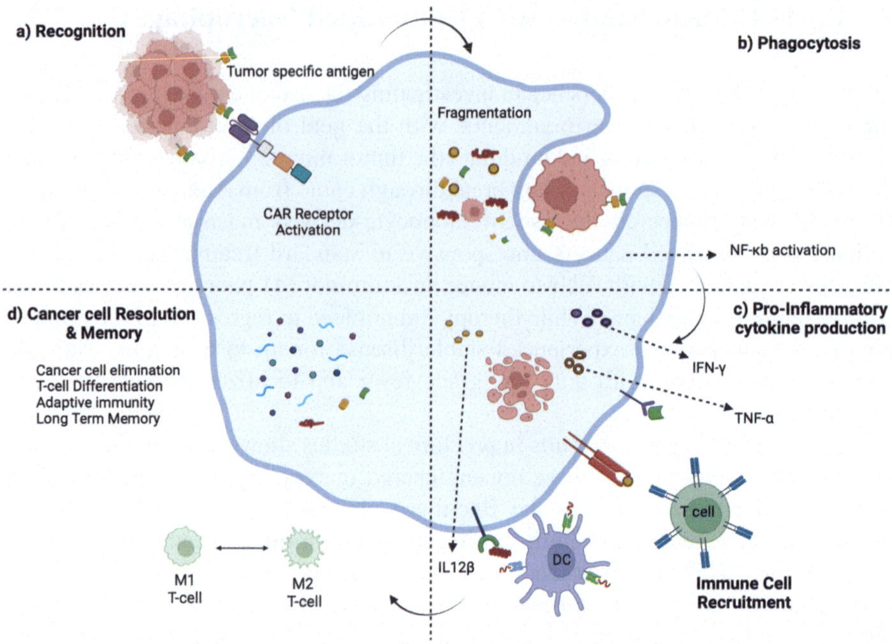

Fig. 2 Mechanisms of CAR receptor activation and the immune response in the tumor microenvironment. This schematic illustrates the sequence of immunological events triggered by CAR receptor activation in response to tumor-specific antigens presented by solid tumor cells. (**a**, **b**) Initial recognition and activation lead to antigen fragmentation and presentation. Reactive oxygen species (ROS), nitric oxide (NO), and antigen fragments drive a proinflammatory microenvironment. This process involves NF-κB activation and the production of key cytokines, including IL-12, TNF-α, and IFN-γ, promoting an immune response. (**c**, **d**) T cells and dendritic cells (DCs) coordinate to eliminate cancer cells, phagocytose debris, and support T-cell differentiation into memory cells, establishing adaptive immunity for long-term cancer resolution. The balance between M1 and M2 macrophages contributes to the regulation of the tumor microenvironment and immune response outcomes

potential against solid tumors. *First-generation CAR macrophages* utilize the CD3ζ-activating domain borrowed from CAR-T cells to enable phagocytosis [33]. However, these macrophages lack the ability to polarize into a durable M1-like proinflammatory state, limiting their ability to sustain robust antitumor responses. In contrast, *second-generation CAR*-T cells contain the intracellular CD3ζ domain with the TIR domain of TLR4, a pattern-recognition receptor critical for immune regulation. This advancement facilitated the induction and maintenance of an M1-like state via the NF-κB pathway while suppressing the M2-like state. Moreover, the second-generation design enhances antigen-dependent phagocytosis and introduces a novel mechanism of CAR-mediated efferocytosis, where tumor cell apoptosis and debris clearance amplify antitumor effects [34]. These innovations underscore the superior therapeutic potential of second-generation CAR macrophages in solid tumor immunotherapy.

5 Engineering CAR Macrophages

The engineering of CAR-Ms involves the integration of synthetic chimeric antigen receptors into macrophages to enable these innate immune cells to recognize and destroy cancer cells. While the fundamental principles of CAR design are derived from CAR-T cells, the unique biology of macrophages necessitates tailored approaches to ensure the effectiveness and safety of CAR-Ms.

Design of CAR Macrophages
The design of CAR-Ms requires careful adaptation of their structural components to align with macrophage-specific functionalities. CARs for macrophages typically consist of three primary elements:

Antigen Recognition Domain
The antigen recognition domain is a single-chain variable fragment (scFv) derived from monoclonal antibodies that bind specifically to tumor antigens such as HER2, EGFR, or mesothelin. This domain allows CAR-Ms to recognize and engage cancer cells displaying these antigens. Tumor-specific antigens are carefully chosen to minimize off-target effects and ensure the selective targeting of cancerous cells.

Signaling Domain
CAR-Ms utilize intracellular signaling domains that are engineered to activate macrophage-specific functions upon antigen binding. Unlike CAR-T cells, which rely on domains such as CD28 or 4-1BB for cytotoxic T-cell activation, CAR-Ms incorporate domains such as CD3ζ or FcRγ. These domains are optimized to trigger key macrophage effector responses, including phagocytosis, reactive oxygen species (ROS) production, and cytokine release. The signaling domains must be robust enough to sustain these activities while avoiding excessive inflammation or systemic toxicity.

5.1 Gene Delivery Methods

The successful generation of CAR-Ms hinges on the efficient delivery of CAR constructs into macrophages. Various methods are employed to achieve this goal, each with distinct advantages and limitations:

Viral Vectors
Lentiviral and retroviral vectors are commonly used for the stable integration of CAR genes into immune cells. These vectors ensure the long-term expression of CAR constructs, making them suitable for persistent therapeutic applications. However, concerns over genotoxicity and manufacturing complexity necessitate careful quality control. Additionally, adenoviral vectors (AdVs), particularly chimeric Ad5f35, are pivotal in engineering CAR-Ms because of their ability to efficiently transduce CD14+ monocytes and macrophages. By leveraging the CD46 receptor,

Ad5f35 ensures high transduction efficiency and stable CAR expression [23, 27, 35].

mRNA Electroporation
The electroporation of mRNAs encoding CAR constructs results in the transient expression of CARs in macrophages. This method reduces the risk of genomic integration-related adverse effects and is ideal for applications where short-term CAR expression is sufficient. The nonintegrative nature of mRNA electroporation also simplifies regulatory approval [36].

Nanoparticles
Emerging nanoparticle-based delivery systems offer a nonviral approach to introduce CAR constructs into macrophages. These systems are being developed to improve transfection efficiency, minimize off-target effects, and facilitate scalable manufacturing. Nanoparticles also allow for the codelivery of other therapeutic agents, potentially enabling combination therapies [37].

5.2 *Challenges in CAR-M Engineering*

Despite its promise, CAR-M technology faces unique challenges that must be addressed to achieve widespread clinical adoption. One major hurdle is the relatively low efficiency of gene transfer into macrophages compared with that into T cells. This is partly due to the intrinsic resistance of macrophages to transduction and their shorter lifespan in culture, which limits the scalability of CAR-M production. Overcoming these obstacles will require advances in gene delivery technologies, optimized culture conditions, and scalable manufacturing platforms. Another challenge lies in ensuring that CAR-Ms maintain their therapeutic efficacy in the immunosuppressive tumor microenvironment. Strategies such as co-engineering resistance to suppressive signals (e.g., IL-10 or TGF-β) and incorporating safety mechanisms, such as inducible CAR expression, are critical for addressing these issues.

6 Mechanisms of CAR Macrophages in Solid Tumors

CAR-Ms are being developed for solid tumor immunotherapy, leveraging direct cytotoxicity, tumor microenvironment (TME) remodeling, and immune response enhancement. In addition to efficiently phagocytosing tumor cells, CAR-Ms secrete proinflammatory cytokines such as IL-12 and TNF-α, which not only exert direct cytotoxic effects on tumor cells but also recruit and activate other immune cells, amplifying the antitumor response. A distinctive feature of CAR-Ms is their ability to modulate the TME by neutralizing suppressive cytokines such as IL-10 and TGF-β, degrading the extracellular matrix through matrix metalloproteinases (MMPs), and recruiting effector immune cells such as T cells and NK cells. These activities convert immunologically "cold" tumors into "hot" tumors, which are more receptive to immunotherapies (Fig. 2). Furthermore, CAR-Ms enhance the efficacy of immune checkpoint inhibitors and exhibit potential for integration into multimodal therapeutic strategies, underscoring their versatility in overcoming challenges associated with solid tumors.

7 Preclinical and Clinical Development

7.1 Preclinical Evidence

Preclinical studies have provided strong support for the development of CAR-Ms, demonstrating their ability to target and combat various solid tumors effectively. These studies underscore the versatility of CAR-Ms in multiple tumor models and highlight their unique mechanisms of action [23, 26, 27, 38, 39].

One of the earliest demonstrations of CAR-M efficacy involved HER2-targeted macrophages in breast cancer models. In these studies, CAR-Ms engineered to express HER2-specific chimeric antigen receptors were able to directly recognize and attack HER2-expressing tumor cells. The treatment significantly reduced the tumor burden in a mouse model and was associated with increased infiltration of T cells into the tumor microenvironment (TME). These findings suggest that CAR-Ms not only exert direct cytotoxic effects but also remodel the TME to support adaptive immune responses, thereby increasing overall antitumor activity [27].

7.2 Clinical Trials

The clinical translation of CAR-Ms began with the initiation of the first human trial, *CT-0508*, which marked a significant milestone in this field. CT-0508 is a phase I clinical trial evaluating the safety, tolerability, and preliminary efficacy of HER2-targeting CAR macrophages in patients with advanced solid tumors (NCT04660929). The trial utilizes macrophages derived from peripheral blood monocytes, which are genetically engineered ex vivo to express HER2-specific CARs. These CAR-Ms are then reintroduced into the patient. Early results from this trial revealed that HER2 CAR-Ms are well tolerated, with minimal adverse effects reported. Importantly, evidence of antitumor activity, including stable disease and tumor shrinkage in some patients, has been reported. These findings support the potential of CAR-Ms as a viable therapeutic option for HER2-positive tumors. The introduction of CT-0508 in human clinical trials has sparked interest in expanding CAR-M clinical trials to include additional tumor types and antigen targets. Ongoing efforts are focused on refining CAR-M-cell manufacturing processes to improve scalability and consistency, as well as exploring combination therapies with immune checkpoint inhibitors to further increase efficacy [39].

8 Challenges and Future Directions

The development of chimeric antigen receptor macrophages (CAR-Ms) faces significant challenges and presents promising avenues for future improvement. One of the primary obstacles is overcoming the immunosuppressive tumor microenvironment (TME), which is characterized by high levels of suppressive factors such as interleukin-10 (IL-10), transforming growth factor-beta (TGF-β), and regulatory immune cells. To counteract these suppressive effects, strategies such as combining CAR-Ms with immune checkpoint inhibitors (e.g., anti-PD-1) have shown potential in enhancing therapeutic efficacy [40].

Manufacturing CAR-Ms at a scale suitable for clinical application also remains crucial. Efforts to improve production processes focus on ex vivo expansion methods that maintain macrophage functionality, leverage bioreactor technologies for automated large-scale manufacturing, and employ synthetic biology to ensure uniformity and enhanced CAR-M performance. Expanding the range of targetable tumor antigens is another critical frontier facilitated by advances in single-cell sequencing, proteomics, and computational modeling. These tools enable the identification of novel antigens uniquely or predominantly expressed in tumors, paving the way for multiantigen-targeting and tunable CAR-M systems to effectively combat antigen heterogeneity in solid tumors [23, 33, 35, 41].

9 Conclusion

The development of CAR-Ms represents a significant advancement in the field of cancer immunotherapy, offering a novel approach to address the unique challenges associated with treating solid tumors. Unlike traditional CAR-T cells, CAR-Ms leverage the inherent properties of macrophages, such as their natural ability to infiltrate the tumor microenvironment (TME) and their plasticity in adapting to diverse conditions [35, 41]. By reprogramming macrophages with chimeric antigen receptors, researchers have created a therapeutic platform capable of not only targeting and destroying tumor cells but also reshaping the TME to increase overall immune activity. CAR-Ms exhibit a multifaceted mechanism of action, encompassing direct tumor killing through phagocytosis, the secretion of proinflammatory cytokines, and the modulation of the immunosuppressive TME. These functions make them particularly well suited to overcoming the barriers that have historically hindered immunotherapy in solid tumors, such as physical obstructions, hypoxia, and immune evasion by cancer cells. Moreover, their potential to recruit and activate other immune cells amplifies the therapeutic response, converting "cold" tumors into "hot" tumors that are more amenable to additional treatments. While CAR-M-cell therapy is still in its early stages, preclinical studies and early clinical trials have demonstrated its promise, with evidence of efficacy and safety in models of breast cancer, glioblastoma, and ovarian cancer, among others. The ongoing clinical evaluation of HER2-targeted CAR-Ms provides insight into their potential to become a transformative therapy for patients with advanced solid tumors. Nevertheless, challenges remain in optimizing CAR-M engineering, scaling up manufacturing processes, and addressing safety concerns such as excessive immune activation. Advances in synthetic biology, gene delivery methods, and combinatorial therapeutic strategies are critical to overcoming these hurdles and unlocking the full potential of CAR macrophages. Additionally, identifying tumor-specific antigens and tailoring CAR-Ms to resist the suppressive signals of the TME will be crucial for ensuring efficacy and minimizing off-target effects [27, 33–35, 38, 39, 41, 42].

Declaration of Interests The authors declare no competing interests.

References

1. Wang G. Cancer immunotherapy and immunomodulation. Curr Med Chem. 2019;26(17):2988.
2. Khalil DN, Smith EL, Brentjens RJ, Wolchok JD. The future of cancer treatment: immunomodulation, CARs and combination immunotherapy. Nat Rev Clin Oncol. 2016;13:394.
3. Mishra AK, Ali A, Dutta S, Banday S, Malonia SK. Emerging trends in immunotherapy for cancer. Diseases [Internet]. 2022;10(3):60. Available from: https://www.mdpi.com/2079-9721/10/3/60.
4. Mishra AK, Banday S, Bharadwaj R, Ali A, Rashid R, Kulshreshtha A, et al. Macrophages as a potential immunotherapeutic target in solid cancers. Vaccines. 2023;11:55.
5. Argyle D, Kitamura T. Targeting macrophage-recruiting chemokines as a novel therapeutic strategy to prevent the progression of solid tumors. Front Immunol. 2018;9:2629.
6. Mantovani A. Macrophages, neutrophils, and cancer: a double edged sword. New J Sci. 2014;2014:1.
7. Chen Y, Song Y, Du W, Gong L, Chang H, Zou Z. Tumor-associated macrophages: an accomplice in solid tumor progression. J Biomed Sci. 2019;26:78.
8. Hirayama D, Iida T, Nakase H. The phagocytic function of macrophage-enforcing innate immunity and tissue homeostasis. Int J Mol Sci. 2018;19:92.
9. Cassetta L, Pollard JW. Targeting macrophages: therapeutic approaches in cancer. Nat Rev Drug Discov. 2018;17:887.
10. Shen M, Du Y, Ye Y. Tumor-associated macrophages, dendritic cells, and neutrophils: biological roles, crosstalk, and therapeutic relevance. Med Rev. 2021;1(2):222.
11. Liu J, Geng X, Hou J, Wu G. New insights into M1/M2 macrophages: key modulators in cancer progression. Cancer Cell Int. 2021;21:389.
12. Balkwill F, Mantovani A. Inflammation and cancer: back to Virchow? Lancet. 2001;357:539.
13. Mantovani A, Allavena P, Marchesi F, Garlanda C. Macrophages as tools and targets in cancer therapy. Nat Rev Drug Discov. Nature Research. 2022;21:799.
14. Mantovani A, Ponzetta A, Inforzato A, Jaillon S. Innate immunity, inflammation and tumor progression: double-edged swords. J Intern Med. 2019;285(5):524.
15. Mosser DM, Edwards JP. Exploring the full spectrum of macrophage activation. Nat Rev Immunol. 2008;8:958.
16. Shapouri-Moghaddam A, Mohammadian S, Vazini H, Taghadosi M, Esmaeili SA, Mardani F, et al. Macrophage plasticity, polarization, and function in health and disease. J Cell Physiol. 2018;233:6425.

17. Mantovani A, Marchesi F, Malesci A, Laghi L, Allavena P. Tumor-associated macrophages as treatment targets in oncology. Nat Rev Clin Oncol. 2017;14:399.
18. Wu WK, Llewellyn OPC, Bates DO, Nicholson LB, Dick AD. IL-10 regulation of macrophage VEGF production is dependent on macrophage polarization and hypoxia. Immunobiology. 2010;215(9–10):796.
19. Liu CY, Xu JY, Shi XY, Huang W, Ruan TY, Xie P, et al. M2-polarized tumor-associated macrophages promoted epithelial–mesenchymal transition in pancreatic cancer cells, partially through TLR4/IL-10 signaling pathway. Lab Investig. 2013;93(7):844.
20. Tamura R, Tanaka T, Yamamoto Y, Akasaki Y, Sasaki H. Dual role of macrophage in tumor immunity. Immunotherapy. 2018;10:899.
21. Bingle L, Brown NJ, Lewis CE. The role of tumor-associated macrophages in tumor progression: implications for new anticancer therapies. J Pathol. 2002;196:254.
22. Pereira JFS, Jordan P, Matos P. A signaling view into the inflammatory tumor microenvironment. Immuno. 2021;1(2):91.
23. Mishra AK, Malonia SK. Advancing cellular immunotherapy with macrophages. Life Sci. 2023;328:121857.
24. Cortese N, Carriero R, Laghi L, Mantovani A, Marchesi F. Prognostic significance of tumor-associated macrophages: past, present and future. Semin Immunol. 2020;48:101408.
25. Choi MR, Stanton-Maxey KJ, Stanley JK, Levin CS, Bardhan R, Akin D, et al. A cellular trojan horse for delivery of therapeutic nanoparticles into tumors. Nano Lett. 2007;7(12):3759.
26. Anderson NR, Minutolo NG, Gill S, Klichinsky M. Macrophage-based approaches for cancer immunotherapy. Cancer Res. 2021;81:1201.
27. Klichinsky M, Ruella M, Shestova O, Lu XM, Best A, Zeeman M, et al. Human chimeric antigen receptor macrophages for cancer immunotherapy. Nat Biotechnol. 2020;38(8):947.
28. Forssell J, Öberg Å, Henriksson ML, Stenling R, Jung A, Palmqvist R. High macrophage infiltration along the tumor front correlates with improved survival in colon cancer. Clin Cancer Res. 2007;13(5):1472.
29. Lee S, Kivimäe S, Dolor A, Szoka FC. Macrophage-based cell therapies: the long and winding road. J Control Release. 2016;240:240.
30. Tan PH, Beutelspacher SC, Xue SA, Wang YH, Mitchell P, McAlister JC, et al. Modulation of human dendritic-cell function following transduction with viral vectors: implications for gene therapy. Blood. 2005;105(10):3824.
31. Hennemann B, Scheibenbogen C, Schiimichen C, Andreesen R. Intrahepatic adoptive immunotherapy with autologous tumorcytotoxic macrophages in patients with cancer. J Immunother. 1995;18(1):19.
32. Andreesen R, Hennemann B, Krause SW. Adoptive immunotherapy of cancer using monocyte-derived macrophages: rationale, current status, and perspectives. J Leukoc Biol. 1998;64:419.
33. Zhang L, Tian L, Dai X, Yu H, Wang J, Lei A, et al. Pluripotent stem cell-derived CAR-macrophages with antigen-dependent anticancer cell functions. J Hematol Oncol. 2020;13:153.
34. Lei A, Yu H, Lu S, Lu H, Ding X, Tan T, et al. A second-generation M1-polarized CAR macrophage with antitumor efficacy. Nat Immunol. 2024;25(1):102.
35. Sloas C, Gill S, Klichinsky M. Engineered CAR-macrophages as adoptive immunotherapies for solid tumors. Front Immunol. 2021;12:783305.
36. Shin MH, Oh E, Kim Y, Nam DH, Jeon SY, Yu JH, et al. Recent advances in CAR-based solid tumor immunotherapy. Cells. 2023;12:1606.
37. Shin S, Lee P, Han J, Kim SN, Lim J, Park DH, et al. Nanoparticle-based chimeric antigen receptor therapy for cancer immunotherapy. J Tissue Eng Regen Med. 2023;20:371.
38. Gabrusiewicz K, Schmierer M, Best A, Zeeman M, Ohtani Y, Gabitova L, et al. Abstract 2180: genetically engineered chimeric antigen receptor (CAR) monocytes demonstrate targeted anti-tumor activity and differentiate into M1-polarized CAR macrophages. Cancer Res. 2020;80(16_Supplement):2180.
39. Gabrusiewicz K, Anderson N, Lu X, Shan X, Shestova O, Petty N, et al. Abstract B65: CT-0508, a novel CAR macrophage product directed against HER2, promotes a proinflammatory tumor microenvironment. Cancer Immunol Res. 2020;8(3_Supplement):B65.

40. Pierini S, Gabbasov R, Worth A, Ramos I, Blumenthal D, Ohtani Y, et al. Abstract 2112: chimeric antigen receptor macrophages (CAR-M) sensitize solid tumors to anti-PD1 immunotherapy. Cancer Res. 2022;82(12_Supplement).
41. Li J, Chen P, Ma W. The next frontier in immunotherapy: potential and challenges of CAR-macrophages. Exp Hematol Oncol. BioMed Central Ltd. 2024;13:76.
42. Liang Y, Xu Q, Gao Q. Advancing CAR-based immunotherapies in solid tumors: CAR- macrophages and neutrophils. Front Immunol. 2023;14:1291619.

Index

A. K. Mishra, S. K. Malonia (eds.), *Cell-based Immunotherapies for Cancer*,
https://doi.org/10.1007/978-3-031-88695-9

The manufacturer's authorised representative in the EU is Springer Nature Customer Service Centre GmbH, Europaplatz 3, 69115 Heidelberg, Germany. If you have any concerns regarding our products, please contact ProductSafety@springernature.com

Printed and bound by CPI Group (UK) Ltd, Croydon, CR0 4YY
30/09/2026
02225564-0015